T0211848

Lecture Notes in Computer Science   12735

More information about this subseries at http://www.palgrave.com/gp/series/7407

Peter J. Stuckey (Ed.)

Integration of Constraint Programming, Artificial Intelligence, and Operations Research

18th International Conference, CPAIOR 2021
Vienna, Austria, July 5–8, 2021
Proceedings

 Springer

Editor
Peter J. Stuckey
Monash University
Melbourne, VIC, Australia

ISSN 0302-9743   ISSN 1611-3349   (electronic)
Lecture Notes in Computer Science
ISBN 978-3-030-78229-0   ISBN 978-3-030-78230-6   (eBook)
https://doi.org/10.1007/978-3-030-78230-6

LNCS Sublibrary: SL1 – Theoretical Computer Science and General Issues

This Springer imprint is published by the registered company Springer Nature Switzerland AG
The registered company address is: Gewerbestrasse 11, 6330 Cham, Switzerland

Preface

This volume contains the papers that were presented at the 18th International Conference on the Integration of Constraint Programming, Artificial Intelligence, and Operations Research (CPAIOR 2021), held in Vienna, Austria as a hybrid physical/virtual conference in response to the COVID-19 pandemic.

The conference received a total of 87 submissions, including 75 regular paper and 12 extended abstract submissions. The regular papers reflect original unpublished work, whereas the extended abstracts contain either original unpublished work or a summary of work that was published elsewhere. Each regular paper was reviewed by at least three Program Committee members. The reviewing phase was followed by an author response period and a general discussion by the Program Committee. The extended abstracts were reviewed for appropriateness for the conference. At the end of the review period, 30 regular papers were accepted for presentation during the conference and publication in this volume, and 6 abstracts were accepted for short presentation at the conference. Among the 30 regular papers, two were published directly in the journal Constraints via a fast-track review process. The abstracts of these papers can be found in this volume.

In addition to the regular papers and extended abstracts, three invited talks, whose abstracts and/or articles can be found in this volume, were given by Maya Gupta (Didero, USA), Adam Elmachtoub (Columbia University, USA), and Nikolaj Bjørner (Microsoft Research, USA).

The conference program included a Master Class on the topic "Explanation and Verification of Machine Learning Models" organized by Alexey Ignatiev and Nina Narodytska with invited talks by Alessio Lomuscio (Imperial College London, UK), Gagandeep Singh (University of Illinois Urbana-Champaign, USA), Guy Katz (Hebrew University of Jerusalem, Israel), Guy Van den Broeck (University of California, Los Angeles, USA), João Marques-Silva (CRNS, France), and Sameer Singh (University of California, Irvine, USA).

Of the regular papers accepted to the conference a committee comprising of myself, Helmut Simonis, and Louis-Martin Rousseau selected for the Best Paper Award the paper "Between Steps: Intermediate Relaxations between big-M and Convex Hull Formulations" by Jan Kronqvist, Ruth Misener, and Calvin Tsay and selected for the Best Student Paper Award the paper "Improving the filtering of Branch-and-Bound MDD Solver" by Xavier Gillard, Vianney Coppé, Pierre Schaus, and André Augusto Cire.

We acknowledge the generous support of our sponsors including, at the time of writing, the Vienna Center for Logic and Algorithms (VCLA), Artificial Intelligence Journal (AIJ), Springer, and TU Wien.

July 2021   Peter J. Stuckey

Organization

Program Chair

Peter J. Stuckey   Monash University, Australia

Conference Chair

Nysret Musliu   TU Wien, Austria

Master Class Chairs

Alexey Ignatiev   Monash University, Australia
Nina Narodytska   VMWare Research, USA

Program Committee

Fahiem Bacchus   University of Toronto, Canada
Chris Beck   University of Toronto, Canada
Nicolas Beldiceanu   LS2N, IMT Atlantique, France
Jeremias Berg   University of Helsinki, Finland
Armin Biere   Johannes Kepler University Linz, Austria
Mats Carlsson   RISE Research Institutes of Sweden, Sweden
Andre Augusto Cire   University of Toronto, Canada
Carleton Coffrin   Los Alamos National Laboratory, USA
Emir Demirović   Delft University of Technology, the Netherlands
Bistra Dilkina   University of Southern California, USA
Ambros Gleixner   HTW Berlin and Zuse Institute Berlin, Germany
Tias Guns   KU Leuven, Belgium
Emmanuel Hebrard   LAAS, CNRS, France
Philip Kilby   Data61 and the Australian National University,
                                    Australia
Joris Kinable   Amazon, USA
Zeynep Kiziltan   University of Bologna, Italy
Lars Kotthoff   University of Wyoming, USA
T. K. Satish Kumar   University of Southern California. USA
Jimmy Lee   The Chinese University of Hong Kong, Hong Kong
Michele Lombardi   University of Bologna, Italy
João Marques-Silva   IRIT, CNRS, France
Ian Miguel   University of St Andrews, Scotland
Nysret Musliu   TU Wien, Austria
Nina Narodytska   VMware Research, USA

Jakob Nordström	University of Copenhagen, Denmark, and Lund University, Sweden
Barry O'Sullivan	University College Cork, Ireland
Justin Pearson	Uppsala University, Sweden
Laurent Perron	Google, France
Claude-Guy Quimper	Université Laval, Canada
Louis-Martin Rousseau	Polytechnique Montréal, Canada
Domenico Salvagnin	University of Padova, Italy
Scott Sanner	University of Toronto, Canada
Pierre Schaus	UCLouvain, Belgium
Thomas Schiex	INRAE, France
Laurent Simon	LaBRI, Bordeaux Institute of Technology, France
Helmut Simonis	University College Cork, Ireland
Guido Tack	Monash University, Australia
Charlotte Truchet	LS2N, Université de Nantes, France
Pascal Van Hentenryck	Georgia Institute of Technology, USA
Petr Vilím	IBM, Czech Republic
Mark Wallace	Monash University, Australia
Roland Yap	National University of Singapore, Singapore

Additional Reviewers

Aghaei, Sina
Akgün, Özgür
Antuori, Valentin
Artigues, Christian
Belov, Gleb
Besançon, Mathieu
Blais, Nicolas
Boudreault, Raphaël
Cappart, Quentin
Eifler, Leon
Espasa Arxer, Joan
Geibinger, Tobias
Gent, Ian
Hendel, Gregor
Hoffmann, Ruth
Hojny, Christopher
Hu, Xinyi
Huang, Isaac
Huguet, Marie-José
Jefferson, Christopher
Jeong, Jihwan

Karahalios, Anthony
Katsirelos, George
Kletzander, Lucas
Le Bodic, Pierre
Leo, Kevin
Maher, Stephen
Meel, Kuldeep S.
Mischek, Florian
Nagarajan, Harsha
Portoleau, Tom
Prestwich, Steve
Pulatov, Damir
Siu, Charles
Spiegel, Christoph
Van Hoeve, Willem-Jan
Vavrille, Mathieu
Wang, Ruiwei
Winter, Felix
Yang, Hojin
Zhang, Han
Zhuowei, Zhong

Extended Abstracts

The following extended abstracts were accepted for presentation at the conference:

- Marleen Balvert. IRELAND: an MILP-based algorithm for learning interpretable input-output relationships from large binary classification data.
- Nick Doudchenko, Miles Lubin, Aditya Paliwal, Pawel Lichocki, and Ross Anderson. MipConfigBench: A dataset for learning in the space of Mixed-Integer Programming algorithms.
- Eleftherios Manousakis, Grigoris Kasapidis, Chris Kiranoudis, and Emmanouil Zachariadis. A matheuristic for the Production Routing Problem: Infeasibility Space Search and Mixed Integer Programming.
- Thibault Prunet, Nabil Absi, Valeria Borodin, and Diego Cattaruzza. Storage Location Assignment Problem in Fast Pick Areas: A novel formulation and decomposition method.
- Jana Koehler, Josef Bürgler, Urs Fontana, Etienne Fux, Florian Herzog, Marc Pouly, Sophia Saller, Anastasia Salyaeva, Peter Scheiblechner, and Kai Waelti. Cable Tree Wiring - Benchmarking Solvers on a Real-World Scheduling Problem with a Variety of Precedence Constraints.
- Mathijs de Weerdt, Robert Baart, and Lei He. Single-Machine Scheduling with Release Times, Deadlines, Setup Times, and Rejection.

Abstracts

Why You Should Constrain Your Machine Learned Models

Maya Gupta ⓘD

Didero, USA
founders@didero.com

Abstract. Common use of machine learning is to gather what training examples one can, train a flexible model with some smoothness regularizers, test it on a held-out set of random examples, and *hope* it works well in practice. But we will show that by adding constraints, we can prepare our models better for their futures, and be more certain of their performance. Based on 8 years of experience at Google researching, designing, training, and launching hundreds of machine-learned models, I will discuss dozens of ways that we found one can constrain ML models to produce more robust, fairer, safer, more accurate models that are easier to debug and that when they fail, do so more predictably and reasonably. This talk will focus on two classes of model constraints: shape constraints, and rate constraints. The most common shape constraint is monotonicity, and it has long been known how to learn monotonic functions over one input using isotonic regression. We will discuss new R&D about 6 different practically useful shape constraints, and how to impose them on flexible, mulit-layer models. The second class of constraints, rate constraints, refers to constraints on a classifiers' output statistics, and is commonly used to make classifiers act responsibly for different groups. For example, we may constrain a classifier used globally to be at least 80% accurate on training examples from India or China, as well as minimizing classification errors on average. We will point listeners to Google's open-source Tensor Flow libraries to impose these constraints, and papers with more technical detail.

Contextual Optimization: Bridging Machine Learning and Operations

Adam Elmachtoub ⓘ

Columbia University, New York, USA
adam@ieor.columbia.edu

Abstract. Many operations problems are associated with some form of a prediction problem. For instance, one cannot solve a supply chain problem without predicting demand. One cannot solve a shortest path problem without predicting travel times. One cannot solve a personalized pricing problem without predicting consumer valuations. In each of these problems, each instance is characterized by a context (or features). For instance, demand depends on prices and trends, travel times depend on weather and holidays, and consumer valuations depend on user demographics and click history. In this talk, we review recent results on how to solve such contextual optimization problems, with a particular emphasis on techniques that blend the prediction and decision tasks together.

Complete Symmetry Breaking Constraints
for the Class of Uniquely Hamiltonian Graphs

Avraham Itzhakov and Michael Codish

Department of Computer Science,
Ben-Gurion University of the Negev,
Beer-Sheva, Israel
{itzhakoa,mcodish}@cs.bgu.ac.il

Abstract. Graph search problems are fundamental in graph theory. Such problems include: existence problems, where the goal is to determine whether a simple graph with certain graph properties exists, enumeration problems, which are about finding all solutions modulo graph isomorphism, and extremal problems, where we seek the smallest/largest solution with respect to some target such as the number of edges or vertices in a solution. Solving graph search problems is typically hard due to the enormous search space and the large number of symmetries.

One common approach to break symmetries in constraint programming is to add symmetry breaking constraints which are satisfied by at least one member of each isomorphism class. A symmetry breaking constraint is called *complete* if it is satisfied by exactly one member of each isomorphism class and *partial* otherwise. A universal measure for the size of a symmetry breaking constraint is the size of its representation in propositional logic. All known techniques to define complete symmetry breaking constraints for graph search problems are based on predicates which are exponential in size. There is no known polynomial size complete symmetry breaking constraint for graph search problems.

This paper introduces, for the first time, a complete symmetry breaking constraint of polynomial size for a significant class of graphs: the class of uniquely Hamiltonian graphs. This is the class of graphs that contain exactly one Hamiltonian cycle. We introduce a canonical form for uniquely Hamiltonian graphs and prove that testing whether a given uniquely Hamiltonian graph is canonical can be performed efficiently. Based on this canonicity test, we construct a complete symmetry breaking constraint of polynomial size which is satisfied only by uniquely Hamiltonian graphs which are canonical. We apply the proposed symmetry breaking constraint to determine the, previously unknown, smallest orders for which uniquely Hamiltonian graphs of minimum degree 3 and girths 3 and 4 exist.

Given that it is unknown if there exist polynomial sized complete symmetry breaking constraints for graphs, this paper makes a first step in the direction of identifying specific classes of graphs for which such constraints do exist.

Supported by the Israel Science Foundation, grant 625/17.

Variable Ordering for Decision Diagrams: A Portfolio Approacho

Anthony Karahalios and Willem-Jan van Hoeve

Carnegie Mellon University, Pittsburgh PA 15213, USA
{akarahal,vanhoeve}@andrew.cmu.edu

Abstract. Relaxed decision diagrams have recently been successfully applied within a range of solution methodologies for discrete optimization, including constraint programming, integer linear programming, integer nonlinear programming, and combinatorial optimization. The variable ordering is often of crucial importance for their effectiveness. For example, Bergman et al. [1, 2] demonstrate that a variable ordering that yields a small exact diagram typically also provides stronger dual bounds from the relaxed diagram. When decision diagrams are built from a single top-to-bottom compilation, dynamic variable orderings can be very effective. For example, a recent work by Cappart et al. [3] deploys deep reinforcement learning to dynamically select the next variable during compilation. Dynamic variable orderings are less applicable, however, to compilation via iterative refinement, in which case the ordering must be specified in advance. In this work, we consider variable ordering strategies for the latter case.

Oftentimes there is no single variable ordering strategy that dominates all others for a given set of problem instances. Selecting the best ordering, or more generally the best algorithm, from a set of alternatives is a well-studied problem in artificial intelligence, in the context of *algorithm portfolios*. There are several ways to construct an algorithm portfolio: using static or dynamic features, formulating predictive models at the algorithm or portfolio level, predicting one algorithm to run per instance or creating a schedule of algorithms to run, using a fixed portfolio or updating it online [4]. We consider several different portfolio mechanisms: an offline predictive model of the single best algorithm using classifiers, an online low-knowledge algorithm selection, a static uniform time-sharing portfolio, and a dynamic online time allocator.

As a case study, we consider the graph coloring problem, for which a decision diagram approach was recently introduced [5, 6]. It uses an iterative refinement procedure much like Benders decomposition or lazy-clause generation, by repeatedly refining conflicts in the diagram until the solution is conflict free. Our experimental results show that predictive methods using classification models or exploration phases can lead to more instances solved optimally. However, these methods may lead to delayed optimality results on problem instances that are easy to solve. Another insight is that a mixed portfolio can outperform a clairvoyant selection of the best individual ordering for each

Partially supported by Office of Naval Research Grant No. N00014-18-1-2129 and National Science Foundation Award #1918102.

instance, by yielding a solution with a unique best upper bound from one ordering and a unique best lower bound from a different ordering.

References

1. Bergman, D., Cire, A.A., van Hoeve, W.-J., Hooker, J.N.: Variable ordering for the application of BDDs to the maximum independent set problem. In: Beldiceanu, N., Jussien, N., Pinson, É. (eds.) CPAIOR 2012. LNCS, vol. 7298, pp. 34–49. Springer, Heidelberg (2012). https://doi.org/10.1007/978-3-642-29828-8_3
2. Bergman, D., Cire, A.A., van Hoeve, W.-J., Hooker, J. N.: Optimization bounds from binary decision diagrams. Inform. J. Comput. **26**(2), 253–268 (2014)
3. Cappart, Q., Goutierre, E., Bergman, D., Rousseau, L.-M.: Improving optimization bounds using machine learning: decision diagrams meet deep reinforcement learning. In: Proceedings of AAAI, pp. 1443–1451. AAAI Press (2019)
4. Kotthoff, L.: Algorithm selection for combinatorial search problems: a survey. In: Bessiere, C., De Raedt, L., Kotthoff, L., Nijssen, S., O'Sullivan, B., Pedreschi, D. (eds). Data Mining and Constraint Programming. LNCS, vol. 10101, pp. 149–190. Springer, Cham (2016).https://doi.org/10.1007/978-3-319-50137-6_7
5. van Hoeve, W.-J.: Graph Coloring with Decision Diagrams. Under review.http://www.optimization-online.org/DB_HTML/2021/01/8215.html
6. van Hoeve, W.-J.: Graph coloring lower bounds from decision diagrams. In: Bienstock, D., Zambelli, G. (eds.) IPCO 2020. LNCS, vol. 12125, pp. 405–419. Springer, Cham (2020). https://doi.org/10.1007/978-3-030-45771-6_31

Contents

Supercharging Plant Configurations Using Z3.......................... 1
  Nikolaj Bjørner, Maxwell Levatich, Nuno P. Lopes,
  Andrey Rybalchenko, and Chandrasekar Vuppalapati

A Computational Study of Constraint Programming Approaches
for Resource-Constrained Project Scheduling with Autonomous
Learning Effects.. 26
  Alessandro Hill, Jordan Ticktin, and Thomas W. M. Vossen

Strengthening of Feasibility Cuts in Logic-Based Benders Decomposition ... 45
  Emil Karlsson and Elina Rönnberg

Learning Variable Activity Initialisation for Lazy Clause
Generation Solvers.. 62
  Ronald van Driel, Emir Demirović, and Neil Yorke-Smith

A*-Based Compilation of Relaxed Decision Diagrams for the Longest
Common Subsequence Problem 72
  Matthias Horn and Günther R. Raidl

Partitioning Students into Cohorts During COVID-19 89
  Richard Hoshino and Irene Fabris

A Two-Stage Exact Algorithm for Optimization of Neural
Network Ensemble 106
  Keliang Wang, Leonardo Lozano, David Bergman,
  and Carlos Cardonha

Heavy-Tails and Randomized Restarting Beam Search in Goal-Oriented
Neural Sequence Decoding................................... 115
  Eldan Cohen and J. Christopher Beck

Combining Constraint Programming and Temporal Decomposition
Approaches - Scheduling of an Industrial Formulation Plant............. 133
  Christian Klanke, Dominik R. Bleidorn, Vassilios Yfantis,
  and Sebastian Engell

The Traveling Social Golfer Problem: The Case of the Volleyball
Nations League.. 149
  Roel Lambers, Laurent Rothuizen, and Frits C. R. Spieksma

Towards a Compact SAT-Based Encoding of Itemset Mining Tasks 163
  Ikram Nekkache, Said Jabbour, Lakhdar Sais, and Nadjet Kamel

A Pipe Routing Hybrid Approach Based on A-Star Search
and Linear Programming   179
   Marvin Stanczak, Cédric Pralet, Vincent Vidal, and Vincent Baudoui

MDDs Boost Equation Solving on Discrete Dynamical Systems   196
   Enrico Formenti, Jean-Charles Régin, and Sara Riva

Two Deadline Reduction Algorithms for Scheduling Dependent Tasks
on Parallel Processors   214
   Claire Hanen, Alix Munier Kordon, and Theo Pedersen

Improving the Filtering of Branch-and-Bound MDD Solver   231
   Xavier Gillard, Vianney Coppé, Pierre Schaus, and André Augusto Cire

On the Usefulness of Linear Modular Arithmetic
in Constraint Programming.   248
   Gilles Pesant, Kuldeep S. Meel, and Mahshid Mohammadalitajrishi

Injecting Domain Knowledge in Neural Networks: A Controlled
Experiment on a Constrained Problem.   266
   Mattia Silvestri, Michele Lombardi, and Michela Milano

Learning Surrogate Functions for the Short-Horizon Planning in Same-Day
Delivery Problems.   283
   Adrian Bracher, Nikolaus Frohner, and Günther R. Raidl

Between Steps: Intermediate Relaxations Between Big-M and Convex
Hull Formulations   299
   Jan Kronqvist, Ruth Misener, and Calvin Tsay

Logic-Based Benders Decomposition for an Inter-modal Transportation
Problem.   315
   Ioannis Avgerinos, Ioannis Mourtos, and Georgios Zois

Checking Constraint Satisfaction   332
   Victor Jung and Jean-Charles Régin

Finding Subgraphs with Side Constraints   348
   Özgür Akgün, Jessica Enright, Christopher Jefferson, Ciaran McCreesh,
   Patrick Prosser, and Steffen Zschaler

Short-Term Scheduling of Production Fleets in Underground Mines Using
CP-Based LNS   365
   Max Åstrand, Mikael Johansson, and Hamid Reza Feyzmahdavian

Learning to Reduce State-Expanded Networks for Multi-activity
Shift Scheduling   383
   Till Porrmann and Michael Römer

SeaPearl: A Constraint Programming Solver Guided
by Reinforcement Learning 392
   Félix Chalumeau, Ilan Coulon, Quentin Cappart,
   and Louis-Martin Rousseau

Learning to Sparsify Travelling Salesman Problem Instances 410
   James Fitzpatrick, Deepak Ajwani, and Paula Carroll

Optimized Item Selection to Boost Exploration for Recommender Systems... 427
   Serdar Kadıoğlu, Bernard Kleynhans, and Xin Wang

Improving Branch-and-Bound Using Decision Diagrams
and Reinforcement Learning................................ 446
   Augustin Parjadis, Quentin Cappart, Louis-Martin Rousseau,
   and David Bergman

Physician Scheduling During a Pandemic........................ 456
   Tobias Geibinger, Lucas Kletzander, Matthias Krainz, Florian Mischek,
   Nysret Musliu, and Felix Winter

Author Index 467

Supercharging Plant Configurations Using Z3

Nikolaj Bjørner[1](✉), Maxwell Levatich[2], Nuno P. Lopes[3],
Andrey Rybalchenko[3], and Chandrasekar Vuppalapati[4]

[1] Microsoft, Redmond, WA, USA
nbjorner@microsoft.com
[2] Columbia University, New York City, NY, USA
[3] Microsoft, Cambridge, UK
[4] Microsoft, Sunnyvale, CA, USA

Abstract. We describe our experiences using Z3 for synthesizing and optimizing next generation plant configurations for a car manufacturing company (The views expressed in this writing are our own. They make no representation on behalf of others). Our approach leverages unique capabilities of Z3: a combination of specialized solvers for finite domain bit-vectors and uninterpreted functions, and a programmable extension that we call constraints as code. To optimize plant configurations using Z3, we identify useful formalisms from Satisfiability Modulo Theories solvers and integrate solving capabilities for the resulting non-trivial optimization problems.

1 Introduction

The *digital transformation* is widely recognized as an ongoing seismic shift in today's industries. It encompasses an integration of software technologies in every aspect of a business. AI advances, overwhelmingly dominated by deep machine learning, are widely hailed as pivotal to this shift. Meanwhile, advances in symbolic reasoning, exemplified by Microsoft's Z3 symbolic solver for automated reasoning, have powered automated programming and analysis engines in the past decade. They have been transforming software engineering life cycles by enabling tools for ensuring strong provable guarantees and automatically synthesizing code and configurations. Likewise, digital transformations in the car industry are powering driving experiences. With new models and factories being churned out at a brisk pace, there is an urgent need for automating and optimizing production plants to increase the pace of production while reducing costs and resource requirements. The organization of production assembly lines involves a combination of hundreds of assembly stations and thousands of operators completing tens of thousands tasks with tens of thousands different tools available. Some tasks must be completed in sequential order, some stations may not be able to service tasks with conflicting requirements, only a subset of available

---

M. Levatich—Work performed while an intern at Microsoft.

P. J. Stuckey (Ed.): CPAIOR 2021, LNCS 12735, pp. 1–25, 2021.
https://doi.org/10.1007/978-3-030-78230-6_1

operators may be able to work on a given task, and all tasks are packed into stringent timing bounds on each station.

Planning large scale production lines is thus becoming a complex *inhuman* puzzle, that at best takes weeks for an extensively trained expert to solve manually. A manual assignment of tasks comes with no automatic assurance of optimality and with no easy way to explore alternatives. Our experiences with Z3 are primarily based on software analysis, verification and synthesis. Z3 is a default tool when it comes to applications around translation validation, symbolic execution, and program verification. It has taken inhuman tasks out of network verification in Azure's operations [21], invoking $O(1B)$ (small) Z3 queries a day, verifying compiler optimizations [23,24], and finding or preventing security vulnerabilities in complex systems code [6,17].

Are techniques that have been tested in the area of software analysis usable for production line scheduling? We have some partial answers that indicate the underlying technologies in Z3 can be put to good use in combinatorial domains. The Dynamics product configurator tool [25] ships using Z3 for solving product configuration tasks. It uses Z3 to enumerate consequences: when an operator fixes a fabric of a sofa, the color choices narrow and Z3 integrates a custom optimized consequence finding module built tightly with it's Conflict Driven Clause Learning, CDCL [34], engine. The production plant design scenario is very different from the Dynamics use case, though. The model is complex and does not fit within a commoditized environment. It involves solving multi-knapsack problems with complex side-constraints. We describe our experiences using Z3 for automating next generation production plant designs of a car manufacturing company. Our journey so far involves a combination of *deep cleaning*, thus the activities involved with formalizing a complex model and in the process identifying and fixing data-entry bugs; and *deep solving*, that is, the combination of solver capabilities used to optimize virtual plant configurations. Within the scope of our experiences we pose and test a hypothesis that solving constraints using uninterpreted functions, a base theory of SMT solvers, together with solving for bit-vectors (that capture finite domains), presents a compelling target for multi-knapsack problems with complex side constraints. We argue that uninterpreted functions can be used effectively to encode assignment constraints. Furthermore, solving for these constraints using decision procedures for uninterpreted functions may have a substantial advantage solving MIP or SAT based formulations. To handle multi-knapsack constraints we were compelled to extend Z3 with an interface for user theories: encoding these constraints as *code* appeared readily more viable than supporting custom global constraints.

1.1 Complexity Without Perplexity

The complexity and difficulty of the problem we are tackling can be characterized along two dimensions: the complexity inherent in capturing production line models and complexity based on the size of production lines that are solved for. In our case, the production line model requires a few dozen different types of data points. Each type is represented as a database table, and each data-point requires at least three and sometimes more than twenty attributes, where each attribute is represented as a database column. The second dimension of complexity can be measured by the size of database instances that are required to capture production line models. In our case, the order of magnitude along main tables are as follows:

- Stations: $O(100)$. A production line is a sequence of stations. Each station is a collection of around 10 operator positions, of which a subset can be used.
- Operator positions: $O(1K)$.
- Processes: $O(1K)$, each process is a collection of tasks.
- Tasks: $O(10K)$, assigned among the processes.
- Tools: $O(1K)$ are assigned to tasks.

The real killer for straight-forward approaches, though, is that production line constraints involve joining several large tables.

We approach the first dimension through the lens of software engineering methodologies: we are creating a formal model of a production plant and synthesizing optimized configurations. We address the second dimension by describing the technological features we found useful for (efficiently) solving the production line automation.

1.2 Domain Engineering - *Deep Cleaning*

Our approach to production line modeling is very much influenced by concepts and methodologies honed and developed in the software engineering and most specifically formal methods communities. Thus, a starting point is to describe using logical notation a set of domains, functions over the domains, and constraints over the signature. At this stage we seek to delay lower level decisions on how constraints are encoded into a solver. Domain engineering produces a mathematically unambiguous and machine checkable account for production line modeling. We also claim our case is distinguished by some level of complexity: it integrates a combination of many rules and constraints that apply only for special cases. Our experiences with domain engineering falls into two categories. *Domain invariants* are global properties of well-formed production line models. A production line model that violates domain invariants does not correspond to a physical production line. *Domain constraints* capture the solution space of virtual plants. They can take advantage of domain invariants by assuming they have been checked and they don't have to be re-enforced in the constraint encoding.

To summarize, we distinguish between the two categories:

1.2.1 Domain Invariants
They describe well-formedness conditions of virtual plant configurations, such as:

- Dependencies between processes are acyclic.
- Stations are connected in a rooted tree comprising of sub-lines.

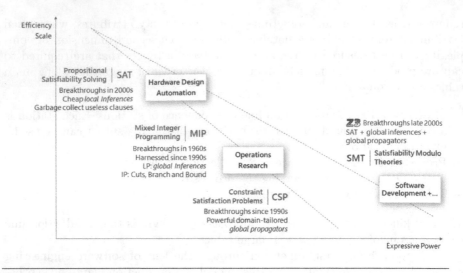

Fig. 1. SAT, MIP, CP and SMT

- Sub-lines may be labeled to force processes within bounds, the same label (called monuments) cannot be used on different branches.

We found that graph visualization tools, in particular MSAGL [26], provided a highly effective way to both communicate assumptions about domain invariants and to uncover violations.

1.2.2 Domain Constraints
They encode constraints on valid solutions, and should be:

- Sufficient to capture the solution space of virtual plants.
  - Every solution to hard constraints is a virtual plant solution.
  - Every virtual plant solution has a solution to hard constraints.
- Usable to map constraint violations back into a root cause analysis for data-entry errors.

1.3   Solver Engineering - *Deep Solving*

As the main tool used so far for solving for production lines is Z3, we are going to mainly describe the approach taken relative to Z3 and SMT technologies. Figure 1 suggests a classification of main techniques pursued in the CPAIOR community.

The way to understand the expressiveness/efficiency trad-off is that expressiveness comes with the benefit of handling increasingly succinct ways of capturing constraints, their propositional encoded counter-parts being impractical for SAT solvers. The lower efficiency means that the more expressive solvers use relatively more overhead per succinct constraint than a SAT solver per clause.

Each class of domain is labeled by distinguishing features of their mainstream state-of-art solvers. Modern SAT solvers are mainly based on a CDCL architecture that alternates a search for a solution to propositional variables with resolution inferences when a search dead-ends. These inferences rely on a limited set of premises. Garbage collection of unused derived clauses is a central ingredient to make CDCL scalable. MIP solvers use interior point methods, primal and dual simplex algorithms. Simplex pivoting performs a Gauss-Jordan elimination, which amounts to globally solving for a selected variable with respect to *all* constraints. CP solvers have perfected the art of efficient propagators for global constraints. Effective propagators narrow the solution space maximally with minimal overhead.

While we have shamelessly positioned SMT solving as reasonable efficient while exceedingly expressive relative to peers, the main message of the illustration is that SMT technologies have been developed, especially since the 2000s, with an emphasis on software applications and borrowing and stealing techniques that have otherwise been perfected by peer technologies. This is exemplified by the fact that Z3's main solver is based on a SAT solver CDCL architecture; it uses dual simplex and related global in-processing techniques from SAT solving to take advantage of global inferences; and finally, our use case illustrates incorporating global propagators. A well-recognized competing foundation for integration of solvers is to leverage a MIP solver instead of a CDCL core. This foundation benefits from global inferences and having strong MIP. Lookahead solvers, developed in the SAT community in the 90s [19] and revitalized for cube and conquer solving in the past decade [20], share some of the same traits as global MIP inferences and are promising methods for partitioning harder problems for a setting with distributed solving.

The solution to virtual plant configurations we are going to describe relies on SMT techniques. In particular, we are going to leverage mostly propositional SAT solving, coupled with a core SMT theory, the theory of uninterpreted functions, and augmented with a CP-inspired plugin for propagating global constraints.

2   Virtual Plant Configurations

In the following we describe virtual plant configurations in sufficient detail to appreciate problem characteristics and the nuances involved. The plant configurations are *virtual*; several points in a design space are explored for planning final physical configurations. We do omit several details that don't introduce crucial different concepts. For example, our full model contains a notion of *sub-line*, which are line segments. It contains constraints that limit how processes can be assigned to common sub-lines. Otherwise, our presentation is purposefully somewhat low level to convey an idea of the number and nature of concepts required for domain engineering. We use the following main domains to describe production lines:

Station, Line, Monument, Process, Task, Zone, Operator

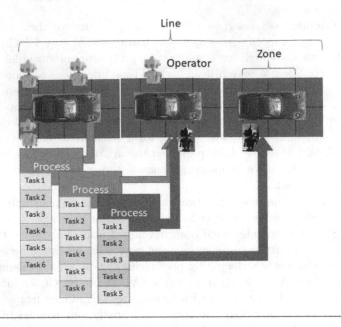

Fig. 2. Elements of a production line

Production lines are specific instantiations of these domains. The set *Task* is instantiated with $O(10K)$ different tasks and *Station* comprises of hundreds of stations (Fig. 2).

The virtual plant optimization problem is, in a nutshell, to assign each *Task* to a station and operator on the selected station. Thus, we are synthesizing two functions:

station: *Task* → *Station*
operator: *Task* → *Operator*

The assignment is subject to timing, capacity, and precedence constraints, and optimization objectives to minimize operator use, minimize station utilization, minimize tool utilization, and minimize height incompatibilities between tasks assigned to the same station.

Let us first describe the relevant domains and then give an idea of the hard constraints and optimization objectives.

2.1 Domains

2.1.1 Stations and Monuments

Each station supports a subset of viable *operators* and has an associated timeout. Tasks assigned to the same operator on a station must be completed within the station timeout. The optional *monument* attribute is used to impose ordering between processes and stations. The reader can think of a monument as a

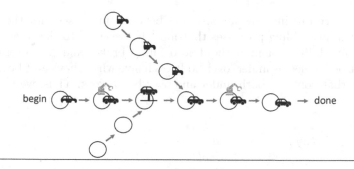

Fig. 3. A fishbone assembly line. Each circle represents a station. Partially completed artifacts move between stations. The main line (in the center) carries partial cars where parts are attached to, while sub-lines flowing to the main line assemble smaller parts into smaller ones so that they reach the main line as a single part. For example, doors can be assembled in a sub-line and attached to the car in the main line. Each station has a different set of tools and machines.

coloring, and ordering constraints can be imposed on a set of stations with the same color. Stations are indexed by unique keys.

$$Station = \langle$$

key :	Id	
$monument$:	$[Monument]$	$Optional\ monument\ tag$
$next$:	$[Station]$	$Next\ station\ in\ line$
$timeout$:	$Numeral$	$Station\ time\ bound$
$line$:	$Line$	$Sub-line\ where\ station\ resides$
$operators$:	$Operator\text{-}\mathbf{set}$	$Viable\ operators\ on\ a\ station$

$$\rangle$$

Stations are organized in a tree structure, also referred to as a *fish-bone* structure. Figure 3 illustrates an abstract production line. The *next* attribute points to the successor station closer to the root of the tree. It is null if the station is last on the line.

For the purpose of this paper, monuments and lines are identified by a unique key.

$$Monument = \langle key : Id \rangle$$
$$Line = \langle key : Id \rangle$$

With each monument, m, the set of stations tagged by m is given by:

$$stations(m) := \{s \in Station \mid s.monument = m \wedge m \neq null\}$$

2.1.2 Processes, Tasks, Zones, and Operators

A process encapsulates a set of related tasks. Processes may be constrained in three ways: The *before* and *after* attributes are used to constrain processes to be assigned to stations before/after stations labeled by the given monuments. The *predecessor* attribute imposes an ordering between processes and the *parallel* attribute identifies sibling processes that must be assigned to the same stations (e.g., one cannot fill coolant without also filling brake fluids). Processes may furthermore be labeled as under/over body exclusive when they can't be assigned to a station that contains both under and over-body work. Thus, we have:

$$
Process \;=\; \langle
$$

key :	*Id*	
ubx :	*Bool*	
before :	*[Monument]*	*Monument to precede*
after :	*[Monument]*	*Monument to succeed*
predecessor :	*[Process]*	*Process to succeed*
parallel :	*Process*-**set**	*Processes to co − assign*

$$
\rangle
$$

Tasks are associated with a host process and characterized by a completion time, the height where the task is completed, a set of viable operators that are capable of servicing the task, and a flag *ub*, indicating whether the task is completed below the car.

$$
Task \;=\; \langle
$$

key :	*Id*	
process :	*Process*	*Process where task belongs*
time :	*Numeral*	*task execution time*
height :	*Numeral*	*work height*
zone :	*Zone*	*area where task takes place*

$$
\rangle
$$

$$
Zone \;=\; \langle
$$

key :	*Id*	
operators :	*Operator*-**set**	*viable operators for zone*
ub :	*Bool*	*is the zone upper or lower body*

$$
\rangle
$$

The set of tasks associated with a process p is therefore given by:

$$
tasks(p) \;:=\; \{t \in Task \mid t.process = p\}
$$

Finally, there is a finite small set of possible operators per station.

$$
Operator \;=\; \{Op_1, \ldots Op_{10}\}
$$

2.2   A Formalization of Domain Constraints

In the following we will describe a representative set of domain constraints. They capture *hard* constraints that must be met by physical or policy requirements, such as, one cannot attach a steering wheel before the dashboard is in place. The formalization comes close to the working model, but leaves out a few details to preserve space. Precedence constraints capture ordering requirements between processes and between stations and processes. Operator constraints capture how tasks can be assigned to operators on assigned stations. Finally, cycle-time constraints bound the number and duration of tasks assignable to a station. To formulate the precedence constraints we will need a predicate that captures the partial order on stations.

2.2.1   Station Precedence Encoding

The relation

$$\preceq: Station \times Station \rightarrow Bool$$

defines a partial (tree) order on stations. The ordering on stations can be encoded by introducing two sequence numbers: The first sequence number $leftOrd$ is obtained by assigning sequence numbers following a depth-first, left-to-right order tree traversal, the other, $rightOrd$ from a depth-first, right-to-left traversal. In other words, the traversal starts with the last station in the line, walks back and either branches to the left-most sub-line whenever two lines join or branches to the right-most sub-line. Then $\preceq$ is defined as:

$$s_2 \preceq s_1 := leftOrd(s_1) \leq leftOrd(s_2) \land rightOrd(s_1) \leq rightOrd(s_2)$$

2.2.2   Process Precedence Constraints

For every task t we have the *monument ordering* constraints that confine tasks between monuments:

$$\min stations(t.process.after) \preceq station(t) \preceq \max stations(t.process.before)$$

In words: a monument m may be associated with a set of stations S. A process $p := t.process$ that has m as the **before** monument, should be assigned to a station that is before, or including, the last station in S. A process p that has m as the **after** monument should take place after, or including, the first station in S. We conveniently assume $\min \emptyset = -\infty$ and $\max \emptyset = +\infty$ to deal with the cases where the *before* or *after* monument attributes are null.

The astute reader will note that it is also possible to have monument attributes, such as $t.process.before$, that are untethered. That is, there is no station tagged with the monument. For these monuments, the rule is that all processes associated with the same untethered monument reside on the same line, so we impose equations of the form

$$station(t_1).line = station(t_2).line$$

for cases such as $t_1.process.before = t_2.process.before \neq null$ with the requirement that $stations(t_1.process.before) = \emptyset$.

Similarly, a precedence relation is imposed by *predecessor* processes:

$$t_1.process = t_2.process.predecessor \Rightarrow station(t_1) \preceq station(t_2)$$

Parallel tasks have to be assigned the same stations:

$t_1.process \in t_2.process.parallel \Rightarrow station(t_1) = station(t_2)$

Tasks belonging to the same process are assigned to the same or at most two neighboring stations:

$$t_1.process = t_2.process \Rightarrow \begin{aligned} &station(t_1) \in \{station(t_2), station(t_2).next\} \\ \vee\; &station(t_2) \in \{station(t_1), station(t_1).next\} \end{aligned}$$

2.2.3   Operator Constraints

Tasks may only be assigned to stations that supports one of the assignable operators:

$operator(t) \in t.zone.operators \cap station(t).operators$

Tasks assigned the same zone on a station must use the same operator.

$$(t_1.zone = t_2.zone \wedge station(t_1) = station(t_2)) \\ \Rightarrow operator(t_1) = operator(t_2)$$

Tasks that are marked as under-body exclusive cannot be assigned to a station with tasks having conflicting zones:

$$(t_1.process.ubx \wedge station(t_1) = station(t_2)) \\ \Rightarrow t_1.zone.ub \Leftrightarrow t_2.zone.ub$$

At most 6 operators (preferably at least 2) can be assigned to a station:

$2 \leq |\{operator(t) \mid t \in Tasks, station(t) = s\}| \leq 6$

The difference between the max and min height used at a station s is bounded (by 200 mm):

$StationHeights(s) := \{t.height \mid t \in Tasks, station(t) = s\}$
$\max StationHeights(s) - \min StationHeights(s) \leq 200$

2.2.4   Cycle-Time Constraints

The time taken by tasks assigned in each operator zone on a station cannot exceed the station completion-time. To formulate this constraint, define the $opTime$ of operator op on stations s as:

$opTime(s, op) := \sum\{t.time \mid t \in Task, station(t) = s, operator(t) = op\}$

Then the cycle time constraints are, for every station s and $op \in s.operators$:

$opTime(s, op) \leq s.timeout$

2.3   Objectives

There is no single objective that governs as a metric for the quality of a production line. Instead there is a collection of objectives that are desirable. They are derived from reducing the cost and maximizing throughput of a production line. Costs are determined by the number of operators and the number of physical assets, stations, and tools; the main cost reduction objectives are therefore:

- Minimize overall number of operators used in a production line.
- Minimize overall number of utilized stations, that is, stations with a non-zero number of operators.
- Minimize overall number of different tools used for the production line.

Other auxiliary objectives are indirectly related to cost. For instance, avoiding lifting and lowering tools and cars between stations, contributes to a smoother operation with reduced risks for accidents.

- Minimize operator congestion on stations. A station is congested if it uses more than four operators.
- Minimize process fragmentation, that is minimize the number of processes that are split.
- Minimize the height differences of tasks within each station and between adjacent stations.
- Minimize operators that are used, but under-utilized on a station, i.e., the operator's assigned tasks can be completed in a small fraction of the station's overall timeout.

2.4   Solvable Formalizations

The formalization we just presented fully describes a set of admissible configurations. It is, however impractical to work with and a much more compact encoding is possible by taking advantage of characteristics of the model and by using specialized code to enforce constraints instead of creating large formulas.

2.4.1   Processes Instead of Tasks

An intrinsic property of the model is that tasks are naturally grouped by processes. The grouping is reflected in the admissible assignments: tasks belonging to the same process can only be assigned to at most two adjacent stations. The number of tasks is furthermore an order of magnitude larger than the number of processes. Thus, by formulating constraints by referring to processes instead of tasks saves roughly an order of magnitude constraints. So instead of solving for assigning tasks to stations we solve for assigning processes to stations, and independently determine whether processes are split. In the modified formulation we are therefore synthesizing the functions:

$$station: \quad Process \rightarrow Station$$
$$operator: \quad Process \times Zone \rightarrow Operator$$

Operator assignment takes a work zone as argument to account for that different tasks within a process are allowed to be assigned different zones.

For processes where all tasks are assigned the same station, all properties of tasks are preserved. But for processes whose tasks are split between stations, the reformulation to processes reduces the solution space from the solver. For splittable processes we partition process tasks into two partitions $p.preTasks$ and $p.postTasks$, such that $p.preTasks \cup p.postTasks = tasks(p)$ and $p.preTasks \cap p.postTasks = \emptyset$.

Committing early on for whether processes can be split is a potential source of fragmentation: a solution may not be able to fully utilize station resources because processes are split while they can still utilize some station time. Characteristics of the production plant models come to the rescue, though. The vast majority of processes are relatively short running compared to station timeouts and any internal fragmentation resulting from restricting how they may be split is a smaller fraction of station timeouts.

To describe the process-based encoding we introduce a predicate:

$isSplit :   Process \rightarrow Bool$ *Is process split between stations*

and require that split processes reside on the same line:

$isSplit(p) \Rightarrow station(p).line = station(p).next.line$

Constraints that are originally formulated using $station(t)$, where t is a task, are now reformulated using processes, using $station(p)$ for process p containing task t. Converting the encoding to use processes is relatively straight-forward, thus we omit it.

2.4.2 Uninterpreted Functions to the Rescue

One approach to encode height constraints is to introduce two functions:

$minHeight :   Station \rightarrow Nat$ *Minimal height of tasks on a station*
$maxHeight :   Station \rightarrow Nat$ *Maximal height of tasks on a station*

and then impose

$$minHeight(station(t)) \leq t.height \leq maxHeight(station(t)) \quad \forall t \in Task$$
$$maxHeight(s) - minHeight(s) \leq 200 \qquad\qquad\qquad \forall s \in Station$$

If we did not have functions to our disposal, and instead used two variables $s.minHeight, s.maxHeight$ per station s, we would have to formulate the bounds on $s.minHeight$ and $s.maxHeight$ using $|Task| \times |Station|$ constraints of the form:

$$station(t) = s \Rightarrow s.minHeight \leq t.height \leq s.maxHeight,$$

for each task t and station s.

By using uninterpreted functions we only assert $|Task|$ constraints to enforce each min-height bound. With hundreds of stations, this saves two orders of magnitudes in the encoding. Furthermore, as we are also only indirectly encoding $station(t)$ by instead using an assignment of stations on processes, we save another order of magnitude in terms of number of constraints. The process-based encoding, thus takes the form:

$$\neg isSplit(p) \Rightarrow minHeight(station(p)) \leq \min\{t.height \mid t \in tasks(p)\}$$
$$isSplit(p) \Rightarrow minHeight(station(p)) \leq \min\{t.height \mid t \in p.preTasks\}$$
$$isSplit(p) \Rightarrow minHeight(station(p).next) \leq \min\{t.height \mid t \in p.postTasks\}$$

and symmetrically for $maxHeight$.

2.4.3 Avoiding Pairwise Constraints

Modeling with uninterpreted functions comes with some useful tricks of the trade. For example, if we wish to enforce that a function f is injective, it can be encoded by requiring for every pair of argument combination x, y:

$$f(x) = f(y) \Rightarrow x = y$$

But a much more succinct encoding uses an auxiliary partial inverse function g with constraints

$$g(f(x)) = x$$

for every x. This can have a dramatic effect if the domain of f is large; say the number of tasks is $O(10K)$, then the pairwise encoding requires $O(100M)$ constraints. A phenomenon related to injectivity surfaces when encoding zone assignments and under-body mutual exclusion. Recall the requirements

$$(t_1.zone = t_2.zone) \wedge (station(t_1) = station(t_2)) \Rightarrow operator(t_1) = operator(t_2)$$
$$t_1.process.ubx \wedge (station(t_1) = station(t_2)) \Rightarrow (t_1.zone.ub \Leftrightarrow t_2.zone.ub)$$

They consider all pairs of tasks. The first requirement can be captured more succinctly by introducing a predicate that tracks which work zones are used on a station and a function $wz2op$ that assigns operators to work zones on stations. The second can be handled using a similar idea that uses a predicate that tracks whether a station is assigned an under-body exclusive task.

Thus, for each process p:

$$\neg isSplit(p) \quad \Rightarrow wzUsed(station(p), z) \qquad \forall z \in \{t.zone \mid t \in tasks(p)\}$$
$$isSplit(p) \quad \Rightarrow wzUsed(station(p), z) \qquad \forall z \in \{t.zone \mid t \in p.preTasks\}$$
$$isSplit(p) \quad \Rightarrow wzUsed(station(p).next, z) \quad \forall z \in \{t.zone \mid t \in p.postTasks\}$$

If the process has $p.ubx$ set to true, we add also:

$$\neg isSplit(p) \quad \Rightarrow wzUbx(station(p), z) \qquad \forall z \in \{t.zone \mid t \in tasks(p)\}$$
$$isSplit(p) \quad \Rightarrow wzUbx(station(p), z) \qquad \forall z \in \{t.zone \mid t \in p.preTasks\}$$
$$isSplit(p) \quad \Rightarrow wzUbx(station(p).next, z) \quad \forall z \in \{t.zone \mid t \in p.postTasks\}$$

Note that since practically all tasks associated with each process share the same zone, there are in the common case only three constraints per process for $wzUsed$, and for $wzUbx$, respectively.

For station s and each work zone z

$$wzUsed(s, z) \;\Rightarrow\; \begin{aligned} & wz2op(s, z) \in z.operators \cap s.operators \\ \wedge\; & wzUbx(s, z) \Rightarrow (ubUsed(s) \Leftrightarrow z.ub) \end{aligned}$$

Note how the predicate $ubUsed(s)$ gets constrained to be true if $z.ub$ is true and $wzUbx(s, z)$ is implied based on some task occupying the workzone z on station s.

2.4.4   Cycle Time Constraints as Code

Finally, we omit encoding cycle time constraints entirely in our formulation. A major issue with fully expanding cycle time constraints is that it requires in the worst case to include the possibility that each task is assigned to every possible station and operator zone. Thus, it requires $|Station| \times |Operators|$ constraints each adding up $|Task|$ terms. Section 4.4 describes our encoding of cycle time constraints as a custom propagator using an API of Z3 that allows encoding constraints as code.

3   Experiences with Domain Engineering

Section 2 described a formalization of virtual plant configurations. Let us describe how the formalization was used to debug virtual plan configurations. Instances of virtual plant configurations are stored in SQL tables. Enforcing the domain constraints is well outside the scope of domain-agnostic database consistency guarantees, but we can take a software-inspired view and treat configurations as code and check invariants as if we are checking assertions of software.

3.1   Model Visualization

The value of model visualization is very well recognized in the CP and model-based development communities [33]. The MSAGL tool [26] was initially developed to support model-based software development using abstract state machines [3], but has since been used broadly, such as in Visual Studio [30]. In our case, graph visualization proved to be an effective way to communicate how a virtual plant model in a database is interpreted in a formal model.

3.2   Checking Global Model Invariants

Initial experiments with visualization suggested that the virtual plant representation in the database did not contain sufficient information to reproduce

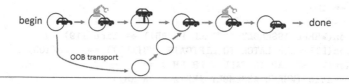

Fig. 4. Out-of-band transportation of parts between the main line and sub-lines. We cannot have a precendence relation between the producer and consumer of these parts as these processes run in parallel lines.

a physically connected production plant. Omitted data-entries or data-entry errors would render product sub-lines disconnected. Similarly, precedence relations between processes could end up being cyclic as a result of data-entry errors. The situation is analogous to software development: a type checker can catch a large class of unsafety bugs cheaply.

A common type of bugs we encountered was in the processes' precedence relations. We found several cases where a process preceding another process was supposed to run in a parallel sub-line. This is not possible as stations in parallel sub-lines have no precedence relation between them (ordering of stations is partial). This was caused by a confusion when the data was entered. These processes effectively run one before the other in a deployed production line if we consider the sub-lines side-by-side. However, there was no process precedence; it was just an artifact of the current solution.

Another kind of precedence bugs we found was related with processes that are not explicitly modeled. For example, some parts are removed from the chassis of the car in one station. Then they are transported on the side to a subsequent station where they get re-attached (Fig. 4). The transportation of these parts is not modeled because we know it can be done in a timely fashion and does not happen in the main conveyor belt, which is what we model. However, initially the processes that receive the removed parts had precedence on the processes that remove these parts, even when the receiving processes were in stations what were not successors of the removing stations. The fix was to remove the precedence relation and consider it on paper only (i.e., the process engineers have to ensure the out-of-band transportation can be done in a timely fashion).

3.3   Root-Cause Analysis Using Unsatisfiable Cores

Global invariants only ensure that solutions to satisfiable constraint encodings correspond to feasible plant configurations. They don't ensure that constraints are feasible. Infeasible constraints are as inevitable as software bugs: they originate from manual data-entry errors that are difficult to avoid because consistency is a global property involving thousands of entries. Bug localization using unsatisfiable cores and program repair using correction sets is already well recognized [22,31,36]. In Fig. 5 we show an example of an unsatisfiable core that was encountered in one of our runs. It involves chaining several equalities and arriving at the equality $1 = 3$.

```
line(103) == 1
line(119) == 3
line(station(WHEEL ASSEMBLY INSTALL FR RH)) == line(119)
line(station(LIFTGATE LATCH TO LIFTGATE INSTALL)) == line(103)
line(station(WHEEL CAP INSTALL - FR LH)) ==
    line(station(WHEEL ASSEMBLY INSTALL FR RH))
line(station(WHEEL CAP INSTALL - FR LH)) ==
    line(station(LIFTGATE SEAL - RIP CORD LH))
line(station(LIFTGATE SEAL - RIP CORD LH)) ==
    line(station(LIFTGATE LATCH TO LIFTGATE INSTALL))
```

Fig. 5. Unsatisfiable core from Z3

```
stations [TR1-240-R-N] are on line [TR]
stations [FN1-150-R-N] are on line [FN]
processes [WHEEL ASSEMBLY INSTALL FR RH] must be assigned to the
  same line as their monuments at [FN1-150-R-N]
processes [WHEEL CAP INSTALL - FR LH] sharing untethered monument
  [WHEEL ASSEMBLY INSTALL FR RH, LIFTGATE SEAL - RIP CORD LH] must
  be assigned to the same line
processes [LIFTGATE LATCH TO LIFTGATE INSTALL] must be assigned to the
  same line as their monuments at [TR1-240-R-N]
processes [LIFTGATE SEAL - RIP CORD LH] sharing untethered monument
  [LIFTGATE LATCH TO LIFTGATE INSTALL] must be assigned to the same line
```

Fig. 6. Explanation from unsat core

To make the inconsistency palatable in terms of concepts used in the data-model, we tracked each assertion by originator information and used this to produce an error report that could be digested at the level of the model, as opposed to the raw encoding. Figure 6 illustrates the same unsatisfiable core, but rendered from the perspective of the database.

Z3 uses MiniSAT's approach [16] by using tracking literals to extract unsatisfiable cores. Cores are optionally minimized using a greedy core minimization algorithm that forms the basis of SAT-based MUS extraction tools [4].

4   Experiences with Solver Engineering

We will be describing the elements used in our current solver. It finds feasible solutions to production lines within a couple of minutes and then yields optimized solutions in a steady stream as the solver explores Pareto fronts. The journey to our current approach took several iterations. During initial iterations, finding just one feasible solution was elusive.

We tried three conceptually different approaches, prior to the eventual solution we describe next. These approaches were differentiated by how they attempted to address the special complexity of cycle-time constraints.

- A first approach created an encoding of all domain constraints, except cycle-time constraints. Then the solver would assign a small batch of processes by adding cycle-time constraints at a time. The approach scaled to less than a dozen processes per batch and it took around 20 h to solve for 10% of all processes.
- In a second experiment we added cycle-time constraints for processes one by one, and greedily assigned them to stations. The experiment relied on auxiliary static analysis to narrow the range of possible station assignments for every process and we would prioritize processes with the narrowest range of feasible stations. With this approach we could assign 80% of processes using 10 h CPU processor time.
- A third experiment aimed to build a CP engine on top of Z3 by augmenting the greedy approach with backtracking so that it could assign all processes. The idea was that the external CP engine would make branching decisions on how to assign processes and also manage backtracking. While engineering this approach was too complex to fully realize, it served as a guide for the approach we arrived at with *constraints as code*. Here, branching decisions remain inside of Z3, but conflict detection and theory propagation is programmed by a CP module for cycle-time constraints.

4.1 SMT Theories and Solvers

Z3 supports a rich collection of formalisms that go well beyond the features used in this work. It supports theories of bit-vectors, uninterpreted functions, arrays, algebraic datatypes, floating points, strings, regular expressions, sequences, bounded recursive function unfolding and partially ordered relations. To support the many formalisms and different classes of formulas Z3 contains a plethora of powerful engines. A CDCL(T) core glues together most supported theories in a combined reasoning engine. The core also integrates with quantifier instantiation engines. Other reasoning cores can be invoked in stand-alone ways, including a core for non-linear real Tarskian arithmetic, decidable quantified theories, and a Horn clause solver [12]. The work described in this paper draws on only a few of the available formalisms and engines: bit-vectors and uninterpreted functions. Central to the art of solver engineering is choosing the best theories and encoding for a particular problem.

4.2 Uninterpreted Functions

We already mentioned that we use the theory of *uninterpreted functions*, also known as EUF. It is basic to first-order logic and treated as a base theory for SMT solvers. EUF admits efficient saturation using congruence closure algorithms [14]. Uninterpreted functions are well recognized in SMT applications as useful for abstracting data [13] and in model checking of hardware designs [2]. Congruence closure consumes a set of equalities over terms with uninterpreted functions and infers all implied equalities over the terms used in the equalities. Consider for example, the two equalities

$$x \;=\; f(g(f(x))), \; x \;=\; g(f(x))$$

We can use the second equality to simplify the first one: by replacing the sub-term $g(f(x))$ in $f(g(f(x)))$ by x, the first equality reduces to $x = f(x)$. This new equality can be used to simplify the second equality by replacing the sub-term $f(x)$ in $g(f(x))$ by x. The resulting equality is $x = g(x)$. Congruence closure algorithms perform such inferences efficiently, without literally substituting terms in equations.

Using EUF instead of encoding directly into SAT is not necessarily without a cost. By default, SMT solvers allow only inferences over EUF that do not introduce new terms. This prevents the solvers from producing short resolution proofs in some cases, but has the benefit of avoiding bloating the search space with needless terms. Efficient solvers seek a middle-ground by introducing transitive chaining of equalities and Ackerman reductions on demand [9,15]. For the use case described in this paper, even these on-demand reductions turn out to be harmful and slow down search. They are disabled for this application. Furthermore, we found it useful to delay restarts to give the solver time to perform model-repair in contrast to producing resolvents. Precisely how to tune SAT solvers for satisfiable instances is a topic [7,27,29] where new insights are currently developed.

4.3   Bit-Vectors

The first few encoding attempts used the theory of arithmetic and integers to represent all domains. While not exclusively responsible for inferior performance, we noticed an order of magnitude speedup on the same formulations when switching to bit-vectors. Finite domains can be encoded directly using bounded integers. The usual ordering $\leq$ on integers can then be used whenever requiring precedence relations or comparing heights. Except for cycle-time constraints that we deal with separately, there is however very little or practically no arithmetic involved with the constraints. By using bit-vectors instead of integer data-types we can force Z3 to use bit-vector reasoning for finite domains. The theory of bit-vectors is used to capture machine arithmetic, with noteworthy applications for analysis of binary code or compiler intermediary languages, thus two-complements arithmetical operations over 32-bit or 64-bit arithmetic found in machine code. Comparison, $\leq$ is defined for both signed and unsigned interpretations of bit-vectors. These operations are used extensively for modeling operator precedence and height constraints. The bit-vector representation and reasoning was order of magnitudes more efficient than using encoding relying on arithmetic. It conforms to common experiences where using arithmetic for finite domain combinatorial problems is rarely an advantage. Mainstream SMT solvers solve bit-vectors by a reduction to propositional SAT. It works well for this domain, in contrast to constraints involving multiplication of large bit-vectors. Handling larger bit-widths is a long standing open challenge for SMT solvers.

4.4　Constraints as Code

Early experiments suggested that adding Pseudo-Boolean inequalities corresponding to cycle-time constraints would be a show-stopper. It is an instance where existing built-in features do not allow for a succinct encoding. These constraints highlighted a need for exposing a flexible approach for extending Z3 with ad-hoc, external, theory solvers. Z3 exposed a way for encoding external solvers more than a decade ago [8]. External theories were subsequently removed from Z3 because not all capabilities of internal theory solvers could be well supported for external solvers. Moreover, with Z3 being open source, the path was prepared for external contributions, such as Z3Str3 [5]. But we found that the cycle-time constraints are not easily amenable to a new theory; the conditions for when they propagate consequences or identify conflicts depend on properties that are highly specific to this particular model. It is thus much easier to represent propagation and conflict detection in code than in constraints.

We will illustrate the user propagator by a simple example borrowed from [10]. It illustrates a Pseudo-Boolean constraint that requires a quadratic size encoding. In contrast, the user propagator does not suffer from this encoding overhead. The example constraint is:

$$3 \left| \{(i,j) \mid i < j \wedge x_i + x_j = 42 \wedge (x_i > 30 \vee x_j > 30)\} \right|$$
$$+ \left| \{(i,j) \mid i < j \wedge x_i + x_j = 42 \wedge x_i \leq 30 \wedge x_j \leq 30\} \right| \quad \leq 100$$

For illustration, we instantiate the example with 8 bit-vectors each with 10 bits over Python:

```
from z3 import *

xs = BitVecs(["x%d" % i for i in range(8)], 10)
```

Then a user-propagator can be initialized by sub-classing to the UserPro pagateBase class that implements the main interface to Z3's user propagation functionality.

```
class UserPropagate(UserPropagateBase):
    def __init__(self, s):
        super(self.__class__, self).__init__(s)
        self.add_fixed(self.myfix)
        self.add_final(self.myfinal)
        self.xvalues = {}
```

```
self.id2x = {self.add(x) : x for x in xs}
self.x2id = {self.id2x[id] : id for id in self.id2x}
self.trail = []
self.lim = []
self.sum = 0
```

The map xvalues tracks the values of assigned variables and id2x and x2id maps tracks the identifiers that Z3 uses for variables with the original variables. The sum maintains the running sum of according to our unusual constraint.

The class must implement methods for pushing and popping backtrackable scopes. We use a trail to record closures that are invoked to restore the previous state and lim to maintain the the size of the trail for the current scope.

```
# overrides a base class method
def push(self):
    self.lim.append(len(self.trail))

# overrides a base class method
def pop(self, num_scopes):
    lim_sz = len(self.lim)-num_scopes
    trail_sz = self.lim[lim_sz]
    while len(self.trail) > trail_sz:
        fn = self.trail.pop()
        fn()
    self.lim = self.lim[0:lim_sz]
```

We can then define the main callback used when a variable tracked by identifier id is fixed to a value e. The identifier is returned by the solver when calling the function self.add(x) on term x. It uses this identifier to communicate the state of the term x. When terms range over bit-vectors and Booleans (but not integers or other types), the client can register a callback with self.add_fixed to pick up a state where the variable is given a full assignment. For our example, the value is going to be a bit-vector constant, from which we can extract an unsigned integer into v. The trail is augmented with a restore point to the old state and the summation is then updated and the Pseudo-Boolean inequalities are then enforced.

```
def myfix(self, id, e):
    x = self.id2x[id]
    v = e.as_long()
    old_sum = self.sum
    self.trail.append(lambda : self.undo(old_sum, x))
    for w in self.xvalues.values():
        if v + w == 42:
            if v > 30 or w > 30:
```

```
            self.sum += 3
    else:
            self.sum += 1
self.xvalues[x] = v
if self.sum > 100:
    self.conflict([self.x2id[x] for x in self.xvalues])
elif self.sum < 10 and len(self.xvalues) > len(xs)/2:
    self.conflict([self.x2id[x] for x in self.xvalues])
```

It remains to define the last auxiliary methods for backtracking and testing.

```
def undo(self, s, x):
    self.sum = s
    del self.xvalues[x]


def myfinal(self):
    print(self.xvalues)


s = SimpleSolver()
for x in xs:
    s.add(x % 2 == 1)
p = UserPropagate(s)
s.check()
print(s.model())
```

4.5 Solving for Multiple Objectives

Z3 supports optimization modulo theories out of the box [11], including weighted MaxSAT and optimization of linear objectives. It can also be instructed to enumerate Pareto fronts or combine objectives through a lexicographic combination. In our case we are not, at present, using these features for optimizing objectives. Instead, we built a custom Pareto optimization mechanism on top of the user propagator. It is inspired by the branch-and-bound method for MaxSAT from [28]. The idea is that each objective function is registered with an independent constraint handler. Each handler maintains a current cost. The current cost is incremented when a variable gets fixed in a way that adds to the running cost. For example, when a task is assigned a station, the tool used by the task is added to the pool of tools used, unless the tool is already used at the station. When then number of used tools exceeds the current running best bound for tools, the handler registers a conflict. Handlers may also cause unit propagation when the current bound is reached. This approach has the benefit from producing partial results as soon as they are available. Several improvements are possible over this scheme, such as neighborhood search around current solutions. We leave this for future explorations, as the current approach is sufficient within the generous time budget for the plant configuration domain.

5   Experiences with MiniZinc

We also developed a plant model in MiniZinc. Following MiniZinc best practices, we used so-called global constraints that deal with functions/relations as first-class values, hence avoiding quantified constraints over individual elements. First, we used a channel constraint to connect a function that assigns a station to a process with its inverse (given that the inverse is used for various aggregations over station's processes). This channel dramatically improved the solving performance, compared to its equivalent formulation using universal quantification. Furthermore we used a bin packing constraint to ensure fit of processes into a station. Finally, we had to address the shortcoming of global constraints that they cannot be driven by decision variables and hence require an eager case distinction as a work-around. To reduce the ranges of decision variables participating the bin packing constraint, i.e., to avoid assuming that any process could be placed in *any* station, we developed an abstract-interpretation style approximation of the set of stations for a given process. Technically, this approximation is computed iteratively as the least-fixpoint of the propagation operator manually derived from the constraints.

We observed that the resulting performance with the Gecode solver backend is comparable with the Z3. We left it for future work to automate the construction of approximation operators.

6   Perspective

We described our experiences with using SMT and CP techniques for solving virtual plant configurations for production plants. The domain shares characteristics of job-shop scheduling and constrained knapsack problems. The scenario integrates a plethora of side constraints. Our perspective in tackling this domain is heavily influenced by methods adapted in the software-engineering, and particularly model-driven engineering and formal methods communities. Several synergies with configuration domains and advances in software engineering communities seem ripe to be explored: Automated software synthesis has gained considerable traction in the software engineering community [1]. SMT and SAT solvers are some of the popular options for handling software synthesis and *program sketching* problems. Super-compilation can be recast as a quantifier instantiation problem and template-based methods use a template space defined by abstract grammars to define a search space for synthesis problems. CVC4 [32] builds in grammar based synthesis as an extension of its quantifier instantiation engine; efficient, custom, synthesis tools such as Prose [18], Rosette [37], and for program sketching [35], integrate specialized procedures.

Our SMT solution is based on Z3 with uninterpreted functions, bit-vectors and user-programmed constraint propagators. The virtual plant configuration

solver is currently actively used for planning next generation production facilities. There are still many exciting avenues to pursue for super-charging virtual
plant configurations, or network cloud configurations and policies for that matter: methodologies and tools developed for programming languages have substantial potential to transform configuration management; configurations can be
improved using feedback measurements from deployments; and symbolic solving
have a central role in checking integrity constraints, and synthesizing solutions
while exploring a design space.

References

1. Alur, R., et al.: Syntax-guided synthesis. In: Dependable Software Systems Engineering. NATO Science for Peace and Security Series, D: Information and Communication Security, vol. 40, pp. 1–25. IOS Press (2015). https://doi.org/10.3233/
   978-1-61499-495-4-1

2. Andraus, Z.S., Liffiton, M.H., Sakallah, K.A.: Reveal: a formal verification tool for
   Verilog designs. In: Cervesato, I., Veith, H., Voronkov, A. (eds.) LPAR 2008. LNCS
   (LNAI), vol. 5330, pp. 343–352. Springer, Heidelberg (2008). https://doi.org/10.
   1007/978-3-540-89439-1_25

3. Barnett, M., Grieskamp, W., Nachmanson, L., Schulte, W., Tillmann, N., Veanes,
   M.: Towards a tool environment for model-based testing with AsmL. In: Petrenko,
   A., Ulrich, A. (eds.) FATES 2003. LNCS, vol. 2931, pp. 252–266. Springer, Heidel-
   berg (2004). https://doi.org/10.1007/978-3-540-24617-6_18

4. Belov, A., Marques-Silva, J.: Muser2: an efficient MUS extractor. J. Satisf. Boolean
   Model. Comput. **8**(3/4), 123–128 (2012). https://doi.org/10.3233/sat190094

5. Berzish, M., Ganesh, V., Zheng, Y.: Z3str3: a string solver with theory-aware
   heuristics. In: FMCAD (2017). https://doi.org/10.23919/FMCAD.2017.8102241

6. Bhargavan, K., et al.: Everest: towards a verified, drop-in replacement of HTTPS.
   In: SNAPL. LIPIcs, vol. 71, pp. 1:1–1:12 (2017). https://doi.org/10.4230/LIPIcs.
   SNAPL.2017.1

7. Biere, A., Fazekas, K., Fleury, M., Heisinger, M.: CaDiCaL, Kissat, Paracooba,
   Plingeling and Treengeling entering the SAT competition 2020. In: Proceedings
   of SAT Competition 2020 - Solver and Benchmark Descriptions. Department of
   Computer Science Report Series B, vol. B-2020-1, pp. 51–53. University of Helsinki
   (2020)

8. Bjørner, N.: Engineering theories with Z3. In: Yang, H. (ed.) APLAS 2011. LNCS,
   vol. 7078, pp. 4–16. Springer, Heidelberg (2011). https://doi.org/10.1007/978-3-
   642-25318-8_3

9. Bjørner, N., de Moura, L.M.: Tractability and modern satisfiability modulo theories
   solvers. In: Tractability: Practical Approaches to Hard Problems, pp. 350–377.
   Cambridge University Press (2014). https://doi.org/10.1017/CBO9781139177801.
   014

10. Bjørner, N., Nachmanson, L.: Navigating the universe of Z3 theory solvers. In:
    Carvalho, G., Stolz, V. (eds.) SBMF 2020. LNCS, vol. 12475, pp. 8–24. Springer,
    Cham (2020). https://doi.org/10.1007/978-3-030-63882-5_2

11. Bjørner, N., Phan, A.-D.: νZ - maximal satisfaction with Z3. In: SCSS. EPiC
    Series in Computing, vol. 30 pp. 1–9. EasyChair (2014). https://easychair.org/
    publications/paper/xbn

12. Bjørner, N., de Moura, L., Nachmanson, L., Wintersteiger, C.M.: Programming Z3. In: Bowen, J.P., Liu, Z., Zhang, Z. (eds.) SETSS 2018. LNCS, vol. 11430, pp. 148–201. Springer, Cham (2019). https://doi.org/10.1007/978-3-030-17601-3_4

13. Burch, J.R., Dill, D.L.: Automatic verification of pipelined microprocessor control. In: Dill, D.L. (ed.) CAV 1994. LNCS, vol. 818, pp. 68–80. Springer, Heidelberg (1994). https://doi.org/10.1007/3-540-58179-0_44

14. Downey, P.J., Sethi, R., Tarjan, R.E.: Variations on the common subexpression problem. J. ACM **27**(4), 758–771 (1980). https://doi.org/10.1145/322217.322228

15. Dutertre, B.: Yices 2.2. In: Biere, A., Bloem, R. (eds.) CAV 2014. LNCS, vol. 8559, pp. 737–744. Springer, Cham (2014). https://doi.org/10.1007/978-3-319-08867-9_49

16. Eén, N., Sörensson, N.: An extensible SAT-solver. In: Giunchiglia, E., Tacchella, A. (eds.) SAT 2003. LNCS, vol. 2919, pp. 502–518. Springer, Heidelberg (2004). https://doi.org/10.1007/978-3-540-24605-3_37

17. Godefroid, P., Levin, M.Y., Molnar, D.A.: SAGE: whitebox fuzzing for security testing. Commun. ACM **55**(3), 40–44 (2012). https://doi.org/10.1145/2093548.2093564

18. Gulwani, S., Polozov, O., Singh, R.: Program synthesis. Found. Trends Program. Lang. **4**(1–2), 1–119 (2017). https://doi.org/10.1561/2500000010

19. Heule, M., van Maaren, H.: Look-ahead based SAT solvers. In: Handbook of Satisfiability, vol. 185, pp. 155–184. IOS Press (2009). https://doi.org/10.3233/978-1-58603-929-5-155

20. Heule, M.J.H., Kullmann, O.: The science of brute force. Commun. ACM **60**(8), 70–79 (2017). https://doi.org/10.1145/3107239

21. Jayaraman, K., et al.: Validating datacenters at scale. In: SIGCOMM (2019). https://doi.org/10.1145/3341302.3342094

22. Jose, M., Majumdar, R.: Bug-assist: assisting fault localization in ANSI-C programs. In: Gopalakrishnan, G., Qadeer, S. (eds.) CAV 2011. LNCS, vol. 6806, pp. 504–509. Springer, Heidelberg (2011). https://doi.org/10.1007/978-3-642-22110-1_40

23. Lopes, N.P., Menendez, D., Nagarakatte, S., Regehr, J.: Provably correct peephole optimizations with Alive. In: PLDI (2015). https://doi.org/10.1145/2737924.2737965

24. Lopes, N.P., Lee, J., Hur, C.-K., Liu, Z., Regehr, J.: Alive2: bounded translation validation for LLVM. In: PLDI (2021). https://doi.org/10.1145/3453483.3454030

25. Microsoft. Microsoft dynamics (2021). https://dynamics.microsoft.com

26. Nachmanson, L.: Microsoft automated graph layout tool (2021). https://github.com/microsoft/automatic-graph-layout

27. Nadel, A., Ryvchin, V.: Chronological backtracking. In: Beyersdorff, O., Wintersteiger, C.M. (eds.) SAT 2018. LNCS, vol. 10929, pp. 111–121. Springer, Cham (2018). https://doi.org/10.1007/978-3-319-94144-8_7

28. Nieuwenhuis, R., Oliveras, A.: On SAT modulo theories and optimization problems. In: Biere, A., Gomes, C.P. (eds.) SAT 2006. LNCS, vol. 4121, pp. 156–169. Springer, Heidelberg (2006). https://doi.org/10.1007/11814948_18

29. Oh, C.: Between SAT and UNSAT: the fundamental difference in CDCL SAT. In: Heule, M., Weaver, S. (eds.) SAT 2015. LNCS, vol. 9340, pp. 307–323. Springer, Cham (2015). https://doi.org/10.1007/978-3-319-24318-4_23

30. Pupyrev, S., Nachmanson, L., Bereg, S., Holroyd, A.E.: Edge routing with ordered bundles. Comput. Geom. **52**, 18–33 (2016). https://doi.org/10.1016/j.comgeo.2015.10.005

31. Reiter, R.: A theory of diagnosis from first principles. Artif. Intell. **32**(1), 57–95 (1987). https://doi.org/10.1016/0004-3702(87)90062-2
32. Reynolds, A., Barbosa, H., Nötzli, A., Barrett, C.W., Tinelli, C.: CVC4SY for sygus-comp 2019. CoRR, abs/1907.10175 (2019). http://arxiv.org/abs/1907.10175
33. Sander, G., Vasiliu, A.: The ILOG JViews graph layout module. In: Mutzel, P., Jünger, M., Leipert, S. (eds.) GD 2001. LNCS, vol. 2265, pp. 438–439. Springer, Heidelberg (2002). https://doi.org/10.1007/3-540-45848-4_35
34. Marques Silva, J.P., Sakallah, K.A.: GRASP: a search algorithm for propositional satisfiability. IEEE Trans. Comput. **48**(5), 506–521 (1999). https://doi.org/10.1109/12.769433
35. Solar-Lezama, A.: Program sketching. Int. J. Softw. Tools Technol. Transfer 475–495 (2012). https://doi.org/10.1007/s10009-012-0249-7
36. Sülflow, A., Fey, G., Bloem, R., Drechsler, R.: Using unsatisfiable cores to debug multiple design errors. In: VLSI (2008). https://doi.org/10.1145/1366110.1366131
37. Torlak, E., Bodík, R.: Growing solver-aided languages with rosette. In: SPLASH (2013). https://doi.org/10.1145/2509578.2509586

A Computational Study of Constraint Programming Approaches for Resource-Constrained Project Scheduling with Autonomous Learning Effects

Alessandro Hill[1]([✉]), Jordan Ticktin[1], and Thomas W. M. Vossen[2]

[1] Industrial and Manufacturing Engineering, California Polytechnic State University,
San Luis Obispo, USA
{ahill29,jticktin}@calpoly.edu
[2] Leeds School of Business, University of Colorado, Boulder, USA
vossen@colorado.edu

Abstract. It is well-known that experience can lead to increased efficiency, yet this is largely unaccounted for in project scheduling. We consider project scheduling problems where the duration of activities can be reduced when scheduled after certain other activities that allow for learning relevant skills. Since per-period availabilities of renewable resources are limited and precedence requirements have to be respected, the resulting optimization problems generalize the resource-constrained project scheduling problem. We introduce four constraint programming formulations that incorporate the alternative learning-based job durations via logical constraints, dynamic interval lengths, multiple job modes, and a bi-objective reformulation, respectively. To provide tight optimality gaps for larger problem instances, we further develop five lower bounding techniques based on model relaxations. We also devise a destructive lower bounding method. We perform an extensive computational study across thousands of instances based on the PSPlib to quantify the impact of project size, potential learning occurrences, and learning effects on the optimal project duration. In addition, we compare formulation strength and quality of the obtained lower bounds using a state-of-the-art constraint programming solver.

Keywords: Resource-constrained project scheduling · Autonomous learning · Constraint programming

1 Introduction

On-the-job learning occurs in most projects that involve human workforce. However, its impact is rarely taken into account during the project planning stage, and has received relatively little attention in the literature. In this paper,

P. J. Stuckey (Ed.): CPAIOR 2021, LNCS 12735, pp. 26–44, 2021.
https://doi.org/10.1007/978-3-030-78230-6_2

we explore the impact of *autonomous* learning [5], which relies on the well-established phenomenon that repeatedly performing similar jobs results in a performance increase that follows the learning curve of the executing personnel. Specifically, we consider a setting where selected pairs of similar project tasks provide a potential for learning. Executing these tasks sequentially leads to increased efficiency in that the succeeding task can be completed faster than when processing the tasks in parallel or inverse order. We assume that the tasks are performed by exogenous personnel that has access to efficient knowledge transfer. In practice, it can be difficult to identify learning relations and quantify their corresponding learning potential. Therefore, we consider the case in which each task can learn from at most one other task. This leads to a more manageable identification strategy for project managers: (I) For each task, find a task that if completed beforehand would be most beneficial if such a task exists. (II) Estimate the corresponding learning effects and quantify the potential time savings. Note, however, that we do allow multiple tasks to benefit from a common predecessor. Moreover, if training on a task helps with a potential successor, then learning is likely to happen in the converse direction in practice.

Related Work. A comprehensive survey of the resource-constrained project scheduling problem (RCPSP), model variants, and corresponding algorithmic approaches is given in [14] and [37]. Mathematical programming formulations have been studied since [32], and time-indexed RCPSP formulations [1] theoretically and computationally outperform less common event-based formulations [21]. Successful implementations incorporate cutting planes and efficient separation algorithms (e.g., [8,42]).

Over the last decade however, Constraint programming (CP) approaches have emerged as the dominant exact solution methods for the RCPSP. The efficient incorporation of lazy clause generation [10] and the cumulative resource constraint [36], followed by several hybrid acceleration techniques (e.g., see [25]) has made CP solvers the state-of-the-art when solving small to medium sized instances. The study of different CP formulations proved CP-based approaches to also be prevailing for different RCPSP variants (e.g., [22,35]). Recently, CP has also shown to produce near-optimal schedules at very large scale when hybridized with mathematical optimization techniques [16].

In [31], learning effects are incorporated in form of both reduced resource utilization and altered job duration by considering the discrete time/cost trade-off problem (DTCTP). The DTCTP is restricted to a single non-renewable resource which is commonly limited by worker availability or budget. Each job is allowed to be executed in one of multiple possible predefined modes. The DTCTP is a special case of the multi-mode RCPSP but neither reduces to the RCPSP nor generalizes it. The setup in [31] uses the alternative modes to represent the execution of a job consuming increased amounts of resources at augmented speed. The relation between resource consumption and duration of the different modes of a job is follows an underlying (resource) learning curve. The authors devise a genetic algorithm, compare its performance to existing approaches for the non-learning case, and conclude that learning effects notably affect the obtained project schedules. A bi-objective variant of the multi-skill RCPSP with project costs and learning benefits that stem from increased resource utilization is considered in [18].

Learning effects have been studied in scheduling problems without resource considerations, such as single-machine scheduling problems [3,27,33]. An overview of scheduling problems that incorporate learning is given in [2,5]. Deterioration is a related concept that also considers dynamic job durations [13], while more general job start time-dependent durations are studied in [40]. For an overview of applications of learning in production and operations management we refer to [11]. In [28], a corresponding review with respect to human factors is provided.

To the best of our knowledge, there is little research that considers the impact of learning in resource-constrained scheduling. In [38], learning effects are incorporated in form of both reduced resource utilization and altered job duration by considering the discrete time/cost trade-off problem (DTCTP). The DTCTP is closely related to the RCPSP but is restricted to a single non-renewable resource which is commonly limited by a worker availability or a budget, and each job can be executed in one of multiple possible predefined modes. The DTCTP is a special case of the multi-mode RCPSP but neither reduces to nor generalizes the RCPSP. The setup in [38] uses the alternative modes to represent the execution of a job consuming increased amounts of resources at augmented speed. The relation between resource consumption and duration of the different job modes follows an underlying (resource) learning curve. For a review on related time/cost trade-off models we refer to [39]. Alternative job durations representing learning effects are related to job crashing in resource-constrained project scheduling [9], and can be interpreted as a rapid variant of plateauing for learning curves [41].

Contribution. There has been considerable research that evaluates the strength and lower bounds of different IP formulation alternatives. However, comparisons of alternative approaches to formulating CPs have received substantially less attention, and we believe this is an important aspect of our work. In addition, we carefully quantify the remarkable lower bounding capabilities of state-of-the-art CP solvers for RCPSPs which are largely underexplored. Overall, our main contributions are as follows.

- We define a novel RCPSP variant that incorporates autonomous learning;
- We propose problem reduction techniques for the resulting problem;
- We introduce four CP formulations, and provide an empirical comparison of their scheduling and lower bounding performance;
- We consider various lower bounding techniques based on model relaxations and destructive lower bounding, and empirically evaluate their strengths;
- We quantify the potential benefits of learning, and perform an extensive computational study based on a comprehensive set of PSPlib-based instances using a broad set of parameters.

The remainder of this paper is organized as follows. In Sect. 2, we formally define and illustrate the learning-enhanced optimization model. Different CP formulations are introduced in Sect. 3, and Sect. 4 suggests various model relaxations as well as a destructive lower bounding method. A comprehensive computational analysis for both the model and the developed approaches is conducted in Sect. 5, followed by our conclusion in Sect. 6.

2 Optimization Model

In the following, we formally define the resource-constrained project scheduling problem with learning, which we refer to as RCPSP+L. Let $T = \{0, ..., z\}$ be a discrete time horizon. We are given a set of non-preemptive jobs $J = \{1, \ldots, n\}$ and each job $j \in J$ has a duration $d_j \in \mathbb{Z}_0^+$. A set of renewable resources is denoted by R and each resource $r \in R$ has a per-period availability of q_r. Each job $j \in J$ consumes $u_{j,r} \in \mathbb{Z}_0^+$ units of resource $r \in R$ in each period that it is processed ($u_{j,r} \leq q_r$). Let $A \subset J \times J$ be a set describing a precedence relation between pairs of jobs. For $(i,j) \in A$, i is called the predecessor and j the successor. We presume that the precedence digraph (J, A) is acyclic, and that job 1 (n) has duration zero, no resource utilization, and is the only job without predecessors (successors).

Furthermore, we are given a learning relation $L \subseteq J \times J$, and *learning potentials* $l_{i,j} \in \{0, \ldots, d_j - 1\}$ for $(i,j) \in L$. We assume that the node in-degree in the learning digraph (J, L) is at most one. The latter will translate to the property that a job can benefit from the experience gained by the execution of at most one preceding job.

A schedule S is an assignment of start and end times in T to the jobs in J. Let $s(j, S)$ denote the start time, $e(j, S)$ the end time, and $d(j, S) = e(j, S) - s(j, S)$ the duration of $j \in J$ in S. Then a feasible schedule, or solution, for the RCPSP+L is a schedule S such that

- the total consumption of resource r in each period does not exceed the corresponding resource availability q_r for $r \in R$,
- job j does not start before job i ends for $(i,j) \in A$, and
- the duration of job j equals $d_j - l_{i,j}$ if job i precedes j, and d_j, otherwise.

The RCPSP+L asks for a feasible schedule S of minimal makespan; i.e., the latest job end time, $\max_{j \in J} e(j, S)$, is minimized. In the following, we denote the alternative duration $d_j - l_{i,j}$ of job j as d_j' and assume that $l_{i,j} = 0$ if $(i,j) \notin L$. Note that the RCPSP+L reduces to the RCPSP in the case that $L = \emptyset$, or $l_{i,j} = 0 \ \forall (i,j) \in L$. Accordingly, the RCPSP+L is strongly NP-hard as the RCPSP is strongly NP-hard [6]. Moreover, the relation L is neither transitive nor symmetric. We define the *learning frequency* ϕ for a problem instance as the relative number of jobs with a learning potential; i.e., $\phi = |L|/|J|$. The *learning intensity* λ is defined as the average of the learning potentials; i.e., $\lambda = \sum_{(i,j) \in L} l_{i,j}/|L|$. Let $i \prec j$ denote that there exists a directed path from i to j in (J, A) for $i \neq j \in J$.

Example. To illustrate the impact of learning, consider an instance of the RCPSP+L with $J = \{1, \ldots, 8\}$ and $R = \{1, 2\}$. Let the precedence digraph

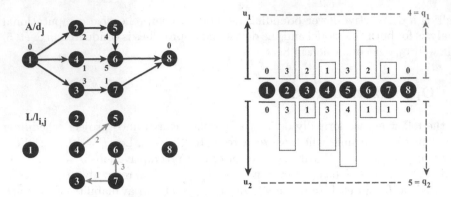

Fig. 1. Precedences with job durations (left, upper), learning digraph with learning potentials (left, lower), and resource details (right), for an example RCPSP+L instance.

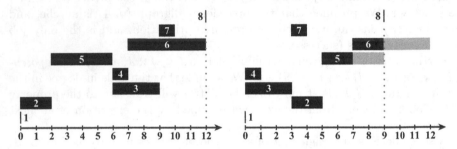

Fig. 2. An optimal solution with makespan 12 for an RCPSP instance (left) and an optimal solution with makespan 9 for the learning-enhanced RCPSP+L instance (right).

with job durations, the learning digraph with learning potentials, and the resource requirements and availabilities be as depicted in Fig. 1. An optimal schedule for the corresponding RCPSP instance (i.e., when $L = \emptyset$) with makespan 12 is illustrated in Fig. 2 (left). However, the integration of learning in the RCPSP+L allows an optimal schedule S' with makespan 9 (Fig. 2, right). It can be seen that job 5 and job 6 benefit from shortened durations $(d(5, S') = d(6, S') = 2)$ since they are scheduled after job 4 and job 7, respectively $(l_{4,5} = 2, l_{7,6} = 3)$. Note that the duration of job 3 remains unchanged since job 7 is not finished before job 3 starts.

To conclude this section, we describe two techniques that focus on preprocessing of learning relations in an RCPSP+L instance.

Precedence-Induced Learning Effects. A learning effect might be implied by the presence of a corresponding (in-)direct precedence requirement. In this case the following implication holds for every feasible schedule S.

$$((i, j) \in L \wedge i \prec j) \implies (d(j, S) = d_j - l_{i,j}) \qquad \forall i \neq j \in J$$

Consequently, we can set $d_j := d'_j$ and $L := L \backslash \{(i,j)\}$. The test can be performed efficiently via depth-first-search in (J, A) for each (i,j) in $\mathcal{O}(|L|(V + E))$.

Directed Cycle Elimination. In both the RCPSP and the RCPSP+L, it is assumed that A is acyclic since circular precedences cannot be implemented in any feasible schedule. Therefore, an arc $(i, j) \in L$ can be removed in an instance of the RCPSP+L if $(J, A \cup \{(i,j)\})$ contains a directed cycle. To efficiently check for these directed cycles, it is sufficient to check whether j is a direct or indirect predecessor of i in A using depth first search. Formally, we obtain:

$$((i,j) \in L \land j \prec i) \implies (d(j, S) = d_j) \qquad \forall i \neq j \in J.$$

3   Constraint Programming Formulations

In this section, we propose four CP formulations for the RCPSP+L that extend an efficient formulation for the RCPSP. Our objective is twofold. First, we demonstrate that using existing constraints with their propagators can lead to notable resolution performance variation. Second, we empirically identify an efficient formulation that can be implemented with minimal effort. Even though we use the term formulation, it is important to emphasize that we are considering constraint programs. As pointed out in [29], the latter generally describes a computer program rather than a mathematical description of the problem as it is done in integer programming (IP). For an in-depth comparison of CP and IP, we refer to [15, 29, 34].

We consider a well-known and efficient CP formulation [23, 25] for the RCPSP that uses an interval variable y_j to represent each job $j \in J$. We access its start time, end time, and duration by $\text{start}(y_j)$, $\text{end}(y_j)$, and $\text{length}(y_j)$.

$$(F_0) \; Min \quad \max_{j \in J} \big(\text{end}(y_j) \big) \tag{1}$$

$$subject \; to \;\; \text{cumulative\_function}\big((y_1, q_{1,r}), \ldots, (y_n, q_{n,r})\big) \; \leq \; q_r \;\; \forall r \in R, \tag{2}$$

$$\text{end\_before\_start}(y_i, y_j) \qquad\qquad \forall (i,j) \in A, \tag{3}$$

$$y_j \; \text{interval variable in } [0, z] \text{ of length } d_j \qquad \forall j \in J. \tag{4}$$

The objective (1) minimizes the latest job end time. Inequalities (2) ensure that the per-period capacity is not exceeded for any resource using a cumulative function which represents the total resource usage by all jobs in each period. The efficient propagation of resource consumption is a key for solver scheduling performance [36]. The end-start precedence relation given in A is enforced by the precedence constraints (3). We assume that the interval variables introduced in (4) have start and end times in T and fixed durations.

Logical Formulation. We first present a logic-based incorporation of learning into formulation (F_0). Then the RCPSP+L can be formulated as follows.

$$(F_1) \; Min \quad \max_{j \in J} \big(\text{end}(y_j) \big) \tag{5}$$

$$subject\ to \quad (2),\ (3), \tag{6}$$

$$\big(\texttt{end}(y_i) \leq \texttt{start}(y_j)\big) \implies \big(\texttt{length}(y_j) = d_j'\big) \qquad \forall (i,j) \in L, \tag{7}$$

$$\big(\texttt{end}(y_i) > \texttt{start}(y_j)\big) \implies \big(\texttt{length}(y_j) = d_j\big) \qquad \forall (i,j) \in L, \tag{8}$$

$$y_j\ \texttt{interval variable in}\ [0,z]\ \texttt{of length}\ [d_j', d_j] \qquad \forall j \in J. \tag{9}$$

Implications in constraints (7) and (8) enforce that the alternative duration is used if the precedence requirement is met; otherwise, the original duration is imposed. The duration of interval variables introduced in (9) is between the job's alternative duration and the original duration. The following RCPSP+L formulations reduce to formulation (F_0) if $L = \emptyset$.

Dynamic Duration Formulation. Instead of using the logical constraints (7) and (8) to implement learning effects, we can dynamically subtract the learning-based reduction from the job durations. This yields the following formulation.

$$(F_2)\ Min \quad \max_{j \in J}\big(\texttt{end}(y_j)\big) \tag{10}$$

$$subject\ to \quad (2),\ (3),\ (9), \tag{11}$$

$$\texttt{length}(y_j) = d_j - l_{i,j} * \big[\texttt{end}(y_i) \leq \texttt{start}(y_j)\big] \qquad \forall (i,j) \in L. \tag{12}$$

The length of an interval variable (Eq. (12)) is set to the original job duration minus the learning potential if a learning precedence requirement is met.

Multi-mode Formulation. The two different durations in case of a learning potential can be interpreted as two modes in which the job can be processed. The multi-mode resource-constrained project scheduling problem (MM-RCPSP) is a well-known generalization of the RCPSP that allows such multiple modes. In addition to the broad applicability of the MM-RCPSP itself, it was shown that MM-RCPSP reformulations of optimization problems yield computationally efficient approaches (e.g., [17]). We formulate the RCPSP+L using multiple modes by introducing two optional interval variables z_j and z_j' for each job j to represent the two modes.

$$(F_3)\ Min \quad \max_{j \in J}\big(\texttt{end}(y_j)\big) \tag{13}$$

$$subject\ to \quad (2),\ (3),\ (9), \tag{14}$$

$$\texttt{alternative}(y_j, [z_i, z_j]) \qquad \forall (i,j) \in L, \tag{15}$$

$$\texttt{presence\_of}(z_j') = \big(\texttt{end}(y_i) \leq \texttt{start}(y_j)\big) \qquad \forall (i,j) \in L, \tag{16}$$

$$z_j, z_j'\ \texttt{optional interval vars in}\ [0,z]\ \texttt{of length}\ d_j, d_j' \quad \forall (i,j) \in L. \tag{17}$$

The alternative constraint (15) effects that exactly one of the two variables z_j and z_j' is present, and that its length equals the length of y_j. The left-hand side of Eq. (16) consists of a `presence_of` expression which returns `true` if variable

z'_j is present in a solution and `false`, otherwise. The latter is equated with the logical precedence expression used in the right-hand side of Eq. (12).

Bi-objective Reformulation. Motivated by providing additional guidance to the CP solver, we reformulate the RCPSP+L as a bi-objective optimization problem that also maximizes the learning utilization. We apply an a priori method that ranks the makespan objective over the secondary objective by modifying formulation (F_2) as follows.

$$(F_4) \ Min \ \max_{j \in J}\big(\text{end}(y_j)\big) \ - \ \frac{l}{|J|+1} \tag{18}$$

$$subject \ to \quad (2), \ (3), \ (9), \ (12), \tag{19}$$

$$\text{int} \ l = \sum_{j \in J} \big[\text{length}(y_j) < d_j\big]. \tag{20}$$

The integer variable l (Eq. (20)) measures the number of active learning effects. In objective function (18), the minimization of the project makespan dominates the maximization of l. However, incrementing the number of learning effects ameliorates the objective function by $1/(|J|+1)$.

4 Relaxations, Restrictions and Lower Bounding

In this section, we introduce various relaxations for the RCPSP+L, and describe a destructive lower bounding method. We also suggest an approach to obtain upper bounds based on imposing a problem restriction. For an overview of lower bounding techniques for the RCPSP and their use, we refer to [7,30,37]. Relaxations strengthened by Lagrangian dualization for the related net present value RCPSP are studied in [12]. The subsequent bounding techniques are implemented as standalone methods and their individual results are analyzed in Sect. 5.

4.1 CP-Based Lower Bounding

State-of-the-art CP solvers store and dynamically update lower bounds in order to prune effectively. Besides lower bounds that are raised during the core branch-and-bound method and through inference, strong bounds are derived via the efficient solution of relaxations (e.g., [24]). Recent developments have led to significant, albeit CP solver and model-dependent, performance improvements. Currently, this strength appears to be largely unknown. We will provide quantitative evidence for the quality of this bound, and use these bounds as an optimality guarantee for unsolved instances.

4.2 Relaxations

Let $opt(P)$ be the optimal objective function value for problem instance P, and $lb(P)$ $(ub(P))$ describe a valid lower (upper) bound for $opt(P)$. Let us begin with a note on the minimal project makespan for an RCPSP+L instance P. Certainly, every job that is shortened due to a learning effect is preceded by an activity that facilitates the learning. Therefore, at least one job has to be executed with its original duration; or, in other words, not all jobs in a schedule can benefit from learning. Assuming that $d_j > 0$, this yields a lower bound: $opt(P) \geq \min_{j \in J} d_j$.

An RCPSP Relaxation. Let RCPSP$^-$(P) denote the RCPSP obtained from an RCPSP+L instance P after dropping the learning effects and setting the static job durations to be the reduced ones (i.e., $L := \emptyset$ and $d_j := d'_j \; \forall (i,j) \in L$). Since the optimal makespan for P never exceeds the optimal makespan for RCPSP$^-$(P), every lower bound for the RCPSP$^-$ is a valid lower bound for RCPSP+L: For an RCPSP+L instance P, it holds that $lb(RCPSP^-(P)) \leq opt(P)$. Consequently, every lower bounding technique for the RCPSP can be applied to derive a lower bound for the RCPSP+L.

A Project Scheduling Relaxation. A classical relaxation for the RCPSP is obtained by the removal of the resources, leading to the well-known project scheduling problem (PSP). In our setting, the obtained lower bound is only valid for an RCPSP+L instance P, if derived from RCPSP$^-$(P). We denote the corresponding PSP obtained for P by PSP$^-$(P). For an RCPSP+L instance P, it holds that $opt(PSP^-(P)) \leq opt(P)$. It is well known that the PSP can be solved efficiently via topological sorting [26] of the acyclic digraph (J, A).

A Learning Project Scheduling Relaxation. Let PSP+L(P) define the project scheduling problem with learning effects derived from an RCPSP+L instance P that does not consider resources; i.e., $R = \emptyset$. Note that this model incorporates our learning concepts into the classical project scheduling problem. For an RCPSP+L instance P, it holds that $lb(PSP+L(P)) \leq opt(P)$.

A Resource-Constrained Scheduling Problem Relaxation. We consider the scheduling problem with learning effects RCSP+L(P) obtained from RCPSP+L instance P after dropping the precedences (i.e., $A := \emptyset$). This relaxation is related to multi-dimensional packing problems when considering every resource in P and the project duration as spatial dimensions. For an RCPSP+L instance P, it holds that $lb(RCSP+L(P)) \leq opt(P)$. The RCSP+L is NP-hard since the RCSP (without learning) generalizes the two-dimensional strip-packing problem which is known to be strongly NP-hard [4].

Algorithm 1: Destructive Lower Bounding for the RCPSP+L

Input: P (RCPSP+L instance), t_{max} (destruction time limit)
Output: lb (valid lower bound for P)
1   $lb \leftarrow 0$;
2   **while** $(infeasible(P, MS \leq lb, t_{max}) = true)$ **do**
3     $\lfloor \; lb \leftarrow lb + 1$;
4   **return** lb;

4.3   Destructive Lower Bounding

Destructive lower bounding (DLB) [19] is known to produce strong lower bounds on the optimal makespan for the RCPSP. DLB is an iterative procedure. For a given valid lower bound lb, the basic idea is to prove that no schedule with makespan lb exists. If this can be done efficiently, we can increment the current lower bound ($lb := lb + 1$), and argue that the initial value for lb was destructed. Algorithm 1 describes the method. Procedure $infeasible(P, MS \leq lb, t_{max})$ returns true if infeasibility can be proven within t_{max} seconds for the RCPSP+L instance P with the additional side constraint that the makespan must not exceed lb; false otherwise. In line 1, lb can be initialized by any valid lower bound on the makespan. In practice, an efficiently computable bound, such as from Sect. 4.2, can be used to accelerate the method. We use formulation (F_2) to detect infeasibility (line 2). Note that Algorithm 1 may be modified to return an optimal schedule when replacing the latter procedure by a black-box solver that is capable of finding an optimal solution. If such an optimal solution is found, then the current lb-value is returned and equals the optimal objective function value.

4.4   A Restriction-Based Upper Bound

Let RCPSP$^+$(P) be the RCPSP derived from an RCPSP+L instance P by dropping the learning relations. Then every schedule S that is feasible for RCPSP$^+$(P) yields a valid upper bound for P, since allowing learning would lead to a makespan that is as least as good as the one of S. For an RCPSP+L instance P, it holds that $opt(P) \leq ub(RCPSP^+(P))$. The obtained optimization problem is NP-hard. However, we may use efficient methods that have been developed for the RCPSP. Furthermore, the described upper bounding method can be used to obtain a feasible schedule for P when computed in a constructive fashion. More detailed, every solution that is found for RCPSP$^+$(P) can be transformed to a solution for P by applying learning effects that may occur. This could be followed by a compacting step which aims at further reducing the current makespan.

5   Computational Study

We present the results of an extensive experimental study to better understand the possible impact of learning and the effectiveness of the scheduling and lower-bounding methods.

Since the RCPSP+L is a new problem, no real-world instances are available literature. We derive RCPSP+L test instances from 480 RCPSP instances with 30 jobs and 600 RCPSP instances with 120 jobs from the well-known PSPlib [20]. For each RCPSP base instance, we randomly choose jobs in an iterative fashion from J for frequency $\phi \in \{0, 0.1, \ldots, 1\}$ ($\phi \in \{0.25, 0.5, 0.75\}$ for $|J| = 120$) and intensity $\lambda \in \{0.1, 0.3, \ldots, 0.9\}$ ($\lambda \in \{0.1, 0.5, 0.9\}$ for $|J| = 120$) until $|L| = \lceil \phi|J| \rceil$. For such a learning successor job j, we randomly draw a learning

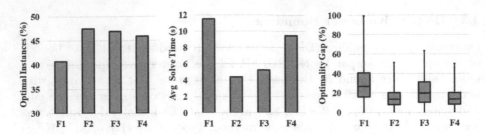

Fig. 3. The relative number of RCPSP+L instances with 120 jobs solved by each CP formulation in 60 s (left), the average run time for the optimally solved instances (center), and the corresponding optimality gaps for the unsolved instances (right).

predecessor i such that i is no (indirect) predecessor of j in A. The learning potential is set to $l_{i,j} = \min\{\lceil \lambda d_j \rceil, \max\{d_j - 1, 1\}\}$ and the arc (i, j) is added to L if $l_{i,j} > 0$. We do this for three different random seeds, resulting in a set of 79,200 RCPSP+L instances (L_{30}) for 30-job RCPSP instances, and a set of 21,600 RCPSP+L instances (L_{120}) for 120-job RCPSP instances (100,800 instances in total). We include the non-learning case (RCPSP) in our study for 30-job instances ($\phi = 0$). Moreover, the target number of arcs in L may not be met in which case we stop the procedure but include the instance. This instance generation scheme ensures that no redundancies with respect to the reduction techniques occur. Computations are conducted on an Intel i5-2320, 3 GHz, 8 GB machine. We use IBM ILOG CP Optimizer 12.9.0 as CP solver, using 4 workers.

5.1   CP Formulation Comparison

We first compare formulations (F_1), (F_2), (F_3) and (F_4) computationally with respect to test set L_{120}. We limit the per-instance computation time to 60 s. Figure 3 shows the percentage of instances that could be solved to optimality by each formulation (left), the average solve time (center) and the optimality gaps obtained for the remaining unsolved instances (right). Formulation (F_2) solves the largest number of test instances to optimality (47.6%). Formulations (F_3) and (F_4) perform almost as well (47.0%, 46.0%) whereas formulation (F_1) produces notably less optimal schedules (40.7%). Regarding the average running

Table 1. Comparison of relative (engine) formulation size, initial #solutions, and optimality gaps from lower bounds w.r.t. formulation (F_2) for 120-job instances.

Formulation	Variables ($\Delta\%$)		Constraints ($\Delta\%$)		#Sols($\Delta\%$)	Gap($\Delta\%$)
	Formulation	Engine	Formulation	Engine		
F_1	0.0	−15.7	20.8	49.7	−0.4	217.6
F_3	200	115.0	35.5	−22.1	−8.0	286.1
F_4	0.0	144.6	0.0	289.5	215.2	0.0

time to solve instances to optimality, formulation (F_2) outperforms formulation (F_3) (4.4 s < 5.3 s), formulation (F_4) (9.5 s), and formulation (F_1) (11.5 s). For 316 instances (1.5%), formulation (F_1) could not find a feasible schedule at all, which never happened for the other formulations. However, formulation (F_1) found optimal schedules for 60 instances that could not be solved by the others. Although formulation (F_2) and formulation (F_3) seem comparable, the superior performance of formulation (F_2) is further observed when considering the achieved optimality gaps for the unsolved instances in Fig. 3 (right). The average optimality gap is 15.2% compared to 22.0%. Similarly, formulation (F_2) outperforms (F_4) in terms of average optimal solution times (4.4 s < 9.5 s), as depicted in Fig. 3 (center). As mentioned in Sect. 3, constraint programs differ from integer programs and they are typically not compared with respect to their theoretical strength. A variety of solver-dependent reduction (preprocessing), extraction (internal model representation), inference (constraint propagation) and branching strategies (no-good learning) are responsible for how efficiently they can be solved. Nevertheless, we provide some empirical insights into the solver's internal behavior. In Table 1, we show the average relative differences $(\Delta\%)$ of formulation/engine variables, formulation/engine constraints, solutions found, and optimality gap (w.r.t. the best upper bound) derived from the initial lower bound for between formulation (F_2) and $(F_1)/(F_3)/(F_4)$. Note that the engine variables (constraints) are the variables (constraints) that are internally used by the solver after extracting the formulation. For example, formulation (F_3) has three times more formulation variables than (F_2), whereas formulations (F_1) and (F_4) use identical variables sets. The solver uses a reduced number (-15.7%) of engine variables for (F_1), but more than twice as many variables are used in the other formulations. Note that (F_4) produces a larger number of solutions within the time limit because schedule perturbations that do not affect the makespan may still lead to alternative solutions of different objective function value due to a change in the secondary objective value. The most significant difference between formulation (F_2) and the inferior formulations (F_1) and (F_3) is the quality of the initial lower bound on the makespan. The former leads to an optimality gap that is more than three times smaller than for the others.

5.2 Lower Bounding Performance

In order to better understand the efficacy of our algorithmic approaches, we apply both lower and upper bounding techniques from Sect. 4 to the large instances in L_{120}. First, we illustrate how effectively the bounding models can be solved using CP in Fig. 4. We use a time limit of 60 s per instance and model. It can be seen that problems PSP and PSP+L can be solved to optimality for all instances. Models RCPSP$^-$ and RCPSP$^+$ can be solved in 52.3% and 45.3%, respectively, of the cases, which is close to the RCPSP+L optimality rate (47.9%) obtained by formulation (F_2). This indicates that our best RCPSP+L formulations find optimal solutions efficiently. Note that formulations (F_1) and (F_4) solve a slightly lower number of instances to optimality (40.2% and 44.6%). Our CP approach has notable difficulties with the RCSP+L model for which only

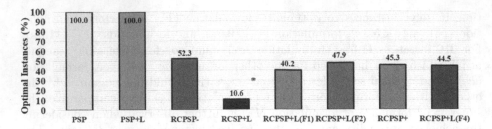

Fig. 4. The relative number of 120-job instances for which the different RCPSP+L relaxations, and formulations (F_1), (F_2) and (F_4), can be solved to optimality by CP.

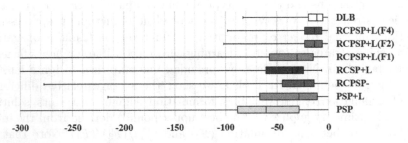

Fig. 5. The distribution of the optimality gaps achieved by the different lower bounding methods with respect to the best upper bounds for unsolved instances in L_{120}.

10.6% of the instances can be solved. The average optimality gap for the latter model is 7.0%. We also analyze the strength of the obtained lower bounds for the 13118 (60.7%) instances that could not be solved to optimality by formulation (F_1). Figure 5 depicts the achieved optimality gap distributions relative to the best upper bound. Note that the time limit in the DLB method is 60 s per bound destruction which allocates more computation time (88 s) to the overall method. Typically, all DLB subproblems can be solved efficiently except for the last one in which either a feasible schedule is found or infeasibility cannot be proven. The best non-DLB lower bound is always achieved by a RCPSP+L CP formulation. However, in 9338 cases (71%), DLB produces an even stronger bound. For (28%) of the instances, it returns the same bound, and for only 114 instances (<1%) the computed lower bound is inferior. We note that, presumably for even larger instances, model PSP+L can be used to compute initial lower bounds very efficiently. Moreover, the makespan computed by RCPSP+L can be improved by RCPSP$^+$ for nine instances.

5.3 Scheduling and Upper Bounding Efficacy

We compute upper bounds using CP for RCPSP$^+$, formulation (F_1), formulation (F_2), and formulation (F_4). Figure 6 depicts the achieved optimality gap distributions relative to the best lower bound. Formulation (F_4) performs similar to formulation (F_2) in terms of average gap (both 13.8%). However, neither one strictly dominates the other. Formulation (F_2) produces a strictly better schedule than formulation (F_4) for 35.3% of the unsolved instances, and 28.4% vice versa. Based on these observations, we suggest using formulation (F_2) since objective function values directly represent the schedules' makespans.

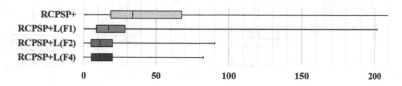

Fig. 6. The distribution of the optimality gaps achieved by the different upper bounding methods with respect to the best lower bounds for unsolved instances in L_{120}.

5.4   Overall Performance

We summarize the achievements of our CP-based approaches in comparison to the initial formulation (F_1) in Fig. 7. It shows the distribution of the optimality gaps for formulation (F_1) versus the best gaps (left), for the 120-job instances that cannot be solved by formulation (F_1). Our intention is to highlight the benefits of investigating alternative formulations, beyond formulation (F_1). We observe an 86.4% average gap reduction (96.6%→13.1%). In some cases the lower bound from formulation (F_1) is extremely poor (363 times ≤ 5), causing a high average gap. We also analyzed the optimality gap reduction for all initially open 120-job instances after applying our lower and upper bounding techniques (right). The average gap reduction is 86.2% (58.1%→8.0%). We tighten gaps for almost all the open instances (99.1%). Finally, we are able to optimally solve 8.3% of initially unsolved instances. For comparison, about 49% of the optimal solutions for 120-job RCPSP base instances are known. The average optimality gap for open instances is 5.7%. The 30-job RCPSP instances are known to be all solved. For 120-job projects, we improve the lower bound for 13 instances (see Appendix A). The optimality gap is reduced by 30% on average. When increasing the time limit to 600 s we could improve another lower bound (instance j1208_6). Note that the focus of this work is on the RCPSP+L and comprehensive computational scenario studies; but in related studies on smaller instance sets, a significantly higher time limit leads to improved results when using CP (see, for instance, [35]).

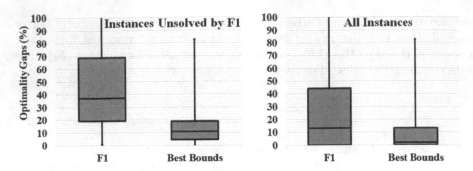

Fig. 7. The distribution of optimality gaps obtained by formulation (F_1) versus best optimality gaps from our lower bounding methods and formulations for 120-job instances; Left: instances not solved to optimality by (F_1); Right: All instances.

5.5 Learning Potential and Benefit

We perform an optimization-based computational model analysis using the cases in L_{30}. Our goal is to obtain insight into how pre-solving parameterization relates to actual makespan benefits. As in [31], we consider a very-large instance set to carefully capture the model behavior. All instances in L_{30} can be solved to optimality by formulation (F_2) in under 600 s. The average solve time is 0.85 s, and for 71094 (89.8%) instances, optimality can be proven within under 1 s. Therefore, we are able to accurately examine the impact of the model parameters on actual learning effects and best possible makespan for these instances. Figure 8 (left) shows the average relative makespan reductions for different values of ϕ and λ. It can be seen that the maximally achievable average makespan reduction (by 49%) is obtained at a learning frequency of $\phi = 1$ and a learning intensity $\lambda = 0.9$. When reducing the intensity to $\lambda = 0.1$ the optimal makespan is reduced by only 7% on average. For increasing ϕ, we observe that the average makespan reduction can be approximated by a linear function. Note that when $\phi \geq 0.8$, then the number of instances differs from other scenarios because the target number of arcs in L cannot not always be generated while asserting instance feasibility. To better comprehend the utilization of learning potentials in optimal schedules, the relative numbers of jobs that are actually performed in reduced time (with respect to all the jobs with learning potential) are shown in Fig. 8 (right). Again, we observe an almost linear growth of actual learning effects when augmenting ϕ. The average learning utilization ranges from 35.5% to 52.2% for the different values of λ. However, the maximal utilization over all the individual instances is 83.3%. In the scenario $\phi = 1$ and $\lambda = 0.9$ the lowest value is 23.3%. In both charts we observe an irregular jump between $\lambda = 0.5$ and $\lambda = 0.7$.

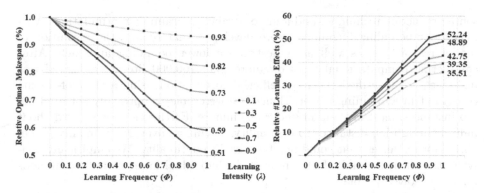

Fig. 8. Average relative optimal makespan (left), and relative #jobs that benefit from learning (right) w.r.t. instance learning frequency and intensity (30-job instances).

5.6 Parameter Performance Impact

In the following, we analyze the impact of learning frequency ϕ and learning intensity λ on the efficacy of our techniques. We use the best upper and lower bounds obtained by all methods. Figure 9 (left) shows the percentage of instances that we solve to optimality with respect to the various learning parameter combinations. It can be seen that more instances can be solved when considering $\lambda = 0.9$. The contrast to the $\lambda \in \{0.1, 0.5\}$ cases increases when augmenting the learning frequency ϕ. As illustrated in Fig. 9 (right), we achieve the best optimality gaps for instances with a low learning frequency ($\phi = 0.25$). The hardest instances for our approaches are the ones with $\lambda = 0.5$ and $\phi = 1$. In this case (1,800 instances), the average optimality gap is 12.0%, and even reaches 82.7% in the worst case.

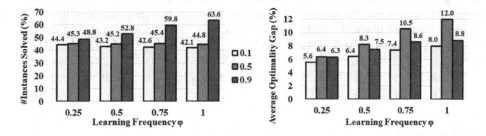

Fig. 9. The relative number of solved instances in L_{120} for the different learning parameters (left), and the corresponding average optimality gaps (right).

6 Conclusion

We introduced and studied a novel variant of the resource-constrained project scheduling problem that incorporates autonomous learning capabilities. We presented reduction techniques and four constraint programming formulations. Various lower bounding techniques were developed that require the resolution of model relaxations, as well as a destructive lower bounding approach. After conducting computational tests on more than 100,000 literature-derived test instances, we identified the most efficient formulation using a state-of-the-art constraint programming solver. Hence, we were able to optimally solve all 30-job instances, most of them in under one minute. Furthermore, we solved about half of the 120-job instances to optimality, leaving an average optimality gap of 13.1% for instances that could not be solved to optimality. We empirically analyzed the efficiency and effectiveness of the individual lower bounding methods for the unsolved problems.

Our study shows that projects can dramatically benefit from considering learning opportunities. Significant makespan reductions ($\leq 50\%$) can be achieved with ample opportunities for learning. The parameter-makespan dependency can be described as near-linear. In sum, we observe that the integration of learning potentials into resource-constrained scheduling leads to problems that can be solved by CP - when properly formulated - as efficiently as the RCPSP itself.

The resulting model represents a first step towards a new direction of research in project scheduling. We see research potential in the exploration of further CP formulations (e.g., using sequence variables) and alternative CP solvers. Furthermore, the analysis of alternative or extended learning concepts, such as for example multi-predecessor learning, could be of interest. Moreover, the development of IP-based approaches could help to better understand the challenges and opportunities of integrated learning benefits in project scheduling.

Appendix A

The improved lower bounds that we obtain with the destructive lover bounding approach described in Subsect. 4.3 are as follows (instance/lb(old)/lb(new)). Lower bounds marked with an asterisk (*) are proven optimal upper bounds. j1201_1/104/105*, j1207_9/84/85, j1208_6/84/85*, j12012_7/115/116, j12012_8/110/111, j12013_10/85/87, j12014_8/108/109, j12019_4/99/101, j12032_7/117/118, j12032_9/123/124, j12033_3/100/101, j12034_5/100/101, j12047_7/111/112, j12059_9/116/117.

References

1. Artigues, C.: On the strength of time-indexed formulations for the resource-constrained project scheduling problem. Oper. Res. Lett. **45**(2), 154–159 (2017)
2. Azzouz, A., Ennigrou, M., Ben Said, L.: Scheduling problems under learning effects: classification and cartography. Int. J. Prod. Res. **56**(4), 1642–1661 (2018)
3. Bai, D., Tang, M., Zhang, Z.H., Santibanez-Gonzalez, E.D.: Flow shop learning effect scheduling problem with release dates. Omega **78**, 21–38 (2018)
4. Baker, B.S., Coffman Jr., E.G., Rivest, R.L.: Orthogonal packings in two dimensions. SIAM J. Comput. **9**(4), 846–855 (1980)

5. Biskup, D.: A state-of-the-art review on scheduling with learning effects. Eur. J. Oper. Res. **188**(2), 315–329 (2008)
6. Blazewicz, J., Lenstra, J., Kan, A.: Scheduling subject to resource constraints: classification and complexity. Discret. Appl. Math. **5**(1), 11–24 (1983)
7. Brucker, P., Knust, S.: Lower bounds for resource-constrained project scheduling problems. Eur. J. Oper. Res. **149**(2), 302–313 (2003)
8. Demassey, S., Artigues, C., Michelon, P.: Constraint-propagation-based cutting planes: an application to the resource-constrained project scheduling problem. INFORMS J. Comput. **17**(1), 52–65 (2005)
9. Dodin, B., Elimam, A.: Integrated project scheduling and material planning with variable activity duration and rewards. IIE Trans. **33**(11), 1005–1018 (2001)
10. Feydy, T., Stuckey, P.J.: Lazy clause generation reengineered. In: Gent, I.P. (ed.) CP 2009. LNCS, vol. 5732, pp. 352–366. Springer, Heidelberg (2009). https://doi.org/10.1007/978-3-642-04244-7_29
11. Glock, C.H., Grosse, E.H., Jaber, M.Y., Smunt, T.L.: Applications of learning curves in production and operations management: a systematic literature review. Comput. Ind. Eng. **131**, 422–441 (2019)
12. Gu, H., Stuckey, P.J., Wallace, M.G.: Maximising the net present value of large resource-constrained projects. In: Milano, M. (ed.) CP 2012. LNCS, pp. 767–781. Springer, Heidelberg (2012). https://doi.org/10.1007/978-3-642-33558-7_55
13. Gupta, J.N., Gupta, S.K.: Single facility scheduling with nonlinear processing times. Comput. Ind. Eng. **14**(4), 387–393 (1988)
14. Hartmann, S., Briskorn, D.: A survey of variants and extensions of the resource-constrained project scheduling problem. Eur. J. Oper. Res. **207**(1), 1–14 (2010)
15. Heipcke, S.: Comparing constraint programming and mathematical programming approaches to discrete optimisation-the change problem. J. Oper. Res. Soc. **50**(6), 581–595 (1999)
16. Hill, A., Brickey, A., Newman, A., Goycoolea, M.: Hybrid optimization strategies for resource constrained project scheduling problems in underground mining (2019, manuscript)
17. Hill, A., Lalla-Ruiz, E., Voß, S., Goycoolea, M.: A multi-mode resource-constrained project scheduling reformulation for the waterway ship scheduling problem. J. Sched. **22**(2), 173–182 (2019)
18. Hosseinian, A.H., Baradaran, V., Bashiri, M.: Modeling of the time-dependent multi-skilled RCPSP considering learning effect. J. Model. Manag. **14**(2), 521–558 (2019)
19. Klein, R., Scholl, A.: Computing lower bounds by destructive improvement: an application to resource-constrained project scheduling. Eur. J. Oper. Res. **112**(2), 322–346 (1999)
20. Kolisch, R., Sprecher, A.: PSPLIB-A project scheduling problem library: OR software-ORSEP operations research software exchange program. Eur. J. Oper. Res. **96**(1), 205–216 (1997)
21. Koné, O., Artigues, C., Lopez, P., Mongeau, M.: Event-based MILP models for resource-constrained project scheduling problems. Comput. Oper. Res. **38**(1), 3–13 (2011)
22. Kreter, S., Schutt, A., Stuckey, P.J.: Using constraint programming for solving RCPSP/max-cal. Constraints **22**(3), 432–462 (2017)
23. Laborie, P.: IBM ILOG CP optimizer for detailed scheduling illustrated on three problems. In: van Hoeve, W.-J., Hooker, J.N. (eds.) CPAIOR 2009. LNCS, vol. 5547, pp. 148–162. Springer, Heidelberg (2009). https://doi.org/10.1007/978-3-642-01929-6_12

24. Laborie, P., Rogerie, J.: Temporal linear relaxation in IBM ILOG CP optimizer. J. Sched. **19**(4), 391–400 (2016)
25. Laborie, P., Rogerie, J., Shaw, P., Vilím, P.: IBM ILOG CP optimizer for scheduling. Constraints Int. J. **23**(2), 210–250 (2018)
26. Lasser, D.J.: Topological ordering of a list of randomly-numbered elements of a network. Commun. ACM **4**(4), 167–168 (1961)
27. Lee, W.C., Wu, C.C., Hsu, P.H.: A single-machine learning effect scheduling problem with release times. Omega **38**(1–2), 3–11 (2010)
28. Lodree, E.J., Geiger, C.D., Jiang, X.: Taxonomy for integrating scheduling theory and human factors: review and research opportunities. Int. J. Ind. Ergon. **39**(1), 39–51 (2009)
29. Lustig, I.J., Puget, J.F.: Program does not equal program: constraint programming and its relationship to mathematical programming. Interfaces **31**(6), 29–53 (2001)
30. Néron, E., et al.: Lower bounds for resource constrained project scheduling problem. In: Józefowska, J., Weglarz, J. (eds.) Perspectives in Modern Project Scheduling. ISOR, vol. 92, pp. 167–204. Springer, Heidelberg (2006). https://doi.org/10.1007/978-0-387-33768-5_7
31. Peteghem, V.V., Vanhoucke, M.: Influence of learning in resource-constrained project scheduling. Comput. Ind. Eng. **87**, 569–579 (2015)
32. Pritsker, A.A.B., Waiters, L.J., Wolfe, P.M.: Multiproject scheduling with limited resources: a zero-one programming approach. Manag. Sci. **16**(1), 93–108 (1969)
33. Qian, J., Steiner, G.: Fast algorithms for scheduling with learning effects and time-dependent processing times on a single machine. Eur. J. Oper. Res. **225**(3), 547–551 (2013)
34. Rossi, F., Van Beek, P., Walsh, T.: Handbook of Constraint Programming. Elsevier (2006)
35. Schutt, A., Chu, G., Stuckey, P.J., Wallace, M.G.: Maximising the net present value for resource-constrained project scheduling. In: Beldiceanu, N., Jussien, N., Pinson, É. (eds.) CPAIOR 2012. LNCS, vol. 7298, pp. 362–378. Springer, Heidelberg (2012). https://doi.org/10.1007/978-3-642-29828-8_24
36. Schutt, A., Feydy, T., Stuckey, P.J., Wallace, M.G.: Explaining the cumulative propagator. Constraints **16**(3), 250–282 (2011)
37. Schwindt, C., Zimmermann, J., et al.: Handbook on Project Management and Scheduling. Springer, Heidelberg (2015)
38. Van Peteghem, V., Vanhoucke, M.: Influence of learning in resource-constrained project scheduling. Comput. Ind. Eng. **87**, 569–579 (2015)
39. Vanhoucke, M., Debels, D.: The discrete time/cost trade-off problem: extensions and heuristic procedures. J. Sched. **10**(4–5), 311–326 (2007)
40. Wei, C.M., Wang, J.B., Ji, P.: Single-machine scheduling with time-and-resource-dependent processing times. Appl. Math. Model. **36**(2), 792–798 (2012)
41. Yelle, L.E.: The learning curve: historical review and comprehensive survey. Decis. Sci. **10**(2), 302–328 (1979)
42. Zhu, G., Bard, J.F., Yu, G.: A branch-and-cut procedure for the multimode resource-constrained project-scheduling problem. INFORMS J. Comput. **18**(3), 377–390 (2006)

Strengthening of Feasibility Cuts in Logic-Based Benders Decomposition

Emil Karlsson[1,2] and Elina Rönnberg[1(✉)]

[1] Department of Mathematics, Linköping University, 581 83 Linköping, Sweden
elina.ronnberg@liu.se
[2] Saab AB, 581 88 Linköping, Sweden

Abstract. As for any decomposition method, the computational performance of a logic-based Benders decomposition (LBBD) scheme relies on the quality of the feedback information. Therefore, an important acceleration technique in LBBD is to strengthen feasibility cuts by reducing their sizes. This is typically done by solving additional subproblems to evaluate potential cuts. In this paper, we study three cut-strengthening algorithms that differ in the computational efforts made to find stronger cuts and in the guarantees with respect to the strengths of the cuts. We give a unified description of these algorithms and present a computational evaluation of their impact on the efficiency of a LBBD scheme. This evaluation is made for three different problem formulations, using over 2000 instances from five different applications. Our results show that it is usually beneficial to invest the time needed to obtain irreducible cuts. In particular, the use of the depth-first binary search cut-strengthening algorithm gives a good performance. Another observation is that when the subproblem can be separated into small independent problems, the impact of cut strengthening is dominated by that of the separation, which has an automatic strengthening effect.

Keywords: Logic-based Benders decomposition · Cut strengthening · Feasibility cuts · Irreducible infeasible subset of constraints

1 Introduction

Logic-based Benders decomposition (LBBD) [9,12] is an extension of Benders decomposition [2,8] in the sense that it allows for a more general type of optimisation problem as a subproblem. In the classical Benders decomposition scheme [2], developed for mixed integer programs (MIPs), the subproblem obtained after fixing master problem variables is a linear program (LP). The dual of this subproblem is solved to find Benders cuts in terms of the master problem variables. The master problem is then resolved, and the procedure is repeated until an optimal solution to the original problem is found. In LBBD, an inference dual of the subproblem is used to find Benders cuts. With an LP subproblem, the inference dual of LBBD reduces to the LP dual as in classical Benders decomposition.

© Springer Nature Switzerland AG 2021
P. J. Stuckey (Ed.): CPAIOR 2021, LNCS 12735, pp. 45–61, 2021.
https://doi.org/10.1007/978-3-030-78230-6_3

The relation to generalized Benders decomposition is that the inference dual of LBBD reduces to the nonlinear convex dual under the conditions described in [8]. In other cases, however, the inference dual and the corresponding Benders cuts must be derived and tailored to the problem structure.

LBBD has since its introduction been applied to a variety of discrete optimisation problems [11]. One reason for the success of LBBD is the possibility to create exact hybrid algorithms that use techniques from both mathematical programming and constraint programming (CP), exploiting the respective strengths of these techniques. A common type of hybrid is to apply a MIP solver to an assignment-type master problem and a CP-based solver to a feasibility-check type of subproblem [7,10,17]. This paper addresses only this particular type of hybrid, even if we recognise that feasibility checks also occur in more general schemes where both feasibility and optimality cuts are generated.

The contributions of this paper are a unified description and overview of three cut-strengthening algorithms for feasibility cuts in the context of LBBD together with a computational evaluation of the algorithms. This evaluation is made for three different problem formulations with instances from five different applications. These problem formulations are chosen because LBBD has been applied to them in previous work, which indicate that they are good candidates for this type of decomposition. The computational evaluation is preliminary in the sense that it only reports the impact that the different algorithms have on the computational time. A detailed evaluation to fully compare the trade-off between the computational effort to strengthen the cuts and the impact these cuts have on the progress of LBBD schemes is left for future work.

The next section gives a literature overview on strengthening of feasibility cuts in LBBD. Section 3 gives a brief introduction to LBBD and presents the cut-strengthening algorithms evaluated in this paper. Section 4 introduces the three problem formulations and their decomposition. Computational results are presented in Sect. 5 and concluding comments are given in Sect. 6.

2   Literature Background

The quality of the cuts is an important aspect to consider when deriving a LBBD scheme and the application of some method to strengthen the cuts is a commonly used acceleration technique [20]. Already in a seminal work on LBBD [9], the aspect of finding minimal no-goods (feasibility cuts) was described in a chapter on search strategies. Finding such feasibility cuts in the context of LBBD can be considered as the problem of finding a minimal set of variables from the master problem whose current values cause an infeasibility in a subproblem. This problem is strongly related to that of finding an irreducible infeasible subset of constraints (IIS) in a mathematical program. Therefore, many algorithms that finds an infeasible subset of constraints can be adapted to strengthen feasibility cuts in LBBD. When choosing what algorithm to use within a particular LBBD scheme, there is a computational trade-off between the quality of cuts and the time spent to obtain this quality.

In the literature on LBBD, there are two extremes in how to deal with feasibility cuts and cut strengthening. One of them is to do no cut strengthening and the other is to find an irreducible set of master problem variables that causes the infeasibility. Between the two extremes, there are greedy approaches that aim at reducing the set of master problem variables, but they give no guarantees that the set is irreducible. When viewing this from the perspective of an infeasibility analysis, both greedy cut strengthening and irreducible cut strengthening can be seen as finding an infeasible subset of constraints, but only irreducible cut strengthening corresponds to finding an IIS.

In [5], the authors use the infeasibility conflict analyser QUICKXPLAIN for CP [14] to find feasibility cuts. First, QUICKXPLAIN is used to find an irreducible set of constraints causing the infeasibility and then this set is used to construct the feasibility cut. The problem studied in their paper is a task-allocation problem, where tasks are assigned to processors in a master problem. The subproblem checks feasibility with respect to communication and task constraints. A drawback of using QUICKXPLAIN to find feasibility cuts is that it identifies constraints rather than variables. If a global constraint (such as DISJUNCTIVE) is the cause of the infeasibility, all variables connected to this constraint are included in the cut, which may result in an unnecessarily large and less efficient cut.

A more problem-agnostic approach to strengthening of feasibility cuts is taken in [3], where the authors solve a series of subproblems to strengthen the cuts. The authors propose an algorithm that greedily strengthens the feasibility cut. They also briefly mention (and implement) an extended cut-strengthening algorithm that embeds their greedy algorithm within the iterative conflict detection algorithm implemented in QUICKXPLAIN (described in [14]) to find irreducible feasibility cuts. The problem studied in [3] is a task-to-core allocation problem handled in a multi-stage LBBD algorithm. Their master problem assigns tasks to cores and the two subproblems check feasibility with respect to memory allocation and scheduling constraints, respectively. The computational results indicate that the cut-strengthening algorithm that finds irreducible feasibility cuts performs better than their greedy algorithm.

The strategy of solving a series of subproblems to strengthen feasibility cuts is also used in [7] and [10], where LBBD is applied to scheduling problems. In [10], they derive a LBBD scheme for a facility allocation and scheduling problem with the objective to minimise either the cost of assigning tasks to facilities, the makespan or the total tardiness. In [7], they address a single facility scheduling problem, both with and without a segmented timeline. They study the feasibility version, minimisation of makespan, and minimisation of total tardiness. In both [7] and [10], any obtained feasibility cut is greedily strengthened by removing one task (corresponds to one master problem variable) in each iteration until the subproblem becomes feasible. The master problem variables corresponding to the smallest number of tasks causing an infeasibility are then used in the cut. In their cut-strengthening algorithm, they evaluate tasks in an order that exploits knowledge of the problem, but without any guarantees that the resulting feasibility cut is irreducible.

In the hybrid branch-and-check-type LBBD solver Nutmeg [17], they also take a problem-agnostic approach to strengthening of feasibility cuts. In each iteration of their cut-strengthening algorithm, they try to remove one variable of the original feasibility cut and then they evaluate the feasibility of a subproblem where the variables of this reduced feasibility cut are fixed. If the subproblem is feasible without this variable, they add it back. This algorithm produces an irreducible feasibility cut and can be viewed as an application of the deletion filter for finding an IIS, see [6] for early work in the context of linear programs. Results from applying this solver to a variety of problems, including facility allocation and scheduling, vehicle routing with location congestion, and satellite scheduling, are presented in [17].

3   Logic-Based Benders Scheme and Cut Strengthening

This section gives an overview of the LBBD scheme used in this paper and presents the algorithms used to strengthen feasibility cuts.

3.1   Logic-Based Benders Decomposition

The decomposition is applied to a problem given on the form

$$
\begin{aligned}
[\text{P}] \quad &\min \ f(x), \\
&\text{s.t.} \ \ C(x), \\
&\quad \ \ C(y), \\
&\quad \ \ x_i \rightarrow C^i(y), \quad i \in I, \\
&\quad \ \ x_i \in \{0,1\}, \quad i \in I, \\
&\quad \ \ y \in D_y,
\end{aligned}
$$

where the binary variables x are chosen as master problem variables and the variables y, that belong to the domain D_y, are chosen as subproblem variables. The objective function $f(x)$ only depends on the master variables x since the applications included in our computational evaluation have a subproblem of feasibility type. The constraints that depend only on x and y, respectively, are referred to as $C(x)$ and $C(y)$. The applications included in the study have an assignment-type master problem where each decision $x_i = 1$ imposes a restriction on y that does not depend on the other decisions $x_j = 1$, $j \in I \backslash \{i\}$. Therefore, the connection between x and y is formulated such that $x_i = 1$ implies a restriction on y by the constraints $C^i(y)$, $i \in I$. The use of such formulation facilitates a direct application of the cut-strengthening algorithms to be presented.

Below, we describe the LBBD scheme along with some additional notation. The master problem in iteration k is

$$
\begin{aligned}
[\mathrm{MP}^k] \quad &\min\ f(x), \\
&\mathrm{s.\,t.}\ \ C(x), \\
&\quad\quad B^k(x), \\
&\quad\quad \text{[Subproblem relaxation]}, \\
&\quad\quad x_i \in \{0,1\}, \quad i \in I,
\end{aligned}
$$

where $B^k(x)$ is the set of feasibility cuts generated in previous iterations. At the first iteration, $k = 1$, $B^1(x) = \emptyset$. By [Subproblem relaxation], we highlight the possibility to strengthen the master problem with a subproblem relaxation component, which can include additional variables and constraints.

Denote a solution to MP^k by $\bar{x}^k$. The subproblem for iteration k is obtained from Problem [P] by the restriction $x = \bar{x}^k$, and is given as

$$
\begin{aligned}
[\mathrm{SP}(\bar{x}^k)] \quad &\min\ 0, \\
&\mathrm{s.\,t.}\ \ C(y), \\
&\quad\quad C^i(y), \quad i \in I(\bar{x}^k), \\
&\quad\quad y \in D_y,
\end{aligned}
$$

where $I(\bar{x}^k) = \{i \in I : \bar{x}_i^k = 1\}$. If $\mathrm{SP}(\bar{x}^k)$ is feasible and has the solution $\bar{y}^k$, we have obtained an optimal solution $(\bar{x}^k, \bar{y}^k)$ to Problem [P]. If $\mathrm{SP}(\bar{x}^k)$ is infeasible, a feasibility cut $B(\bar{x}^k)$ is added to the master problem by the update $B^{k+1}(x) = B^k(x) \cup B(\bar{x}^k)$, where

$$
B(\bar{x}^k) = \sum_{i \in I(\bar{x}^k)} (1 - x_i) \geq 1. \tag{1}
$$

To obtain convergence of the LBBD scheme, a feasibility cut must be chosen such that $\bar{x}^k$ becomes infeasible in MP^{k+1}, which will always hold if the added cut is $B(\bar{x}^k)$. However, if a subset $I(\bar{x}) \subseteq I(\bar{x}^k)$ for which $\mathrm{SP}(\bar{x})$ is infeasible is found, this gives a possibly stronger cut $B(\bar{x})$ that can replace $B(\bar{x}^k)$. This is the property that is explored in the cut-strengthening algorithms presented in the next section.

3.2 Cut-Strengthening Algorithms

The cut-strengthening algorithms presented in this section attempt to strengthen a feasibility cut by reducing the number of variables included in the constraint. This is accomplished by solving the subproblem $\mathrm{SP}(x)$ for different values of x.

To categorise cuts, we use the following definition.

Definition 1. *A feasibility cut $B(\bar{x})$ is irreducible if subproblem $\mathrm{SP}(\bar{x})$ is infeasible and if subproblem $\mathrm{SP}(\tilde{x})$ is feasible for each $\tilde{x}$ such that $I(\tilde{x}) \subset I(\bar{x})$ holds.*

Note that the number of variables in an irreducible cut does not need to be minimal and that it can be possible to derive more than one irreducible cut from one feasibility cut.

In the following sections, we present three algorithms to strengthen cuts. These algorithms differ both in how they search over different subsets of variables to include in the cut and with respect to the guarantees of the strengths of the resulting cuts. The deletion filter and the depth-first binary search (DFBS) algorithm will ensure that the strengthened feasibility cut is irreducible, while the greedy algorithm will not. Note that most cut-strengthening algorithms can be tailored to a specific problem by selecting in what order to evaluate different variables. To keep a generality in our comparison, we do however not exploit this possibility and rely on using a random order.

Greedy. The greedy cut-strengthening algorithm attempts to strengthen a feasibility cut $B(\bar{x})$ by, in each iteration, selecting an index $i \in I(\bar{x})$, making the assignment $\bar{x}_i = 0$, and solving the subproblem to evaluate the feasibility. This is repeated until the subproblem becomes feasible, and then infeasibility is restored by the assignment $\bar{x}_i = 1$ for the last selected index i. Thereafter, the resulting cut, which is not guaranteed to be irreducible, is returned. The pseudo-code is given in Algorithm 1. Cut-strengthening algorithms similar to this greedy algorithm are used in [7] and [10].

    Data: A feasibility cut $B(\bar{x})$
    Result: An (improved) feasibility cut $B(\bar{x})$
1 **while** *True* **do**
2 | Select an index $i \in I(\bar{x})$;
3 | $\bar{x}_i \leftarrow 0$;
4 | **if** $\mathrm{SP}(\bar{x})$ *is feasible* **then**
5 | | $\bar{x}_i \leftarrow 1$;
6 | | return $B(\bar{x})$;
7 | **end**
8 **end**

Algorithm 1: Pseudo-code of the greedy cut-strengthening algorithm

---

Deletion Filter. The deletion filter cut-strengthening algorithm is based on the deletion filter for finding an IIS [6]. There is one iteration for each $i \in I(\bar{x})$ where the feasibility of a subproblem with the assignment $\bar{x}_i = 0$ is evaluated. If the subproblem is infeasible, the assignment $\bar{x}_i = 0$ is made permanent in the remaining iterations and in the final cut. If the subproblem is feasible, the assignment is permanently changed to $\bar{x}_i = 1$, both in the remaining iterations and in the final cut. The deletion filter algorithm finds an irreducible feasibility cut and its pseudo-code is given in Algorithm 2. The algorithm is used to strengthen feasibility cuts in [17].

Data: A feasibility cut $B(\bar{x})$
Result: An irreducible feasibility cut $B(\bar{x})$

```
1 for i ∈ I(x̄) do
2 │   x̄ᵢ ← 0;
3 │   if SP(x̄) is feasible then
4 │   │   x̄ᵢ ← 1;
5 │   end
6 end
7 return B(x̄);
```

Algorithm 2: Pseudo-code of the deletion filter cut-strengthening algorithm

Data: A feasibility cut $B(\bar{x})$
Result: An irreducible feasibility cut $B(\bar{x})$

```
 1 T ← I(x̄); S ← ∅; x̄ᵢ ← 0,   i ∈ I;
 2 while True do
 3 │   if |T| ≤ 1 then
 4 │   │   x̄ᵢ ← 1,   i ∈ T;
 5 │   │   if SP(x̄) is infeasible then
 6 │   │   │   return B(x̄);
 7 │   │   end
 8 │   │   T ← S; S ← ∅;
 9 │   │   if |T| ≥ 2 then
10 │   │   │   go to Line 3;
11 │   │   end
12 │   │   T₂ ← T; T₁ ← ∅
13 │   else
14 │   │   Split T into T₁ and T₂;
15 │   end
16 │   x̄ᵢ ← 1,   i ∈ S ∪ T₁;
17 │   if SP(x̄) is feasible then
18 │   │   S ← S + T₁; T ← T₂;
19 │   else
20 │   │   T ← T₁;
21 │   end
22 │   x̄ᵢ ← 0,   i ∈ S ∪ T₁;
23 end
```

Algorithm 3: Pseudo-code of the DFBS cut-strengthening algorithm

Depth-First Binary Search. The DFBS cut-strengthening algorithm is similar to the deletion filter algorithm, but instead of evaluating only a single index at a time, subsets of indices are evaluated. In each major iteration, the algorithm evaluates subproblems to iteratively reduce a subset of indices into a single index i, for which the permanent assignment $\bar{x}_i = 1$ is made and used in the final cut. Each time a permanent assignment is made, a subproblem is solved with the assignment $\bar{x}_i = 0$ for all i for which there is no permanent assignment. If this subproblem is infeasible, no more variable needs to be included in the cut. If the subproblem is feasible, the complete procedure is repeated and an additional variable to include in the final cut is found by exploring a new subset of indices. As for the deletion filter algorithm, the final cut will be irreducible. The difference is that, by not only exploring one individual index at a time there is a possibility to decrease the number of subproblems that needs to be solved. The pseudo-code for this algorithm is given in Algorithm 3 and it is based on the presentation in [1] for finding an IIS for a mathematical program. This type of algorithm is one of the components of the infeasibility analyser QUICKXPLAIN, described in [15], which is used to strengthen cuts in [5].

4   Problems and Modelling

In this section, the problems that we apply LBBD on to evaluate the cut-strengthening algorithms are presented. For each problem, we give a brief problem statement, present a mathematical model and its decomposition, and relate this formulation to the general problem formulation [P].

4.1   Cumulative Facility Scheduling with Fixed Costs

A LBBD scheme for cumulative facility scheduling with fixed costs was introduced in [10]. Computational results for this problem were later also given for the branch-and-check-type LBBD solver Nutmeg [17].

The problem formulation includes a set of facilities $\mathcal{F}$ and a set of tasks $\mathcal{I}$. Each task $i \in \mathcal{I}$ must be assigned to a facility $f \in \mathcal{F}$, where it is to be performed for the duration of its processing time p_{if}, using a resource at the rate c_{if} per time unit. The maximum total rate of resource consumption on facility $f \in \mathcal{F}$ is limited to C_f per time unit. Each task $i \in \mathcal{I}$ must be scheduled in the interval between its release time r_i and deadline d_i. To assign task $i \in \mathcal{I}$ to facility $f \in \mathcal{F}$ incurs the cost F_{if}, and the objective is to minimise the cost of assigning each task to a facility.

The master problem variables x_{if} are binary and indicate if task $i \in \mathcal{I}$ is assigned to facility $f \in \mathcal{F}$ or not. The subproblem checks feasibility with respect to the scheduling of each facility, where the continuous variable $y_{if} \geq 0$ equals the start time of task $i \in \mathcal{I}$ on facility $f \in \mathcal{F}$.

A formulation of this problem, given on the same form as [P], is

$$\min \ \sum_{i \in \mathcal{I}} \sum_{f \in \mathcal{F}} F_{if} x_{if}, \tag{2}$$

$$\text{s.t.} \ \sum_{f \in \mathcal{F}} x_{if} = 1, \quad i \in \mathcal{I}, \tag{3}$$

$$\text{CUMULATIVE}((y_{if}|i \in \mathcal{I}), (p_{if}|i \in \mathcal{I}), (c_{if}|i \in \mathcal{I}), C_f), \quad f \in \mathcal{F}, \tag{4}$$

$$x_{if} \to r_i \leq y_{if} \leq d_i - p_{if}, \quad i \in \mathcal{I}, f \in \mathcal{F}, \tag{5}$$

$$[\text{Energy relaxation}]. \tag{6}$$

The objective (2) is to minimise the cost for assigning tasks to facilities. Constraints (3) assign each task to a facility and correspond to constraints $C(x)$

in [P]. The cumulative constraints (4) sequence the tasks while respecting the available number of resources. The corresponding constraints in [P] are $C(y)$. Constraints (5) make tasks respect their release times and deadlines if they are assigned to a facility. Thereby they connect the master problem decisions with which constraints to include in the subproblem, and these connections correspond to $x_{if} \rightarrow C^i(y), i \in I$, in [P]. The [Energy relaxation], introduced in [10], is a relaxation of constraints (3)–(5) and it is used to strengthen the master problem. It can be written as

$$\frac{1}{C_f} \sum_{i \in \mathcal{I}(t_1, t_2)} p_{if} c_{if} x_{if} \leq t_2 - t_1, \quad (t_1, t_2) \in T, \tag{7}$$

where $T = \{(r_i, d_{i'}) : (i, i') \in I \times I, d_{i'} > r_i\}$ is the set of release time and deadline pairs for which the energy relaxation is used. Further, the set $\mathcal{I}(t_1, t_2) = \{i \in \mathcal{I} : t_1 \leq r_i, d_i \leq t_2\}$ gives the tasks that have a time window that starts after t_1 and ends before t_2.

4.2 Single Machine Scheduling with Sequence-Dependent Setup Times and Multiple Time Windows

In [7], a LBBD scheme was proposed for solving the feasibility version of a single machine scheduling problem with a segmented timeline. In this paper we study a somewhat more general problem formulation which we refer to as single machine scheduling with sequence-dependent setup times and multiple time windows. For this formulation, it is assumed that there can be more tasks than can be feasible scheduled, and the objective is to maximise the sum of prizes obtained for scheduling tasks. The model we use is in essence taken from [13]. It is a generalisation of the model used in [7] since it allows for unique time-windows for each task and because it includes sequence-dependent setup times. To solve the feasibility version from [7], the prize-collecting objective is used to determine the maximum number of tasks that can be feasibly scheduled.

The given set of tasks is $\mathcal{I}$ and a prize q_i is collected if task $i \in \mathcal{I}$ is scheduled in the interval between the release time r_{iq} and deadline d_{iq} of one of its time-windows $q \in \mathcal{Q}_i$. Each scheduled task $i \in \mathcal{I}$ must be given exclusive uninterrupted access to the machine for the duration of its processing time p_i. If task i is performed before task j, there is a minimum setup time s_{ij} between the end of task i and the start of task j, $i, j \in \mathcal{I} : i \neq j$.

The master problem variables x_{iq} are binary and indicate if task $i \in \mathcal{I}$ is scheduled in time window $q \in \mathcal{Q}_i$ or not. The subproblem checks if the selected tasks can be scheduled within their assigned time windows. In the subproblem, the continuous variable $y_i \geq 0$ equals the start time of task $i \in \mathcal{I}$.

A formulation of this problem, given on the same form as [P], is

$$\min \sum_{i \in \mathcal{I}} \sum_{q \in \mathcal{Q}_i} q_i x_{iq}, \tag{8}$$

$$\text{s.t.} \quad \sum_{q \in \mathcal{Q}_i} x_{iq} \leq 1, \quad i \in \mathcal{I}, \tag{9}$$

$$\text{DISJUNCTIVE}((y_i | i \in \mathcal{I}), (p_i | i \in \mathcal{I}), (s_{ij} | i, j \in \mathcal{I})), \tag{10}$$

$$x_{iq} \rightarrow r_{iq} \leq y_i \leq d_{iq} - p_i, \quad q \in \mathcal{Q}_i, i \in \mathcal{I}, \tag{11}$$

$$[\text{Segment relaxtion}]. \tag{12}$$

The objective (8) is to maximise the total prize collected by assigning tasks to time-windows. Constraints (9) make sure that each task is scheduled in at most one of its time windows and they correspond to constraints $C(x)$ in [P]. The disjunctive constraint (10) ensures that no tasks overlap and that the tasks respect their sequence-dependent setup times. The corresponding constraints in [P] are $C(y)$. Constraints (11) ensure that if a task is assigned to a time-window, the task is performed within this time window. Thereby they connect the master problem decisions with which constraints to include in the subproblem, and these connections correspond to $x_{if} \rightarrow C^i(y), i \in I$, in [P]. To strengthen the master problem, we use the [Segment relaxation] described in [16]. The [Segment relaxation] is derived from constraints (9)–(11) and formulated as

$$\sum_{i \in \mathcal{I}} \sum_{q \in \mathcal{Q}_i(t_1, t_2)} p_i x_{iq} \leq t_2 - t_1, \quad (t_1, t_2) \in T, \tag{13}$$

where $T = \{(r_{iq}, d_{i'q'}) : (q, q') \in \mathcal{Q}_i \times \mathcal{Q}_{i'}, (i, i') \in I \times I, d_{i'q'} > r_{iq}\}$ is the set of release time and deadline pairs for which the segment relaxation is used. Further, the set $\mathcal{Q}_i(t_1, t_2) = \{q \in \mathcal{Q}_i : t_1 \leq r_{iq}, d_{iq} \leq t_2\}$ gives the time windows for task $i \in \mathcal{I}$ that that starts after t_1 and ends before t_2.

4.3   Vehicle Routing with Location Congestion

The vehicle routing problem with location congestion was introduced in [18] and a LBBD scheme for solving it was derived in [17]. In this problem, vehicles that originate from a depot are to deliver goods, referred to as requests, at different locations. For this purpose, routes that has a minimal total transportation cost are to be constructed such that all requests are delivered while respecting vehicle capacity and location congestion constraints. The set of requests and locations are denoted by R and $\mathcal{L}$, respectively. Each request $i \in R$ is associated with a specific location $l_i \in \mathcal{L}$ and the set $R_l = \{i \in R : l_i = l\}$ includes all requests at location $l \in \mathcal{L}$. When a vehicle arrives at request $i \in R$, it requires a processing time p_i to deliver the goods. At each location $l \in \mathcal{L}$, a maximum of C_l vehicles can deliver goods at a given time.

To represent the routing aspect of the problem, a graph $G = (N, A)$ is defined with a set of nodes $N = R \cup \{0^-, 0^+\}$ and a set of arcs $A = \{(i, j) \in N \times N : i \neq$

j}, where 0^- and 0^+ correspond to artificial start and end nodes at the depot. A vehicle must arrive to each node $i \in N$ between a release time r_i and a deadline d_i, and all vehicles must return to the depot before the time T. If a node $i \in N$ is visited by a vehicle, a weight q_i is added to the vehicle at the depot. For each route the maximum added weight for a vehicle is Q. For the artificial nodes 0^- and 0^+ the release times are 0, the deadlines are T and the weights are 0. The cost for transportation on arc $(i, j) \in A$ is denoted by c_{ij}. There is no cost for using a vehicle and there is no upper bound on the number of used vehicles.

The master problem variables x_{ij} are binary and indicate if arc $(i, j) \in A$ is used in a route or not. The subproblem checks the feasibility of the suggested routes with respect to location congestion, vehicle weight, and time windows. In the subproblem, the continuous variables $y_i^{\text{start}} \in [r_i, d_i]$ and $y_i^{\text{weight}} \in [q_i, Q]$ equal, for each node $i \in N$, the time a vehicle arrives and its total accumulated weight, respectively.

A formulation of this problem, given on the same form as [P], is

$$\min \quad \sum_{(i,j) \in A} c_{ij} x_{ij}, \tag{14}$$

$$\text{s.t.} \quad \sum_{i:(i,j) \in A} x_{ij} = 1, \quad j \in R, \tag{15}$$

$$\sum_{j:(i,j) \in A} x_{ij} = 1, \quad i \in R, \tag{16}$$

$$\textsc{Cumulative}((y_i^{\text{start}} | i \in R_l), (p_i | i \in R_l), (1 | i \in R_l), C_l), \quad l \in L, \tag{17}$$

$$x_{ij} \rightarrow y_i^{\text{weight}} + q_j \leq y_j^{\text{weight}}, \quad (i, j) \in A, \tag{18}$$

$$x_{ij} \rightarrow y_i^{\text{start}} + p_i + c_{ij} \leq y_j^{\text{start}}, \quad (i, j) \in A, \tag{19}$$

[Weight and time relaxation]. $\tag{20}$

The objective (14) is to minimise the total transportation cost. Constraints (15)–(16) ensure that exactly one vehicle enters and leaves each request and correspond to $C(x)$ in [P]. The cumulative constraints (17) make sure that, at each point in time and for each location, the number of vehicles delivering goods does not exceed the maximum limit. Constraints (17) correspond to $C(y)$ in [P]. Constraints (18) and (19) keep track of the vehicle weights and arrival times, respectively, at each node. Constraints (18)–(19) connect the master problem decisions with which constraints to include in the subproblem, and these connections correspond to $x_{if} \rightarrow C^i(y), i \in I$, in [P]. Note that the number of vehicles used in a solution is given by the number of used arcs from (or to) the depot. Also, because the vehicles are identical and all requests have exactly one incoming and outgoing arc, there is no need to explicitly represent the vehicles in the model, but only which arcs that are used. The master problem is strengthened by the [Weight and time relaxation] component that contains the variables y_i^{start}, y_i^{weight}, $i \in N$, and the constraints (18)–(19).

5   Computational Evaluation

This section provides results from applying the presented cut-strengthening algorithms and compare these to not using cut strengthening. Below, the different approaches are, for short, referred to as DFBS, deletion filter, greedy, and none (no cut strengthening), respectively. In the case when the subproblem can be separated into small independent problems, we also make a comparison between solving these small problems individually and solving them together as one large problem; referred to as solving with or without subproblem separation. Note that subproblem separation has an automatic strengthening effect since it results in a cut being generated for each infeasible separated subproblem.

The LBBD scheme was implemented in Python 3.8, and the MIP and CP models were solved using Gurobi Optimizer version 9.0.3 and IBM ILOG CP Optimizer version 12.10, respectively. All tests were carried out on a computer with two Intel Xeon Gold 6130 processors (16 cores, 2.1 GHz each) and 96 GB RAM. Each instance was given a total time of 20 min and the MIP-gaps were set to 0 for the master problems.

5.1   Instances

The instances are either taken from previous work or generated in line with descriptions in previous work, but with new parameter settings. All instances can be accessed, either directly or via reference, from our repository[1]. For cumulative facility scheduling with fixed costs, referred to as Problem 4.1, we use 336 instances from [10]. For vehicle routing with location congestion, referred to as Problem 4.3, we use 450 instances from [17].

For the single machine scheduling problem with sequence-dependent setup times and multiple time windows, referred to as Problem 4.2, we use instances from two different sources. Instance set I contains 480 instances that we have generated based on the description in [7]. We did, however, change a few parameter settings to make the instances harder, since those generated according to the parameter settings in [7] were solved within seconds and did not facilitate an interesting comparison. Instance set II contains 900 instances introduced in [13]; they originate from two different applications, namely avionics scheduling [4] and particle therapy patient scheduling [19].

For two of the instance sets, it is possible to apply subproblem separation. For Problem 4.1, there is one subproblem for each facility, and for Instance set I of Problem 4.2, there is one subproblem for each time segment.

5.2   Percentage of Solved Instances

The results of our computational evaluation are presented in Fig. 1. In each subplot, the horisontal axis gives the time and the vertical axis gives the percentage

---

[1] https://gitlab.liu.se/eliro15/lbbd_instances.

of solved instances. We begin with describing the results for the instances that do not have an inherent subproblem separation structure.

Figure 1a and Fig. 1b give the results for Instance set II for Problem 4.2 and Problem 4.3, respectively. For the former, the algorithms that find irreducible cuts give the best performance. With DFBS, 75.3% of the instances are solved to optimality and with the deletion filter, 72.4% of the instances are solved to optimality. The corresponding percentages for greedy and none are 53.7% and 26.4%, respectively.

For Problem 4.3, using DFBS gives the best performance with 100.0% of the instances solved to optimality. Using the deletion filter also gives a good performance with 91.1% of the instances solved to optimality. The corresponding percentages for greedy and none are 66.0% and 55.1%, respectively. A peculiarity of the instances for Problem 4.3 is that many of them are infeasible. This is detected for many of these instances already in the first LBBD iteration, irrespective of the cut-strengthening approach used. The infeasibility is detected either because the master problem is infeasible without any feasibility cuts or because the subproblem is infeasible even without a restriction on x (such a test is implemented as part of the initialisation of our LBBD scheme). All instances solved without cut strengthening belong to this category. For the remaining instances, feasibility cuts must be generated to prove optimality or infeasibility. A few of these instances are solved using greedy, but the results from using greedy are still weak in comparison with using deletion filter or DFBS.

Results from solving Problem 4.1 with and without subproblem separation are given in Fig. 1c and Fig. 1d, respectively. In the case without subproblem separation, the use of DFBS gives the best performance with 56.8% of the instances solved to optimality. The corresponding percentages for deletion filter, greedy, and none are 54.8%, 51.2%, and 46.4%, respectively. When subproblem separation is applied, the impact of cut strengthening is small and the percentages of solved instances range between 76.5% and 78.9% for all approaches. Note that the results when applying subproblem separation are consistently and significantly better than those when subproblem separation is not applied.

Instance set I for Problem 4.2 can be solved both with and without subproblem separation. The results for solving with separation and without separation are given in Fig. 1e and Fig. 1f, respectively. When subproblem separation is not applied, the use of DFBS gives the best performance with 54.2% of the instances solved to optimality. The corresponding percentages for deletion filter, none, and greedy are 40.8%, 37.7%, and 31.5%, respectively. With subproblem separation, the percentages of solved instances when applying deletion filter, DFBS, greedy, and none are 73.1%, 72.9%, 70.8%, and 63.1%, respectively. Again, subproblem separation has a significant impact on the number of solved instances. Also, it is clear in this case that using some cut-strengthening algorithm is beneficial both with and without separation.

When summarising the results, the following can be observed. In the view of all instance sets and different types of problems, the best performance is obtained when DFBS is applied for cut strengthening. This suggests that DFBS

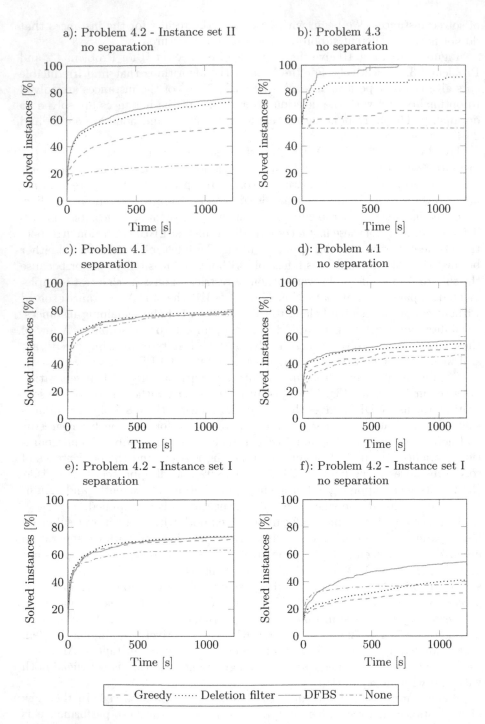

Fig. 1. Percentage of solved instances for different cut-strengthening algorithms

is a good candidate when choosing an algorithm for strengthening of feasibility cuts in a general LBBD scheme. As expected, exploiting problem structure and applying subproblem separation is very beneficial, and this effect dominates that of applying cut strengthening in the cases when separation is possible. We believe that the reason for this is two-fold. Firstly, the separation itself strengthens the cuts, since cuts are generated for each subproblem. Secondly, solving the separated subproblem is easier than solving one large subproblem.

6   Concluding Remarks

This paper presents how different algorithms for strengthening of feasibility cuts can be applied in a LBBD scheme and provides results that illustrate the different algorithms' impact on the computational performance of the LBBD scheme. To make the comparison generic, the cut strengthening is made without exploiting knowledge about problem structure when deciding in which order to evaluate variables to include in a cut. The computational evaluation is based on three different problem formulations, using over 2000 instances from five different applications. For some of these instances, there is an inherent structure that makes is possible to separate the subproblem into smaller independent problems. For such instances, the benefits from this separation dominates that of applying cut strengthening where no problem structure is exploited.

The general conclusion is that for the tested instances, using the DFBS cut-strengthening algorithm gives the best computational performance. To gain a deeper understanding of our results, we suggest further work that includes the following considerations. Firstly, both the sizes of the cuts and the total number of LBBD iterations can be compared to the number of subproblems solved during the cut strengthening. These quantities can then be analysed in the light of how computationally challenging the subproblem is in comparison with the master problem. Secondly, the difference in performance between DFBS and deletion filter can be further analysed. Both algorithms find irreducible cuts, but they differ in their search for finding them. Do the computational results differ because they tend to find different cuts or because of their different efficiencies in finding what is most often the same cut? Lastly, it would be relevant to investigate the impact of exploiting problem structure, especially variable sorting, and integrate this with the considered cut-strengthening algorithms. In the case when subproblem separation can be applied, it is interesting to understand how much of the computational gain from the separation that follows from sorting it implies.

Acknowledgement. Emil Karlsson is funded by the Research School in Interdisciplinary Mathematics at Linköping University. The work is also partly funded by the Center for Industrial Information Technology (CENIIT), Project-ID 16.05. Computational experiments were performed on resources provided by the Swedish National Infrastructure for Computing (SNIC) at National Supercomputer Centre (NSC).

References

1. Atlihan, M.K., Schrage, L.: Generalized filtering algorithms for infeasibility analysis. Comput. Oper. Res. **35**, 1446–1464 (2008). https://doi.org/10.1016/j.cor.2006.08.005
2. Benders, J.F.: Partitioning procedures for solving mixed-variables programming problems. Numerische Mathematik **4**, 238–252 (1962). https://doi.org/10.1007/BF01386316
3. Benini, L., Lombardi, M., Mantovani, M., Milano, M., Ruggiero, M.: Multi-stage benders decomposition for optimizing multicore architectures. In: Perron, L., Trick, M.A. (eds.) CPAIOR 2008. LNCS, vol. 5015, pp. 36–50. Springer, Heidelberg (2008). https://doi.org/10.1007/978-3-540-68155-7_6
4. Blikstad, M., Karlsson, E., Lööw, T., Rönnberg, E.: An optimisation approach for pre-runtime scheduling of tasks and communication in an integrated modular avionic system. Optim. Eng. **19**(4), 977–1004 (2018). https://doi.org/10.1007/s11081-018-9385-6
5. Cambazard, H., Hladik, P.-E., Déplanche, A.-M., Jussien, N., Trinquet, Y.: Decomposition and learning for a hard real time task allocation problem. In: Wallace, M. (ed.) CP 2004. LNCS, vol. 3258, pp. 153–167. Springer, Heidelberg (2004). https://doi.org/10.1007/978-3-540-30201-8_14
6. Chinneck, J.W., Dravnieks, E.W.: Locating minimal infeasible constraint sets in linear programs. ORSA J. Comput. **3**, 157–168 (1991). https://doi.org/10.1287/ijoc.3.2.157
7. Coban, E., Hooker, J.N.: Single-facility scheduling by logic-based Benders decomposition. Ann. Oper. Res. **210**, 245–272 (2013). https://doi.org/10.1007/s10479-011-1031-z
8. Geoffrion, A.M.: Generalized Benders decomposition. J. Optim. Theory Appl. **10**, 237–260 (1972). https://doi.org/10.1007/BF00934810
9. Hooker, J.N.: Logic-Based Methods for Optimization: Combining Optimization and Constraint Satisfaction. Wiley, Hoboken (2000). https://doi.org/10.1002/9781118033036
10. Hooker, J.N.: Planning and scheduling by logic-based Benders decomposition. Oper. Res. **55**, 588–602 (2007). https://doi.org/10.1287/opre.1060.0371
11. Hooker, J.N.: Logic-based benders decomposition for large-scale optimization. In: Velásquez-Bermúdez, J.M., Khakifirooz, M., Fathi, M. (eds.) Large Scale Optimization in Supply Chains and Smart Manufacturing. SOIA, vol. 149, pp. 1–26. Springer, Cham (2019). https://doi.org/10.1007/978-3-030-22788-3_1
12. Hooker, J.N., Ottosson, G.: Logic-based Benders decomposition. Math. Program. **96**, 33–60 (2003). https://doi.org/10.1007/s10107-003-0375-9
13. Horn, M., Raidl, G.R., Rönnberg, E.: A* search for prize-collecting job sequencing with one common and multiple secondary resources. Ann. Oper. Res. (2020). https://doi.org/10.1007/s10479-020-03550-7
14. Junker, U.: QuickXPlain: conflict detection for arbitrary constraint propagation algorithms. In: IJCAI01 Workshop on Modeling and Solving Problems with Constraints (CONS-1) (2001)
15. Junker, U.: QuickXPlain: preferred explanations and relaxations for over-constrained problems. In: Proceedings of AAAI 2004, pp. 167–172 (2004)
16. Karlsson, E., Rönnberg, E., Stenberg, A., Uppman, H.: A matheuristic approach to large-scale avionic scheduling. Ann. Oper. Res. (2020). https://doi.org/10.1007/s10479-020-03608-6

17. Lam, E., Gange, G., Stuckey, P.J., Van Hentenryck, P., Dekker, J.J.: Nutmeg: a MIP and CP hybrid solver using branch-and-check. SN Oper. Res. Forum **1**, 22:1–22:27 (2020). https://doi.org/10.1007/s43069-020-00023-2
18. Lam, E., Van Hentenryck, P.: A branch-and-price-and-check model for the vehicle routing problem with location congestion. Constraints **21**, 394–412 (2016). https://doi.org/10.1007/s10601-016-9241-2
19. Maschler, J., Riedler, M., Stock, M., Raidl, G.R.: Particle therapy patient scheduling: first heuristic approaches. In: Proceedings of the 11th International Conference of the Practice and Theory of Automated Timetabling, PATAT 2016, pp. 223–244 (2016)
20. Rahmaniani, R., Crainic, T.G., Gendreau, M., Rei, W.: The Benders decomposition algorithm: a literature review. Eur. J. Oper. Res. **259**, 801–817 (2017). https://doi.org/10.1016/j.ejor.2016.12.005

Learning Variable Activity Initialisation for Lazy Clause Generation Solvers

Ronald van Driel, Emir Demirović⬤, and Neil Yorke-Smith⁽✉⁾⬤

Algorithmics, Delft University of Technology, Delft, Netherlands
R.A.vanDriel@student.tudelft.nl, {e.demirovic,n.yorke-smith}@tudelft.nl

Abstract. Contemporary research explores the possibilities of integrating machine learning (ML) approaches with traditional combinatorial optimisation solvers. Since optimisation hybrid solvers, which combine propositional satisfiability (SAT) and constraint programming (CP), dominate recent benchmarks, it is surprising that the literature has paid limited attention to machine learning approaches for hybrid CP–SAT solvers. We identify the technique of *minimal unsatisfiable subsets* as promising to improve the performance of the hybrid CP–SAT lazy clause generation solver Chuffed. We leverage a graph convolutional network (GCN) model, trained on an adapted version of the MiniZinc benchmark suite. The GCN predicts which variables belong to an unsatisfiable subset on CP instances; these predictions are used to initialise the activity score of Chuffed's Variable-State Independent Decaying Sum (VSIDS) heuristic. We benchmark the ML-aided Chuffed on the MiniZinc benchmark suite and find a robust 2.5% gain over baseline Chuffed on MRCPSP instances. This paper thus presents the first, to our knowledge, successful application of machine learning to improve hybrid CP–SAT solvers, a step towards improved automatic solving of CP models.

1 Introduction

Neuro-symbolic approaches to combinatorial optimisation problems include improving optimisation solver performance or robustness by incorporating machine learning (ML). This trend shows successful promise in integer programming [2,10,14,26], propositional satisfiability (SAT) [22,25] as well as constraint programming (CP) [1,9,24]. Hybrid CP–SAT solvers are the state of the art for CP according to recent MiniZinc Challenge competitions [19]. Such solvers, labelled as *Lazy Clause Generation* (LCG) solvers [21], combine the conflict learning ability from SAT solvers with finite domain propagation from CP solvers.

However to the best of our knowledge there have not been research to date on combining machine learning to improve the performance of hybrid CP–SAT solvers. For example, Song et al. [24] show that machine learning can be used to automatically learn variable ordering heuristics for traditional constraint satisfaction solving. Portfolio approaches have shown excellent performance [1,13], but such methods select one of the solving strategies in a portfolio rather than

© Springer Nature Switzerland AG 2021
P. J. Stuckey (Ed.): CPAIOR 2021, LNCS 12735, pp. 62–71, 2021.
https://doi.org/10.1007/978-3-030-78230-6_4

directly modify a LCG solver. Similarly, applications that combine machine learning and constraint programming have been studied (e.g., [5]), but the machine learning part does not influence the internal hybrid CP–SAT algorithm.

We aim to utilise machine learning to improve a single component of hybrid CP–SAT solvers, namely the activity-based variable selection heuristic (VSIDS). Our approach is motivated by *Neurocore* [23], a method that uses ML to influence the variable selection of SAT solvers. Given that LCG solvers use SAT solvers in their inner-workings, a natural question to ask is whether an approach such as Neurocore can be employed in constraint programming. While related ideas may be exploited, a direct application of Neurocore in CP is not possible. Neurocore trains its learned model on clauses derived from the proof of unsatisfiability. However, unsatisfiability proofs are not an established concept in CP. Current SAT techniques are not easily extendable, as CP considers *optimisation* problems, possibly using integer variables, and complex constraints that may require an exponential number of clauses when encoding into SAT. There has been progress in this direction using cutting planes reasoning, but only for specific problems or constraints [7,11,12]. For similar reasons, the machine learning features used in SAT may not directly translate to (LCG-based) CP.

This paper provides a first demonstration of the value of using ML within the LCG solver Chuffed [4]. We develop a modified version of Neurocore for constraint programming and employ it to learn initialisation values for the activities used in the variable-selection heuristic. We benchmark our ML-aided approach on problems from the MiniZinc benchmark suite and find a statistically-significant 2.5% average gain over the baseline Chuffed on MRCPSP instances.

2   Background

The Satisfiability problem (SAT) is concerned with deciding whether or not there exists an assignment of truth values to variables such that a given propositional logic formula is satisfied. A *SAT solver* is an algorithm that explores the space of possible assignments with the aim of either finding a satisfying assignment or proving that the formula is unsatisfiable. For the purposes of this paper, the search may be viewed as a backtracking algorithm over the variable assignments.

The *Variable-State Independent Decaying Sum* (VSIDS) [17,20], originally developed as a variable selection heuristic for SAT solver Chaff, is commonly used in LCG solvers. When using VSIDS in a SAT solver during search, variables are selected according to their *activity*. Intuitively, the activity score indicates the likelihood that the variable will quickly lead to a conflict, and selecting such variables early in the search is beneficial. Initially, the activity value of each variable is initialised to zero. Once the solver encounters a *conflict*, i.e., it is detected that the current partial assignment is infeasible, analysis is performed to determine the reason for the conflict. The reason is recorded as a *learned clause*, which consists of a subset of the variables from the partial assignment. Each time a variable is involved in a conflict, its activity is increased. To emphasise recent conflicts, the activity scores of all variables are periodically non-linearly decreased. As a result, variables recently involved in conflicts have the highest scores. LCG solvers make use of VSIDS in their internal SAT solver.

SAT solvers, upon concluding that a problem is unsatisfiable, may provide a *certificate of unsatisfiability*. Intuitively, the certificate consists of a set of clauses and a sequence of logic derivation steps result that in the empty clause, i.e., unsatisfiability. A related concept in CP is a *minimal unsatisfiable subset*, which is a set of constraints that unsatisfiable together, but are not unsatisfiable if any constraint is removed from the set. Conceptually, SAT solvers operate on the low-level of propositional logic, whereas CP solvers consider a more expressive CP setting, e.g., complex constraints over integer variables. Hybrid CP–SAT solvers [21] maintain a dual view of the problem: in addition to the CP view, a portion of the problem is converted into propositional logic. An internal SAT solver is invoked on the propositional logic formula, augmented with CP propagators to infer variable assignments based on the current partial assignment. Once a conflict is encountered, the conflict analysis procedure from SAT operates as usual, with the exception that variables set by propagators are queried to provide the reason for their propagation in the form of a *clause*. Since all reasons are clausal, this allows the solver to use the SAT conflict analysis procedure while still retaining the benefit of CP. In this way, hybrid CP–SAT solvers combine SAT and CP solving techniques.

The *Neurocore* [23] approach uses machine learning to influence the variable selection heuristic of a SAT solver. Since a SAT solver may make thousands of decisions per second using VSIDS, a possible replacement of variable selection is expected to run with a tight time budget. Hence Neurocore does not *directly* use ML to replace the variable selection heuristic, but instead *indirectly* influences the selection procedure by periodically modifying the activity values of the variables. The ML model, represented as a graph convolutional network (GCN), is trained to assign a confidence value between zero and one for each variable depending on its features. The estimate represents the probability that the variable is part of an unsatisfiable core. The first assumption is that variables that are used in the proof of unsatisfiability are likely to quickly lead to conflicts during search, and therefore the solver should aim to select these variables as soon as possible. The second assumption is that, even though unsatisfiable cores do not exist in satisfiable instances, the GCN predictions will nevertheless be valuable even for satisfiable instances to identify highly conflict-inducing variables.

3   Approach

Recall our goal is to predict the initial values of variable activity for a CSP instance. Since it is difficult to formulate directly learning VSIDS initialisations as a feasible learning problem (as discussed later), instead we leverage the analogous precedent in SAT solving discussed above [23].

By default, the activity values in LCG solvers are set to zero or to random values at the start of the search. The scores do not provide any meaningful information to the solver in the beginning but they gradually become more useful as search proceeds. By providing useful initial values we posit that the solver performance can be improved; improvements at the start of search are particularly valuable. In the absence of meaningful VSIDS values, *Chuffed* [4], the LCG solver used in this work, typically uses (user-specified) search annotations if provided, before switching to VSIDS for making branching decisions.

Our approach is to train a graph convolutional network model on *unsatisfiable* instances, to make a prediction on which variables belong to an unsatisfiable subset. The trained model is then used as part of the LCG solver to classify the variables of input instances at the start of the search. The classification is done by assigning a value between zero and one for each variable, which may be interpreted as the probability that the variable is in an unsatisfiable subset. These values are used to initialise the activity values for VSIDS.

Whereas training is done on unsatisfiable instances, the target instances used afterwards do not necessarily need to be unsatisfiable, e.g., it is expected the instances represent optimisation problems for which a feasible solution exists. Note that for satisfiable instances, no unsatisfaible subset exists, but the predictions made by the network are still valuable since, intuitively, higher predicted values indicate variables that are more likely to engage in a conflict.

It is important to note that, similar to the approach proposed by Selsam and Bjørner [23] – and with works in the predict-and-optimise paradigm [6,8] – our ambition is not to achieve the best possible ML predictions. The reason for this is that more accurate predictions do not necessarily imply that they are more useful for the solver; rather the metric to optimise is the runtime of the solver. The hypothesis is that, even though satisfiable instances do not have unsatisfiable cores, the *confidence* of classifying a variable to be part of an unsatisfiable set correlates with the effectiveness of branching on that variable. This can be seen as a surrogate for the runtime. The *true* metric that directly optimises the runtime remains an open question.

An alternative could be to learn based on the final VSIDS scores. However such scores are biased towards the last few conflicts before termination even though many other conflicts were needed to prove optimality. On a related note, in core-boosted MAxSAT [3], after the core-guided (lower bounding) phase, it was beneficial to *nullify* the VSIDS scores before switching to the linear search (upper bounding) phase, as opposed to keeping the final VSIDS scores of the lower bounding method, indicating that VSIDS scores that are good for one phase of the search may not be good for another phase.

3.1 Machine Learning Model

We adopt the Graph Convolution Network (GCN) model of Kipf and Welling [15].[1] A GCN learns a function of the features on a graph: in our case the constraint graph. The features we choose pertain to the variables: 1. Categorical features indicating if a variable is declared as a Boolean, integer, float or set. 2. Minimum value within the variable domain. 3. Maximum value within the

---

[1] Code available at https://github.com/tkipf/gcn; we use their default settings.

variable domain. 4. The range of the variable domain. 5. A set of identifiers of variables which co-occur in some constraint. Then the input of the GCN is:

1. **A feature matrix of size** $N \times D$. Here N represents the number of variables and D the number of selected features.
2. **An adjacency matrix of size** $N \times N$. In this matrix variables are considered adjacent if they co-occur in a constraint.
3. **The labels in an** $N \times C$ **matrix.** Here C represents the number of output classes, in our case two: one for variables which are part of a minimal unsatisfiable subset (MUS) and the other for variables which are not.

The output of the model is a $N \times C$ matrix, the softmax outputs – which can be interpreted as the probability for each variable to belonging to each class. Because we consider two classes only, it is possible to express the output of the ML predictions with a single value, which is the prediction confidence of a variable belonging to a MUS.

4   Empirical Study

We now examine experimentally the effectiveness of the proposed approach. We compiled from source three different versions of Chuffed: CHUFFED0_OG, CHUFFED1_EX and CHUFFED1_INC. All three versions were configured to switch to VSIDS as soon as 100 conflicts have been encountered.[2] While all three versions have an identical configuration, they are different in the way the ML was integrated. CHUFFED0_OG was otherwise left completely unmodified, and serves as a baseline. CHUFFED1_EX was modified to have the VSIDS scores initialised with the predictions obtained after being trained on a training set which contained only instances from *other* problem types. Similarly, CHUFFED1_INC was modified to initialise the VSIDS scores with predictions after being trained on *all* training instances, including from the same problem type.

4.1   Data Sets

We require two different datasets containing CP instances. One of these datasets should only contain unsatisfiable instances to train on; the other should contain satisfiable instances to solve for evaluation. The MiniZinc benchmark suite [18] supplies over 13,000 satsifiable instances for evaluation. Since we found no public CP dataset contained sufficiently many unsatisfiable instances for training a ML model, the constraint optimisation problem (COP) instances from the MiniZinc benchmark suite were modified to become unsatisfiable. This was done by first solving them for their optimal value; then the original instance was modified by bounded the objective variable to be strictly better than the optimal value.

Using this procedure allows the creation both the satisfiable datset as well as the unsatisfiable dataset. For the unsatisfiable dataset the labels were generated

---

[2] This is lower than the Chuffed default, in order to ensure that VSIDS is used.

Table 1. Experiments on MRCPSP benchmarks

Instances	Chuffed0_OG	Chuffed1_Ex	Chuffed1_Inc
	Avg. runtime (s)	Avg. runtime (s)	Avg. runtime (s)
mrcpsp10900	4.507	4.356	4.461
mrcpsp36	2.399	2.428	2.410
mrcpsp4425	311.565	296.139	302.595
mrcpsp4777	5274.736	5153.284	5155.367
mrcpsp4871	892.922	865.954	865.404
mrcpsp4960	32.713	32.241	32.099
mrcpsp7051	16.091	15.884	16.028
mrcpsp896	0.152	0.155	0.189
mrcpsp9880	0.236	0.241	0.240
mrcpsp9994	0.033	0.034	0.035
Total(s)	6535.354	6370.715	6378.829
Standard Deviation	282.493	273.983	271.103
Relative(%)	100.0%	97.5%	97.6%

using MiniZinc's *findMUS* command [16]. Note findMUS often returns multiple different MUS combinations and a variable is deemed being part of a MUS if it is in *any* one of them. The datasets contained 13,667 problem instances for which features were available and 8,057 instances for which labels could be extracted. These latter instances contain 1,532,444 variables, of which 623,293 (40.7%) are part of at least one MUS. The dataset is dominated by a single problem type, namely the Multi-mode Resource-Constrained Project Scheduling Problem (MRCPSP): over 90% of total instances. Additionally, over 80% of the instances in the dataset could be solved in less than 0.1 s. These non-challenging instances were excluded as being of limited use for training and testing.

4.2 Experimental Configuration and Results

In training the two learning Chuffed variants, the parameters of the GCN model were set as follows, based on initial trial runs: Learning rate: 0.3; Number of epochs: 200; Number of units in the first hidden layer: 16; Dropout rate: 0.1; Weight decay: $5e^{-4}$; Tolerance for early stopping: 10; Prediction accuracy at the point of early stopping was between 0.7 and 0.8.

The three Chuffed versions were used to solve test-sets containing instances respectively from the four largest problem types: MRCPSP, bin-packing, price-collecting and fastfood. The experiments were run on a Linux machine with a 16-core 2.50 GHz Xeon Gold 6248 CPU and 32 GB RAM. The code and datasets are available at doi.org/10.4121/14259635.

The box-plot in Fig. 1 shows the resulting distribution of the the total runtimes of all instances from the each of the four largest problem types, averaged

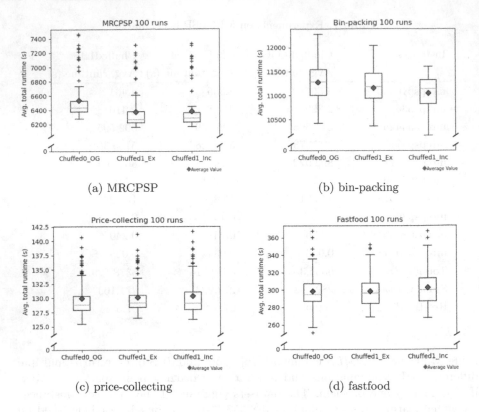

(a) MRCPSP (b) bin-packing

(c) price-collecting (d) fastfood

Fig. 1. Box-plots of total runtime of all test instances averaged over 100 runs.

over a total of 100 runs. A more detailed summary of the results is presented for the two larger domains in Tables 1 and 2, which show the average runtime over 100 runs for each of the instances from the test-set as well as statistics on the total runtime. Table 3 reports the outcome of two-tailed t-tests.

The t-test analysis shows that the machine learning enhanced version significantly outperform the unmodified version for both MRCPSP and bin-packing instances. The gain is about 2.5% for MRCPSP and 1–2% for bin-packing. There is no sufficient statistical evidence to conclude any significant difference between the results obtained with CHUFFED1_INC and CHUFFED1_EX for MRCPSP. However, for bin-packing, there is a statistically significant difference between CHUFFED1_EX and CHUFFED1_INC, of about 1%. This may indicate that bin-packing shares less 'learn-able' concepts with other problem types than MRCPSP. For price-collecting and fastfood there is insufficient evidence to conclude statistically-significant differences between any of the different Chuffed versions. The most likely explanation is not about the dis-similarity of these instances to other problem types, but because the tested instances were not sufficiently large.

Table 2. Experiments on bin-packing benchmarks

Instances	Chuffed0_OG	Chuffed1_Ex	Chuffed1_Inc
	Avg. runtime (s)	Avg. runtime (s)	Avg. runtime (s)
2DLevelPacking238	171.700	151.000	152.580
2DLevelPacking23	1563.956	1499.611	1512.328
2DLevelPacking492	1221.866	1275.854	1237.965
2DPacking13	5065.462	5037.534	5025.021
2DPacking165	683.933	708.044	641.285
2DPacking168	2511.413	2430.075	2431.017
2DPacking62	58.744	57.180	57.587
Total(s)	11277.074	11159.298	11057.783
Standard Deviation	381.016	359.230	347.639
Relative(%)	100.0%	99.0%	98.1%

Table 3. Pairwise t-test analysis

MRCPSP	t-stat	p-value	bin-packing	t-stat	p-value
Chuffed0_OG − Chuffed1_Ex	4.163	$4.693e^{-5}$	Chuffed0_OG − Chuffed1_Ex	2.238	0.026
Chuffed0_OG − Chuffed1_Inc	3.978	$9.761e^{-5}$	Chuffed0_OG − Chuffed1_Inc	4.230	$3.577e^{-5}$
Chuffed1_Ex − Chuffed1_Inc	-0.209	0.834	Chuffed1_Ex − Chuffed1_Inc	-2.020	0.045

price-collecting	t-stat	p-value	fastfood	t-stat	p-value
Chuffed0_OG − Chuffed1_Ex	-0.226	0.821	Chuffed0_OG − Chuffed1_Ex	-1.316	0.190
Chuffed0_OG − Chuffed1_Inc	-1.506	0.134	Chuffed0_OG − Chuffed1_Inc	-1.907	0.058
Chuffed1_Ex − Chuffed1_Inc	-1.390	0.166	Chuffed1_Ex − Chuffed1_Inc	-0.648	0.518

5   Conclusion

This paper shows that it is possible to use machine learning approaches designed for solving SAT instances to improve lazy clause generation solving techniques. Specifically, we have shown how to use unsatisfiable core learning in its CP flavour as minimal unsatisfiable subsets, to improve the performance of the LCG solver Chuffed. We do this by learning the probability a variable is involved in a MUS, as a proxy for initial values of Chuffed's VSIDS scores. With CP–SAT approaches dominating recent MiniZinc benchmarks it is noteworthy that the proposed approach is able to consistently achieve an improved performance on sizeable instances. Although the relative margin of improvement is small (up to 2.5% on MRCPSP scheduling benchmarks), it is statistically significant in the largest two tested problem domains. This suggests that the *similarity* of a variable with variables from MUSs seen during training is a proxy for determining the conflicting nature of a variable.

Our work demonstrates the first, to our knowledge, successful application of machine learning to aid a CP–SAT optimisation solver. This paper thus opens the door to further research. For instance, integrating the classification part directly into the solver can be investigated; this would require embedding the feature extraction part directly into the solver together with additional computational resources, e.g., a GPU as in the Neurocore approach. Moreover, one could consider alternative surrogates other than MUS membership to learn important variables for branching in CP–SAT solvers.

Acknowledgement. We thank the anonymous reviewers of CPAIOR. Thanks to S. van der Laan, K. Leo and P. J. Stuckey. This research was partially supported by TAILOR, a project funded by EU Horizon 2020 research and innovation programme under grant number 952215.

References

1. Amadini, R., Gabbrielli, M., Mauro, J.: SUNNY-CP: a sequential CP portfolio solver. In: Proceedings of the 30th ACM Symposium on Applied Computing, pp. 1861–1867 (2015)
2. Bengio, Y., Lodi, A., Prouvost, A.: Machine learning for combinatorial optimization: a methodological tour d'Horizon. Eur. J. Oper. Res. **290**(2), 405–421 (2021)
3. Berg, J., Demirović, E., Stuckey, P.J.: Core-boosted linear search for incomplete MaxSAT. In: Rousseau, L.-M., Stergiou, K. (eds.) CPAIOR 2019. LNCS, vol. 11494, pp. 39–56. Springer, Cham (2019). https://doi.org/10.1007/978-3-030-19212-9_3
4. Chu, G., Stuckey, P.J., Schutt, A., Ehlers, T., Gange, G., Francis, K.: Chuffed, a lazy clause generation solver (2018). https://github.com/chuffed/chuffed
5. De Uña, D., Rümmele, N., Gange, G., Schachte, P., Stuckey, P.J.: Machine learning and constraint programming for relational-to-ontology schema mapping. In: Proceedings of IJCAI 2018, pp. 1277–1283 (2018)
6. Demirović, E., et al.: An investigation into prediction + optimisation for the Knapsack problem. In: Rousseau, L.-M., Stergiou, K. (eds.) CPAIOR 2019. LNCS, vol. 11494, pp. 241–257. Springer, Cham (2019). https://doi.org/10.1007/978-3-030-19212-9_16
7. Elffers, J., Gocht, S., McCreesh, C., et al.: Justifying all differences using pseudo-boolean reasoning. In: Proceedings of AAAI 2020, pp. 1486–1494 (2020)
8. Elmachtoub, A.N., Grigas, P.: Smart 'predict, then optimize'. CoRR abs/1710.08005 (2017). http://arxiv.org/abs/1710.08005
9. Galassi, A., Lombardi, M., Mello, P., Milano, M.: Model agnostic solution of CSPs via deep learning: a preliminary study. In: van Hoeve, W.-J. (ed.) CPAIOR 2018. LNCS, vol. 10848, pp. 254–262. Springer, Cham (2018). https://doi.org/10.1007/978-3-319-93031-2_18
10. Gasse, M., Chételat, D., Ferroni, N., Charlin, L., Lodi, A.: Exact combinatorial optimization with graph convolutional neural networks. In: Proceedings of NeurIPS 2019, pp. 15554–15566 (2019)
11. Gocht, S., McBride, R., McCreesh, C., Nordström, J., Prosser, P., Trimble, J.: Certifying solvers for clique and maximum common (connected) subgraph problems. In: Simonis, H. (ed.) CP 2020. LNCS, vol. 12333, pp. 338–357. Springer, Cham (2020). https://doi.org/10.1007/978-3-030-58475-7_20
12. Gocht, S., McCreesh, C., Nordström, J.: Subgraph isomorphism meets cutting planes: solving with certified solutions. In: Proceedings of IJCAI 2020, pp. 1134–1140 (2020)

13. Guerri, A., Milano, M.: Learning techniques for automatic algorithm portfolio selection. In: Proceedings of ECAI 2004, pp. 475–479 (2004)
14. Khalil, E., Le Bodic, P., Song, L., Nemhauser, G., Dilkina, B.: Learning to branch in mixed integer programming. In: Proceedings of AAAI 2016, pp. 724–731 (2016)
15. Kipf, T.N., Welling, M.: Semi-supervised classification with graph convolutional networks. CoRR abs/1609.02907 (2016). http://arxiv.org/abs/1609.02907
16. Leo, K., Tack, G.: Debugging unsatisfiable constraint models. In: Salvagnin, D., Lombardi, M. (eds.) CPAIOR 2017. LNCS, vol. 10335, pp. 77–93. Springer, Cham (2017). https://doi.org/10.1007/978-3-319-59776-8_7
17. Liang, J.H., Ganesh, V., Zulkoski, E., Zaman, A., Czarnecki, K.: Understanding VSIDS branching heuristics in conflict-driven clause-learning SAT solvers. In: Piterman, N. (ed.) HVC 2015. LNCS, vol. 9434, pp. 225–241. Springer, Cham (2015). https://doi.org/10.1007/978-3-319-26287-1_14
18. MiniZinc: The MiniZinc benchmark suite (2016). https://github.com/MiniZinc/minizinc-benchmarks
19. MiniZinc: Minizinc challenge 2020 (2020). https://www.minizinc.org/challenge2020/results2020.html
20. Moskewicz, M.W., Madigan, C.F., Zhao, Y., Zhang, L., Malik, S.: Chaff: engineering an efficient SAT solver. In: Proceedings of 38th Annual Design Automation Conference, pp. 530–535 (2001)
21. Ohrimenko, O., Stuckey, P.J., Codish, M.: Propagation via lazy clause generation. Constraints 14(3), 357–391 (2009)
22. Selsam, D., Bjørner, N.: Guiding high-performance SAT solvers with unsat-core predictions. In: Janota, M., Lynce, I. (eds.) SAT 2019. LNCS, vol. 11628, pp. 336–353. Springer, Cham (2019). https://doi.org/10.1007/978-3-030-24258-9_24
23. Selsam, D., Bjørner, N.: Neurocore: guiding high-performance SAT solvers with unsat-core predictions. CoRR abs/1903.04671 (2019). http://arxiv.org/abs/1903.04671
24. Song, W., Cao, Z., Zhang, J., Lim, A.: Learning variable ordering heuristics for solving constraint satisfaction problems. CoRR abs/1912.10762 (2019). http://arxiv.org/abs/1912.10762
25. Soos, M., Kulkarni, R., Meel, K.S.: CrystalBall: gazing in the black box of SAT solving. In: Janota, M., Lynce, I. (eds.) SAT 2019. LNCS, vol. 11628, pp. 371–387. Springer, Cham (2019). https://doi.org/10.1007/978-3-030-24258-9_26
26. Yilmaz, K., Yorke-Smith, N.: A study of learning search approximation in mixed integer branch and bound: node selection in SCIP. AI 2(2), 150–178 (2021). https://doi.org/10.3390/ai2020010

A*-Based Compilation of Relaxed Decision Diagrams for the Longest Common Subsequence Problem

Matthias Horn$^{(\boxtimes)}$ and Günther R. Raidl

Institute of Logic and Computation, TU Wien, Vienna, Austria
{horn,raidl}@ac.tuwien.ac.at

Abstract. We consider the longest common subsequence (LCS) problem and propose a new method for obtaining tight upper bounds on the solution length. Our method relies on the compilation of a relaxed multi-valued decision diagram (MDD) in a special way that is based on the principles of A* search. An extensive experimental evaluation on several standard LCS benchmark instance sets shows that the novel construction algorithm clearly outperforms a traditional top-down construction (TDC) of MDDs. We are able to obtain stronger and at the same time more compact relaxed MDDs than TDC and this in shorter time. For several groups of benchmark instances new best known upper bounds are obtained. In comparison to existing simple upper bound procedures, the obtained bounds are on average 14.8% better.

Keywords: Longest common subsequence problem · Multi-valued decision diagram · A* search

1 Introduction

In the last 10–15 years *decision diagrams* (DDs) have shown to be a powerful tool in combinatorial optimization with which for a wide range of problems new state-of-the-art approaches could be obtained [1,4,14]. This includes prominent problems such as minimum independent set, set covering, maximum cut, maximum 2-satisfiability [3,5] as well as variants of the traveling salesman problem and other sequencing and scheduling problems [14,24]. In particular can DD-based methods be superior where traditional mixed integer linear programming (MIP) or constraint programming (CP) approaches suffer, e.g., from weak dual bounds?

In essence, DDs are data structures that provide graphical representations of the solution space of a combinatorial optimization problem. Restricted DDs represent a subset of feasible solutions and can be used to obtain heuristic solutions

This project is partially funded by the Doctoral Program "Vienna Graduate School on Computational Optimization", Austrian Science Foundation (FWF) Project No. W1260-N35.

P. J. Stuckey (Ed.): CPAIOR 2021, LNCS 12735, pp. 72–88, 2021.
https://doi.org/10.1007/978-3-030-78230-6_5

and primal bounds [3,6] whereas relaxed DDs represent a superset of all feasible solutions in a compact way and can therefore be seen as a discrete relaxation of the problem. Relaxed DDs can be used to obtain dual bounds and provide, for instance, promising new branching schemes [3]. The more general form of a DD in which a node may have more than two outgoing arcs to successor nodes is called multi-valued DD (MDD), and MDDs have proven particularly useful for sequencing and scheduling problems.

Recently, a new A*-based construction (A*C) scheme was presented to compile relaxed MDDs [21]. The authors demonstrate on a prize-collecting scheduling problem that with A*C it is possible to compile in shorter running times relaxed MDDs that provide stronger bounds and are at the same time smaller than relaxed MDDs constructed by traditional top-down or incremental refinement methods. As the name A*C suggests, this method is inspired by A* search [20] and utilizes some fast but not necessarily that strong problem-specific bounding procedure during the construction. However, the prize-collecting scheduling problem from [21] is rather new. The goal of the current work therefore is to investigate the applicability of A*C on the prominent *longest common subsequence* (LCS) problem in order to see if this construction method has the potential to lead to superior results also on this already deeply investigated kind of problem. In our experimental evaluation we will compare the A*C approach for the LCS problem not only to a top-down MDD construction but also to several upper bounding procedures for the LCS from the literature.

The goal of the LCS problem [27] is to find the longest string which is a common subsequence of a set of m input strings $S = \{s_1, s_2, \ldots, s_m\}$ over an alphabet Σ. We denote the length of a string s by $|s|$, and let n be the maximum length of the input strings, i.e., $n = \max_{i=1,\ldots,m} |s_i|$. A subsequence is a string that can be derived from another string by deleting zero or more characters. A common subsequence can be derived from all input strings. For instance, for the input strings ABCDBA and ACBDBA, an LCS is ABDBA. Determining the length of an LCS is a way to measure the similarity of strings and has a wide range of applications, for example in computational biology where strings often represent segments of RNA or DNA [23,30]. Other applications can be found in text editing, file comparison, data compression, and the production of circuits in field programmable gate arrays, to just name a few [2,12,25]. If m is fixed then the LCS problem can be solved by *dynamic programming* (DP) based algorithms in polynomial time $O(n^m)$ [19]. For an arbitrary number of input strings, however, the problem is known to be NP-hard [27].

In the literature plenty of exact approaches have been proposed for solving the LCS problem. Besides the already mentioned DP based approaches, Blum and Festa [10] investigated a MIP model, which is however not competitive and cannot be practically applied to any of the commonly used benchmark sets in the literature due to its excessive size. Further exact methods are for instance based on dominant point approaches and/or parallelization [13,26,28,31] or on a transformation to the max clique problem [9], but they are still not applicable to practical instances with a large number of long input strings. Solving LCS instances of practical relevance to proven optimality is still a challenging task in terms of computation time and memory consumption. Therefore, heuristic approaches are used for larger m and n. Fast construction heuristics are, e.g., the expansion algorithm [11] or the best next heuristic [18,22]. Among the more advanced search strategies, in particular *beam search* (BS) based approaches have been frequently proposed differing in various details such as the heuristic guidance and filtering. This culminated in a general BS-based framework by Djukanovic at el. [15] which can express essentially all heuristic state-of-the-art approaches from the literature by respective configuration settings. They authors proposed also a novel heuristic guidance function, which approximates the expected length of a LCS for random strings. The BS framework in combination with this novel guidance dominates the other existing approaches on most of the available benchmark instances. The same authors further described novel A* based anytime algorithms by interleaving A* search with BS or anytime column search, respectively [16]. Thereby the novel search guidance from before plays again a crucial role.

Before we proceed let us define further notation. We denote the character at position j in a string s by $s[j]$, and $s[j, j']$, $j \leq j'$, refers to the continuous subsequence of s starting at position j and ending at position j'. For $j > j'$, substring $s[j, j']$ is the empty string denoted by ε. Last but not least, let $|s|_a$ be the number of occurrences of character $a \in \Sigma$ in string s.

The next section gives a formal definition of MDDs for the LCS problem. Section 3 reviews two known procedures to obtain upper bounds for the length of an LCS and presents a new one that extends one of those. Section 4 explains how relaxed MDDs are compiled for the LCS problem with A*C. Results of computational experiments are discussed in Sect. 5. Finally, Sect. 6 concludes this work.

2   Multi-valued Decision Diagrams for the LCS Problem

In the context of the LCS problem a MDD is a directed acyclic multi-graph $\mathcal{M} = (V, A)$ with one root node $\mathbf{r}$. For classical layer-based MDDs, all nodes are partitioned into at most $n + 1$ layers $L_1, \ldots, L_{n+1}$, where L_1 is a singleton containing only $\mathbf{r}$ and L_i, $i > 0$ contains only nodes that are reachable from $\mathbf{r}$ over exactly $i - 1$ arcs. An arc $\alpha = (u, v) \in A(\mathcal{M})$ in such a MDD is always directed from a source node u in some layer L_i to a target node v in a subsequent layer L_{i+1}. In this work we will also construct MDDs that do not follow this layer structure. In all cases, each arc α is associated with a character $c(\alpha) \in \Sigma$ s.t. any directed path originating from $\mathbf{r}$ identifies a sequence of characters and thus a (partial) solution. The length of a path is defined as the number of its arcs and corresponds to the number of characters of the represented sequence. For non-layered MDDs, there is only one node that has no further outgoing arcs which we denote by $\mathbf{t}$.

An *exact* MDD encodes precisely the set of all feasible solutions, i.e., the set of all feasible common subsequences, and a longest path encodes a longest common subsequence. Due to the NP-hardness of the LCS problem such exact MDDs will in general have exponential size. Therefore we consider more compact relaxed MDDs which encode supersets of all feasible solutions. In such a relaxed MDD nodes of an exact MDD are superimposed (*merged*) s.t. new paths may emerge representing sequences that are no feasible common subsequences. The length of a longest path then represents an upper bound on the LCS length. In contrast, restricted MDDs approximate exact MDDs by removing nodes and arcs from the exact one s.t. a longest path represents a heuristic solution and its length a lower bound on the LCS length. We remark that compiling a restricted MDD to obtain a heuristic solution essentially corresponds to the well-known beam search, and the leading heuristics for the LCS problem are diverse beam search variants as already pointed out in the previous section.

Each node $u \in V(\mathcal{M})$ is associated with a state which is a *position vector* $\mathbf{p}(u)$, with $p_i(u) \in \{1, \ldots, |s_i|\}$, $i = 1, \ldots, m$. On the basis of this position vector it is possible to define a subproblem $S[\mathbf{p}(u)]$ of S by considering the substrings $s_i[p_i(u), |s_i|]$, $i, \ldots, m$. Thus, $S[\mathbf{p}(u)]$ consists of the right part of each string from S starting from the position indicated in position vector $\mathbf{p}(u)$. The root state represents the original problem S, indicated by $S[\mathbf{p}(\mathbf{r}) = (1, \ldots, 1)]$. An arc $\alpha = (u, v) \in A(\mathcal{M})$ represents the transition from state $\mathbf{p}(u)$ to state $\mathbf{p}(v)$ by appending character $c(\alpha)$ to the sequences of characters encoded by the paths from $\mathbf{r}$ to u. The transition function to obtain successor state $\mathbf{p}(v)$ by considering character $c(\alpha)$ is defined as

$$
\tau(\mathbf{p}(u), c(\alpha)) = \begin{cases} (p_{1,c(\alpha)}(u) + 1, \ldots, p_{m,c(\alpha)}(u) + 1) & \text{if } c(\alpha) \in \Sigma^{\text{nd}}(u) \\ (n + 1, \ldots, n + 1) & \text{else}, \end{cases} \tag{1}
$$

where $p_{i,a}(u)$, $i = 1, \ldots, m$ denotes for each character $a \in \Sigma$ the position of the first occurrence of a in $s_i[p_i(u), |s_i|]$ and set $\Sigma^{\text{nd}}(u) \subseteq \Sigma$ contains all letters that can be feasibly appended at state $\mathbf{p}(u)$, thus letters that occur at least once in each string in $S[\mathbf{p}(u)]$, and are non-dominated. A character $a \in \Sigma$ dominates character $b \in \Sigma$ iff $p_{i,a}(u) \leq p_{i,b}(u)$ for all $i = 1, \ldots, m$, and therefore it never can be better to append a dominated letter next. States that have no further feasible transition, i.e., where $\Sigma^{\text{nd}} = \emptyset$, are mapped to state $(n + 1, \ldots, n + 1)$ of target node $\mathbf{t}$.

To create relaxed MDDs we have to define a state merger which computes the state of merged nodes. Let U be a set of nodes that should be merged. An appropriate state merger is

$$
\oplus (U) = \left(\min_{u \in U} p_i(u) \right)_{i=1,\ldots,m}. \tag{2}
$$

Since we take always the minimum of each position, each feasible solution of any subproblem $S[\mathbf{p}(u)]$, $u \in U$, will also be a feasible solution of the subproblem $S[\mathbf{p}(\oplus(U))]$. Hence, no feasible solution will be lost in the relaxed MDD, but new paths corresponding to infeasible solutions may emerge.

3   Independent Upper Bounds

To compile MDDs based on A* search we need a fast-to-calculate independent upper bound on the solution length of LCS subproblems to guide the construction mechanism. We use two well known upper bounds from the literature as well as a third bound which is an adaption of one of the former. The first upper bound from Fraser [18] was tightened by Blum et al. [8] and is based on the number of occurrences of each character. Given a node u and the associated position vector $\mathbf{p}(u)$, this bound calculates the sum of the minimal number of occurrences of each character over all the strings of the corresponding subproblem $S[\mathbf{p}(u)]$, i.e.,

$$\mathrm{UB}_1(u) = \mathrm{UB}_1(\mathbf{p}(u)) = \sum_{a \in \Sigma} \min_{i=1,\ldots,m} |s_i[p_i(u), |s_i|]|_a. \qquad (3)$$

By using a suitable data structure prepared in pre-processing, UB_1 can be efficiently computed in $O(m|\Sigma|)$ time.

The second upper bound is based on DP and was introduced by Wang et al. [31]. Since the LCS for two input strings can be efficiently computed, for each pair $\{s_i, s_{i+1}\} \subseteq S$, $i = 1, \ldots, m-1$ a so-called scoring matrix $\mathbf{M}_i^2$ is computed, where an entry $M_i^2[p,q]$ with $p = 1, \ldots, |s_i|$ and $q = 1, \ldots, |s_{i+1}|$, stores the length of the LCS of strings $s_i[p, |s_i|]$ and $s_{i+1}[q, |s_{i+1}|]$. The scoring matrices are determined in a pre-processing step. Then

$$\mathrm{UB}_2(u) = \mathrm{UB}_2(\mathbf{p}(u)) = \min_{i=1,\ldots,m-1} \mathbf{M}_i^2[p_i(u), p_{i+1}(u)] \qquad (4)$$

is an upper bound for the subproblem $S[\mathbf{p}(u)]$ of a given node u and the associated position vector $\mathbf{p}(u)$.

The third upper bound we consider adapts the above one as follows. For UB_2, $m-1$ scoring matrices are computed, one for each pair of input strings $\{s_i, s_{i+1}\}$, $i = 1, \ldots, m-1$. However, the pairs of input strings are just chosen according to their natural order given by the instance specification. We are aiming now to choose pairs of input strings in a more controlled and more promising way by utilizing as guidance the version of the first upper bound function for two strings, i.e., $\mathrm{UB}_1(s_i, s_{i'}) = \Sigma_{a \in \Sigma} \min(|s_i|_a, |s_{i'}|_a)$, $s_i, s_{i'} \in S$, $s_i \neq s_{i'}$. Pairs of strings for which this value is small can be expected to typically also have shorter LCSs, possibly leading to an overall tighter bound. The subset of pairs of input strings for which we will compute corresponding scoring matrices, denoted by P, is determined as follows. We iterate over all pairs of input strings $\{(s_i, s_{i'}) \in S \times S \mid i < i'\}$ sorted according to $\mathrm{UB}_1(\cdot, \cdot)$ in non-decreasing order and add each string pair for which not both strings already appear in some string pair earlier added to P. In this way it is ensured that each input string is used at least once and $|P| = O(m)$. The upper bound of a given node u is then

$$\mathrm{UB}_3(u) = \mathrm{UB}_3(\mathbf{p}(u)) = \min_{(s_i, s_{i'}) \in P} \mathbf{M}_{s_i, s_{i'}}^3[p_i(u), p_{i'}(u)], \qquad (5)$$

where $\mathbf{M}_{i,i'}^3$ is the scoring matrix for string pair $(s_i, s_{i'}) \in P$.

Finally, let $\mathrm{UB}(u) = \min\{\mathrm{UB}_1(u), \mathrm{UB}_2(u), \mathrm{UB}_3(u)\}$ be the strongest upper bound we can obtain.

4   A*-Based Construction of MDDs

To construct relaxed MDDs we essentially follow the A*C approach from [21] using the principles of A* search. We maintain an open list Q of nodes that need to be (re-)expanded. This list is sorted according to a priority function

$$f(u) = Z^{\mathrm{lp}}(u) + \mathrm{UB}(u), \tag{6}$$

where $Z^{\mathrm{lp}}(u)$ denotes the length of the so far best path from the root node $\mathbf{r}$ to node u and $\mathrm{UB}(u)$ is the upper bound described in Sect. 3. To break ties, we prefer the node with higher Z^{lp}-value. Initially, Q contains the root node. At each step, A* search and consequently A*C always take a node $u \in Q$ from the open list that maximizes $f(u)$ and expands u by considering all feasible successor states using transition function τ from Eq. (1). Newly created nodes are inserted into Q; nodes that are reached in better ways via the expanded node are updated and reinserted into Q. If Q finally gets empty then A*C has compiled a complete *exact* MDD, since all encountered nodes and arcs corresponding to feasible transitions are stored. However, since UB never underestimates the length of the LCS, the classical A* optimality condition can be applied by terminating early as soon as the target node $\mathbf{t}$ gets selected for expansion the first time. In this case we get in general an incomplete MDD, but due to the optimality condition of A* search at least one optimal path is contained in the MDD.

4.1   Relaxed MDDs

To create a relaxed MDD we limit the size of Q by some threshold value ϕ. As soon as the size of the open list $|Q|$ exceeds ϕ the algorithm starts to merge nodes from Q. If the MDD construction process is carried out until the open list becomes empty a complete relaxed MDD is obtained. Alternatively, we may again terminate as soon as $\mathbf{t}$ is the first time selected for expansion. Then we obtain in general an incomplete relaxed MDDs where not all feasible solutions may be contained, however, due to the optimality condition of A* search the length of the longest path from $\mathbf{r}$ to $\mathbf{t}$—that is $Z^{\mathrm{lp}}(\mathbf{t})$—is a valid upper bound to the length of the LCS. We denote this upper bound by $Z^{\mathrm{ub}}_{\mathrm{min}}$. Note that this bound cannot further be improved by continuing the MDD construction.

Merging. If $|Q|$ exceeds ϕ then nodes are selected in a pairwise fashion for merging. This must be done carefully since we have to ensure that no cycles emerge and that the open list gets empty after a finite number of expansions. Furthermore we do not merge nodes which are already expanded since this would require to update all successor states from the expanded node onward. Note that, since nodes are selected from Q for merging, this approach is able to merge nodes across layers, by introducing so called "long arcs" that skip certain layers. Moreover, to compile relaxed MDDs with A*C the recursive problem formulation does not necessarily have to be based on layers at all.

To do the selection for pairing, we label each node $u \in V(\mathcal{M})$ by a labeling function $\mathcal{L}(u)$ that maps the state $\mathbf{p}(u)$ to a simpler label of a restricted finite domain $\mathcal{D}_{\mathcal{L}}$. The idea is that nodes with the same label are considered similar s.t. the merged state is still a reasonable representative for both nodes. Hence, we only merge nodes with the same label. Moreover, labels are chosen in such way that no cycles will emerge through merging and the open list gets empty in a finite number of steps.

To efficiently select partner nodes for merging we use a global set of so-called collector nodes V^c which is realized as a dictionary indexed by the labels and is initially empty. As long as Q is too large, nodes that are not yet expanded are selected from it in increasing Z^{lp}-order. If for a selected node already a collector node V^c with the same label exists then the two nodes get merged s.t. all incoming arcs from the two nodes will be redirected to the new merged node. The two original nodes are removed from Q and V^c and the new merged node is integrated into $V(\mathcal{M})$ and becomes a new collector node in V^c. Note that during the whole construction process we never allow multiple nodes for the same state. For more details we refer to [21].

Static Labeling Function. For the LCS problem we label all nodes $u \in Q$ by

$$\mathcal{L}_1(u) = (p_{i_{\mathrm{lcs1}}}(u), p_{i_{\mathrm{lcs2}}}(u)) \tag{7}$$

where i_{lcs1}, i_{lcs2} are two specific indexes that refer to a pair of input strings $\{s_{i_{\mathrm{lcs1}}}, s_{i_{\mathrm{lcs2}}}\} \in S$ with smallest $\mathbf{M}^3_{s_i,s_{i'}}[0,0]$ over all $(s_i, s_{i'}) \in P$. Hence, we merge only nodes whose states have the same positions in strings $s_{i_{\mathrm{lcs1}}}$ and $s_{i_{\mathrm{lcs2}}}$ and thus partially represent the same subproblems. Consequently, the longest path in the relaxed MDD will never be worse than the upper bound obtained from the corresponding scoring matrix and each path originating from $\mathbf{r}$ will be a feasible common subsequence w.r.t. input strings $\{s_{i_{\mathrm{lcs1}}}, s_{i_{\mathrm{lcs2}}}\} \subseteq S$. Since any merged node will have the same values for $p_{i_{\mathrm{lcs1}}}$ and $p_{i_{\mathrm{lcs2}}}$ as the original nodes, and each transition from a state to a corresponding successor state increases the values from $p_{i_{\mathrm{lcs1}}}$ and $p_{i_{\mathrm{lcs2}}}$, the values $p_{i_{\mathrm{lcs1}}}$ and $p_{i_{\mathrm{lcs2}}}$ strictly increase along each path in the relaxed MDD. Consequently, no cycles can occur and the open list gets empty within an finite number of iterations.

Dynamic Labeling Function. To derive stronger relaxed MDDs we investigate further the static labeling function

$$\mathcal{L}_2(u) = (p_{i_{\mathrm{lcs1}}}(u), p_{i_{\mathrm{lcs2}}}(u), p_{i_{\mathrm{lcs3}}}(u), p_{i_{\mathrm{lcs4}}}(u)) \tag{8}$$

where $\{s_{i_{\mathrm{lcs3}}}, s_{i_{\mathrm{lcs4}}}\} \in S$ is the additional pair of input strings with smallest $\mathbf{M}^3_{s_i,s_{i'}}[0,0]$ over all $(s_i, s_{i'}) \in P \setminus \{(i_{\mathrm{lcs1}}, i_{\mathrm{lcs2}})\}$. Note that the convergence speed of A*C depends on the size of the domain $|\mathcal{D}_{\mathcal{L}}|$ of the used labeling function $\mathcal{L}$. If the domain size is large then nodes can be grouped into many subgroups and it may be harder to keep the open list size under the desired threshold value ϕ since there are fewer possibilities to merge nodes. If the domain size is small then nodes are merged more aggressively, which makes it easier to keep the open list size under ϕ. However, the finally compiled relaxed MDD will in general be weaker than a relaxed MDD compiled with a labeling function of a larger domain size. For $\mathcal{L}_1$ the domain size is $|\mathcal{D}_{\mathcal{L}_1}| = \sum_{a \in \Sigma} |s_{i_{\mathrm{lcs1}}}|_a |s_{i_{\mathrm{lcs2}}}|_a$. Preliminary results showed that the domain size of $\mathcal{L}_2$ is already too large to let A*C finish in reasonable time on our benchmark instances, but the obtained relaxed MDDs have the potential to be stronger than relaxed MDDs compiled with $\mathcal{L}_1$. Therefore we follow a different strategy: Instead of using a labeling function that is static over the whole compilation process we use a function that adapts its domain size depending on the current situation. We propose the labeling function

$$\mathcal{L}_{2,\Delta}(u) = (p_{i_{\mathrm{lcs1}}}(u),\, p_{i_{\mathrm{lcs2}}}(u),\, \lfloor p_{i_{\mathrm{lcs3}}}(u)/\Delta \rfloor,\, \lfloor p_{i_{\mathrm{lcs4}}}(u)/\Delta \rfloor) \qquad (9)$$

which discretizes the values for $p_{i_{\mathrm{lcs3}}}$ and $p_{i_{\mathrm{lcs4}}}$ by discretization factor Δ. A*C starts with $\Delta = 1$ and doubles this parameter after every k consecutive failures of reducing the open list size below ϕ. If the open list size could be reduced to size ϕ then Δ is reset to one. Each time Δ is adapted, the set of collector nodes V^c is cleared.

4.2 Further Details

Similar to [21], in we merge an already expanded node $u \in V(\mathcal{M})$ and a not yet expanded node $v \in Q$ if $\mathbf{p}(v) \oplus \mathbf{p}(u) = \mathbf{p}(u)$, $Z^{\mathrm{lp}}(v) \leq Z^{\mathrm{lp}}(u)$, and $\mathcal{L}_1(u) = \mathcal{L}_1(v)$ holds since we do not need to update the state of node u. This is efficiently done by indexing each expanded node by labeling function $\mathcal{L}_1$ and checking the condition after each node expansion for each newly created node.

5   Experimental Results

To test our approaches we use six benchmark sets from the literature.

BL instance set from [10]: 450 instances grouped by different values for m, n, and $|\Sigma|$. For each combination there are ten uniform random instances.

Rat, Virus, and Random instance set from [29]: Three benchmark sets consisting of 20 instances each. The Rat and Virus benchmark sets have a biological background whereas instances of the Random benchmark sets are randomly generated.

ES instance set from [17]: 600 instances grouped by different values for m, n and $|\Sigma|$, where each group includes 50 instances.

BB instance set from [7]: 800 instances that were artificially generated in a way s.t. input strings have a large similarity to each other. There are ten instances for each combination of m and $|\Sigma|$.

We used all of these instances for the experimental evaluation but report here only some due to the lack of space. In particular, the main results table to come in Sect. 5.3 omits data for sets Virus and Random since they are similar to those obtained for Rat, and from set BL only instances with $n = 100$ are considered as also done in [16]. However, all instances from all mentioned benchmark sets are considered in all the boxplots to come. Complete results over all benchmark instances are available from https://www.ac.tuwien.ac.at/files/resources/results/LCS/cpaior21_mdds.zip. The algorithms were implemented using GNU C++ 7.5.0, and all experiments were performed on a single core of an Intel Xeon E5649 with 2.53 GHz and 32 GB RAM.

To evaluate the A*C algorithm we use the standard *top-down construction* (TDC) as baseline, which compiles a relaxed MDD layer by layer starting with the root node $\mathbf{r}$. All nodes of the current layer L_i, $i = 1, \dots, n$, are expanded and the newly created nodes are inserted into layer L_{i+1}. The layer size is limited by parameter β. If the size of L_{i+1} exceeds β then L_{i+1} is reduced, after all nodes of L_i are expanded, by sorting the nodes according to priority function f in non-increasing order and replacing the last nodes with smallest f-values from position β onward into a single merged node. Note that TDC in general yields an MDD with be multiple target nodes at different layers. In this case the notation $Z^{lp}(\mathbf{t})$ refers to the length of the longest path from $\mathbf{r}$ to any target node.

5.1   Comparison of Independent Upper Bounds

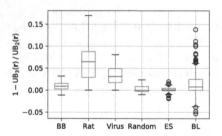

Fig. 1. Relative differences of upper bounds $UB_2(\mathbf{r})$ and $UB_3(\mathbf{r})$.

We start with a comparison of the upper bounds $UB_2(\mathbf{r})$ and $UB_3(\mathbf{r})$ from Sect. 3. Figure 1 shows boxplots for the relative differences $1 - UB_3(\mathbf{r})/UB_2(\mathbf{r})$ over the different benchmark sets. Over all instances, tighter upper bounds can be obtained from $UB_3(\mathbf{r})$ than from $UB_2(\mathbf{r})$ in 62.2% of the cases, and in these the relative difference is on average 1.6%. Both upper bounds are equal in 17.6% of all instances. Overall, upper bound $UB_3(\mathbf{r})$ has on average a relative difference to $UB_2(\mathbf{r})$ of 0.9%. However, differences vary significantly with the type of benchmarks as the figure shows. The largest relative differences could be observed on benchmark sets Rat and Virus. For randomly generated instances, the relative differences seems to be smaller in general. Overall, we conclude that UB_3 provides in general slightly tighter upper bounds than UB_2 but does not dominate it. As both bounding procedures are relatively fast, we conclude that their joint application makes sense.

5.2   Impact of Parameters ϕ and β

Next we investigate the impact of parameter ϕ as well as the choice of the labeling function on the quality of the obtained relaxed MDDs. For this purpose we compile MDDs for middle size instances from benchmark set BB with $m = 100$, $n = 1000$, and $|\Sigma| = 8$. Figure 2 depicts aggregated characteristics of the relaxed

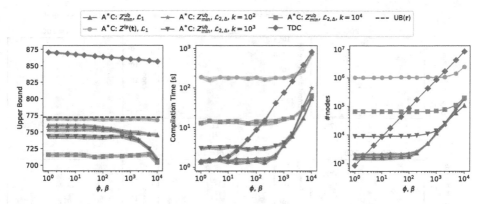

Fig. 2. Relaxed MDDs obtained by A*C and TDC for different settings of ϕ and β for benchmark set BB, $n = 1000$, $m = 100$, $|\Sigma| = 8$

MDDs created by A*C and TDC, respectively. The diagram to the left shows obtained upper bounds, i.e., average lengths of longest **r-t** paths, for different values of ϕ and β in the range of 1 to 10^4. The different solid lines represent different choices of labeling functions for A*C as well as results obtained from TDC. The small tubes around the lines indicate corresponding standard deviations. For A*C we generally report the upper bound values Z_{min}^{ub} obtained when **t** was selected the first time for expansion, and in case of labeling function $\mathcal{L}_1$ additionally the longest path lengths in the complete relaxed MDDs. The dashed line indicates the combined bound UB(**r**) from Sect. 3. The diagrams in the middle and to the right report the corresponding average computation times in seconds and average numbers of nodes of the relaxed MDDs, respectively.

In general we can observe that tighter upper bounds can be obtained when choosing larger values for ϕ or β. Naturally, this comes at the cost of larger compilation times and lager relaxed MDDs. In comparison to TDC, A*C provides consistently much better results in terms of tightness of obtained upper bounds and for larger values of ϕ and β also in terms of compilation time and compactness of obtained relaxed MDDs. A*C with the dynamic labeling function $\mathcal{L}_{2,\Delta}$ yields stronger bounds than with $\mathcal{L}_1$, requires, however, more time than $\mathcal{L}_1$. This is not surprising since domain $\mathcal{D}_{\mathcal{L}_{2,\Delta}}$ is larger than $\mathcal{D}_{\mathcal{L}_1}$ and thus leads less frequently to merges. The tightest upper bounds can be obtained with function $\mathcal{L}_{2,\Delta}$ where the discretization factor Δ is doubled after every $k = 10^4$ consecutive failures of reducing Q below ϕ. Again this can be explained due to less merges than with other parameter settings. For the same reason these settings need in general more computation time and produce larger relaxed MDDs. Note also that, even for small values of ϕ, upper bounds obtained from A*C are substantially smaller than UB(**r**).

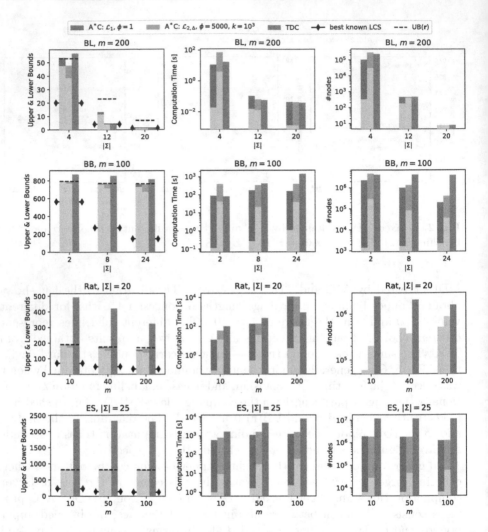

Fig. 3. Lower and upper bounds, respective compilation times, and sizes of obtained relaxed MDDs for selected benchmark sets.

5.3   Main Comparison of A*C and TDC

We start with a graphical comparison for a selected subset of instance classes in Fig. 3. Shown are upper bounds obtained from relaxed MDDs compiled with A*C and TDC, respectively, corresponding compilation times, and the sizes of the obtained MDDs. Each group of bars corresponds to a specific instance class and shows average results, except for instance class Rat which contains only one instance per instance class. The first two bars from the left to right always correspond to relaxed MDDs obtained from A*C with parameters $\{\mathcal{L}_1, \phi = 1\}$ and $\{\mathcal{L}_{2,\Delta}, \phi = 5000, k = 10^3\}$, respectively. The first parameter setting is the case where A*C merges nodes most aggressively whereas the latter setting lets A*C select nodes for merging more carefully, but still with a reasonable total compilation time. The third bar corresponds to relaxed MDDs obtained from TDC with $\beta = 5000$. The brighter parts of the bars indicate the results for the in general incomplete relaxed MDDs obtained when A*C terminates as soon as t is selected for expansion whereas the darker parts show the results for the completed relaxed MDDs. For instance, the brighter part of the bars in the diagrams on the left side show average $Z_{\min}^{ub}$ values. Diamond markers indicate the average lengths of the best known LCSs from the literature obtained from [16]. The black dashed lines show the independent upper bounds UB(**r**).

We can see that if A*C terminates as soon as t is selected for expansion then we obtain in all considered cases MDDs yielding significantly tighter bounds than the MDDs obtained from TDC. Moreover, compilation times are shorter and the obtained MDDs are smaller in case of A*C. Note that although these relaxed MDDs are incomplete in the sense that not all feasible solutions are covered, they can still be further used, e.g., for the DD-based branch-and-bound approach as described by Bergman et al. [3]. It is still possible to derive an exact cut set of nodes to branch on by considering nodes that are not expanded yet, too. If we consider complete relaxed MDDs from A*C then the obtained upper bounds are still tighter or equal than those from relaxed MDDs obtained from TDC, however the compilation with A*C is not faster anymore. Note that TDC was not able to compile relaxed MDDs with $\beta = 5000$ within the time limit of three hours for instances from set ES with $n = 5000$. Also the A*C approach could not compile a complete relaxed MDD for instances of set Rat with $m = 200$, $|\Sigma| = 20$, and $n = 600$ within the three hours time limit. However, with the stopping condition of selecting t for expansion, A*C terminated much earlier. As the length of the longest path of the incomplete relaxed MDD when A*C aborts after three hours is also a feasible upper bound, we show these values in these cases, too.

Finally, Table 1 presents more detailed main results of our computational experiments. Here, A*C is always terminated when t is selected for expansion. Each row contains aggregated results of one instance class. The characteristics of the instance classes can be seen in the first four columns whereas column UB(**r**) shows the average independent upper bound. The next eight columns belong to results obtained from relaxed MDDs compiled with A*C and TDC, respectively. Hereby, columns $\overline{Z_{\min}^{ub}}$ and $\overline{Z^{lp}}(\mathbf{t})$ state the average lengths of the longest paths obtained from the compiled MDDs. Columns $\sigma(\cdot)$ report corresponding standard deviations. Average compilation times in seconds are listed in columns t. Finally, columns gap report the remaining optimality gaps (ub−obj)/ub·100% in relation to the objective values of so far best known solutions obtained from [16] and listed in column obj; value ub refers to the upper bound obtained from the considered approach, i.e., $Z_{\min}^{ub}$ or Z^{lp}. We remark that [16] shows experimental results for two parameter settings, one tailored to obtain as good as possible heuristic solutions, and one targeted towards smallest possible remaining optimality gaps. While we use the better objective values from the former results, the gaps listed in our table for [16] are those of the latter.

Table 1. Main results for A*C and TDC and comparison to the anytime A* search from [16].

	n	$\|\Sigma\|$	m	UB(r)	A*C $\overline{Z^{ub}_{min}}$	$\sigma(Z^{ub}_{min})$	t[s]	gap[%]	TDC $\overline{Z^{lp}(\mathbf{t})}$	$\sigma(\overline{Z^{lp}(\mathbf{t})})$	t[s]	gap[%]	lit. best [16] obj	gap[%]
BB 1000		2	10	807.4	**781.6**	9.1	8.9	**13.4**	882.7	4.4	18.7	23.3	676.7	16.2
			100	792.7	**767.3**	4.5	43.2	**26.5**	871.8	4.3	74.2	35.4	563.6	30.6
		4	10	796.5	**759.5**	6.7	6.4	**28.2**	879.5	4.1	27.7	38.0	545.5	29.4
			100	779.0	**739.4**	8.2	22.5	**47.2**	868.2	4.7	181.3	55.1	390.2	50.9
		8	10	794.8	**732.8**	11.3	7.7	**36.9**	874.9	5.7	45.5	47.1	462.7	38.0
			100	772.3	**708.2**	5.3	23.1	**61.4**	857.6	3.3	386.4	68.1	273.4	65.0
		24	10	786.1	**689.1**	14.5	12.5	44.0	846.9	3.5	131.8	54.5	385.6	**40.5**
			100	768.4	**669.8**	9.9	42.0	**77.7**	818.3	1.8	1261.4	81.7	149.5	79.5
Rat 600		4	10	345.0	**319.0**	–	4.8	35.4	570.0	–	27.7	63.9	206.0	38.0
			15	347.0	**331.0**	–	5.2	42.9	564.0	–	20.6	66.5	189.0	44.5
			20	293.0	**277.0**	–	9.4	37.2	494.0	–	34.4	64.8	174.0	39.5
			25	344.0	**327.0**	–	5.4	47.1	557.0	–	44.1	68.9	173.0	47.4
			40	315.0	**300.0**	–	10.6	48.7	455.0	–	26.2	66.2	154.0	**48.1**
			60	343.0	**323.0**	–	5.1	52.3	548.0	–	62.5	71.9	154.0	53.1
			80	281.0	**261.0**	–	12.6	44.8	466.0	–	36.6	69.1	144.0	47.6
			100	279.0	**263.0**	–	5.5	47.1	497.0	–	118.5	72.0	139.0	49.6
			150	**222.0**	**222.0**	–	13.2	41.0	443.0	–	142.3	70.4	131.0	**40.2**
			200	231.0	**228.0**	–	40.3	44.7	436.0	–	223.7	71.1	126.0	44.9
		20	10	191.0	**167.0**	–	19.6	56.9	493.0	–	101.3	85.4	72.0	58.7
			15	198.0	**169.0**	–	45.1	62.7	467.0	–	268.2	86.5	63.0	62.9
			20	190.0	**159.0**	–	101.7	65.4	456.0	–	278.2	87.9	55.0	**65.2**
			25	173.0	**145.0**	–	18.5	64.1	417.0	–	158.6	87.5	52.0	68.1
			40	176.0	**143.0**	–	53.0	65.0	421.0	–	379.6	88.1	50.0	70.3
			60	195.0	**161.0**	–	439.7	70.8	431.0	–	284.2	89.1	47.0	**70.3**
			80	180.0	**145.0**	–	518.7	69.7	376.0	–	269.5	88.3	44.0	**69.1**
			100	173.0	**138.0**	–	103.5	71.0	359.0	–	545.3	88.9	40.0	71.8
			150	172.0	**145.0**	–	128.0	73.8	323.0	–	609.6	88.2	38.0	**71.5**
			200	170.0	**133.0**	–	195.7	73.7	324.0	–	897.6	89.2	35.0	70.2
ES 1000		2	10	795.3	**783.6**	4.3	5.6	**21.0**	987.5	1.3	19.7	37.3	618.9	21.2
			50	791.0	**779.4**	3.0	12.8	30.6	982.7	1.2	40.8	45.0	540.9	**30.6**
			100	788.7	**777.3**	3.0	18.4	32.8	980.8	0.9	77.6	46.8	522.1	32.9
		10	10	477.6	**462.2**	2.9	4.9	55.6	951.8	2.7	138.4	78.5	205.0	**54.9**
			50	473.7	**455.7**	1.8	15.4	69.8	928.7	2.1	339.4	85.2	137.5	**69.1**
			100	472.2	**454.0**	2.0	28.9	72.7	919.5	2.1	591.8	86.5	124.1	**71.9**
ES 2500		25	10	820.1	**800.1**	2.4	11.5	70.4	2389.2	4.3	1453.7	90.1	236.6	**70.1**
			50	816.5	**791.0**	1.7	39.1	82.3	2332.4	4.5	4367.0	94.0	140.4	**81.9**
			100	814.4	**788.3**	1.4	74.2	84.3	2309.5	3.6	7514.3	94.7	123.4	**84.0**
ES 5000		100	10	888.3	**853.9**	2.6	62.7	82.9	–	–	–	–	145.7	**82.9**
			50	883.5	**835.9**	1.7	152.1	91.4	–	–	–	–	72.0	**91.3**
			100	882.3	**829.5**	1.6	373.3	92.7	–	–	–	–	60.8	**92.6**
BL 100		4	10	58.8	**47.5**	1.6	0.5	28.2	75.6	2.0	3.0	54.9	34.1	**10.8**
			50	56.2	**41.7**	1.4	2.1	42.0	65.0	1.2	6.1	62.8	24.2	**18.7**
			100	54.7	**40.6**	1.1	3.2	45.8	61.0	1.8	9.6	63.9	22.0	**20.4**
			150	53.8	**38.7**	1.2	3.9	46.8	58.0	1.4	11.6	64.5	20.6	**18.1**
			200	53.0	**38.3**	0.8	5.0	47.8	56.8	1.8	15.9	64.8	20.0	**20.2**
		12	10	37.4	**21.2**	1.7	0.2	40.1	36.3	4.1	3.7	65.0	12.7	**0.0**
			50	34.4	**8.7**	2.1	0.2	20.7	9.6	3.0	0.3	28.1	6.9	**0.0**
			100	28.8	**5.2**	0.4	<0.1	**0.0**	**5.2**	0.4	<0.1	**0.0**	5.2	**0.0**
			150	23.8	**4.7**	0.5	<0.1	**0.0**	**4.7**	0.5	<0.1	**0.0**	4.7	**0.0**
			200	22.8	**4.1**	0.3	<0.1	**0.0**	**4.1**	0.3	<0.1	**0.0**	4.1	**0.0**
		20	10	29.2	**9.5**	1.0	<0.1	16.8	10.5	2.2	0.3	24.8	7.9	**0.0**
			50	17.5	**3.0**	0.0	<0.1	**0.0**	**3.0**	0.0	<0.1	**0.0**	3.0	**0.0**
			100	12.1	**2.1**	0.3	<0.1	**0.0**	**2.1**	0.3	<0.1	**0.0**	2.1	**0.0**
			150	7.2	**1.9**	0.3	<0.1	**0.0**	**1.9**	0.3	<0.1	**0.0**	1.9	**0.0**
			200	6.8	**1.1**	0.3	<0.1	**0.0**	**1.1**	0.3	<0.1	**0.0**	1.1	**0.0**

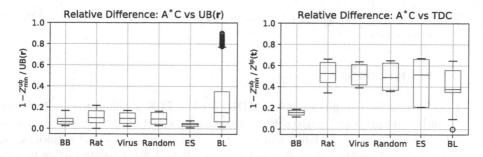

Fig. 4. Relative differences of upper bounds between $Z_{\min}^{\mathrm{ub}}$ and $\mathrm{UB}(\mathbf{r})$ as well as between $Z_{\min}^{\mathrm{ub}}$ and $Z^{\mathrm{lp}}(\mathbf{t})$.

For the compilation of MDDs we set $\beta = 5000$ for TDC and $\phi = 5000$ for A*C with labeling function $\mathcal{L}_{2,\Delta}$ and $k = 10^4$ for all instance except for benchmark set ES where k is set to 10^3.

We observe that in all considered cases the obtained upper bounds $Z_{\min}^{\mathrm{ub}}$ are tighter than $\mathrm{UB}(\mathbf{r})$ as well as the upper bounds obtained from relaxed MDDs compiled with TDC. Only in one single case, for benchmark set Rat with $|\Sigma| = 4$, $m = 150$, $n = 600$, the upper bound $\mathrm{UB}(\mathbf{r})$ is equal to $Z_{\min}^{\mathrm{ub}}$. We notice an average relative difference between $Z_{\min}^{\mathrm{ub}}$ and $\mathrm{UB}(\mathbf{r})$ of 14.8% over all instances. Considering $Z_{\min}^{\mathrm{ub}}$ and $Z^{\mathrm{lp}}(\mathbf{t})$ from relaxed MDDs compiled with TDC we get an average relative difference of 43.7%. The boxplots shown in Fig. 4 give deeper insight on the relative differences between $Z_{\min}^{\mathrm{ub}}$ and $\mathrm{UB}(\mathbf{r})$ as well as the differences between $Z_{\min}^{\mathrm{ub}}$ and $Z^{\mathrm{lp}}(\mathbf{t})$. The largest relative difference between upper bounds obtained from relaxed MDDs compiled by A*C and TDC occurs for instance sets Rat, Virus, Random, and ES. For these benchmark sets the median of the obtained relative differences is about 50%. Regarding instances of the BB benchmark set, substantially smaller relative differences are obtained. The fact that BB instances are created in a way s.t. input strings have a large similarity to each other seems to be an explanation for this discrepancy. The median of the relative differences between upper bounds $Z_{\min}^{\mathrm{ub}}$ and $\mathrm{UB}(\mathbf{r})$ is about 10% for all benchmark sets. Only results from benchmark set ES exhibit a median relative difference of about 4%, which can be explained by the longer input strings of ES instances, e.g., $n = 5000$. Finally, BL instances exhibit some outliers, e.g., instances with a relative difference between $Z_{\min}^{\mathrm{ub}}$ and $\mathrm{UB}(\mathbf{r})$ of 80% and differences between $Z_{\min}^{\mathrm{ub}}$ and $Z^{\mathrm{lp}}(\mathbf{t})$ (TDC) of 0%. This is not surprising, since benchmark set BL contains small instances that could be solved to proven optimality by exact methods, and both construction methods, A*C as well as TDC, are able to compile relaxed MDDs that yield the optimal solution values as upper bounds. This is also documented in Table 1 for instance classes of set BL with $n = 100$ and $|\Sigma| \in \{12, 20\}$ where the average optimality gap is 0%. In comparison to [16], we can observe that A*C is able to obtain even smaller optimality gaps in 315 cases and equal optimality gaps in 73 cases. Most of the gaps from

[16] were only obtained after a time limit of 15 min, while A*C created the MDDs in much shorter time.

6   Conclusions

In this work we compiled relaxed MDDs for the LCS problem to obtain upper bounds. The proposed construction algorithm A*C is not layer-oriented such as TDC and utilizes fast independent upper bounds on subproblems for guidance in the spirit of A* search. As independent upper bound we suggested using a combination of two fast-to-calculate bounds from the literature and the new variant UB_3 that is approximately equally fast to compute but occasionally stronger than the former bounds. To control the size of the relaxed MDD, A*C merges nodes when the list of open nodes exceeds a certain size. To determine suitable partner nodes for merging, we investigated different LCS-specific labeling functions. The better performing dynamic labeling function adapts the domain dynamically during the compilation process s.t. depending on the current situation nodes are merged more or less aggressively. When rigorously comparing A*C with a classical TDC on several benchmark instance sets from the literature, we observed that A*C is able to provide more compact relaxed MDDs that are significantly stronger than relaxed MDDs obtained from TDC in shorter time. For several instance classes relaxed MDDs compiled with A*C yielded stronger bounds than the best known upper bounds from the literature.

For future work it seems promising to embed the compilation of relaxed MDDs into a branch-and-bound approach that branches over exact nodes of the relaxed MDD, as already done in the literature for other kinds of problems, where, however, the classical TDC was used instead of A*C to compile relaxed MDDs. To obtain also high quality heuristic solutions for subproblems within such a branch-and-bound approach, ideas from the leading beam search methods can further be adopted. Further interesting research directions will be to investigate different strategies for the novel dynamic labeling functions mechanism and to perform more detailed analysis of the ability of A*C to reduce the size of relaxed MDDs, e.g. comparing for small instances the size of exact reduced MDDs with relaxed MDDs compiled with A*C.

References

1. Andersen, H.R., Hadzic, T., Hooker, J.N., Tiedemann, P.: A constraint store based on multivalued decision diagrams. In: Bessière, C. (ed.) CP 2007. LNCS, vol. 4741, pp. 118–132. Springer, Heidelberg (2007). https://doi.org/10.1007/978-3-540-74970-7_11
2. Beal, R., Afrin, T., Farheen, A., Adjeroh, D.: A new algorithm for "the LCS problem" with application in compressing genome resequencing data. BMC Genom. **17**(4), 544 (2016). https://doi.org/10.1186/s12864-016-2793-0
3. Bergman, D., Cire, A.A., van Hoeve, W.J., Hooker, J.N.: Discrete optimization with decision diagrams. INFORMS J. Comput. **28**(1), 47–66 (2016)

4. Bergman, D., Cire, A.A., von Hoeve, W.J., Hooker, J.N.: Optimization bounds from binary decision diagrams. INFORMS J. Comput. **26**(2), 253–268 (2014)
5. Bergman, D., Cire, A.A., van Hoeve, W.J., Hooker, J.N.: Decision Diagrams for Optimization. Artificial Intelligence: Foundations, Theory, and Algorithms. Springer, Heidelberg (2016). https://doi.org/10.1007/978-3-319-42849-9
6. Bergman, D., Cire, A.A., van Hoeve, W.-J., Yunes, T.: BDD-based heuristics for binary optimization. J. Heuristics **20**(2), 211–234 (2014). https://doi.org/10.1007/s10732-014-9238-1
7. Blum, C., Blesa, M.J.: Probabilistic beam search for the longest common subsequence problem. In: Stützle, T., Birattari, M., Hoos, H.H. (eds.) SLS 2007. LNCS, vol. 4638, pp. 150–161. Springer, Heidelberg (2007). https://doi.org/10.1007/978-3-540-74446-7_11
8. Blum, C., Blesa, M.J., López-Ibáñez, M.: Beam search for the longest common subsequence problem. Comput. Oper. Res. **36**(12), 3178–3186 (2009)
9. Blum, C., et al.: Solving longest common subsequence problems via a transformation to the maximum clique problem. Comput. Oper. Res. **125**, 105089 (2021). https://doi.org/10.1016/j.cor.2020.105089
10. Blum, C., Festa, P.: Longest common subsequence problems. In: Metaheuristics for String Problems in Bioinformatics, chap. 3, pp. 45–60. Wiley (2016)
11. Bonizzoni, P., Della Vedova, G., Mauri, G.: Experimenting an approximation algorithm for the LCS. Discret. Appl. Math. **110**(1), 13–24 (2001)
12. Brisk, P., Kaplan, A., Sarrafzadeh, M.: Area-efficient instruction set synthesis for reconfigurable system-on-chip designs. In: Proceedings of DAC 2004 - the 41st Annual Design Automation Conference, pp. 395–400. IEEE Press (2004)
13. Chan, H.T., Yang, C.B., Peng, Y.H.: The generalized definitions of the two-dimensional largest common substructure problems. In: Proceedings of the 33rd Workshop on Combinatorial Mathematics and Computation Theory, pp. 1–12. National Taiwan University (2016)
14. Cire, A.A., van Hoeve, W.J.: Multivalued decision diagrams for sequencing problems. Oper. Res. **61**(6), 1411–1428 (2013)
15. Djukanovic, M., Raidl, G.R., Blum, C.: A beam search for the longest common subsequence problem guided by a novel approximate expected length calculation. In: Nicosia, G., Pardalos, P., Umeton, R., Giuffrida, G., Sciacca, V. (eds.) LOD 2019. LNCS, vol. 11943, pp. 154–167. Springer, Cham (2019). https://doi.org/10.1007/978-3-030-37599-7_14
16. Djukanovic, M., Raidl, G.R., Blum, C.: Finding longest common subsequences: new anytime A* search results. Appl. Soft Comput. **95**, 106499 (2020). https://doi.org/10.1016/j.asoc.2020.106499
17. Easton, T., Singireddy, A.: A large neighborhood search heuristic for the longest common subsequence problem. J. Heuristics **14**(3), 271–283 (2008). https://doi.org/10.1007/s10732-007-9038-y
18. Fraser, C.B.: Subsequences and supersequences of strings. Ph.D. thesis, University of Glasgow, UK (1995)
19. Gusfield, D.: Algorithms on Strings, Trees, and Sequences: Computer Science and Computational Biology. Cambridge University Press, Cambridge (1997)
20. Hart, P., Nilsson, N., Raphael, B.: A formal basis for the heuristic determination of minimum cost paths. IEEE Trans. Syst. Sci. Cybern. **4**(2), 100–107 (1968)
21. Horn, M., Maschler, J., Raidl, G.R., Rönnberg, E.: A*-based construction of decision diagrams for a prize-collecting scheduling problem. Comput. Oper. Res. **126**, 105125 (2021). https://doi.org/10.1016/j.cor.2020.105125

22. Huang, K., Yang, C., Tseng, K.: Fast algorithms for finding the common subsequences of multiple sequences. In: Proceedings of the IEEE International Computer Symposium, pp. 1006–1011. IEEE Press (2004)

23. Jiang, T., Lin, G., Ma, B., Zhang, K.: A general edit distance between RNA structures. J. Comput. Biol. **9**(2), 371–388 (2002)

24. Kinable, J., Cire, A.A., van Hoeve, W.J.: Hybrid optimization methods for time-dependent sequencing problems. Eur. J. Oper. Res. **259**(3), 887–897 (2017)

25. Kruskal, J.B.: An overview of sequence comparison: time warps, string edits, and macromolecules. SIAM Rev. **25**(2), 201–237 (1983)

26. Li, Y., Wang, Y., Zhang, Z., Wang, Y., Ma, D., Huang, J.: A novel fast and memory efficient parallel mlcs algorithm for long and large-scale sequences alignments. In: 2016 IEEE 32nd International Conference on Data Engineering (ICDE), pp. 1170–1181. IEEE Press (2016)

27. Maier, D.: The complexity of some problems on subsequences and supersequences. J. ACM **25**(2), 322–336 (1978)

28. Peng, Z., Wang, Y.: A novel efficient graph model for the multiple longest common subsequences (MLCS) problem. Front. Genet. **8**, 104 (2017)

29. Shyu, S.J., Tsai, C.Y.: Finding the longest common subsequence for multiple biological sequences by ant colony optimization. Comput. Oper. Res. **36**(1), 73–91 (2009)

30. Smith, T., Waterman, M.: Identification of common molecular subsequences. J. Mol. Biol. **147**(1), 195–197 (1981)

31. Wang, Q., Korkin, D., Shang, Y.: A fast multiple longest common subsequence (MLCS) algorithm. IEEE Trans. Knowl. Data Eng. **23**(3), 321–334 (2011)

Partitioning Students into Cohorts During COVID-19

Richard Hoshino[1]([✉]) and Irene Fabris[2]

[1] Northeastern University, Vancouver, BC, Canada
r.hoshino@northeastern.edu
[2] Quest University Canada, Squamish, BC, Canada
irene.fabris@questu.ca

Abstract. The COVID-19 pandemic has forced educational institutions to make significant changes to safeguard the health and safety of their students and teachers. One of the most effective measures to reduce virus transmission is partitioning students into discrete cohorts.

In primary and middle schools, it is easy to create these cohorts (also known as "learning groups"), since students in each grade take the same set of required courses. However, in high schools, where there is much diversity in course preferences among individual students, it is extremely challenging to optimally partition students into cohorts to ensure that every section of a course only contains students from a single cohort.

In this paper, we define the Student Cohort Partitioning Problem, where our goal is to optimally assign cohorts to students and course sections, to maximize students being enrolled in their desired courses. We solve this problem by modeling it as an integer linear program, and apply our model to generate the Master Timetable for a Canadian all-boys high school, successfully enrolling students in 87% of their desired courses, including 100% of their required courses. We conclude the paper by explaining how our model can benefit all educational institutions that need to create optimal student cohorts when designing their annual timetable.

Keywords: School timetabling · Integer programming · Optimization

1 Introduction

The COVID-19 virus has led to the worst global pandemic in over a hundred years. Since the first case was identified in December 2019, the disease has spread worldwide, leading to 131 million cases and 2.8 million deaths as of April 1, 2021 [13]. In addition to destabilizing world economies, the pandemic has also had a profound impact on education, with nearly 87% of the world's students, i.e., 1.5 billion learners in over 170 countries, affected by school closures [23]. The switch to Remote Learning has been overwhelming for many students who live in conditions that are not suitable for home study, and has further exacerbated social inequalities such as access to technology [14,20].

© Springer Nature Switzerland AG 2021
P. J. Stuckey (Ed.): CPAIOR 2021, LNCS 12735, pp. 89–105, 2021.
https://doi.org/10.1007/978-3-030-78230-6_6

In many countries, governments and school boards have invested considerable resources to safely bring students back to school. There are numerous measures to mitigate COVID-19 transmission in schools and the Center for Disease Control [8] divides these mitigation measures into three categories: *personal* controls (e.g. hand hygiene, masks, and physical distancing), *engineering* controls (e.g. air ventilation systems, plexiglass barriers), and *administrative* controls (e.g. scheduling breaks and meals at different times).

According to the CDC, the most important administrative control is to ensure that the same group of students learn together each day, avoiding interactions with other student groups. These learning groups, also known as *cohorts*, are already in place at many primary schools, since the same group of students stay together for the entire year, learning from a single teacher. Creating cohorts is straightforward for students in many middle schools and junior high schools, since every student in each grade takes the same set of required courses.

However, creating cohorts is far more complex for students in senior high school (usually 16- and 17-year-olds), since these students have a much larger set of course options available. While some schools allow students to self-select into a particular stream (e.g. Arts, Sciences) where they only take courses with students from that stream, many schools encourage or require students to take courses from all disciplines, with each discipline having numerous options.

To illustrate the challenge of partitioning students into cohorts, consider a scenario where each of 9 students (S_1 to S_9) wishes to enroll in four courses chosen among 12 course offerings (C_1 to C_{12}).

Suppose we have the following set of desired courses.

Student	Desired courses
S_1	C_1, C_4, C_7, C_{10}
S_2	C_1, C_5, C_8, C_{11}
S_3	C_1, C_6, C_9, C_{12}
S_4	C_2, C_4, C_9, C_{11}
S_5	C_2, C_5, C_7, C_{12}
S_6	C_2, C_6, C_8, C_{10}
S_7	C_3, C_4, C_8, C_{12}
S_8	C_3, C_5, C_9, C_{10}
S_9	C_3, C_6, C_7, C_{11}

We can construct a timetable that grants each of the 9 students all four of their requests, with each course taking place in one time slot.

Course	Enrolled students	Time slot	Course	Enrolled students	Time slot
C_1	S_1, S_2, S_3	1	C_7	S_1, S_5, S_9	3
C_2	S_4, S_5, S_6	1	C_8	S_2, S_6, S_7	3
C_3	S_7, S_8, S_9	1	C_9	S_3, S_4, S_8	3
C_4	S_1, S_4, S_7	2	C_{10}	S_1, S_6, S_8	4
C_5	S_2, S_5, S_8	2	C_{11}	S_2, S_4, S_9	4
C_6	S_3, S_6, S_9	2	C_{12}	S_3, S_5, S_7	4

For example, student S_1's timetable is C_1, C_4, C_7, C_{10}. If we give a score of 1 point whenever a student is enrolled in a desired course, we see that the above timetable achieves the best possible score of $9 \times 4 = 36$.

Now suppose we need to split the 9 students and 12 courses into 3 cohorts. Suppose we partition our students and courses into these cohorts of equal size:

Cohort	Students	Courses
1	S_1, S_2, S_6	C_1, C_4, C_8, C_{10}
2	S_4, S_5, S_9	C_2, C_5, C_7, C_{11}
3	S_3, S_7, S_8	C_3, C_6, C_9, C_{12}

Since each course is only available to students from that cohort, we can no longer enroll students in all of their desired courses. For example, student S_1 is enrolled in three desired courses (C_1, C_4, C_{10}) but not in C_7 because this course is only offered to students in Cohort 2, and S_1 is assigned to Cohort 1. We can show that the above cohort partition yields a timetable scoring 21 total points.

Assuming that each of the three cohorts must contain 3 students and 4 courses, we can show that 21 points is optimal. Thus, the best possible cohort partition produces a 41.7% reduction from the maximum score of 36.

This simple example illustrates the challenge of cohort partitioning. At large high schools, where there are thousands of students with heterogeneous course preferences, school administrators are pressured to ensure that their students can enroll in their desired courses. Due to COVID-19, the additional requirement of student cohorts makes timetabling even harder.

In the Canadian province of British Columbia, where both authors reside, the government mandated cohorts of size at most 120, for every high school in the province. Thus, a small high school with 400 students (100 students for each of Grades 9, 10, 11, 12) could treat each grade as a single cohort, and would only need to forbid students from taking courses outside of their cohort (e.g. a Grade 11 student taking Grade 12 math). However, for a large high school with 2000 or more students, creating these cohorts is incredibly challenging.

The provincial government announced the 120-student cohort limit on July 29, 2020, as the centerpiece of their Back to School plan [1]. Thus, schools had just over a month to implement this policy, to ensure their students and staff could return in September under the new guidelines. Numerous solutions were reported: restricting student choice by eliminating courses, scheduling certain courses outside of school hours by making them virtual, and hiring more teachers to teach additional "sections" of a course (e.g. one section of Calculus per cohort). Canadian High Schools usually define a course *section* as an offering distinguished from other course sections by time slot, classroom, and teacher (e.g. offering multiple sections of AP Calculus on different days and times).

Some schools simply ignored the cohort policy, as they would have been forced to re-design their Master Timetable. For example, at one high school, several sections of a course have students from *six* different cohorts, with the students in each cohort required to sit together in the same part of the classroom [21].

One school hired the authors of this paper to apply Linear Programming techniques to maximize the students' ability to take their desired courses while ensuring that each section of every course only consists of students from a single cohort. We recently developed an algorithm for optimizing student course preferences in school timetabling [12], and in this paper, we expand upon this work by introducing and solving the Student Cohort Partitioning Problem (SCPP).

This paper proceeds as follows. In Sects. 2 and 3, we define the SCPP and provide a brief literature review on related work that involves partitioning students into cohorts. In Sect. 4, we describe our solution to the SCPP by formulating it as an integer linear program. In Sect. 5, we apply our model (with over 1.5 million binary decision variables) to generate the Master Timetable for a Canadian high school, partitioning 328 students and 196 course sections into three cohorts. In Sect. 6, we discuss the limitations of cohort-based timetabling on student choice, and in Sect. 7, we conclude the paper with questions and directions for future research.

2   Problem Definition

To avoid confusion in how we label our sets, we now rename timeslots as *blocks* and cohorts as *learning groups*. Let I be the set of individual students, T be the set of teachers, C be the set of courses, S be the set of sections, B be the set of blocks, and L be the set of learning groups.

Each course has one or more sections, and each course section is represented by the pair (c, s), where $c \in C$ and $s \in S$.

In the School Timetabling Problem (STP), our goal is to find a feasible assignment of course sections to teachers and blocks. The more general version of the STP is a combinatorial optimization problem, which asks for the best assignment satisfying all of the hard constraints while maximizing the satisfaction of the teachers being assigned their desired courses in specific blocks.

The Post-Enrollment Course Timetabling Problem (PECTP) was introduced just over a decade ago [17], as part of the second International Timetabling Competition. In the PECTP, points are awarded for enrolling students in any section of a desired course. In addition to all of the constraints in the STP (e.g. no teacher can be assigned to two courses in the same block), the PECTP involves additional student-related hard constraints, such as ensuring that no student is enrolled in multiple sections of the same course.

In this paper, we define the Student Cohort Partitioning Problem (SCPP) to be identical to the PECTP, with these three additional requirements:

(i) Each student $i \in I$ is assigned a learning group $l \in L$.
(ii) Each course section $(c, s) \in C \times S$ is assigned a learning group $l \in L$.
(iii) Student i can be enrolled in section s of course c only if both i and (c, s) are assigned to the same learning group l.

The SCPP can be modeled as a 0-1 integer linear program, where the binary variable $X_{t,s,c,b,l}$ $(Y_{i,s,c,b,l})$ equals 1 if teacher t (student i) is assigned to section s of course c in block b and learning group l, and is equal to 0 otherwise.

In our earlier example with 9 students and 12 cohorts, our optimal solution placed student 8 in learning group 3, assigning this student to courses $3, 6, 9, 12$, which are offered in blocks $1, 2, 3, 4$, respectively. Since each of these courses has a single section $(s = 1)$, we have $Y_{8,1,3,1,3} = Y_{8,1,6,2,3} = Y_{8,1,9,3,3} = Y_{8,1,12,4,3} = 1$.

Let $D_{t,c,b}$ be the *desirability* of teacher t being assigned to course c in block b. This coefficient will be a function of teacher t's ability and willingness to teach course c, combined with their availability in block b. Let $P_{i,c,b}$ be the *preference* of student i being enrolled in course c in block b. (We assume that the coefficients $D_{t,c,b}$ and $P_{i,c,b}$ are independent of the section s and learning group l.)

Then, subject to all of the hard constraints, our integer linear program aims to maximize the following objective function:

$$\sum_{t \in T} \sum_{s \in S} \sum_{c \in C} \sum_{b \in B} \sum_{l \in L} D_{t,c,b} \cdot X_{t,s,c,b,l} + \sum_{i \in I} \sum_{s \in S} \sum_{c \in C} \sum_{b \in B} \sum_{l \in L} P_{i,c,b} \cdot Y_{i,s,c,b,l}.$$

In Sect. 4, we present the hard constraints for this timetabling optimization problem. But first, we present a brief summary of related work.

3   Related Work

Partitioning students into learning groups is a complex challenge faced by school administrators. This explains why scholars in Operations Research have devised innovative techniques to create these student partitions, and apply them to real-life timetabling instances to serve educational institutions all over the world.

Computer Science students at Boston University are optimally matched with peers to form learning groups that increase collective student learning [2] [3]. The practice of creating these cohorts, known as *team formation*, is an NP-Hard combinatorial optimization problem that has been tackled in the last two decades using techniques such as simulated annealing [5], branch and cut [9], and genetic algorithms [24].

Researchers have applied resolvable complete block designs to pre-assign students to groups, ensuring that no pair of students works together on more than one group assignment. This is an application of the Social Golfer Problem [22] to education. For example, Baker et al. [4] deploy both linearized IP and Constraint

Logic Programming (CLP) models to maximize exposure of MBA students to each other at Dartmouth College's Tuck School of Business.

As we do in our paper, many educational institutions want to *section students*, i.e. they want to partition students into non-overlapping groups while maximizing the percentage of students being enrolled in their requested courses. A comprehensive introduction to *student sectioning* is found in Kristiansen et al. [15], which presents the High School Student Sectioning (HSSS) problem, formulating it as an integer program, and solving fifteen real-life instances at Danish Schools using Gurobi, a state-of-the-art MIP solver [16].

Over the last decade, there have been two main approaches to solve instances of the Student Sectioning problem: sectioning during course timetabling and batch sectioning after a complete timetable is developed [18].

In the first approach, students who request similar combinations of courses are grouped into the same course sections to minimize potential student conflicts. Thus, the initial sectioning phase precedes the assignment of course sections to time slots. In this sense, sectioning becomes a pre-processing stage prior to timetabling. Two examples of this practice are Carter's homogeneous sectioning [7] and Schindl's regular sub-division of the students using a conflict graph [19].

In the second approach, the task of assigning course sections to time slots is performed first, and only after are students assigned to sections of their desired courses. For example, Müller et al. [18] solve a batch sectioning problem for Purdue University using a heuristic based on an iterative forward search, which progressively searches different neighborhoods of the solution space finding feasible yet incomplete solutions which are optimized in subsequent iterations.

Our Student Cohort Partitioning Problem (SCPP) most closely resembles the batch sectioning formulation, specifically the work by Goebbels et al. [10] which models a batch sectioning problem at Niederrhein University as an IP and finds an optimal solution using IBM's ILOG CPLEX 12.80 solver. To construct the solution, they first optimally partition students into groups of homogeneous sizes and then match these groups to courses previously assigned to time slots.

Despite some similarities, the SCPP differs from the literature in two important ways. First, we partition both students *and* courses into learning groups. Second, each student's timetable comprises of a personalized set of requested courses, rather than the same set of courses as everyone else in their cohort.

In the next two sections, we present the key results of this paper. First, we model the SCPP, an extension of the Post-Enrollment Course Timetabling Problem that was made necessary by the COVID-19 pandemic. Second, we introduce a multi-stage heuristic that finds a close-to-optimal solution of the SCPP, and apply it to generate the Master Timetable for a Canadian high school, partitioning 328 students and 196 course sections into 3 learning groups.

4 Mathematical Model

Let I be the set of individual students, T be the set of teachers, C be the set of courses, S be the set of sections, B be the set of blocks, and L be the set of learning groups.

Our integer linear program (ILP) aims to maximize

$$\sum_{t \in T} \sum_{s \in S} \sum_{c \in C} \sum_{b \in B} \sum_{l \in L} D_{t,c,b} \cdot X_{t,s,c,b,l} + \sum_{i \in I} \sum_{s \in S} \sum_{c \in C} \sum_{b \in B} \sum_{l \in L} P_{i,c,b} \cdot Y_{i,s,c,b,l},$$

where $D_{t,c,b}$ is the desirability coefficient of teacher t being assigned to course c in block b, and $P_{i,c,b}$ is the preference coefficient of student i being assigned to course c in block b.

We define the following four binary variables, two of which appear in our objective function.

(i) For each $t \in T, s \in S, c \in C, b \in B$, and $l \in L$, let $X_{t,s,c,b,l}$ equal 1 if teacher t is assigned to section s of course c in block b and learning group l, and is equal to 0 otherwise.

(ii) For each $i \in I, s \in S, c \in C, b \in B$, and $l \in L$, let $Y_{i,s,c,b,l}$ equal 1 if student i is assigned to section s of course c in block b and learning group l, and is equal to 0 otherwise.

(iii) For each $s \in S, c \in C$, and $l \in L$, let $\widehat{X}_{s,c,l}$ equal 1 if section s of course c is assigned to learning group l, and is equal to 0 otherwise.

(iv) For each $i \in I$ and $l \in L$, let $\widehat{Y}_{i,l}$ equal 1 if student i is assigned to learning group l, and is equal to 0 otherwise.

We now present the hard constraints.

Each section of a course can belong to at most one learning group.

$$\sum_{l \in L} \widehat{X}_{s,c,l} \leq 1 \qquad \forall s \in S, c \in C \tag{1}$$

For every learning group, each section of a course is assigned to exactly one teacher and is scheduled in exactly one block.

$$\sum_{t \in T} \sum_{b \in B} X_{t,s,c,b,l} = \widehat{X}_{s,c,l} \qquad \forall s \in S, c \in C, l \in L \tag{2}$$

No teacher can be assigned to two different courses in the same block.

$$\sum_{s \in S} \sum_{c \in C} \sum_{l \in L} X_{t,s,c,b,l} \leq 1 \qquad \forall t \in T, b \in B \tag{3}$$

Course c must be timetabled exactly O_c times, where O_c is the number of sections of course c that will be offered.

$$\sum_{t \in T} \sum_{s \in S} \sum_{b \in B} \sum_{l \in L} X_{t,s,c,b,l} = O_c \qquad \forall c \in C \tag{4}$$

Each student must belong to exactly one learning group. (Note that teachers may belong to multiple learning groups.)

$$\sum_{l \in L} \widehat{Y}_{i,l} = 1 \qquad \forall i \in I \tag{5}$$

No learning group can contain more than M_l students, where M_l is the maximum size of a learning group.

$$\sum_{i \in I} \widehat{Y}_{i,l} \leq M_l \qquad \forall l \in L \tag{6}$$

If a student is enrolled in a course section in learning group l, then *both* the student *and* the course section must belong to learning group l.

$$Y_{i,s,c,b,l} \leq \widehat{Y}_{i,l} \qquad \forall i \in I, s \in S, c \in C, b \in B, l \in L \tag{7}$$

$$Y_{i,s,c,b,l} \leq \widehat{X}_{s,c,l} \qquad \forall i \in I, s \in S, c \in C, b \in B, l \in L \tag{8}$$

No student can be enrolled in more than one course in the same block.

$$\sum_{s \in S} \sum_{c \in C} \sum_{l \in L} Y_{i,s,c,b,l} \leq 1 \qquad \forall i \in I, b \in B \tag{9}$$

No student can be enrolled in multiple sections of the same course.

$$\sum_{s \in S} \sum_{b \in B} \sum_{l \in L} Y_{i,s,c,b,l} \leq 1 \qquad \forall i \in I, c \in C \tag{10}$$

Section s of course c can have at most $M_{s,c}$ students, where $M_{s,c}$ is the maximum enrollment for this course section.

$$\sum_{i \in I} \sum_{l \in L} Y_{i,s,c,b,l} \leq M_{s,c} \qquad \forall s \in S, c \in C, b \in B \tag{11}$$

This is our model for the Student Cohort Partitioning Problem (SCPP). Our solution is found by maximizing the objective function of this integer linear program subject to these eleven constraints.

In practice, the large majority of these variables $X_{t,s,c,b,l}$ and $Y_{i,s,c,b,l}$ will be pre-set to 0, since teachers are qualified to only teach a small subset of the offered courses, and students will only want to enroll in a small subset of these courses. By fixing these variables to be 0, we can solve the SCPP whenever our sets $|T|, |I|, |S|, |C|, |B|, |L|$ are of reasonable size.

While our ILP model is guaranteed to output an optimal solution, the computing time grows exponentially as the problem size increases. Thus, for a large school with hundreds of students and course offerings, we might not be able to solve the ILP. This motivates the need for approximation algorithms.

We conclude this section of the paper by proposing two heuristics: a multi-step approach called "Progressive Assignment" that orders the courses by the preference coefficient $P_{i,c,b}$, and a Large Neighbourhood Search (LNS). Both heuristics are inspired by the principle that when tackling an NP-complete problem, it is good practice to reduce the initial intractable problem to a series of simpler tractable subproblems [6].

In our first heuristic, we divide the course sections into k different groups, sorted by course priority. In other words, courses with the highest preference coefficients $P_{i,c,b}$ are partitioned first. We solve the SCPP on this smaller subset of courses in C, and lock in the assignments $\widehat{X}_{s,c,l}$ found in the optimal solution. We then include these assignments as hard constraints in the following step, where we solve the SCPP on the next subset of courses in C.

Thus, we build the timetable progressively, assigning only a subset of the course sections at a time. While the $\widehat{X}_{s,c,l}$ variables stay fixed throughout our Progressive Assignment algorithm, the values of $\widehat{Y}_{i,l}$ and $Y_{i,s,c,b,l}$ change in each of the k steps.

We allow this because once the $\widehat{X}_{s,c,l}$ assignments are set, our ILP quickly finds the best possible assignment of students to learning groups, and students to course sections, to generate the optimal solution at each step. This ensures that the final solution produced by the Progressive Assignment is close (although most certainly not equal) to the optimal solution for the original SCPP.

In our second heuristic, we employ a Large Neighbourhood Search to iteratively improve our solution. Given any solution to the SCPP (e.g. the solution found in our Progressive Assignment), we lock in all but h of the variables $\widehat{X}_{s,c,l}$, set them as hard constraints, and then re-calculate the SCPP to generate a new solution where some of these h course sections may be re-assigned to different learning groups. Since our initial solution was produced by one of the $|L|^h$ assignments of learning groups to these h course sections, our new SCPP solution cannot be worse than our input solution. Thus, our LNS is analogous to "hill climbing" because it performs iterative and incremental changes on an arbitrary initial solution to find a better solution.

We may stop the search at any time, either after a fixed time limit or when the algorithm appears to have converged. This heuristic, like all local search algorithms, may get stuck in a local minimum, especially if the value of h is small. However, when h is sufficiently large, the results of the LNS get better at each step, until a close-to-optimal solution is found.

When these two algorithms are combined, we can rapidly generate a nearly-optimal (or possibly optimal) solution to complex SCPP problems. We now apply these algorithms on a real-world instance, an all-boys high school in Canada.

5 Application

St. George's School (SGS) is located in Vancouver, the most populous city in the Canadian province of British Columbia. SGS is one of Canada's leading independent schools, with an enrollment of 1200 students. Founded in 1930, the school's mission is to "inspire their students to become fine young men who will shape positive futures for their families and the global community".

For the SGS administration, the biggest challenge is creating the timetable for the students in Grades 11 and 12. In 2020–2021 this represented 328 students, with 165 juniors and 163 seniors at this all-boys high school. Unlike students in the lower grades who take mostly required (core) courses, there are numerous elective courses in the final two years, and each student wants to enroll in a different combination of courses from the over *one hundred* options available.

The majority of these course options are offered to students in both Grades 11 and 12, and traditionally the school has viewed students from these two grades as a single cohort. Due to the government's mandate of a 120-person cohort limit, these 328 students needed to be partitioned into $|L| = 3$ learning groups.

The $|I| = 328$ individual students requested a total of 2303 courses, which is fewer than the maximum total of $|I| \times |B| = 328 \times 8 = 2624$. This occurred because the students could take a "self-study period", i.e., a spare block.

Certain courses were canceled due to low enrollment. Based on the student requests, the school decided to offer $|C| = 89$ different courses, of which 41 were single-section courses. Of the 48 multi-section courses, many had two or three sections, though one course (Social Studies 11) had $|S| = 8$ sections, since this course was mandatory for all of the Grade 11 students. In all, there were 196 total course sections.

The SGS timetable has $|B| = 8$ blocks, and each of the 196 course sections needed to be scheduled in one of these 8 blocks.

Each school day consists of four 70-minute class periods and one lunch break, with a four-day "tumbling timetable". Students alternate between their four blocks (A, B, C, D) on Days 1 and 3 and their four blocks (E, F, G, H) on Days 2 and 4, repeating this pattern for the entire academic year.

Class	Day 1	Day 2	Day 3	Day 4
Period 1	A	E	C	G
Period 2	B	F	D	H
Lunch	Lunch	Lunch	Lunch	Lunch
Period 3	C	G	A	E
Period 4	D	H	B	F

Since the assignment of blocks to days and time slots is fixed by the school administration, creating the optimal timetable is equivalent to optimally assigning course sections to blocks.

The school leadership team also *pre-assigned* each of the 196 course sections to one of the $|T| = 50$ teachers at the senior school. Pre-assigning teachers to course sections reduces our ILP's objective function to maximizing student preferences.

Mathematically this is equivalent to setting the desirability coefficient $D_{t,c,b}$ to 0 if teacher t is assigned to at least one section of course c in some block b, and ensuring a hard constraint of $X_{t,s,c,b,l} = 0$ whenever teacher t is not assigned to section s of course c. The pre-assignment of course sections to teachers reduced the problem size from $(|T| + |I|) \cdot |S| \cdot |C| \cdot |B| \cdot |L|$ to a smaller problem with $|I| \cdot |S| \cdot |C| \cdot |B| \cdot |L| = 328 \times 196 \times 8 \times 3 = 1542912$ total binary variables.

The school set the following weights for the preference coefficients $P_{i,c,b}$:

5 points if c is a high-priority elective course
3 points if c is a medium-priority elective course
1 point if c is a low-priority elective course

Of the 196 course sections, 149 were elective courses while the remaining 47 represented *required* courses – i.e., a student desiring a required course had to be registered in at least one section of that course.

We modeled high-priority courses as soft constraints assigning them a large weight in our student course preference matrix, whereas *required* courses were treated as hard constraints which we hard coded in our ILP.

For their entire history, St. George's School has created their timetable manually. Given the challenges of solving a combinatorial optimization by hand, the school has always pre-assigned each course section to one of the eight blocks *before* creating each student's timetable.

While this pre-assignment ensures that the educators know their exact teaching schedule before they go on their summer holidays, this of course has a significant impact on the Objective Function, since each course section is locked into a block rather than optimized to maximize students getting into their desired courses. (The school has hired us to create their 2021–2022 timetable, which will not include teacher pre-assignments or block pre-assignments.)

Our optimization program, written in Python, inputs an Excel sheet consisting of all the course data and the individual student requests. For the actual optimization, we use COIN-OR Branch and Cut (CBC), an open-source MIP solver, with the Google OR-Tools linear solver wrapper [11].

We first generated the optimal timetable for just $|L| = 1$ learning group, which was trivial since course sections were pre-assigned to blocks. Using the model described in Sect. 4, our ILP computed the solution in just 8.5 s on an 8 GB Lenovo laptop running Windows 10 with a 2.1 GHz processor. The Python code and input files deployed to design the optimal timetable are found in a repository at https://github.com/ifabrisarabellapark/CPAIOR2021.

The results are presented below, grouped by course priority.

Priority type	Total requested	Total enrolled
Required	807	807
High	541	533
Medium	166	155
Low	789	727
TOTAL	2303	2222

When there is only one learning group (i.e., Pre-COVID), the objective function of our optimal timetable has value $533 \times 5 + 155 \times 3 + 727 \times 1 = 3857$, with students being enrolled in $2222/2303 = 96\%$ of their desired courses.

In the $2303 - 2222 = 81$ instances where a student was not assigned a section of a desired course, the majority of them were due to over-capacity, including all $8 + 11 = 19$ of the unassigned High and Medium Priority requests.

Fortunately, all 19 of these requests had a reasonable alternative – for example, the five students not getting into AP Chemistry 12 could instead take Chemistry 12, which was offered in the same block and had plenty of available seats.

We then applied our ILP model to partition the students and course sections, to find a solution to our Student Cohort Partitioning Problem (SCPP). By definition, we knew that we could not exceed a success rate of 96%.

Since SGS requested learning groups to be evenly balanced, we set $M_l = 110$ as the maximum cohort size since we had $|I| = 328$ students and $|L| = 3$ learning groups. As expected, our Python program could not solve the ILP within our pre-set limit of 12 h, and so we applied our approximation algorithms.

We first solved the ILP for just the Required courses, and locked in the learning groups for these 47 course sections. We then performed our Progressive Assignment for the remaining course sections in order of priority: High, then Medium, then Low. This entire process took less than five minutes, giving us an initial assignment of course sections to learning groups $(\widehat{X}_{s,c,l})$, students to learning groups $(\widehat{Y}_{i,l})$ and students to course sections $(Y_{i,s,c,b,l})$.

We then applied the Large Neighbourhood Search (LNS), which locked in the learning groups for $196 - h$ course sections, while allowing the remaining h course sections to be re-assigned by our ILP. We found that reshuffling $h = 30$ was ideal for our problem size, so that each step of the search computed in an average time of 60 seconds. Our local search algorithm converged to the same solution within two hours for every single trial we ran.

This solution scores $518 \times 5 + 147 \times 3 + 525 \times 1 = 3556$ points, with students getting into $1997/2303 = 87\%$ of their desired courses, including 100% of their required courses. The results are presented below.

Priority type	Total requested	Progressive assignment	LNS
Required	807	807	807
High	541	514	518
Medium	166	144	147
Low	789	524	525
TOTAL	2303	1989	1997

Our Master Timetable, with students being enrolled in 87% of their desired courses, was delivered to SGS on August 20, 2020, less than two weeks after the authors were introduced to this project. This gave plenty of time for the school administrators to announce the learning groups of each student and each course section, and set up each student's on-campus activities (co-curriculars, lunch, entrance times, and exit times) to occur exclusively within their learning group.

Each student was given a coloured ID card (Orange, Pink, Grey) corresponding to their learning group, in addition to their 8-block timetable. In the 13% of cases where the students were not given a desired course, the Registrar's Office made simple adjustments, such as enrolling students in a similar course (e.g. AP Chem 12 to Chem 12), and inviting students to register for a different elective course. Within a few days, each student had a satisfactory timetable.

St. George's School has hired the authors to build the 2021–2022 Master Timetable and has decided to no longer pre-assign course sections to teachers and blocks. By doing this, the students will be able to get into a higher percentage of their desired courses, regardless of whether the number of learning groups stays at 3, or returns to 1 in a post-COVID world.

6 Discussion

Our final model for the Student Cohort Partitioning Problem (SCPP) has a total of $|I||S||C||B||L| = 328 \times 196 \times 8 \times 3 = 1542912$ total binary variables. The size of this model prevented us from confirming that our solution was optimal. Naturally, we wondered how close our implemented solution, with an 87% success rate and an objective value of $518 \times 5 + 147 \times 3 + 525 \times 1 = 3556$ points, was to the optimal solution for $|L| = 3$ learning groups.

To answer this question, we considered the easier problem of $|L| = 2$ learning groups with at most $M_l = 220$ students in a cohort. If we restrict the set of elective courses to just one priority class (High, Medium, Low) at a time, then we can solve each of these three separate ILPs, each in just a few minutes.

Any solution to the SCPP with $|L| = 3$ and $M_l = 110$ is automatically a solution to the easier SCPP with $|L| = 2$ and $M_l = 110 + 110 = 220$. Therefore, our results for $|L| = 2$ and $M_l = 220$, marked in bold, provide a theoretical upper bound to our optimization problem for St. George's School with $|L| = 3$.

Priority type	Total requested	$|L| = 2, M_l = 220$	$|L| = 3, M_l = 110$
Required	807	**807**	807
High	541	**518**	518
Medium	166	**155**	147
Low	789	**682**	525
Total assignments	2303	**2162**	1997
Objective value	3992	**3737**	3556

The implemented solution scoring 3556 points is less than 5% below the theoretical upper bound which scored $518 \times 5 + 155 \times 3 + 682 \times 1 = 3737$ points. Our solution is provably optimal for the set of Required and High priority courses.

We presume that the actual optimal solution is much closer to 3556, since our model mandates we enroll students in all of the elective courses *simultaneously*, rather than treating each priority class *separately*, as we did when computing the upper bound. Thus, our construction is an overestimate of the actual optimal solution for our problem with $|L| = 3$ and $M_l = 110$.

We were fortunate to find a solution satisfying 87% of student course requests, since partitioning students into groups could have led to a terrible outcome for the school, had the students selected a more heterogeneous set of courses.

To illustrate this point, consider a small school with $|I| = 32$ students, where each student is required to select five courses, one from each of the five sets: $\{A_1, A_2\}$, $\{B_1, B_2\}$, $\{C_1, C_2\}$, $\{D_1, D_2\}$, and $\{E_1, E_2\}$. Furthermore, suppose that each of the $2^5 = 32$ students selects a different set of 5 courses.

Assume each classroom can hold at most 16 students. Since there are 16 students who request each course, there is only the need to offer one section of each course. If learning groups do not exist (i.e., $|L| = 1$), then it is trivial to enroll all 32 students into all 5 of their courses. We simply assign each pair of courses (e.g. A_1 and A_2) to be taught in the same time slot, which guarantees a timetable where all $32 \times 5 = 160$ course requests are satisfied.

Now suppose $|L| = 2$, and we need to partition the $|I| = 32$ students and $|C| = 10$ courses into two learning groups. It is straightforward to see that each course partition is identical, by symmetry. For example, if we assign courses $\{A_1, B_1, C_1, D_1, E_1\}$ to the first learning group and courses $\{A_2, B_2, C_2, D_2, E_2\}$ to the second learning group, then each student is simply assigned to the learning group that gives them the most number of desired courses. For example, the student who wants $\{A_2, B_1, C_2, D_1, E_2\}$ is assigned to the second learning group since this student would prefer a timetable with 3 desired courses instead of 2.

We can show that 2 students get into all five courses, 10 students get into four courses, and 20 students get into three courses. Assuming each desired course assignment has a score of 1 point, the objective value is $2 \times 5 + 10 \times 4 + 20 \times 3 = 110$, resulting in a success rate of $\frac{110}{160} = 0.6875$. By making the small switch from $|L| = 1$ learning groups to $|L| = 2$, our success rate drops from 100% to 68.75%.

This construction generalizes, from 2^5 students and 10 courses to 2^n students and $2n$ courses, showing that when there are $|L| = 2$ learning groups, it is possible for the students to select their courses so that the optimal solution to the SCPP results in a success rate that converges to $\frac{1}{2} = 50\%$ as $n \to \infty$.

For all $|L| > 2$, we conjecture that a similar construction with $|L|^n$ students and $|L|$ learning groups results in the optimal SCPP solution having a success rate converging to $\frac{1}{|L|}$. We have verified this result for $|L| = 3$ and $|L| = 4$ using Python. In other words, creating learning groups has a significant impact on students being assigned their desired courses, especially when there are many single-section courses and the students have heterogeneous course preferences.

We were fortunate that the $|I| = 328$ students at St. George's School selected their courses in such a way that enabled us to enroll them in 87% of their desired courses (1997 out of 2303). This success percentage is notable given that our theoretical upper bound shows that at most 2162 of these 2303 course requests can be fulfilled, even with $|L| = 2$ learning groups.

7 Conclusion

In this paper, we defined the Student Cohort Partitioning Problem (SCPP) and modeled it using an integer linear program. We then applied our model on a real-world problem instance at a Canadian high school, partitioning $|I| = 328$ students into $|L| = 3$ learning groups while enrolling these students in 87% of their desired courses, including 100% of their required courses.

Our collaboration with St. George's School was a success, and the implementation of the learning groups was done smoothly and effectively. This is part of the reason why the school has reported *zero* COVID-19 cases among the students and staff of the Senior School thus far in the 2020–2021 academic year.

We recognize that our research on the SCPP has only begun, and further research will need to be conducted to assess the scalability of our work to larger data sets. A natural first step is to take the set of benchmark instances for the Post-Enrollment Course Timetabling Problem (PECTP) and amend them with two or more learning groups, to create benchmark instances for the SCPP.

The authors have signed contracts with five different high schools in British Columbia, to build each school's 2021–2022 Master Timetable. Here are the questions we will be asking ourselves as we proceed with our work in the coming months:

(a) Could SCPP solutions be improved by adding a few more sections of certain courses? And if so, what would be the cost to the school? If the school had enough money to add X extra sections, which sections should be added to maximize the number of students being assigned their desired courses?

(b) How will our SCPP model handle additional requirements, such as requiring *teachers* to belong to a single learning group? Should our model assign a penalty whenever a teacher is assigned to multiple learning groups, and include this penalty in the objective function?

(c) Given rising COVID-19 numbers in Canada (especially with new variants), some schools may decide to restrict the number of days that each learning group attends the school. How would this policy change affect our timetabling solutions?

(d) What are the best techniques to find fast nearly-optimal solutions to larger SCPP instances? For example, would Constraint Programming techniques and/or Benders decomposition outperform the methods presented in this paper?

In British Columbia, schools will not "return to normal" until they reach the final phase of the provincial Back to School plan, which is conditional on wide vaccination and immunity among the population. As a result, schools in our province, as well as in other areas of Canada and the world, will likely need to design their 2021–2022 timetables to include multiple Learning Groups.

While the global pandemic continues, our SCPP formulation is extremely applicable for educational institutions, as our model enables schools to create cohorts to maximize students being able to take their desired courses, while simultaneously reducing the spread of COVID-19 through optimal partitioning.

Acknowledgments. The authors thank the reviewers for their insightful comments that significantly improved the presentation of this paper. The authors also thank the administrators at St. George's School for making this collaboration possible. Specifically, we acknowledge Sarah Coates (Associate Principal of Academics), Andrew Shirkoff (Director of Risk Management), Jan Chavarie (Head of Applications Support), and Jessie Bahia (Registrar).

References

1. B.C'.s Back to School Plan. https://www2.gov.bc.ca/gov/content/education-training/k-12/covid-19-return-to-school#learning-group. Accessed 12 Apr 2021
2. Bahargam, S., Erdos, D., Bestavros, A., Terzi, E.: Personalized education; solving a group formation and scheduling problem for educational content. In: Proceedings of the 8th International Conference on Educational Data Mining, pp. 488–492. International Educational Data Mining Society, Madrid (2015)
3. Bahargam, S., Erdos, D., Bestavros, A., Terzi, E.: Team formation for scheduling educational material in massive online classes. arXiv preprint arXiv:1703.08762 (2017)
4. Baker, K.R., Magazine, M.J., Polak, G.G.: Optimal block design models for course timetabling. Oper. Res. Lett. **30**(1), 1–8 (2002)
5. Baykasoglu, A., Dereli, T., Das, S.: Project team selection using fuzzy optimization approach. Cybern. Syst.: Int. J. **38**(2), 155–185 (2007)
6. Bessiere, C., Carbonnel, C., Hebrard, E., Katsirelos, G., Walsh, T.: Detecting and exploiting subproblem tractability. In: IJCAI International Joint Conference on Artificial Intelligence, pp. 468–474. AAAI Press, California (2013)
7. Carter, M.W.: A comprehensive course timetabling and student scheduling system at the University of Waterloo. In: Burke, E., Erben, W. (eds.) PATAT 2000. LNCS, vol. 2079, pp. 64–82. Springer, Heidelberg (2001). https://doi.org/10.1007/3-540-44629-X_5
8. CDC Operational Considerations for Schools. https://www.cdc.gov/coronavirus/2019-ncov/global-covid-19/schools.html. Accessed 12 Apr 2021
9. Chen, S.J., Lin, L.: Modeling team member characteristics for the formation of a multifunctional team in concurrent engineering. IEEE Trans. Eng. Manag. **51**(2), 111–124 (2004)
10. Goebbels, S., Pfeiffer, T.: Optimal student sectioning at Niederrhein University of Applied Sciences. In: Neufeld, J.S., Buscher, U., Lasch, R., Möst, D., Schönberger, J. (eds.) Operations Research Proceedings 2019. ORP, pp. 167–173. Springer, Cham (2020). https://doi.org/10.1007/978-3-030-48439-2_20
11. Google OR-Tools: fast and portable software for combinatorial optimization. https://developers.google.com/optimization. Accessed 12 Apr 2021
12. Hoshino, R., Fabris, I.: Optimizing student course preferences in school timetabling. In: Hebrard, E., Musliu, N. (eds.) CPAIOR 2020. LNCS, vol. 12296, pp. 283–299. Springer, Cham (2020). https://doi.org/10.1007/978-3-030-58942-4_19
13. Johns Hopkins University & Medicine. https://coronavirus.jhu.edu/map.html. Accessed 12 Apr 2021
14. Khlaif, Z.N., Salha, S.: The unanticipated educational challenges of developing countries in Covid-19 crisis: a brief report. Interdisc. J. Virtual Learn. Med. Sci. **11**(2), 130–134 (2020)

15. Kristiansen, S., Sørensen, M., Stidsen, T.R.: Student sectioning at high schools in Denmark. In: 6th Multidisciplinary International Conference on Scheduling: Theory and Applications, pp. 628–632. Springer, Belgium (2013)
16. Kristiansen, S., Sørensen, M., Stidsen, T.R.: Integer programming for the generalized high school timetabling problem. J. Sched. **18**(4), 377–392 (2014). https://doi.org/10.1007/s10951-014-0405-x
17. Lewis, R., Paechter, B., McCollum, B.: Post enrolment based course timetabling: a description of the problem model used for track two of the second international timetabling competition (2007)
18. Müller, T., Murray, K.: Comprehensive approach to student sectioning. Ann. Oper. Res. **181**(1), 249–269 (2010). https://doi.org/10.1007/s10479-010-0735-9
19. Schindl, D.: Student sectioning for minimizing potential conflicts on multi-section courses. In: Proceedings of the 11th International Conference of the Practice and Theory of Automated Timetabling (PATAT 2016), pp. 327–337. Springer, Udine (2016)
20. The Hill: Coronavirus shining light on internet disparities in rural America. https://thehill.com/blogs/congress-blog/technology/488848-coronavirus-outbrea k-shining-an-even-brighter-light-on. Accessed 12 Apr 2021
21. The Squamish Chief: concern about mixed-cohort classrooms. https://www. squamishchief.com/news/local-news/amid-covid-19-worries-concern-emerges-abo ut-mixed-cohort-classrooms-1.24199269. Accessed 12 Apr 2021
22. Triska, M., Musliu, N.: Solving the social golfer problem with a GRASP. In: Proceedings of the 7th International Conference on the Practice and Theory of Automated Timetabling, (PATAT 2008). Springer, Montréal (2008)
23. UNESCO: COVID-19 impact on education. https://en.unesco.org/covid19/ educationresponse. Accessed 12 Apr 2021
24. Wi, H., Oh, S., Mun, J., Jung, M.: A team formation model based on knowledge and collaboration. Expert Syst. Appl. **36**(5), 9121–9134 (2009)

A Two-Stage Exact Algorithm
for Optimization of Neural Network
Ensemble

Keliang Wang[1]($\boxtimes$), Leonardo Lozano[2]($\boxtimes$), David Bergman[1]($\boxtimes$),
and Carlos Cardonha[1]($\boxtimes$)

[1] Department of Operations and Information Management,
University of Connecticut, Mansfield, USA
{keliang.wang,david.bergman,carlos.cardonha}@uconn.edu
[2] Operations, Business Analytics and Information Systems,
University of Cincinnati, Cincinnati, USA
leolozano@ucmail.uc.edu

Abstract. We study optimization problems where the objective function is modeled through feedforward neural networks. Recent literature has explored the use of a single neural network to model either uncertain or complex elements within an objective function. However, it is well known that ensembles can produce more accurate and more stable predictions than single neural network. We therefore study how neural network ensemble can be incorporated within an objective function, and propose a two-stage optimization algorithm for solving the ensuing optimization problem. Preliminary computational results applied to a global optimization problem and a real-world data set show that the two-stage model greatly outperforms a standard adaptation of previously proposed MIP formulations of single neural network embedded optimization models.

Keywords: Ensemble learning · Two-stage optimization · Neural network · Surrogate model · Embedded predictive models

1 Introduction

Neural networks (NNs) are particularly useful as surrogate models for functions that are either unknown (i.e., functions that rely on the outcome of non-trivial simulation models) or that cannot be easily represented by simple (linear) expressions. Namely, given a function f with domain $\Omega \in \mathbb{R}^n$, a surrogate NN model $\mathcal{N}$ of f produces estimates $\mathcal{N}(x)$ of $f(x)$ for every x in Ω. NNs have attracted considerable attention recently in the optimization community, and several techniques to represent and optimize NNs as linear programs have been proposed in the literature (see e.g., [1,9,17]). Our work is placed in this stream of research.

NNs typically do not deliver exact representations of their associated functions, i.e., it is not necessarily true that $f(x) = \mathcal{N}(x)$; actually, it is not even true

© Springer Nature Switzerland AG 2021
P. J. Stuckey (Ed.): CPAIOR 2021, LNCS 12735, pp. 106–114, 2021.
https://doi.org/10.1007/978-3-030-78230-6_7

that $f(x) \geq f(x') \iff \mathcal{N}(x) \geq \mathcal{N}(x')$ for every $x, x' \in \Omega$. As a consequence, optimizing f over a surrogate neural network may be problematic, as imprecise representations may lead to solutions that are far from optimal.

One alternative approach to mitigate this issue consists of employing ensembles of neural networks. The idea is that several different neural network representations typically make different mistakes (e.g., in different regions of the search space), so a solution that is optimal for an ensemble is more likely to be robust, i.e., closer to solutions that are indeed optimal for f. The machine learning literature has shown that ensemble models typically perform very well in practice [14]. Similar results have been shown for neural networks, i.e., ensembles of neural networks frequently provide better predictions than a single large neural network [11, 23].

Given f, let $\mathcal{E}_f = \{\mathcal{N}^1, \mathcal{N}^2, \ldots, \mathcal{N}^e\}$ be an ensemble of e neural networks representing f. An estimate of f for any point x in Ω produced by $\mathcal{E}_f$ is given by the average of the individual estimates of each neural network in $\mathcal{E}_f$, that is,

$$\mathcal{E}_f(x) = \frac{1}{e} \sum_{i=1}^{e} \mathcal{N}^i(x).$$

We study the following problem:

$$\max_{x \in \Omega} \mathcal{E}_f(x) \qquad (1)$$

In this work, we adapt existing big-M integer programming formulation and propose a two-stage optimization algorithm to solve Problem 1. We assess the performance of our algorithms using two benchmark instances: the *peaks* function, which has been traditionally used by the global optimization community, and the *concrete* data set, which is introduced by [21] and has also been used in the optimization literature [14]. The results exhibit a superiority of our two-stage algorithm over benchmark model in terms of computational performance.

The manuscript is organized as follows. After providing a literature review (Sect. 2), we introduce the notation and a baseline model for Problem 1 in Sect. 3. The two-stage optimization algorithm is introduced in Sect. 4. We present the results of our experiments in Sect. 5 and conclude with directions for future work (Sect. 6).

2   Literature Review

The ensemble learning method combines predictions of multiple base estimators in order to improve generalization ability and prediction performance over a single estimator [7, 11, 22]. Dietterich [7] provides statistical, computational and representational arguments to show that an ensemble is always able to outperform each of its individual base estimators. Ensembles of neural networks were introduced by Hanse and Salamon [11] and have gained substantial development over the last years. As a result, recent work has shown that ensembles of neural networks are being adopted in many scenarios of practical relevance, such as financial decision applications [20] and time series forecasting [12].

Over the last few years, the interplay of neural networks and optimization started to attract the interest of the scientific community. Bartolini et al. [3] train neural networks to approximate the behavior of thermal controllers in a multi-core CPU and embed the estimator into a Constrained Programming model through Neural Constraints [2]. Schweidtmann et al. [16] embed neural network into a deterministic global optimization problem using McCormick relaxations in a reduced space which employs convex and concave envelopes of the nonlinear activation function.

Recent works in the area have focused on the representation of neural network with rectified linear unit (ReLU) activation functions into mixed-integer programming (MIP) formulation [9]. Anderson et al. [1] propose a generic framework that constructs sharp or ideal MIP formulation for ReLU neural network. Bergman et al. [4] devise a modeling framework that integrates the training process of neural networks with MIP formulation leveraging the resulting estimator. In the machine learning community, MIP formulations have been used to study the properties of neural networks [6,8,18,19]. To the best of our knowledge, optimization over neural network ensemble has not been studied in the literature.

3   Notation and Baseline Formulation

A feedforward neural network $\mathcal{N}$ is a weighted acyclic directed graph composed by a set of nodes (or neurons) $V(\mathcal{N})$, a set of arcs $A(\mathcal{N})$, and a weight function $W : A(\mathcal{N}) \to \mathbb{R}$; if $\mathcal{N}$ is clear from the context, we refer to these elements simply as V, A, and W, respectively.

Set V contains a subset V_0 of n_0 neurons without incoming arcs, which are referred to as input neurons; input neurons are associated with the features (or simply input) of f. The distance (in number of arcs) from nodes in V to the nodes in V_0 induce the partition $V = V_0 \cup V_1 \cup \ldots \cup V_L$, where V_l contains all the n_l neurons reached from some input neuron after the traversal of exactly l arcs, $l \in \{0\} \cup [L]$, where $[n]$ represents the set $\{1, 2, \ldots, n\}$ for n in $\mathbb{N}$. Each V_l can be interpreted as the l-th layer of $\mathcal{N}$; V_0 is the input layer, whereas V_L is the output layer. We assume that $n_L = |V_L| = 1$, i.e., there is only one neuron in the output layer. We use vector $(n_0, n_1, \ldots, n_L)$ in order to succinctly represent the architecture of $\mathcal{N}$.

Each arc $a = (u, v)$ in $A(\mathcal{N})$ is directed from a neuron in V_i to another in V_{i+1}, $0 \le i < L$. There is an arc $a = (u, v)$ for each pair $(u, v) \in V_i \times V_{i+1}$, $0 \le i < L$. For each $l \in \{1, 2, \ldots, L\}$, we let $W^l \in \mathbb{R}^{n_l \times n_{l-1}}$ denote the sub-matrix of W containing the weights of all arcs directed from V_{l-1} to V_l.

Let $x \in \mathbb{R}^{n_0}$ be an input for function f. In order to use $\mathcal{N}$ to evaluate f on x, we define output vectors $y^l \in \mathbb{R}^{n_l}$ and bias vectors $b^l \in \mathbb{R}^{n_l}$ for each layer l in $\{0, 1, \ldots, L\}$, with elements y^l_j and b^l_j being associated with the j-th neuron in layer l. Note that this definition requires orderings for each set V_l; we assume that the neurons in V_0 are ordered in a way that the j-th neuron is associated with the j-th feature of f.

The values of y^l for each layer l are defined as follows. First, we set vector $y^0 = x$, i.e., y^0 contains the input of f. The values of y_j^l for $l \in \{1, 2, \ldots, L-1\}$ and $j \in \{1, 2, \ldots, n_l\}$ is given by

$$y_j^l = ReLU((W_j^l)'y^{l-1} + b_j^l),$$

where $ReLU : \mathbb{R} \to \mathbb{R}^+$ is the Rectified Linear Unit (ReLU) activation function, defined as $ReLU(x) := \max(0, x)$. More succinctly, we can just write $y^l = ReLU(W^l y^{l-1} + b^l)$ for $l = 1, \ldots, L-1$. Finally, the value y_1^L of the single neuron in V_L is the affine combination of previous layer's output without applying the ReLU function, i.e., $y_1^L = W^L y^{L-1} + b^L$. The value delivered by $\mathcal{N}$ given some input x is denoted by $\mathcal{N}(x)$.

Let $\mathcal{E} = \{\mathcal{N}^1, \mathcal{N}^2, \ldots, \mathcal{N}^e\}$ be an ensemble of neural networks representing f. We assume that the architecture and the weights of the neural networks used in the formulations below are given as input (i.e., the models do not compute or readjust the structure of the networks). Below we propose E-NN, a mixed-integer programming formulation that solves Problem 1 given $\mathcal{E}$.

$$\max_x \quad \frac{1}{e} \sum_{i=1}^{e} y_1^{i, L_i} \tag{2a}$$

$$\text{s.t.} \quad y^{i,0} = x \qquad\qquad\qquad i \in [e] \tag{2b}$$

$$y_1^{i, L_i} = h_{n_{L_i}}^{i, L_i} \qquad\qquad i \in [e] \tag{2c}$$

$$h_j^{i,l} = (W_j^{i,l})'y^{i,l-1} + b_j^{i,l} \qquad i \in [e], l \in [L_i], j \in [n_l] \tag{2d}$$

$$h_j^{i,l} \leq y_j^{i,l} \leq h_j^{i,l} + M(1 - z_j^{i,l}) \quad i \in [e], l \in [L_i - 1], j \in [n_l] \tag{2e}$$

$$0 \leq y_j^{i,l} \leq M z_j^{i,l} \qquad\qquad i \in [e], l \in [L_i - 1], j \in [n_l] \tag{2f}$$

$$z_j^{i,l} \in \{0, 1\} \qquad\qquad i \in [e], l \in [L_i - 1], j \in [n_l] \tag{2g}$$

$$h_j^{i,l}, y_j^{i,l} \in \mathbb{R} \qquad\qquad i \in [e], l \in [L], j \in [n_l] \tag{2h}$$

$$x \in \Omega. \tag{2i}$$

Variable x represents the input vector in $\mathbb{R}^{n_0}$; x is restricted to elements of Ω by a set of problem-dependent constraints, which are represented succinctly in the model above in (2i). In order to evaluate $\mathcal{N}^i(x)$, the model employs a binary variable $z_j^{i,l}$ and continuous variables $h_j^{i,l}$ and $y_j^{i,l}$ for the j-th neuron in $V(\mathcal{N}^i)_l$, $l = 1, \ldots, L_i - 1$. With a slight abuse of notation we use variables $y_j^{i,l}$ with the same meaning as before, i.e., $y_j^{i,l} = ReLU((W_j^{i,l})'y^{i,l-1} + b_j^{i,l})$.

Each variable $h_j^{i,l}$ contains the parameter of the ReLU function used in the estimation of $y_j^{i,l}$, i.e., $h_j^{i,l} = (W_j^{i,l})'y^{i,l-1} + b_j^{i,l}$; the value of $h_j^{i,l}$ is computed in constraint (2d). The value of the variables $z_j^{i,l}$ depend on the signal of $h_j^{i,l}$;

if $h_j^{i,l} < 0$, then $z_j^{i,l} = 0$ and $y_j^{i,l}$ must be set to 0; conversely, if $h_j^{i,l} \geq 0$, then $z_j^{i,l} = 1$ and $y_j^{i,l} = h_j^{i,l}$. These two scenarios are modeled by constraints (2e) and (2f). Finally, constraint (2b) ensures that the values of the neurons composing the input layers are equal to the input feature described by x, and constraint (2c) set y_1^{i,L_i} to $h_{n_{L_i}}^{i,L_i}$ for all the NNs in the ensemble (recall that the ReLU function is not applied to the last layer).

4 Two-Stage Optimization Algorithm

We propose a two-stage optimization algorithm for Problem 1. The first stage solves a relaxation of E-NN to obtain upper bounds on y_1^{i,L_i} for all $\mathcal{N}^i$ in $\mathcal{E}$. The second stage solves a strengthened version of E-NN, which includes the bounds from the first stage. Let Ψ be the space defined by constraints (2c)–(2h) and consider the following reformulation of E-NN, which includes e copies of the x-variables:

$$\max_{x} \quad \frac{1}{e} \sum_{i=1}^{e} y_1^{i,L_i} \tag{3a}$$

$$\text{s.t.} \quad x^1 = x^2 = \ldots = x^e \tag{3b}$$

$$y^{i,0} = x^i \qquad i = 1,\ldots,e \tag{3c}$$

$$x^i \in \Omega \qquad i \in [e] \tag{3d}$$

$$(h,y,z) \in \Psi. \tag{3e}$$

Constraints (3b) ensure that all the copies of the x-variables are equal and are often referred to as *non-anticipativity* constraints. Following a similar approach to [5] in the context of two-stage stochastic programming, we relax constraints (3b) to obtain a relaxation of E-NN. Moreover, after removing constraints (3b) our relaxation displays a block-diagonal structure which allows us to decompose it into the following simpler problems, one for each neural network $\mathcal{N}^i$ in $\mathcal{E}$, denoted by R-E-NN(i):

$$\max_{x\,\in\,\Omega} \quad y_1^{i,L_i} \tag{4a}$$

$$\text{s.t.} \quad y^{i,0} = x^i \tag{4b}$$

$$x^i \in \Omega \tag{4c}$$

$$(h^i,y^i,z^i) \in \Psi^i, \tag{4d}$$

where Ψ^i is the space defined by the subset of constraints (2c)–(2h) corresponding to $\mathcal{N}^i \in \mathcal{E}$. Let u^i denote the optimal objective value of R-E-NN(i), and note that $y_1^{i,L_i} \leq u^i$ for all $\mathcal{N}^i$ in $\mathcal{E}$. Algorithm 1 presents our proposed two-stage approach. In the first stage we solve relaxations R-E-NN(i) for all the networks in the ensemble. In the second stage we solve E-NN strengthened by the constraints $y_1^{i,L_i} \leq u^i$ for every neural network in the ensemble.

Algorithm 1. Two-stage optimization algorithm for an ensemble of neural networks

---

1: Solve relaxation R-E-NN(i) for all $i = 1, \ldots, e$ and obtain each optimal value u_i.
2: Solve E-NN with the additional constraints $y_1^{i,L_i} \leq u^i$, $\forall i = 1, \ldots, e$.

---

5   Computational Study

We evaluate the computational performance of Algorithm 1 against E-NN, a direct big-M formulation of Problem 1. We train the neural networks in Python 3.7 and scikit-learn 0.23.2 [15] and use a 2.6 GHz 6-Core Intel i7-9750H CPU with 32 GB of RAM for training. All the algorithms are implemented in Java, and we use Gurobi 9.0.0 to solve the integer programming formulations [10]. The experiments are executed on an Intel Xeon E5–1650 CPU (six cores) running at 3.60 GHz with 32 GB of RAM on Windows 10. Each execution is restricted to a time limit of 600 s.

5.1   Instances

We use two instances in our computational study. One is a traditional global optimization benchmark functions, for which the optimal solutions are known. Additionally, we use a real-world problem for which regression models have been proposed in the literature [14]; for this problem, the original goal is to predict a target value given a set of input features, and there is no global optimal solution.

Peaks: The peaks function $z : [-3,3]^2 \to \mathbb{R}$ is defined in (5) and the global minimum for z is -6.551, which is attained at $x = 0.228$ and $y = -1.626$.

$$z(x,y) := 3(1-x^2)^2 e^{-x^2-(y+1)^2} - 10\left(\frac{x}{5} - x^3 - y^5\right)e^{-x^2-y^2} - \frac{e^{-(x+1)^2-y^2}}{3}. \quad (5)$$

Concrete: This data set contains a list with different types of concrete, described by eight input features. The data set is introduced and first studied by [21], who use neural networks trained on this data to predict compressive strength. The optimization problem in this case consists of the identification of an "ideal concrete", with maximum predicted compressive strength.

For our experiments, we first train neural network ensemble that work as regression models for these problems, and then we use our algorithms to find an optimal solution. In order to generate training data sets, we sample 1000 data points for **peaks** and obtain a data set containing 1030 records for **concrete** via the AppliedPredictiveModeling package in R ([13]). We apply min-max normalization to scale each input feature into range $[-1, 1]$.

The configuration of an ensemble model is determined by the number of neural networks and the topology of each of its individual estimators. We investigate ensembles with 5, 10, 15 and 20 neural networks. Each individual neural network has one hidden layer, containing from 20 to 200 neurons, in a 20 step size. We generate 5 instances for each configuration, so there are 200 instances for each problem.

5.2 Results and Analysis

We first present a scatter plot comparing the execution time of the algorithms for both problems in Fig. 1. Each dot represents an instance, whose coordinates x and y indicate the execution time of the algorithms (two-stage Algorithm 1 and E-NN formulation, respectively), and the size and the color correspond to the number of neurons in the hidden layer and the number of neural networks, respectively. The results show that the Algorithm 1 outperforms the E-NN formulation in virtually all cases for both problems, especially for large-sized configurations.

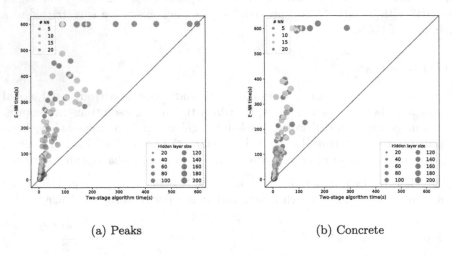

(a) Peaks   (b) Concrete

Fig. 1. Solution time comparison between Algorithm 1 and E-NN.

In Fig. 2 we report the execution times and the optimality gaps using cumulative frequency distribution plots. In the left half of each plot, the y-axis represents the number of instances solved within the time indicated by x-axis; the right half shows the number of instances solved within the optimality gap given in the x-axis after 600 s. Figures 2a and 2b show that Algorithm 1 solves more instances than E-NN after any amount of time with much smaller gaps at time limit for both problems. The advantage of the two-stage algorithm is highlighted for the concrete problem, where all instances are solved to optimality within 300 s.

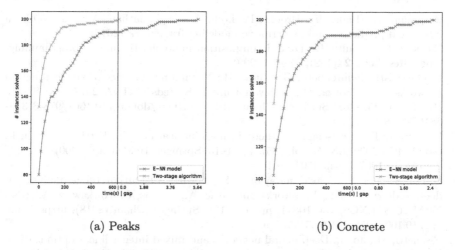

(a) Peaks   (b) Concrete

Fig. 2. Cumulative distribution plots comparing Algorithm 1 and E-NN.

6   Conclusion and Future Work

In this paper we presented a novel algorithmic approach to optimize over ensemble of single-layered neural networks. Our experiments show that the proposed approach reduces the solution times of similar problems relying on the optimization over E-NN by orders of magnitude. Ensemble methods have been show in the machine learning literature to provide more robust results than single estimators, so this work provides a first probe into integrating an ensemble of neural networks as part of an optimization problem.

At the moment, the interaction between mixed-integer programming and neural network is a fast-moving field and there are many interesting open research questions. For example, we do not employ any techniques to compress the neural networks (such as those presented by [17,19]), and it is not clear whether the application of such techniques could lead to computational improvements (especially because the time spent with compression techniques may be non-negligible). Nevertheless, this is an interesting direction for future research.

References

1. Anderson, R., Huchette, J., Ma, W., Tjandraatmadja, C., Vielma, J.P.: Strong mixed-integer programming formulations for trained neural networks. Math. Program. **183**, 3–39 (2020). https://doi.org/10.1007/s10107-020-01474-5
2. Bartolini, A., Lombardi, M., Milano, M., Benini, L.: Neuron constraints to model complex real-world problems. In: Lee, J. (ed.) CP 2011. LNCS, vol. 6876, pp. 115–129. Springer, Heidelberg (2011). https://doi.org/10.1007/978-3-642-23786-7_11
3. Bartolini, A., Lombardi, M., Milano, M., Benini, L.: Optimization and controlled systems: a case study on thermal aware workload dispatching. In: AAAI (2012). http://www.aaai.org/ocs/index.php/AAAI/AAAI12/paper/view/5042

4. Bergman, D., Huang, T., Brooks, P., Lodi, A., Raghunathan, A.U.: JANOS: an integrated predictive and prescriptive modeling framework (2019)
5. Carøe, C.C., Schultz, R.: Dual decomposition in stochastic integer programming. Oper. Res. Lett. **24**(1–2), 37–45 (1999)
6. Cheng, C.-H., Nührenberg, G., Ruess, H.: Maximum resilience of artificial neural networks. In: D'Souza, D., Narayan Kumar, K. (eds.) ATVA 2017. LNCS, vol. 10482, pp. 251–268. Springer, Cham (2017). https://doi.org/10.1007/978-3-319-68167-2_18
7. Dietterich, T.G.: Ensemble methods in machine learning. In: Kittler, J., Roli, F. (eds.) MCS 2000. LNCS, vol. 1857, pp. 1–15. Springer, Heidelberg (2000). https://doi.org/10.1007/3-540-45014-9_1
8. Dutta, S., Jha, S., Sankaranarayanan, S., Tiwari, A.: Output range analysis for deep feedforward neural networks. In: Dutle, A., Muñoz, C., Narkawicz, A. (eds.) NFM 2018. LNCS, vol. 10811, pp. 121–138. Springer, Cham (2018). https://doi.org/10.1007/978-3-319-77935-5_9
9. Fischetti, M., Jo, J.: Deep neural networks and mixed integer linear optimization. Constraints **23**(3), 296–309 (2018). https://doi.org/10.1007/s10601-018-9285-6
10. L Gurobi Optimization: Gurobi optimizer reference manual (2018). http://www.gurobi.com
11. Hansen, L.K., Salamon, P.: Neural network ensembles. IEEE Trans. Pattern Anal. Mach. Intell. **12**(10), 993–1001 (1990)
12. Kourentzes, N., Barrow, D.K., Crone, S.F.: Neural network ensemble operators for time series forecasting. Expert Syst. Appl. **41**(9), 4235–4244 (2014)
13. Kuhn, M., Johnson, K.: Appliedpredictivemodeling: functions and data sets for 'applied predictie modeling' (2014). https://cran.r-project.org/web/packages/AppliedPredictiveModeling/index.html
14. Mišić, V.V.: Optimization of tree ensembles. Oper. Res. **68**(5), 1605–1624 (2020)
15. Pedregosa, F., et al.: Scikit-learn: machine learning in Python. J. Mach. Learn. Res. **12**, 2825–2830 (2011)
16. Schweidtmann, A.M., Mitsos, A.: Deterministic global optimization with artificial neural networks embedded. J. Optim. Theory Appl. **180**(3), 925–948 (2018). https://doi.org/10.1007/s10957-018-1396-0
17. Serra, T., Kumar, A., Ramalingam, S.: Lossless compression of deep neural networks. arXiv preprint arXiv:2001.00218 (2020)
18. Serra, T., Tjandraatmadja, C., Ramalingam, S.: Bounding and counting linear regions of deep neural networks. In: International Conference on Machine Learning, pp. 4558–4566. PMLR (2018)
19. Tjeng, V., Xiao, K., Tedrake, R.: Evaluating robustness of neural networks with mixed integer programming. In: 7th International Conference on Learning Representations, ICLR 2019, pp. 1–21 (2019)
20. West, D., Dellana, S., Qian, J.: Neural network ensemble strategies for financial decision applications. Comput. Oper. Res. **32**(10), 2543–2559 (2005)
21. Yeh, I.C.: Modeling of strength of high-performance concrete using artificial neural networks. Cem. Concr. Res. **28**(12), 1797–1808 (1998)
22. Zhou, Z.H.: Ensemble Methods: Foundations and Algorithms. CRC Press, Boco Raton (2012)
23. Zhou, Z.H., Wu, J., Tang, W.: Ensembling neural networks: many could be better than all. Artif. Intell. **137**(1–2), 239–263 (2002). https://doi.org/10.1016/S0004-3702(02)00190-X

Heavy-Tails and Randomized Restarting Beam Search in Goal-Oriented Neural Sequence Decoding

Eldan Cohen[✉] and J. Christopher Beck

Department of Mechanical and Industrial Engineering, University of Toronto, Toronto, Canada
{ecohen,jcb}@mie.utoronto.ca

Abstract. Recent work has demonstrated that neural sequence models can successfully solve combinatorial search problems such as program synthesis and routing problems. In these scenarios, the beam search algorithm is typically used to produce a set of high-likelihood candidate sequences that are evaluated to determine if they satisfy the goal criteria. If none of the candidates satisfy the criteria, the beam search can be restarted with a larger beam size until a satisfying solution is found. Inspired by works in combinatorial and heuristic search, we investigate whether heavy-tailed behavior can be observed in the search effort distribution of complete beam search in goal-oriented neural sequence decoding. We analyze four goal-oriented decoding tasks and find that the search effort of beam search exhibits fat- and heavy-tailed behavior. Following previous work on heavy-tailed behavior in search, we propose a randomized restarting variant of beam search. We conduct extensive empirical evaluation, comparing different randomization techniques and restart strategies, and show that the randomized restarting variant solves some of the hardest instances faster and outperforms the baseline.

Keywords: Beam search · Neural sequence models · Randomized restarts

1   Introduction

Neural sequence models are commonly used in the modeling of sequential data and are the state-of-the-art approach for tasks such as machine translation [10], text summarization [6], and image captioning [37]. Beam search is the most commonly used algorithm for decoding neural sequence models by (approximately) finding the most likely output sequence conditioned on the input. To do so, beam search generates sequences token-by-token, extending a fixed number of active candidate sequences (beam size) at each step.

Recently, neural sequence models have been successfully applied to different combinatorial search problems such as program synthesis and routing problems. Unlike machine translation and image captioning, such problems often have a

© Springer Nature Switzerland AG 2021
P. J. Stuckey (Ed.): CPAIOR 2021, LNCS 12735, pp. 115–132, 2021.
https://doi.org/10.1007/978-3-030-78230-6_8

goal criteria that can be used to evaluate candidate solutions and require solutions that satisfy the goal criteria. For example, in resource-constrained combinatorial routing problems, we may wish to find a tour that satisfies some resource constraint (e.g., limited fuel or budget). In such scenarios, beam search is used to produce a set of promising (high-likelihood) candidate sequences that are evaluated to determine if they satisfy the goal criteria. If none of them satisfy the criteria, the beam search can be restarted with a larger beam size until a satisfying solution is found.

Previous work on heuristic and combinatorial search algorithms found they tend to exhibit a fat- and heavy-tailed behavior that can be exploited to boost their performance by incorporating randomized restarts in the search (e.g., [8,12]). In this work, we investigate whether a heavy-tailed behavior can also be observed for goal-oriented beam search. We consider four goal-oriented neural sequence decoding tasks, each with a goal criteria that enforces bounded suboptimality with respect to a chosen evaluation metric. We focus on complete anytime beam search (CAB), a complete variant of beam search commonly used in goal-oriented neural sequence decoding, and perform an extensive empirical study of the heavy-tailed behavior and the impact of randomized restarts. Specifically, we make the following contributions:

1. We show that for goal-oriented neural sequence problems, complete anytime beam search exhibits a fat- or heavy-tailed behavior on ensembles of relaxed problems, similar to the behavior observed for CSPs and SAT.
2. We consider a randomized variant of beam search that is based on noise injection to the inputs of the neural network and show that randomized complete anytime beam search exhibits fat- or heavy-tailed behavior on ensembles of multiple runs on a single instance.
3. Inspired by previous work on heavy-tailed behavior in combinatorial and heuristic search problems, we introduce a randomized restarting variant of complete anytime beam search and show that it outperforms the baseline by solving some of the hardest problems faster.
4. We conduct extensive empirical evaluation and analyze the impact of different parameters including the constrainedness of the goal criteria, the restart policy, and the type of randomization.

2 Background

2.1 Beam Search for Goal-Oriented Neural Sequence Decoding

A neural sequence model learns a probability distribution over sequences by being trained to predict the probability of the next token in a sequence, $p(y_t|x; y_{1:t-1})$, conditioned on the input x and the partial sequence $y_{1:t-1}$ [5]. The total probability of a (partial) sequence $y_{1:t}$ follows from the chain rule of probability:

$$p(y_{1:t}|x) = p(y_t|x; y_{1:t-1}) \cdot p(y_{1:t-1}) = \prod_{t'=1}^{t} p(y_{t'}|x; y_{1:t'-1}). \tag{1}$$

It is common to model $p(y_t|x; y_{1:t-1})$ using a Recurrent Neural Network [16], where the input x and the partial sequence $y_{1:t-1}$ we condition on are expressed by a fixed length representation h_t. This representation is updated each step using a non-linear function f: $h_t = f(h_{t-1}, y_{t-1})$ with h_0 being a representation of the input x and y_0 being a special token that represents the start of the sequence. The conditional probability over the next token y_t can then be computed using the softmax function,

$$p(y_t = v_i|x; y_{1:t-1}) = \frac{\exp(w_i h_t)}{\sum_{j=1}^{|\mathcal{V}|} \exp(w_j h_t)},$$

where $\mathcal{V} = \{v_1, v_2, ...\}$ is the set of all possible tokens and w_i are model weights.

Beam search is a limited-width breadth-first search. In the context of sequence models, it is often used as an approximation to finding the (single) sequence y that maximizes Eq. (1), or as a way to obtain a set of high-probability sequences from the model. At the first step, $t = 0$, we only have one (empty) sequence. At each of the following steps, $t \geq 1$, we consider all one-token extensions of the beam sequences from step $t - 1$ and retain (at most) $\mathcal{B}$ partial sequences with the highest probability. In the last step, we return the $\mathcal{B}$ highest probability complete sequences, which we assume to be of equal length (as they can be padded). $\mathcal{B}$ is called the beam width (or, alternatively, beam size) and the probabilities of (partial) sequences are estimated by the neural network.

In goal-oriented neural sequence decoding, we are not looking for the most-likely sequence according to the learned model. Instead, we are looking for a solution that satisfies the goal criteria. In such scenarios, we use beam search to generate a set of $\mathcal{B}$ high-quality candidates that are then evaluated to determine if they satisfy the goal criteria. Once a candidate satisfies the goal criteria, it is returned as the solution of the beam search.

Previous work on goal-oriented neural sequence decoding considered a variant of the complete anytime beam search (CAB) [42] in which failing to find a satisfying solution results in doubling the beam width and re-running the beam search [2,25,43]. As the beam width increases, a larger portion of the hypotheses space is explored and the search is guaranteed to find a solution, if one exists. Algorithm 1 shows pseudo-code for this variant of complete anytime beam search.

Algorithm 1. Complete Anytime Beam Search

---

   function CAB(goalCriteria)
      $beamWidth \leftarrow 1$
      while not solved **do**
         $candidates \leftarrow BeamSearch(beamWidth)$
         for $cand \in candidates$ **do**
            if $Satisfy(cand, goalCriteria)$ **then**
               return $cand$
         $beamWidth \leftarrow 2 \cdot beamWidth$

2.2 Heavy-Tailed Behavior and Randomization in Heuristic and Combinatorial Search Algorithms

Analyzing the empirical distribution of search effort over an ensemble of problems, rather than just the mean or median, can often help design better search algorithms. Previous work has found fat- or heavy-tailed behavior in the distribution of search effort for different search algorithms on NP-complete problems, e.g., the number of backtracks in CSPs, on ensembles of random problems [7,12,13]. This behavior tends to appear in ensembles of relaxed problems, i.e., problems with high density of solutions. In these ensembles, the median search effort is low, however the hardest instances can require orders-of-magnitude higher effort. Interestingly, Gomes et al. [12] also found heavy-tailed behavior in the search effort distribution of a randomized search procedure on a *single* instance, suggesting that some of the hardest problems can be solved easily by minor changes in the search procedure. This result has motivated significant work on reducing heavy-tailed behavior using randomized restarts, portfolios, etc. [11].

Fat- and heavy-tailed distributions have a long tail containing a considerable concentration of mass. Formally, a random variable X is considered heavy-tailed if it has a Pareto-like decay of its tail above some threshold x_l, i.e., there exists some $x_l > 0$, $c > 0$, $\alpha > 0$ such that $P[X > x] = cx^{-\alpha}$ for $x > x_l$ [32]. An approximately linear behavior over several orders of magnitude in the log-log plot of $1 - CDF(x)$ (i.e., the survival function) is a clear sign of heavy-tailed behavior with the slope providing an estimate of the stability index α [14].

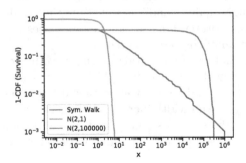

Fig. 1. Heavy and non-heavy tailed behavior [14].

To demonstrate heavy-tailed behavior, we present an example from Gomes et al. [14]. Figure 1 shows the log-log plot of $1 - CDF(x)$ for two normally distributed random variables with a mean of 2 and different standard deviation. It also shows a random variable that represents the number of steps it takes for a symmetric random walk on a line to get back to the starting point. The normal distributions exhibit a fast-decay behavior, while the random walk exhibits a clear heavy-tailed behavior indicated by the approximately linear behavior on the log-log plot.

3   Goal-Oriented Benchmark Problems

In our analysis, we use a set of four goal-oriented benchmark problems. Following is a description of each problem and its goal criteria.

3.1   Combinatorial Routing Problems

Several recent works have demonstrated the potential of using deep learning to solve combinatorial optimzation problems [9, 22, 23, 30]. A recent work [23] proposed an architecture based on attention layers and trained using REINFORCE [41] to generate solutions for combinatorial routing problems that minimize the solution cost. The authors use this architecture to generate solutions to the Travelling Salesman Problem (TSP), two variants of the Vehicle Routing Problem (VRP), the Orienteering Problem (OP), and the Prize Collecting TSP (PCTSP) and show it outperforms a wide range of baselines. Decoding can be done using sampling or beam search, and the best solution among the generated candidates is returned. To eliminate infeasible solutions, e.g., revisiting the same node in TSP, the authors use masking (setting the log-probabilities of infeasible solutions to $-\infty$). In our work, we use Kool et al.'s [23] architecture[1] and problem instances and run experiments on two combinatorial routing problems:

- The Travelling Salesman Problem (TSP) consists of constructing a tour that starts at the depot, visits all nodes exactly once, and returns to the depot.
- The Capacitated Vehicle Routing Problem (CVRP) consists of constructing multiple routes, each starting and ending at the depot, such that the total demand of the nodes in each route does not exceed the vehicle capacity.

The cost of solution in both problems is the sum of pairwise Euclidean distances of consecutive nodes in the solution path (including the depot).

Goal Criteria. As the current model is trained to minimize the solution cost, we consider the goal-oriented problem of finding a solution with a bounded optimality gap. Assuming a minimization problem with cost function $\mathcal{C}$, our goal criteria for a candidate solution x is $\frac{\mathcal{C}(x) - \mathcal{C}(x^*)}{\mathcal{C}(x^*)} \leq \varepsilon$, where x^* is an optimal solution and ε controls the constrainedness of problems (increasing ε leads to a higher expected number of feasible solutions).[2] Following Kool et al. [23], we compute optimal solutions for TSP using Concorde [1] and approximate optimal solutions for CVRP using KLH3 [17] (Kool et al. [23] note CVRP problems with more than 20 location were intractable for an exact solver).

---

[1] Obtained from github.com/wouterkool/attention-learn-to-route.
[2] This notion of constrainedness matches the notion of resource-constrainedness previously used to study planning in resource-constrained environments [29].

3.2   Visual Program Synthesis

Several recent works have considered the problem of synthesizing programs for images using deep neural networks [27,33,36]. These networks take an image as input and output a program that generates the image. The quality of a candidate program can be evaluated using a metric of projection loss, typically a distance measure between the generated image and the input. In our experiments, we use CSGNet[3] [33], a neural architecture that takes in a 2D or 3D shape image and outputs a program to generate the shape using instructions based on constructive solid geometry (CSG). CSGNet is trained using a combination of supervised learning and reinforcement learning (using REINFORCE [41]) to minimize the visual distance between the generated solutions and the input images.

Goal Criteria. Our goal criteria is based on Chamfer Distance ($\mathcal{CD}$), a measure of visual similarity between two shapes that is used by Sharma et al. [33] to evaluate CSGNet. Let $\mathcal{CD}(a, b)$ denote the (non-negative) Chamfer distance between shape a and shape b. We define our goal criteria for a candidate solution x to be $\mathcal{CD}(x, i) \leq \gamma$ where i is the input shape and the parameter γ controls the constrainedness of problems.

3.3   Conditional Molecular Design

A recent line of work focuses on generating molecules with specific properties [18–20], such as the molecular weight, the Wildman-Crippen partition coefficient [40], and a quantitative estimation of drug-likeness (QED) [3]. Kang and Cho [20] proposed a semi-supervised variational autoencoder that is trained on a set of existing molecules from the ZINC dataset [35] with only a partial annotation (i.e., only a fraction of the molecules are labelled with the property values).[4]

The model represents a generative process in which the input molecule x is generated from the distribution $p(x|z, y)$ that is conditioned on the molecule properties y and a latent variable z. The molecules are represented using SMILES strings [39] and are generated character-by-character. For the conditional generation of molecules with a specific property, we sample z from its prior and y from its prior conditioned on the specific property. A molecule representation $\hat{x}$ is obtained from y and z using the decoder's conditional probability $p(x|y, z)$,

$$\hat{x} = \arg\max p(x|y, z), \tag{2}$$

where Eq. (2) is approximated by a beam search.

Goal Criteria. We focus on the QED property [3], a measure of drug-likeness in the range $[0, 1]$ that is based on desirability functions for several molecular properties. We compute QED using RDKit [26] and evaluate the generated

---

[3] Obtained from github.com/Hippogriff/CSGNet.
[4] Obtained from github.com/nyu-dl/conditional-molecular-design-ssvae.

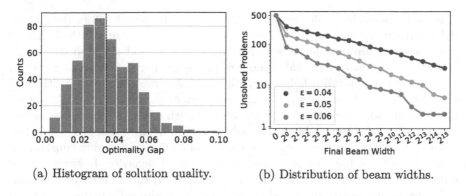

(a) Histogram of solution quality.   (b) Distribution of beam widths.

Fig. 2. TSP (100 nodes): Results for 500 random instances.

molecules based on the absolute difference between their QED and the desired QED. Formally, we define our goal criteria for a candidate solution x to be $|QED(x) - q| \leq \rho$ where q is the desired value of QED and the parameter ρ represents a bound on the deviation from the desired QED value and controls the constrainedness of the criteria.

4   Fat- and Heavy-Tailed Behavior in Goal-Oriented Neural Sequence Decoding

In this section we demonstrate the existence of heavy-tailed behavior in goal-oriented neural sequence decoding. Due to space, we only present results for one benchmark problem, the Travelling Salesman Problem (TSP), however in Appendix A, we present similar results for the other three benchmarks.[5]

We consider a collection of 500 randomly generated TSP problem instances with 100 nodes solved using beam search with a beam width of 10. Figure 2a shows the distribution of solution quality presented as optimality gap $(\frac{\mathcal{C}(x)-\mathcal{C}(x^*)}{\mathcal{C}(x^*)})$ to match our goal criteria. The center of the distribution is around 0.03 with the mean (marked in a dashed line) at approximately 0.034. However, there is a small number of problems for which the optimality gap can be much higher (up to approximately 0.1).

Next, we consider the case of solving the goal-oriented problem where solutions must satisfy a bound on the optimality gap denoted as ε (as discussed in Sect. 3.1). We use complete anytime beam search (Algorithm 1) to solve the problems with the given bound as goal criteria. We start with a beam width of 1, and double the beam width in each iteration if no solution that satisfies the goal criteria is found. We record the beam width for which a satisfying solution was found representing the required search effort.

Figure 2b shows the search effort distribution for three different goal criteria $\varepsilon = 0.04$, $\varepsilon = 0.05$, $\varepsilon = 0.06$. The y-axis represents the number of unsolved

---

[5] All appendices appear in tidel.mie.utoronto.ca/pubs/rr-beam-appendix.pdf.

problems in log-scale, while the x axis represents the search effort (i.e., beam width) in discrete log_2-scale (i.e., in steps of $2^i, i = 0, 1, ...$) to match the behavior of the complete anytime beam search. We artificially add the step 0 (i.e., no search effort) to denote the total number of problems. For $\varepsilon = 0.05$ and $\varepsilon = 0.06$, there is a clear heavy-tailed behavior with a very low median (beam width of 1) and a slow decay of the tail over multiple orders of magnitude. In fact, not all problems were solved for the maximum beam width of 32, 768. Note that when $\varepsilon = 0.05$, 332 of the 500 problems are solved with a beam width of 1, while five problems could not be solved for a beam width of 32, 768. For a more constrained goal criteria of $\varepsilon = 0.04$, we still observe a fat-tailed behavior, however we see a noticeable increase in the difficulty of problems and the number of problems that could not be solve in the search effort limits is significantly higher. We could not analyze more constrained goal criteria due to the high computational cost, however we hypothesize that problems will become significantly harder and the heavy-tailed behavior will reduce, consistent with previous work [7,12].

The above results suggest that goal-oriented beam search exhibits a heavy-tailed behavior in ensembles of random problems, similar to the one observed for other combinatorial and heuristic search algorithms. In these algorithms, much of the large variability in the search effort for ensembles of random problems was found to be associated with the algorithm, rather than the problem instances [12]. To isolate the variability of the search algorithm, in the next section we analyze the search effort distribution of a randomized variant of complete anytime beam seach on a single instance.

4.1   Fat- and Heavy-Tailed Behavior on a Single Instance

In order to introduce randomization into beam search decoding of neural sequence models, we inject random noise in the inputs of the neural network that is being decoded using beam search. Injecting random noise in the inputs of a neural network is a known technique in the training of neural networks in order to improve their robustness [16].[6] Note that the noise injected to the network's inputs does not impact the goal test that is still based on the original input, i.e., the noise does not change the problem we are solving. The sole purpose of the noise is to introduce some randomness in the network's predicted probabilities and, as a result, in the beam search decoding.

For TSP instances, the inputs to the network consist of the locations of all nodes, expressed as two-dimensional coordinates normalized in the range [0, 1]. We inject noise to the network inputs by *adding* random noise drawn from a uniform distribution, $U(-0.01, 0.01)$. Figure 3a shows the distribution of search effort for 500 randomized runs (i.e., runs with different random injected noise) for different values of ε. We can see a fat-tailed behavior that indicates a significant

---

[6] Note that we are not aware of any direct connection between noise injection in training to increase robustness and our use of noise injection in testing to introduce randomness in the decoding process. However, it might be interesting to consider whether there is some underlying connection.

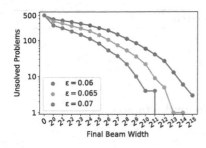

CVRP	Multiplicative uniform noise on locations and demand
Visual Prog. Synthesis	We flip, with small probability, pixels close to the edges of the shape
Molecule Generation	Additive Gaussian noise on random seeds

Fig. 3. (a) TSP (100 nodes): Distribution of beam widths for 500 randomized runs on a single instance. (b) Problem-specific noise injection to network's inputs. See Appendix B for detailed description.

variability is associated with the search method. Note that the results in Fig. 3a were observed for a single, arbitrarily chosen instance. Experiments with other instances also yielded fat- and heavy-heavy tailed behavior, however we found large differences among instances: different instances exhibited different levels of fat- and heavy-tailedness for different levels of goal criteria constrainedness.

Figure 3b briefly summarizes the problem-specific noise injection used for the other three benchmarks. A detailed description of the random noise injection and experimental results for these benchmarks appear in Appendix B.

The above results indicate that significant variability can be associated with the search algorithm itself. Previous works have exploited the large variability associated with the search algorithm to improve problem solving performance by introducing randomized restarts (see Sect. 2.2). In the next section, we propose a complete variant of beam search that incorporates randomized restarts and evaluate its impact on the distribution of search effort.

5   Randomized Restarting Neural-Guided Beam Search for Goal-Oriented Combinatorial Problems

We present randomized-restarting complete anytime beam search (RR-CAB), a variant of complete anytime beam search (Algorithm 1) that uses randomized beam search and a custom restart strategy. Algorithm 2 presents the pseudocode of RR-CAB, where the goal criteria and the restart strategy are passed as parameters. In each iteration the algorithm runs a randomized beam search (using a random seed) with a beam width that is determined by the restart strategy. The algorithm returns when one of the candidate solutions generated by the beam search satisfies the goal criteria.

In order to randomize the results of a beam search, we consider the following two options.

Beam Search with Injected Input Noise. Following the methodology in Sect. 4.1, we inject random noise to the inputs of the neural networks.

Algorithm 2. Randomized Complete Beam Search

---

   function RR-CAB(goalCriteria, restartStrategy)
      iteration ← 1
      while not solved **do**
         beamWidth ← *restartStrategy(iteration)*
         seed ← *RandomSeed()*
         candidates ← *RandomizedBeamSearch(beamWidth, seed)*
         for *cand* ∈ *candidates* **do**
            if *Satisfy(cand, goalCriteria)* **then**
               return *cand*
         iteration ← *iteration* + 1

Stochastic beam search (SBS) [24]. SBS is a stochastic variant of beam search that samples k sequences without replacement from a sequence model and therefore produces randomized output. The level of diversity in SBS is controlled by the softmax temperature that modifies the conditional probability of each token during the decoding process. The probability of token y_t conditioned on the partial sequence $y_{1:t-1}$, is defined as a softmax normalization of the unnormallized log-probabilities, $\phi(y_t|x; y_{1:t-1})$, with a temperature T [24]:

$$p(y_t|x; y_{1:t-1}) = \frac{exp(\phi(y_t|x; y_{1:t-1})/T)}{\sum_{y'} exp(\phi(y'|x; y_{1:t-1})/T)}.$$

The temperature $T > 0$ and higher T leads to higher diversity. The default temperature is $T = 1$, where the predicted probabilities are not modified. In our experiments, we considered two temperature configurations: the default temperature $T = 1.0$ and a higher diversity temperature $T = 1.5$.

Note that we could not perform the analysis in Sect. 4.1 using SBS since, unlike input noise injection, we cannot guarantee that repeated runs with different beam widths will maintain similar conditional probability distributions (see discussion in Sect. 7). However, in RR-CAB, we are not interested in maintaining the same probability distributions across runs and therefore SBS can be used as a randomized variant of beam search.

5.1 Restart Strategies

A restart strategy is a sequence $(t_1, t_2, t_3, ...)$ of run lengths after which the search restarts. In goal-oriented neural sequence decoding, the sequence length is either fixed (e.g., in TSP and CVRP) or predicted by the network (e.g., in visual program synthesis or conditional molecule generation). If we want to allocate more search effort, we simply extend the beam width thus allowing more sequences to be tested against the goal criteria.

In each iteration, we run a beam search with a given beam width until a solution if found. In deterministic complete anytime beam search (Algorithm 1), the beam width is increased in each iteration. In RR-CAB, running a search with the same beam width multiple times leads to different results and can sometimes be more efficient than increasing the beam width. We therefore employ a custom restart strategy to determine the beam width in each iteration. We consider two popular restart strategies from the literature.

Fixed-Cutoff Strategy. Fixed-cutoff strategies [15] are simple strategies of the form $(t_c, t_c, ...)$ where t_c is a constant. This strategy is often not robust enough: a small t_c value might not be sufficient to solve all problems, while a larger value will be computationally inefficient.

Geometric Strategy. Geometric strategies [38] take the form $(r^0, r^1, r^2, r^3, ...)$ where the geometric factor r controls how fast the cutoff values grow. When $r = 2$ and randomization is not applied, this strategy has a similar behavior to the complete beam search procedure described in Sect. 2.1.

6   Empirical Results

In this section, we present empirical analysis of the performance of RR-CAB on the goal-oriented benchmarks. We compare results for the two randomization techniques (input noise injection and SBS) and the two restart strategies (geometric and fixed-cutoff) described in Sect. 5.

6.1   Results for the Travelling Salesman Problem (TSP)

We consider the same collection of 500 randomly generated TSP problems with 100 nodes used in Sect. 4. We analyze the results of RR-CAB with random noise injection and the two restart strategies: geometric with $r = 2$ and fixed-cutoff with beam width $\mathcal{B} = 8$. In order to directly compare the performance of a fixed-cutoff strategy and a geometric strategy, we organize the results of fixed-cutoffs beam search in batches of multiple beam searches with a constant beam width, such that they sum to the beam width of the corresponding beam search with geometric restarts. For example, we present results for a geometric restart policy for the beam width thresholds 1, 2, 4, 8, 16, 32, etc. In comparison, for fixed-cutoff restarts, the result for a threshold of 16 represents a batch of two beam searches, each with a constant beam width of 8.

Figure 4 compares the distribution of search effort of standard CAB and RR-CAB in the configurations described above. In general, the randomized variants tend to under perform for the very small beam width: problems that were easily solved without randomization do not benefit, and even suffer, from adding randomization. In particular, since we use a beam width $\mathcal{B} = 8$ for the fixed-cutoff strategy, solutions are only found starting from a threshold of 8. However, as we increased the search effort, we see that the randomized variants outperform standard CAB. For the more constrained problems, we see that the fixed-cutoff strategy significantly outperforms the geometric restarts strategy. This could be due the use of relatively large r chosen for fair comparison with CAB. For $\varepsilon = 0.6$, geometric restarts seem to have similar performance to fixed cut-offs.

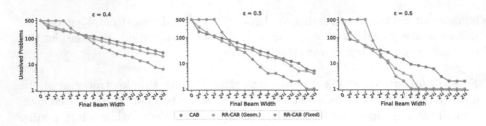

Fig. 4. TSP (100 nodes): Distribution of beam widths for 500 random instance for RR-CAB with input noise injection.

The inherent differences between the two restart strategies result in an apparent inferiority of fixed-cutoffs in smaller beam widths: in addition to having no solutions for beam widths smaller than 8, even for a beam width of 8 it underperforms since RR-CAB with geometric restarts has already made three randomized runs (for beam width 1, 2, and 4) that can lead to solutions. In practice, this is easily mitigated by using a restart policy that starts with geometric restarts before changing to fixed-cutoffs: $1, 2, 4, 8, 8, \ldots$ To maintain simple and clear comparison we do not adopt this enhancement in our evaluation.

Figure 5 shows similar comparison to Fig. 4 where the beam search is randomized using SBS with a softmax temperature of $T = 1$ (top) and $T = 1.5$ (bottom). Again, we see that introducing randomization to CAB leads to better performance. Using softmax temperature of $T = 1.5$ exhibits better performance and manages to solve more hard instances faster. Interestingly, for SBS we find that geometric restarts are approximately as good as fixed-cutoff strategy.

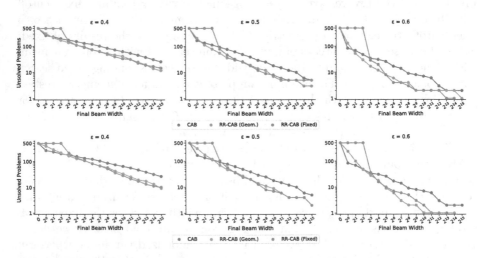

Fig. 5. TSP (100 nodes): Distribution of beam widths for 500 random instances for RR-CAB with SBS using $T = 1.0$ (top) and $T = 1.5$ (bottom).

The above results show that introducing randomization in the search can help solve some of the hardest instances faster. Consistent with previous work on CSPs and SAT, the impact on more relaxed instances tends to be more significant [12]. However, note that we cannot analyze the impact of RR-CAB on more constrained instances due to computational limitations and even for $\varepsilon = 0.4$, using randomization seems to have positive impact on the performance.

6.2   Results for the Other Benchmarks

Figure 6, Fig. 7, and Fig. 8 show the results for CVRP, visual program synthesis and conditional molecule generation, respectively. For CVRP and molecule generation, we found that, similar to TSP, a temperature of $T = 1.5$ yields better results when using SBS. In visual program synthesis, higher temperature did not lead to better results and we present results for $T = 1$.

In the visual program synthesis problem, the number of potential expansions of each of the beam candidates is much higher than the other problems (approximately 400, compared to 36–100 in the other problems). Therefore, when using SBS for this problem, we only consider the top 50 extensions of each candidate. Practically, it is unlikely that an extension of partial hypothesis that is not in the most likely 50 extensions will lead to a hypothesis that will be returned by the beam search. However, when applying randomization it may have the undesired outcome of promoting very low-ranked hypotheses and we therefore consider only the top 50 hypotheses.

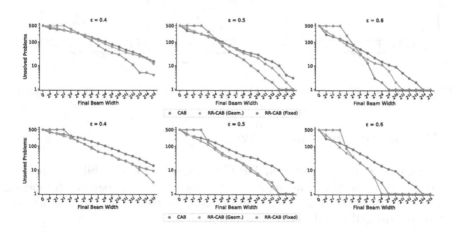

Fig. 6. CVRP (50 nodes): RR-CAB with noise injection (top), SBS (bottom).

Consistent with our results for TSP, we find that RR-CAB solves some the hardest problems faster and outperforms the baseline. As in TSP, when using random noise injection, the fixed cut-offs strategy tends to outperforms the geometric strategy.

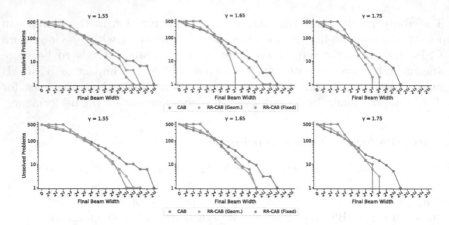

Fig. 7. Visual Program Synthesis: RR-CAB with noise injection (top), SBS (bottom).

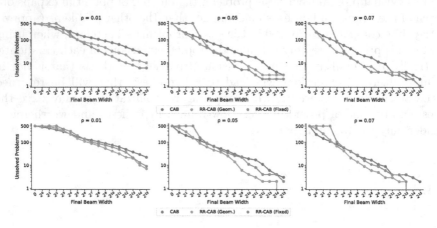

Fig. 8. Molecule Generation: RR-CAB with noise injection (top), SBS (bottom).

7   Discussion and Future Work

Our empirical results suggest that RR-CAB exploits the variability associated with the search procedure and significantly outperforms the baseline by solving some of the hardest problems faster. In this section, we discuss different aspects related to RR-CAB and directions for future work.

Randomization Techniques. We consider two techniques that can randomize the results of a beam search: input noise injection and SBS. While both techniques introduce randomization to the predicted probabilities, there are some important differences between them. A key limitation of the noise injection technique is that it needs to be tailored for each problem. In our work, we had to manually try different randomization approaches in order to find one that would generate sufficient variability on a single instance without making the problem significantly harder across different runs. Alternatively, an inherent limitation of SBS is that we are unable to guaranteed that repeated runs with different beam widths will maintain similar conditional probability distributions. The implication of this limitation is that we cannot analyze the search effort distribution of SBS on a single problem instance, as we do for beam search with noise injection in Sect. 4.1. As future work, it is interesting to investigate other generic ways of introducing noise into the decoding process. Potential directions include applying noise to hidden units [4,31] or using dropout [34] in inference.

Restart Strategies and Parallelization. We focused on two well known restart strategies: fixed-cutoff and geometric restarts. Previous works in combinatorial optimization has considered more advanced restart strategies such as Luby's universal strategy [28] and dynamic and learning restart strategies (e.g., [21]). Investigating ways to incorporate such strategies in RR-CAB is an interesting direction for future work.

A key challenge in designing restart strategies for beam search is their ability to be parallelized on a GPU. In our experiments, we present results for the fixed-cutoff restart strategy by batching together beam searches and comparing these results to the corresponding final beam width of a geometric strategy. As we start investigating more complicated restart strategies, such as Luby's universal strategy [28], we will not be able to batch the results together to maintain comparability. Furthermore, even in our comparison, it is not clear that a set of four beam search instances, each with a beam width of 8 and executed together on a GPU, is comparable to one beam search with a beam width of 32. Our work, therefore, raises the need for well-defined evaluation metrics that can be used to compare the results of parallelized complete beam searches with different restart strategies, even when it not possible to batch together runs as we currently do.

8 Conclusion

In this work we show that fat- and heavy-tailed behavior, that was previously observed for several combinatorial and heuristic search algorithms, can be observed for complete anytime beam search in goal-oriented neural sequence decoding. We perform an extensive empirical analysis, across four goal-oriented benchmarks, and find fat- and heavy-tailed behavior in the distribution of search efforts of beam search. Inspired by previous work on combinatorial and heuristic search, we propose a randomized restarting variant of complete anytime beam search, RR-CAB, and study the impact of different randomization techniques and restart strategies. Our experiments show that RR-CAB solves some of the hardest problems faster and outperforms the baseline. Our work raises interesting questions on the impact of parallelization on the development and evaluation of randomized restarting beam search algorithms and highlights directions for future work.

Acknowledgements. We thank the anonymous reviewers for their valuable feedback. This work was supported by the Natural Sciences and Engineering Research Council of Canada.

References

1. Applegate, D., Bixby, R., Chvatal, V., Cook, W.: Concorde TSP solver (2006)
2. Balog, M., Gaunt, A., Brockschmidt, M., Nowozin, S., Tarlow, D.: Deepcoder: learning to write programs. In: International Conference on Learning Representations (ICLR) (2017)
3. Bickerton, G.R., Paolini, G.V., Besnard, J., Muresan, S., Hopkins, A.L.: Quantifying the chemical beauty of drugs. Nat. Chem. **4**(2), 90 (2012)
4. Cho, K.: Noisy parallel approximate decoding for conditional recurrent language model. arXiv preprint arXiv:1605.03835 (2016)
5. Cho, K., et al.: Learning phrase representations using RNN encoder-decoder for statistical machine translation. In: EMNLP (2014)
6. Chopra, S., Auli, M., Rush, A.M.: Abstractive sentence summarization with attentive recurrent neural networks. In: North American Chapter of the Association for Computational Linguistics: Human Language Technologies (NAACL-HLT), pp. 93–98 (2016)
7. Cohen, E., Beck, J.C.: Fat- and heavy-tailed behavior in satisficing planning. In: AAAI Conference on Artificial Intelligence (AAAI), pp. 6136–6143 (2018)
8. Cohen, E., Beck, J.C.: Local minima, heavy tails, and search effort for GBFS. In: International Joint Conferences on Artificial Intelligence (IJCAI), pp. 4708–4714 (2018)
9. Deudon, M., Cournut, P., Lacoste, A., Adulyasak, Y., Rousseau, L.-M.: Learning heuristics for the TSP by policy gradient. In: van Hoeve, W.-J. (ed.) CPAIOR 2018. LNCS, vol. 10848, pp. 170–181. Springer, Cham (2018). https://doi.org/10.1007/978-3-319-93031-2_12
10. Gehring, J., Auli, M., Grangier, D., Yarats, D., Dauphin, Y.N.: Convolutional sequence to sequence learning. In: International Conference on Machine Learning (ICML), pp. 1243–1252 (2017)
11. Gomes, C.: Randomized backtrack search. In: Milano, M. (ed.) Constraint and Integer Programming: Toward a Unified Methodology, vol. 27, pp. 233–291. Springer, Heidelberg (2003). https://doi.org/10.1007/978-1-4419-8917-8_8
12. Gomes, C.P., Fernández, C., Selman, B., Bessière, C.: Statistical regimes across constrainedness regions. Constraints **10**(4), 317–337 (2005). https://doi.org/10.1007/s10601-005-2807-z
13. Gomes, C.P., Selman, B., Crato, N.: Heavy-tailed distributions in combinatorial search. In: Smolka, G. (ed.) CP 1997. LNCS, vol. 1330, pp. 121–135. Springer, Heidelberg (1997). https://doi.org/10.1007/BFb0017434
14. Gomes, C.P., Selman, B., Crato, N., Kautz, H.: Heavy-tailed phenomena in satisfiability and constraint satisfaction problems. J. Autom. Reason. **24**(1), 67–100 (2000). https://doi.org/10.1023/A:1006314320276
15. Gomes, C.P., Selman, B., Kautz, H., et al.: Boosting combinatorial search through randomization. In: National Conference on Artificial Intelligence (AAAI), vol. 98, pp. 431–437 (1998)
16. Goodfellow, I., Bengio, Y., Courville, A.: Deep Learning. MIT Press (2016). http://www.deeplearningbook.org

17. Helsgaun, K.: An extension of the Lin-Kernighan-Helsgaun TSP solver for constrained traveling salesman and vehicle routing problems. Roskilde University, Roskilde (2017)
18. Jin, W., Barzilay, R., Jaakkola, T.: Junction tree variational autoencoder for molecular graph generation. In: International Conference on Machine Learning (ICML), pp. 2323–2332 (2018)
19. Jin, W., Yang, K., Barzilay, R., Jaakkola, T.: Learning multimodal graph-to-graph translation for molecule optimization. In: International Conference on Learning Representations (ICLR) (2018)
20. Kang, S., Cho, K.: Conditional molecular design with deep generative models. J. Chem. Inf. Model. **59**(1), 43–52 (2018)
21. Kautz, H., Horvitz, E., Ruan, Y., Gomes, C., Selman, B.: Dynamic restart policies. In: National Conference on Artificial Intelligence (AAAI), pp. 674–681 (2002)
22. Khalil, E., Dai, H., Zhang, Y., Dilkina, B., Song, L.: Learning combinatorial optimization algorithms over graphs. In: Conference on Neural Information Processing Systems (NeurIPS), pp. 6348–6358 (2017)
23. Kool, W., van Hoof, H., Welling, M.: Attention, learn to solve routing problems! In: International Conference on Learning Representations (ICLR) (2019)
24. Kool, W., Van Hoof, H., Welling, M.: Stochastic beams and where to find them: The gumbel-top-k trick for sampling sequences without replacement. In: International Conference on Machine Learning (ICML), pp. 3499–3508 (2019)
25. Lample, G., Charton, F.: Deep learning for symbolic mathematics. In: International Conference on Learning Representations (2019)
26. Landrum, G.: RDKit: open-source cheminformatics. http://www.rdkit.org
27. Liu, Y., Wu, Z., Ritchie, D., Freeman, W.T., Tenenbaum, J.B., Wu, J.: Learning to describe scenes with programs. In: International Conference on Learning Representations (ICLR) (2018)
28. Luby, M., Sinclair, A., Zuckerman, D.: Optimal speedup of Las Vegas algorithms. Inf. Process. Lett. **47**(4), 173–180 (1993)
29. Nakhost, H., Hoffmann, J., Müller, M.: Resource-constrained planning: a Monte Carlo random walk approach. In: International Conference on Automated Planning and Scheduling (ICAPS) (2012)
30. Nazari, M., Oroojlooy, A., Snyder, L., Takác, M.: Reinforcement learning for solving the vehicle routing problem. In: Conference on Neural Information Processing Systems (NeurIPS), pp. 9839–9849 (2018)
31. Poole, B., Sohl-Dickstein, J., Ganguli, S.: Analyzing noise in autoencoders and deep networks. arXiv preprint arXiv:1406.1831 (2014)
32. Resnick, S.I.: Heavy-Tail Phenomena: Probabilistic and Statistical Modeling. Springer, Heidelberg (2007). https://doi.org/10.1007/978-0-387-45024-7
33. Sharma, G., Goyal, R., Liu, D., Kalogerakis, E., Maji, S.: CSGNet: neural shape parser for constructive solid geometry. In: Conference on Computer Vision and Pattern Recognition (CVPR), pp. 5515–5523 (2018)
34. Srivastava, N., Hinton, G., Krizhevsky, A., Sutskever, I., Salakhutdinov, R.: Dropout: a simple way to prevent neural networks from overfitting. J. Mach. Learn. Res. **15**(1), 1929–1958 (2014)
35. Sterling, T., Irwin, J.J.: ZINC 15-ligand discovery for everyone. J. Chem. Inf. Model. **55**(11), 2324–2337 (2015)
36. Tian, Y., et al.: Learning to infer and execute 3D shape programs. In: International Conference on Learning Representations (ICLR) (2019)

37. Vinyals, O., Toshev, A., Bengio, S., Erhan, D.: Show and tell: lessons learned from the 2015 MSCOCO image captioning challenge. IEEE Trans. Pattern Anal. Mach. Intell. **39**(4), 652–663 (2017)
38. Walsh, T.: Search in a small world. In: International Joint Conference on Artificial Intelligence (IJCAI), pp. 1172–1177 (1999)
39. Weininger, D.: Smiles, a chemical language and information system. 1. Introduction to methodology and encoding rules. J. Chem. Inf. Comput. Sci. **28**(1), 31–36 (1988)
40. Wildman, S.A., Crippen, G.M.: Prediction of physicochemical parameters by atomic contributions. J. Chem. Inf. Comput. Sci. **39**(5), 868–873 (1999)
41. Williams, R.J.: Simple statistical gradient-following algorithms for connectionist reinforcement learning. Mach. Learn. **8**(3–4), 229–256 (1992). https://doi.org/10.1007/BF00992696
42. Zhang, W.: Complete anytime beam search. In: National Conference on Artificial Intelligence (AAAI), pp. 425–430 (1998)
43. Zohar, A., Wolf, L.: Automatic program synthesis of long programs with a learned garbage collector. In: Conference on Neural Information Processing Systems (NeurIPS), pp. 2094–2103 (2018)

Combining Constraint Programming and Temporal Decomposition Approaches - Scheduling of an Industrial Formulation Plant

Christian Klanke[1]([⊠]) [iD], Dominik R. Bleidorn[2] [iD], Vassilios Yfantis[3] [iD], and Sebastian Engell[1] [iD]

[1] Process Dynamics and Operations Group, Faculty of Biochemical and Chemical Engineering, TU Dortmund University, Emil-Figge-Straße 70, 44229 Dortmund, Germany
christian.klanke@tu-dortmund.de
[2] INOSIM, Joseph-von-Fraunhofer-Str. 20, 44227 Dortmund, Germany
[3] Chair of Machine Tools and Control Systems, Department of Mechanical and Process Engineering, Technische Universität Kaiserslautern, Gottlieb-Daimler-Straße 42, 67663 Kaiserslautern, Germany

Abstract. This contribution deals with the development of a Constraint Programming (CP) model and solution strategy for a two-stage industrial formulation plant with parallel production units for crop protection chemicals. Optimal scheduling of this plant is difficult: a high number of units and operations have to be scheduled while at the same time a high degree of coupling between the operations is present due to the need for synchronizing charging and discharging operations.

In the investigated problem setting the formulation lines produce several intermediates that are filled into a variety of types of final containers by filling stations. Formulation lines and filling stations each consist of parallel, non-identical sets of equipment units. Buffer tanks are used to decouple the two stages, to increase the capacity utilization of the overall plant.

The CP model developed in this work solves small instances of the scheduling problem monolithically. To deal with large instances a decomposition algorithm is developed. The overall set of batches is divided into subsets which are scheduled iteratively. The algorithm is designed in a moving horizon fashion, in order to counteract the disadvantages of the limited lookahead that order-based decomposition approaches typically suffer from. The results show that the complex scheduling problem can be solved within acceptable solution times and that the proposed moving horizon strategy (MHS) yields additional benefits in terms of solution quality.

Keywords: Constraint Programming · Moving-horizon · Decomposition algorithm · Batch process scheduling

This work was partially funded by the European Regional Development Fund (ERDF) in the context of the project OptiProd.NRW.

© Springer Nature Switzerland AG 2021
P. J. Stuckey (Ed.): CPAIOR 2021, LNCS 12735, pp. 133–148, 2021.
https://doi.org/10.1007/978-3-030-78230-6_9

1   Introduction

The goal of the manufacturing and process industries is the profitable production of goods. For this purpose, the production has to be efficient and effective, ensuring that the company remains competitive. One determining factor to succeed is the optimal utilization of machines and material to cut production costs and remain resource-efficient. One means to achieve both, without modifying the production environment, and thus without extensive investments, is optimal scheduling.

While scheduling for simple processes can be done manually, most real-life processes and plants are very complex and automated, optimization-based procedures become necessary. This gave rise to a variety of solution approaches, of which many can be found in [2].

Among the exact approaches, i.e. among approaches that are able to find a provable globally optimal solution, Constraint Programming (CP) has shown to be a promising solution technique on a variety of benchmark scheduling problems and industrial cases [4,5,11]. In comparison to Mixed-Integer Linear Programming (MILP), for which an overview of different modeling approaches for the scheduling of batch processes is provided in [8], CP exhibits several advantages. Those advantages include the richer mathematical syntax of CP and the ability to exploit domain-specific knowledge more easily to improve the search process.

In any case computer-aided scheduling also has to cope with the curse of dimensionality, because almost all scheduling problems belong in the class of NP-hard problems, i.e. problems for which the solution time grows exponentially, when the problem size increases [6]. Therefore much research went into the development of methods to efficiently create schedules, trading-off solution time and solution quality [7].

One of these is temporal decomposition, where the orders are split into several sets of orders, the sets are scheduled sequentially for each subproblem and in every subproblem timing constraints are introduced to take into account the solution of the previous subproblem. The solutions of the subproblems are then concatenated to yield the overall schedule. In [1] temporal decomposition of a scheduling problem formulated as a MILP is complemented by spatial decomposition, i.e. decomposition based on groups of units, and close to optimum solutions are obtained consistently. It was possible to reduce the computational time to a fraction of the time needed to solve the monolithic problem. A similar approach was described in [12] where temporal decomposition approaches were applied to a scheduling problem formulated as an MILP.

In this paper, a real-life industrial batch scheduling problem is modeled and optimized with a CP-based approach. Due to the size of the problem a monolithic solution is computationally not tractable and a moving horizon decomposition approach, that yields acceptable overall solution times, is developed. At the core of the decomposition strategy, temporal decomposition is applied. However, temporal decomposition suffers from myopia with respect the orders that are scheduled in subsequent iterations and unnecessarily large changeover times might occur. Our approach remedies this issue by dynamically reassigning orders to subproblems, based on the most recent partial solution. To the best of the authors knowledge, no such combination of a CP model that is solved with a moving horizon decomposition approach or temporal decomposition for a case study of industrial complexity has been reported yet.

A scheduling problem similar to the one solved in this work was considered in [10] with the constraint logic programming system CHIP. The investigated production process consists of a batch formulation and a continuous packaging part, where a storage tank separates each batch formulation line from the subsequent packaging stage. However, the industrial case study we investigate in this paper is of significantly higher complexity regarding the process constraints that are modeled, which makes the use of a decomposition approach necessary.

This paper is structured as follows: in Sect. 2 the industrial formulation plant for crop protection chemicals that is considered in this work is introduced. Section 3 presents the solution methodology, including the CP model and the moving horizon strategy (MHS). Results are presented in Sect. 4. A conclusion and an outlook are provided in Sect. 5.

2   Case Study

Formulation of crop protection chemicals refers to the preparation of their active ingredients in a way to make them suitable for use. Active ingredients are highly chemically and biologically active substances. Therefore they have to be diluted or suspended in a liquid, in order to ensure the safety of the user. Finally, through the formulation different physicochemical properties of the products can be controlled.

A schematic depiction of the formulation plant is shown in Fig. 1. The plant consists of three main sections, the formulation lines, the filling stations and the buffer tanks. The plant is operated sequentially, i.e. intermediate products are first produced in the formulation lines and then filled into their final containers by the filling stations. The buffer tanks are used for storage of the intermediate products, in order to decouple the two production stages. Formulation lines, buffer tanks and filling stations are fully connected via a transfer panel.

Each formulation line consists of a pre-processing line and three identical standardization tanks, i.e. stirred tank reactors. Standardization refers to a mixing operation of the solvent and the active ingredient. Only a single pre-processing line is connected to all standardization tanks in a formulation line, only one tank can be filled at the same time. After a tank has been filled, a changeover operation has to take place in the pre-processing line prior to the formulation of the next product. This changeover includes cleaning and setting up the equipment for the next product. Only full batches are produced, i.e. the standardization tanks are always filled up to their full capacity. After the standardization of a batch has been completed, the tank is ready to be emptied. However, it is not required to empty the tank immediately. The standardization tanks can be used for intermediate storage. When a tank is emptied, its content is transferred either to a filling station or to a buffer tank.

The filling stations can be connected to every standardization and buffer tank and are used to fill the intermediate products into their final containers. They do not posses a dedicated buffer. Instead they directly drain the tank they are connected to (standardization or buffer tank) in a continuous manner. A filling operation is time consuming and depends on the filled amount and on the flowrate of the filling station. The tank connected to a filling station is therefore gradually emptied. Once the filling operation has been completed, a changeover, i.e. cleaning and preparation, has to be performed both for the tank and the filling station. An additional constraint of the filling stations is that they are not operated during night shifts, in contrast to the formulation lines. In order to decouple the two production stages and to utilize the full production capacity of the formulation lines, primarily the buffer tanks are used for intermediate storage. A buffer tank can store multiple batches of the same intermediate product from the formulation lines. The transfer time between a standardization and a buffer tank is negligible. The buffer tanks are emptied by the filling stations in the same way as the standardization tanks. After a buffer tank has been emptied, a changeover has to be performed before new intermediates can be stored.

Each intermediate product can only be processed in a subset of the available formulation lines and filling stations. The changeover times between the uses of pieces of equipment depend on the product sequence.

The scheduling task for the formulation plant consists of the allocation of the different batches of intermediate products to the standardization tanks and to the filling stations and the timing and sequencing of these operations. The use of the buffer tanks also constitutes a degree of freedom for the scheduling algorithm. The goal is to minimize the production time, i.e. the makespan or the total completion time, needed to meet the specified demand of the products. Due dates and deadlines are not considered, as the plant operates according to a make-to-stock policy. It is assumed that the set of orders is known a priori, rendering this scheduling problem deterministic and static. For the real application, due to uncertainties in the production process and newly incoming orders, frequent rescheduling has to take place with updated information.

The same scheduling problem has previously been addressed in [14], where a decomposition-based MILP strategy was employed. In the following section the CP-based solution strategy is introduced.

3   Methodology

The complexity of the presented case study is such that it is computationally intractable to solve large problem instances in a monolithic fashion. For this reason a temporal decomposition-based approach is proposed. At first the set of all batches is obtained from the set of orders. Then this set of batches is divided into subsets of constant size. Afterwards each of these subsets is scheduled iteratively, fixing the decisions made in previous iterations and considering the timing constraints that are imposed on the current iteration. The approach

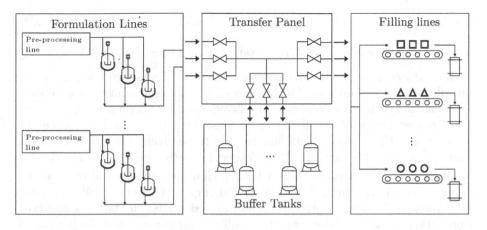

Fig. 1. Schematic depiction of the industrial formulation plant.

is extended to a moving horizon scheme: from the batches considered for a sub-problem, only a certain number are fixed in the schedule, while the remaining batches are withheld to be scheduled in the subsequent iterations.

First the CP model is presented which solves small-scale instances mono-lithically and is the base for the decomposition approach. This is followed by a description of the moving horizon strategy. The model was implemented in Python 3.6, using Google's software suite for combinatorial optimization, OR-Tools (version 7.6.7570) [9]. The corresponding CP-SAT Solver was used.

3.1   Variables

CP is based upon handling discrete variables. The three types of variables that are used within the CP model are integer variables, Boolean variables and interval variables. The latter are explained in more detail: Interval variables enable a higher-level representation of time intervals. Within this work we use interval variables to represent the different activities that are performed on the units (i.e. a single piece of equipment), therefore we use the words activity and interval variable synonymously. The time horizon $\mathcal{T} = [0, t_{Hor}] \subset \mathbb{N}_0$ is discretised into a finite set of timepoints and all durations are rounded up to a multiple of the discretization interval (one hour in our case) to match these integrality requirements. An activity I is defined in terms of integer variables: the start of the activity $Start(I)$, the end of the activity $End(I)$ and its duration $duration(I)$.

Furthermore, activities can be declared as optional, in which case a Boolean variable $I.isPresent$ is attached to activity I. $I.isPresent$ takes the value *False* or 0 if the activity is excluded from the schedule and takes the value *True* or 1 if the activity is included. For activities that are not optional, all constraints assume $I.isPresent = 1$. If an optional activity is not present, no constraint will act on the related variables, unless explicitly specified.

Inherently, all activities are constrained by Eq. (1). $\mathcal{I}$ denotes the set of activities.

$$I.\text{isPresent} \implies \text{End}(I) - \text{Start}(I) = \text{duration}(I) \quad \forall I \in \mathcal{I} \tag{1}$$

At least one activity is introduced per unit $j \in \mathcal{J}$ and batch $b \in \mathcal{B}$, where multiple interval variables may be introduced per batch and unit, to accurately represent the synchronization between different transfer operations, changeovers or the discretization of the filling operations. The sets $\mathcal{J}$ and $\mathcal{B}$ denote the complete set of processing units and batches respectively. On the formulation tanks $\mathcal{J}^T$ an interval variable $I_{j,b}^l$ is defined for each tank, batch and type of operation $l \in \mathcal{L}^T = \{$Charge, Mix, Dis$\}$, i.e. charging, mixing and discharging. Charging and discharging refers to filling and emptying a tank. Similarly for the buffer tanks $\mathcal{J}^{St}$ interval variables are introduced per batch and tank and type of operation $l \in \mathcal{L}^{St} = \{$stor, empt$\}$, i.e. filling and storing, and discharging of the tank. In the case of the filling operations a discretization into n individual intervals represented by the interval variables $I_{j,b,i}^{\text{Fillstep}}$ is necessary, to account for the continuous operation of the filling stations $\mathcal{J}^{Fill}$. The following sets of interval variables are therefore introduced for the formulation stage, the filling stage and the buffer tanks:

$$
\begin{aligned}
&I_{j,b}^l && \forall b \in \mathcal{B}, j \in \mathcal{J}^T, l \in \mathcal{L}^T = \{\text{Charge, Mix, Dis}\} \\
&I_{j,b}^l && \forall b \in \mathcal{B}, j \in \mathcal{J}^{St}, l \in \mathcal{L}^{St} = \{\text{Stor, Empt}\} \\
&I_{j,b,i}^{\text{Fillstep}} && \forall b \in \mathcal{B}, j \in \mathcal{J}^{\text{Fill}}, i = 1, ..., n \\
&I_{j,b}^{\text{Fill}} && \forall b \in \mathcal{B}, j \in \mathcal{J}^{\text{Fill}}
\end{aligned}
$$

For all of the aforementioned activities, the domains of the start and end of the interval variables have to be defined and are initialised as: $\text{Start}(I) = 0, \text{End}(I) = t_{Hor}$.

Except for the intermediate and buffer tank storage times, the duration of the activities is given a priori by the product-dependent charging, discharging and processing rates of the equipment.

3.2   Constraints

The constraints (2) enforce that exactly one charging and mixing activity is present for each batch. Furthermore, at most one discharging operation from the formulation tanks to the filling stations and one storing and emptying operation for the buffer tanks is present. These storing and emptying operations only occur in pairs on the buffer tanks. Since discharging the content of a formulation tank into a buffer tank takes much less time than one discrete time interval, these transfers are not considered via activities and hence there might be less discharging operations from the formulation tanks than charging and mixing operations.

$$\sum_{I \in I_b^l} I.\text{isPresent} = 1, \forall b \in \mathcal{B}, \forall l \in \{\text{Charge, Mix}\},$$

$$\sum_{I \in I_b^l} I.\text{isPresent} \leq 1, \forall b \in \mathcal{B}, \forall l \in \{\text{Dis, Stor, Empt}\} \tag{2}$$

$$I_{j,b}^{\text{Stor}}.\text{isPresent} = I_{j,b}^{\text{Empt}}.\text{isPresent}, \forall b \in \mathcal{B}, j \in \mathcal{J}^{St}$$

Since discharging from a formulation tank or buffer tank into a filling station takes place synchronously, equal start and end times of discharging or filling are enforced via constraints (3), depending on whether the filling stations are fed from the formulation tanks directly or from the buffer tanks:

$$(I_{j',b}^l.\text{isPresent} = 1 \wedge I_{j,b}^{\text{Fill}}.\text{isPresent} = 1) \implies \left\{ \begin{array}{l} \text{Start}(I_{j',b}^l) = \text{Start}(I_{j,b}^{\text{Fill}}) \\ \text{End}(I_{j',b}^l) = \text{End}(I_{j,b}^{\text{Fill}}) \end{array} \right\}$$

$$\forall b \in \mathcal{B}, j \in \mathcal{J}^{\text{Fill}}, (l, j') \in \{\{\text{Empt}\} \times \mathcal{J}^{St}\} \cup \{\{\text{Dis}\} \times \mathcal{J}^T\} \tag{3}$$

Constraints (4) ensure, that the basic sequences within the formulation tanks are satisfied. This means, that the discharging, storing and filling operations have to take place after the mixing operations in the formulation tank have ended, and that at most the maximal intermediate storage time t_{inter} elapses before discharging from the formulation tank to the filling station takes place. Moreover, the end of the charging operation is synchronized with the start of the mixing operation in the formulation tanks.

$$\begin{aligned} \text{End}(I_{j,b}^{\text{Mix}}) &\leq \text{Start}(I_{j',b}^{\text{Stor}}) \; \forall b \in \mathcal{B}, j \in \mathcal{J}^T, j' \in \mathcal{J}^{St} \\ \text{End}(I_{j,b}^{\text{Mix}}) &\leq \text{Start}(I_{j',b}^{\text{Fill}}) \; \forall b \in \mathcal{B}, j \in \mathcal{J}^T, j' \in \mathcal{J}^{\text{Fill}} \\ \text{End}(I_{j,b}^{\text{Mix}}) + t_{inter} &\geq \text{Start}(I_{j,b}^{\text{Dis}}) \; \forall b \in \mathcal{B}, j \in \mathcal{J}^T \\ \text{Start}(I_{j,b}^{\text{Dis}}) &\geq \text{End}(I_{j,b}^{\text{Mix}}) \; \forall b \in \mathcal{B}, j \in \mathcal{J}^T \\ \text{End}(I_{j,b}^{\text{Charge}}) &= \text{Start}(I_{j,b}^{\text{Mix}}) \; \forall b \in \mathcal{B}, j \in \mathcal{J}^T \end{aligned} \tag{4}$$

The disjunctive *NoOverlap*-constraint enables a straightforward formulation of the allocation of resources only in mutually exclusive time periods. In this model it is used to model that a single pre-processing line can only supply one formulation tank at a time during the charging operation on one of the formulation lines in $\mathcal{J}^{\text{Form}}$. Similarly, the interval variables are constrained to not overlap for all operations that can be performed in a formulation tank or on a filling station, to ensure that only one batch is processed at a time on a unit. Beyond that, it is not possible to store intermediates of different types $p, p' \in \mathcal{P}^{inter}$ in the buffer tanks. $\mathcal{I}_{p,j}^{St}$ denotes the set of storage activities associated with intermediate p on unit j and $\mathcal{I}_j$ denotes all activities that can be performed on unit j. As the non-working periods for the filling stations are represented by the interval variables $I \in \mathcal{I}^{\text{Shift}}$, the tasks on the filling stations are also constrained to not overlap with these.

$$\text{NoOverlap}\left(\left\{\mathcal{I}_j^{\text{Charge}}|j \in \mathcal{J}_{fl}^T\right\}\right) \quad \forall fl \in \mathcal{J}^{\text{Form}}$$
$$\text{NoOverlap}(\mathcal{I}_j) \qquad\qquad\qquad \forall j \in \mathcal{J}^T$$
$$\text{NoOverlap}(\{I, I'\}) \qquad\qquad \forall j \in \mathcal{J}^{\text{St}}, \forall I \in \mathcal{I}_{p,j}^{\text{St}}, I' \in \mathcal{I}_{p',j}^{\text{St}}, \qquad (5)$$
$$p \neq p'$$
$$\text{NoOverlap}(\mathcal{I}_j \cup \mathcal{I}^{\text{Shift}}) \qquad \forall j \in \mathcal{J}^{\text{Fill}}$$

Because of the additional efforts associated with changing the connectivity, the filling of a batch only takes place on a single unit. For this purpose the Boolean variable $V_{b,j}^{\text{Fill}}$ is used to indicate that a filling activity (and all related intervals) of batch b are assigned to filling station j. The corresponding constraints are given by Eqs. (6):

$$\sum_{i\in\{1,\dots,n\}} I_{j,b,i}^{\text{Fillstep}}.\text{isPresent} \geq 1 \implies V_{b,j} \quad \forall j \in \mathcal{J}^{\text{Fill}}, b \in \mathcal{B}$$
$$V_{b,j} \implies \neg V_{b,j'} \quad \forall j \in \mathcal{J}^{\text{Fill}}, j' \in \mathcal{J}^{\text{Fill}} \setminus j, b \in \mathcal{B} \qquad (6)$$

For the buffer tanks, a cumulative constraint, given by Eq. (7), is defined. The amount of stored material must not exceed the capacity of the buffer tank at any given time point. The first argument of the *AddCumulative* constraint specifies a set of interval variables, the second argument specifies the corresponding contributions to the filling, i.e. the individual batchsizes C_b, and the third argument specifies the maximum capacity of the unit C_j. Due to the *NoOverlap* constraints on the buffer tank, it is avoided, that different intermediates are stored in the same buffer tank at the same time.

$$\text{AddCumulative}(\mathcal{I}_j, \{C_{b=\mathcal{B}_I}|\forall I \in \mathcal{I}_j\}, C_j) \quad \forall j \in \mathcal{J}^{\text{St}} \qquad (7)$$

To ensure that the order demands are met by the filling stations, constraint (8) is introduced, where $C_{b,i}$ represents the amount filled in a single, discretized step of a filling operation. Due to the errors introduced by the discretization of the scheduling horizon and the fixed batch sizes, constraining the amount to the exact batch size is not possible. $\mathcal{I}_b^{\text{Fillstep}}$ denotes the set of discretized filling intervals of batch b.

$$\sum_{i=1}^{|\mathcal{I}_b^{\text{Fillstep}}|} \sum_{j\in\mathcal{J}^{\text{Fill}}} I_{j,b,i}^{\text{Fillstep}}.\text{isPresent} \cdot C_{b,i} \geq C_b \quad \forall b \in \mathcal{B} \qquad (8)$$

Changeovers, as described in Sect. 2 occur in the following units:

(i) in the pre-processing feed,
(ii) in the formulation tanks,
(iii) in the filling stations,
(iv) and in the buffer tanks.

The duration of changeovers depend on where they occur (pre-processing line, standardization tank, filling station or buffer tank) and which two tasks are adjacent to the changeover, in particular whether the two tasks process batches of the same or of different (intermediate) products.

We handle the first dependency by defining the set of constraints for each set of units separately. The second dependency requires to know the sequence of the tasks.

To model the changeovers, Boolean variables are used to define a graph $G = (N, E)$ formed by all activities on a unit (or line). The set of nodes N consists of all activities considered for the changeover plus a start- and endnode. The set of edges E contains all pairwise combinations of all nodes, represented by the Boolean variables. A path through a graph defines a sequence of tasks on a unit. To trace these sequences, the variable $L_{II'j}$ is introduced, which is 1, if activity I is preceding I', on unit j. An edge from a node to itself can be used to exclude an activity from the graph. This accounts for batches that are not produced on a unit in the schedule or batches that do not affect the changeover times. A circuit constraint defines the directional graph in the model and enforces that each node is visited exactly once (forming a Hamiltonian path). Equation (9) contains the circuit constraints and Eq. (10) contains the constraints necessary to ensure that activities in the path are part of the schedule. If an edge is part of the path, the two corresponding tasks/nodes are constrained to be present in the schedule and a minimum waiting duration between their start- and endtime is enforced.

$$\begin{aligned} &\text{CircuitConstraint}(G_{\text{Preprocess}}) \\ &\text{CircuitConstraint}(G_{\text{Formulation tanks}}) \\ &\text{CircuitConstraint}(G_{\text{filling stations}}) \\ &\text{CircuitConstraint}(G_{\text{Buffer tanks}}) \end{aligned} \tag{9}$$

The path allows the solver to trace the relative position of activities. This is used to check if a solution candidate complies with the changeover constraints as given by Eq. (11):

$$L_{II'j} \implies \left\{ \begin{matrix} I.\text{isPresent} \\ I'.\text{isPresent} \end{matrix} \right\}$$
$$\forall i \in \mathcal{I}_j^l, j \in \mathcal{J}, l \in \{\text{Span, Charge, Fill, Stor, Empt}\} \tag{10}$$

$$L_{II'j} \implies \text{End}(I) + co_{II'j} \leq \text{Start}(I') \quad \forall I \in \mathcal{I}_j^l,$$
$$I' \in \mathcal{I}_j^l, j \in \mathcal{J}, l \in \{\text{Span, Charge, Fill, Stor, Empt}\} \tag{11}$$

Changeovers for cleaning activities on a formulation tank take place between the end of the processing activity and the start of the charging activity of a subsequent batch. The last processing activity can either be the mixing step or the discharging step. This depends on whether the intermediate storage is transferred to a storage tank or filling station. In the buffer tanks changeovers take place between the end of a discharging operation and the start of the subsequent filling and storing operation. To consider these changeovers, additional spanning interval variables $I_j^{\text{Span}} \in \mathcal{I}^{\text{Span}}$ are necessary, which are considered in the circuit constraint (9) instead of the individual activities. They are given by constraints (12):

$$
\begin{aligned}
\text{Start}(I_j^{\text{Span}}) &= \text{Start}(I_j^{\text{Charge}}) && \forall I \in \mathcal{I}_j^T, j \in \mathcal{J}^T \\
\text{End}(I_j^{\text{Span}}) &= \text{End}(I_j^{\text{Mix}}) && \forall I \in \mathcal{I}_j^T, j \in \mathcal{J}^T \\
\text{Start}(I_j^{\text{Span}}) &= \text{Start}(I_j^{\text{Stor}}) && \forall I \in \mathcal{I}_j^T, j \in \mathcal{J}^{\text{St}} \\
\text{End}(I_j^{\text{Span}}) &= \text{End}(I_j^{\text{Empt}}) && \forall I \in \mathcal{I}_j^T, j \in \mathcal{J}^{\text{St}}
\end{aligned}
\tag{12}
$$

A special case are the changeovers within the buffer tanks. In contrast to the other units, they can hold multiple batches of the same order at the same time. A changeover in the buffer tanks only occurs, when a new product is stored. This changeover happens between the end of removing the last batch from a buffer tank, and the start of filling a new batch of a different product in the same buffer tank. Because batches can arrive and leave the buffer that belong to different orders, storage intervals may partially or fully overlap each other. To take this into account additional constraints are necessary. For each pair of storage activities, a Boolean variable $L_{II'j}^{overlap}$ has been defined, which is *True* if the activity overlaps with another activity. The constraints are given by Eq. (13). They are a set of implications, that ensure that changeovers between activities which are partially or completely overlapped by other activities are considered correctly, or excluded from the Hamiltonian path.

$$
\begin{aligned}
L_{II'j}^{overlap} &\implies \text{Start}(I) \geq \text{Start}(I') \ \forall I \in \mathcal{I}_j^{\text{St}}, I' \in \mathcal{I}_j^{\text{St}}, j \in \mathcal{J}^{\text{St}} \\
L_{II'j}^{overlap} &\implies \text{End}(I) < \text{End}(I') \quad \forall I \in \mathcal{I}_j^{\text{St}}, I' \in \mathcal{I}_j^{\text{St}}, j \in \mathcal{J}^{\text{St}} \\
L_{II'j} &\implies \text{Start}(I') \geq \text{Start}(I) \ \forall I \in \mathcal{I}_j^{\text{St}}, I' \in \mathcal{I}_j^{\text{St}}, j \in \mathcal{J}^{\text{St}} \\
L_{II'j} &\implies \text{End}(I') > \text{End}(I) \quad \forall I \in \mathcal{I}_j^{\text{St}}, I' \in \mathcal{I}_j^{\text{St}}, j \in \mathcal{J}^{\text{St}}
\end{aligned}
\tag{13}
$$

The objectives makespan F_{MK} and completion time F_{CT} are defined as given by Eq. (14):

$$
\begin{aligned}
F_{MK} &= \max_{I \in \mathcal{I}} \text{End}(I) \\
F_{CT} &= \sum_{I \in (\mathcal{I}^T \cup \mathcal{I}^{\text{Fill}})} \text{End}(I)
\end{aligned}
\tag{14}
$$

Further constraints have been added to remove symmetries. Constraints in Eq. (15) greatly improve the performance of the search process, as the start time, end time and duration variables of activities on units, that are not used by the activity, would otherwise be branched or propagated on.

$$
\neg I.\text{isPresent} \implies \left\{
\begin{array}{ll}
\text{Start}(I) &= 0 \quad \forall I \in \mathcal{I} \\
\text{End}(I) &= 0 \quad \forall I \in \mathcal{I} \\
\text{duration}(I) &= 0 \quad \forall I \in \mathcal{I}
\end{array}
\right\}
\tag{15}
$$

Batches of the same product on the same line are indistinguishable, yet their permutations increase the search space, as they are considered separate entities by the solver. To remove the symmetries, all batches of the same product are constrained to be scheduled in order (if present) on the formulation lines, as given by Eq. (16). The discrete filling steps of a single batch on the filling stations are treated analogously.

$$\text{Start}(I_{j,b-1}^{\text{Charge}}) \leq \text{Start}(I_{j,b}^{\text{Charge}}) \quad \forall b = 2, \dots, |B_o|, j \in \mathcal{J}^{Fill}, o \in \mathcal{O} \tag{16}$$

3.3 Moving Horizon Strategy

The complexity of the scheduling problem in terms of the interactions between the decisions, as well as its sheer scale, in terms of number of operations to be scheduled, necessitates the use of a decomposition algorithm to obtain optimised schedules in a reasonable amount of time.

In this work a batch-based decomposition strategy is chosen, since a decomposition based on complete orders would lead to prohibitively large subproblems. The complete set of batches $\mathcal{B}$ is evenly decomposed into smaller subsets $\mathcal{B}^k$ that are scheduled in iteration k of the solution algorithm. A final schedule is obtained by concatenation of the partial solutions that are computed in the iterations k. The number of batches that are scheduled in each iteration is chosen by balancing the solution time against the schedule quality.

The MHS is displayed in Algorithm 1. For each subset $\mathcal{B}^k$ a set of corresponding interval variables $\mathcal{I}^k$ is generated.

In every iteration, a two-step solution procedure is employed. The first step is a makespan optimization by solving the model $\pi(F_{MK})$, which can be solved computationally efficiently in the CP setting. This is followed by an optimization of the total completion time $\pi(F_{CT})$ in which the makespan is bounded by the previously obtained value by adding the constraints $F_{MK} \leq F_{MK}^*$. A solution hint (initial solution) $\mathcal{I}^{k,*}$ that is equal to the solution that was obtained in the makespan optimization step is passed to the completion time optimization step.

It is necessary to optimise both objectives sequentially because of makespan-equivalent subsolutions, that are detrimental to the overall solution. When omitting the completion time optimization, units may show periods of inactivity between activities. Constraint propagation is known to perform better on max-type objectives, such as makespan, than on sum-type objectives such as completion time [3]. Therefore the preliminary makespan optimization speeds up the solution procedure noticeably. This is the reason why a sequential optimization was chosen over an objective that consists of a weighted sum of the makespan and the total completion time.

The decomposition algorithm has been implemented in a moving horizon fashion. In each iteration, a number of batches is scheduled, but the batches for which the completion times on the filling stations are the largest are removed from the solution (Algorithm 1, lines 14–21). Only the allocation and timing decisions for the remaining IH batches are fixed on the formulation tanks, filling stations, and buffer tanks. In this strategy IH refers to the number of batches in the implementation horizon and PH to the number of batches in the prediction horizon. The removed $(PH - IH)$ batches are added to the set of batches to be scheduled in the next iteration. This algorithm design aims to decrease the myopia of scheduling only a subset of orders in each iteration, as Fig. 2 demonstrates. On the left of Fig. 2, four iterations of the solution strategy are displayed, where $IH = PH$. This leads to a large changeover time between I_1 and I_2 being incurred in the second iteration. However, when the prediction horizon is increased to $PH = 3$, a better sequence is found. While unnecessarily large changeover times are avoided, an optimization with $|\mathcal{I}^k| = 4$ did not need to be solved, which leads to a decrease in computation times.

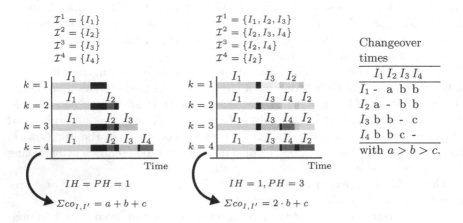

Fig. 2. Illustration of how the moving horizon Strategy can reduce the myopia of order-based decomposition approaches on a single unit.

4   Results

The results section is structured as follows: First results for a case with 78 batches are presented to demonstrate, that reductions in the computational time have been achieved, while the solution quality is equal to the one presented in [14]. Secondly, the influence of different lengths of prediction and implementation horizons is investigated on a set of larger cases with 253 batches each.

All computations have been conducted on a FUJITSU ESPRIMO P920 Workstation with a 4-core Intel i7 processor (3.40 GHz) and 32 GB of RAM. The considered formulation plant consists of 7 formulation lines, each with 3 standardization tanks, 8 filling stations, and 5 buffer tanks.

In Fig. 3 an illustrative schedule for the case study with 78 batches, corresponding to the production of about one week, is shown. Colored, horizontal bars represent operations. Sets of operations that belong to one batch, i.e. charging, mixing, discharging (Formulation tanks) or discretized filling (Filling stations) or filling and discharging (Buffer tanks) are separated by vertical, black lines.

Both the formulation and the filling stations exhibit periods of underutilization, e.g. FL5, FS5 and FS6. Therefore, neither the formulation nor the filling stage is the clear bottleneck stage. The utilization of a line heavily depends on its flexibility, i.e. the set of orders that can be produced on a certain formulation

Algorithm 1. Moving horizon strategy

1: $K = \lceil |\mathcal{B}|/N \rceil$ ▷ Determine number of iterations
2: $\mathcal{I}^{1,*}, F^*_{MK} = \pi(F_{MK})$ ▷ Solve CP min. MK
3: $\mathcal{I}^{1,*}, F^*_{CT} = \pi(F_{CT})$ ▷ Solve CP min. CT
4: **for** $k = 2, \ldots, K$ **do**
5:      **for all** $j \in \mathcal{J}$ **do**   ▷ Define horizon shift constraints
6:          $EST = \max_{I \in \mathcal{I}^{k-1,*}_j} \text{End}(I)$ ▷ Compute Earliest Starting Time
7:          $I' = \{I \in \mathcal{I}^{k-1,*}_j | \text{End}(I) = EST\}$
8:          $\mathcal{D}(\text{Start}(I)) = \{\text{End}(I) + co_{I'Ij}, \ldots, |T_H|\} \; \forall I \in \mathcal{I}^k_j$
9:                                    ▷ Reduce domain of $Start(I)$
10:      **end for**
11:      $\mathcal{I}^{k,*}, F^*_{MK} = \pi(F_{MK})$
12:      $\mathcal{I}^{k,*}, F^*_{CT} = \pi(F_{CT}, \mathcal{I}^{k,*}, F_{MK} \leq F^*_{MK})$
13:                 ▷ Solve with solution hint $\mathcal{I}^{k,*}$ and MK constraint
14:      **for** $k = 1, \ldots, (PH - IH)$ **do**   ▷ Keep IH batches
15:          $I^{k,*}_{last} = \{I \mid \text{End}(I) = \max_{I \in \mathcal{I}^{k,*}_j, j \in \mathcal{J}^{\text{Fill}}} \text{End}(I)\}$ ▷ Retrieve last activity
16:          **for all** $b \in \mathcal{B}^k$ **do**
17:              **if** $I^{k,*}_{last} \in \mathcal{I}^{k,*}_b$ **then**   ▷ Check if the last activity belongs to batch b
18:                 $\mathcal{I}^{k,*} = \mathcal{I}^{k,*} \setminus \mathcal{I}^{k,*}_b$ ▷ Remove all activities of that batch
19:              **end if**
20:          **end for**
21:      **end for**
22:      $\mathcal{I}^* = \mathcal{I}^* \cup \mathcal{I}^{k,*}$ ▷ Append to final solution
23: **end for**

or filling station. Overall a good balancing of the loads of the processing lines is achieved with the exception of FS7, but the orders that are processed on this line cannot be processed elsewhere.

The makespan of this exemplary schedule is 133 h. This is the same value that was reported in [14] for the same set of orders. The schedule was obtained within 23 min, with $IH = 5$ and $PH = 5$. In [14] up to 5 batches were scheduled per iteration, which took 38 min in total. This is reduction in computational time by 40%. Beyond that, an optimality gap of 5% was used in each iteration, while in our approach each iteration is solved to optimality.

A measure to improve the quality of solutions when a decomposition approach is used, is to sort the list of orders prior to scheduling. Sorting for a criterion that reflects the flexibility of an order, similar to the algorithm proposed in [13], could reduce the gaps. However, the list of orders already represents priorities of the demand satisfaction and therefore no presorting was implemented.

In order to evaluate the benefits of the MHS compared to a temporal decomposition that fixes all operations in a single iteration, i.e. where $IH = PH$, a set of 10 random instances was generated and optimised with different values of IH and PH. All random instances consist of 253 batches, but the types of intermediate and final products are randomised. This randomisation of the types

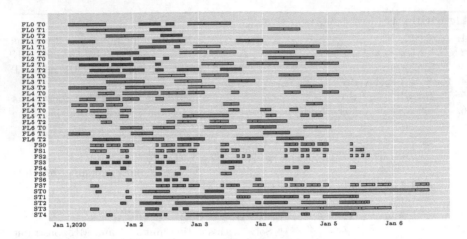

Fig. 3. Illustrative schedule for the case study with 78 batches, IH = 5 and PH = 5. One colour may include multiple orders.

Table 1. Results of the MHS for a set of 10 randomised instances with 253 batches each for varying implementation and prediction horizons.

IH	PH	PH/IH	Runtime [s]	Av. Makespan [h]	Av. Completion time [h]
1	5	5	14550	404	12811
2	5	2.5	8171	399	12823
3	5	1.67	5702	405	12922
5	5	1	3359	399	12955
1	3	3	1799	403	12852
2	3	1.5	1430	408	12978
3	3	1	801	396	13131
1	2	2	1127	409	13020
2	2	1	622	400	13232
1	1	1	549	404	13334

of intermediate and final product impacts the changeover times, the processing times, and which units are eligible for the batches of an order.

In Table 1 the average runtime, average makespan and average completion time results for the 10 instances are given. For the completion time it can be seen, that it tends to decrease with higher values for the ratio PH/IH. However, for the makespan no clear trend can be identified. This might be due to the symmetries implied by the makespan objective which are difficult to account for across the different subproblems. Although better objective functions are achieved with an increasing PH compared to IH, a trade off between the improvements in solution quality against the increased computation times has to be made.

5 Summary, Conclusion and Outlook

In this work a CP model and a moving horizon decomposition strategy were developed to solve a large-scale scheduling problem for a real-life formulation plant. Noteworthy features of the scheduling problem are the necessity to model the allocation and the capacity utilization of the buffer tanks explicitly, the continuous operating mode of the filling stations as well as the non-identity of the processing units in terms of processing rates and eligibility. The MHS that was developed decomposes the scheduling problem into manageable subproblems, while alleviating the limitations of simple order-based decomposition schemes. As a result, acceptable solution times, especially when considering the problem size, have been obtained. Lower solution times than with a MILP approach were achieved for the same solution quality.

Soft metrics are also worth to be discussed. A special challenge in modeling the plant operations of the case study under consideration, for a MILP formulation, is the representation of the continuously operated filling stations. A discretization of the processing time was necessary to represent the variable filling time and to model a set of independent filling operations. In [14] a discrete-time model was suggested because of the presence of mass balances, that leads to variables with up to four indices (three for order, unit and timepoint and one for the processing time discretization). Considering model design and maintenance, CP requires less expert knowledge to represent such features, as interval variables inherently enable for variable durations and mass balances can be easily represented with cumulative constraints. This results in a better overall comprehensibility of the CP model compared to MILP models in this case.

Taking into account the possibility of employing high-level formulations in CP models which improves the ease of application and the evidence that has recently accumulated that CP outperforms MILP formulations in many combinatorial optimization problems [4,5,11], it can be said that CP has received too little attention for practical applications yet. A further practical consideration is that the software suite OR-tools and its CP-SAT solver currently are open source products, while the performance of MILP-based approaches heavily relies on the use of commercial solvers, such as Gurobi or CPLEX.

Future work will include the modelling of the up- and downstream logistics as well as a more detailed scheduling of personnel, because both aspects can lead to bottlenecks in the production process under consideration. Furthermore, a sorting algorithm for the list of orders, that still considers for the fixed priorities of certain orders, will be developed to investigate if the solution quality can be improved further.

References

1. Elkamel, A., Zentner, M., Pekny, J.F., Reklaitis, G.V.: A decomposition heuristic for scheduling the general batch chemical plant. Eng. Optim. **28**(4), 299–330 (1997). https://doi.org/10.1080/03052159708941137

2. Harjunkoski, I., et al.: Scope for industrial applications of production scheduling models and solution methods. Comput. Chem. Eng. **62**, 161–193 (2014). https://doi.org/10.1016/j.compchemeng.2013.12.001
3. Kovács, A., Beck, J.C.: A global constraint for total weighted completion time for unary resources. Constraints **16**(1), 100–123 (2011). https://doi.org/10.1007/s10601-009-9088-x
4. Ku, W.Y., Beck, J.C.: Mixed integer programming models for job shop scheduling: a computational analysis. Comput. Oper. Res. **73**, 165–173 (2016). https://doi.org/10.1016/j.cor.2016.04.006
5. Laborie, P.: An update on the comparison of MIP, CP and hybrid approaches for mixed resource allocation and scheduling. In: van Hoeve, W.-J. (ed.) CPAIOR 2018. LNCS, vol. 10848, pp. 403–411. Springer, Cham (2018). https://doi.org/10.1007/978-3-319-93031-2_29
6. Lenstra, J.K., Rinnooy Kan, A.H., Brucker, P.: Complexity of machine scheduling problems. Ann. Disc. Math. **1**(C), 343–362 (1977). https://doi.org/10.1016/S0167-5060(08)70743-X
7. Leung, J.Y.: Handbook of Scheduling: Algorithms, Models, and Performance Analysis. CRC Press (2004)
8. Méndez, C.A., Cerdá, J., Grossmann, I.E., Harjunkoski, I., Fahl, M.: State-of-the-art review of optimization methods for short-term scheduling of batch processes. Comput. Chem. Eng. **30**(6–7), 913–946 (2006). https://doi.org/10.1016/j.compchemeng.2006.02.008
9. Perron, L., Furnon, V.: Or-tools https://developers.google.com/optimization/
10. Simonis, H., Cornelissens, T.: Modelling producer/consumer constraints. In: Montanari, U., Rossi, F. (eds.) CP 1995. LNCS, vol. 976, pp. 449–462. Springer, Heidelberg (1995). https://doi.org/10.1007/3-540-60299-2_27
11. Wari, E., Zhu, W.: A constraint programming model for food processing industry: a case for an ice cream processing facility. Int. J. Prod. Res. **57**(21), 6648–6664 (2019). https://doi.org/10.1080/00207543.2019.1571250
12. Wu, D., Ierapetritou, M.G.: Decomposition approaches for the efficient solution of short-term scheduling problems. Comput. Chem. Eng. **27**(8–9), 1261–1276 (2003). https://doi.org/10.1016/S0098-1354(03)00051-6
13. Yfantis, V., Corominas, F., Engell, S.: Scheduling of a consumer goods production plant with intermediate buffer by decomposition and mixed-integer linear programming. IFAC-PapersOnLine **52**(13), 1837–1842 (2019). https://doi.org/10.1016/j.ifacol.2019.11.469
14. Yfantis, V., Siwczyk, T., Lampe, M., Kloye, N., Remelhe, M., Engell, S.: Iterative medium-term production scheduling of an industrial formulation plant. Comput. Aided Chem. Eng. **46**, 19–24 (2019). https://doi.org/10.1016/B978-0-12-818634-3.50004-7

The Traveling Social Golfer Problem: The Case of the Volleyball Nations League

Roel Lambers, Laurent Rothuizen, and Frits C. R. Spieksma[✉]

Eindhoven University of Technology, Eindhoven, The Netherlands
f.c.r.spieksma@tue.nl

Abstract. The Volleyball Nations League is the elite annual international competition within volleyball, with the sixteen best nations per gender contesting the trophy in a tournament that spans over 6 weeks. The first five weeks contain a single round robin tournament, where matches are played in different venues across the globe. As a result of this setup, there is a large discrepancy between the travel burdens of meeting teams, which is a disadvantage for the teams that have to travel a lot. We analyse this problem, and find that it is related to the well-known Social Golfer Problem. We propose a decomposition approach for the resulting optimization problem, leading to the so-called Venue Assignment Problem. Using integer programming methods, we find, for real-life instances, the fairest schedules with respect to the difference in travel distance.

Keywords: Social Golfer problem · Volleyball Nations League · Integer programming

Prologue

It is the beginning of June 2018 when the Italian men volleyball team go undefeated in the first round of the inaugural Volleyball Nations League. They played their three first-round matches in Kraljevo, Serbia, and for the next round of three matches they have to travel, via Belgrade, Rome, Buenos Aires, to reach the next venue in San Juan, Argentina, after more than 24 h of traveling. Playing only a few days after this trip, their momentum seems lost and they lose two out of their three games, all played within a week, upon which they immediately need to fly to Japan for the third round.

Ultimately, the Italian team had to travel literally across the globe within a time span of four weeks, playing matches against the best volleyball teams in the world. Even though they started off with three victories, they ended eighth and did not qualify for the final stages. In comparison, the French team played all their matches within Europe and emerged as winner of the main event. Later that year however, during the World Championship, the Italian team outperformed the French team.

This work is based on [13].

© Springer Nature Switzerland AG 2021
P. J. Stuckey (Ed.): CPAIOR 2021, LNCS 12735, pp. 149–162, 2021.
https://doi.org/10.1007/978-3-030-78230-6_10

1   Introduction

The above example is just one of many that highlights the importance of travel times in the Volleyball Nations League (VNL). It is well established within the scientific literature that extensive travelling has a negative impact on sport performance. Although we do not intend to survey the literature on this subject, this finding is reported for various sports ranging from rugby [11] to baseball ([15], and [18]) and from basketball [9] to triathletes [16]; see also the references contained in these papers. We close this paragraph by a quote from [14] who concludes: "Jet lag and travel fatigue are considered by high-performance athletic support teams to be a substantial source of disturbance to athletes." This finding is illustrated in the prologue of this paper, and serves as its motivation.

The Volleyball Nations League is a tournament organized every year by the FIVB (Fédération Internationale de Volleyball), for both men and women (see https://www.volleyball.world/en/vnl/2021). This tournament was first organized in 2018 to replace the World League/World Grand Prix as annual volleyball tournament. Every tournament contains 16 teams, and consists of multiple phases. In the first phase, lasting for five weeks, all 16 teams play in a single round robin. The best 6 teams then qualify for the second phase, where out of two groups of three, four teams emerge to play cross finals. Our interest is exclusively on the first phase.

In the VNL tournament, teams play in rounds. In each round, each team is in a poule consisting of 4 teams, meeting all teams in their poule once. After 5 rounds, all teams have played all the other teams exactly once, and a ranking is made based on the results in this single round robin tournament. All the matches in a single poule are held at the same venue, however, every round has its 4 poules played out in different venues. As it is a disadvantage to have traveled more than your opponent going into a match, our main interest lies in minimizing a measure that captures the imbalance in travel times between opposing teams.

A priori, it is not clear how a round robin schedule that can be decomposed in poules is obtained. In fact, finding a schedule that fits this VNL-format is related to the so-called Social Golfer Problem (SGP). In this problem we are given gp golfers and w rounds, and the SGP-question is whether it is possible to let the gp golfers play in g groups of p golfers in each of the w rounds, in such a way that every pair of golfers plays in the same group in at most one round, see [4,10,17]. This question is far from innocent: only for restricted sets of values for g, p, w the answer to this question is known. For instance, when $g = p = w - 1$, solutions are known to exist when g is a prime power - and no other solution to these type of instances has been found, nor has it been proven that these are the only instances for which a solution can exist [8].

Of course, in the context of the Volleyball Nations League, each golfer corresponds to a team, and a group corresponds to a poule. Since the Volleyball Nations League has $g = p = 4$ and $w = 5$, it follows that the answer to the SGP-question is affirmative, and hence a schedule for the VNL that consists of 5 rounds, each round consisting of 4 poules, is known to exist. In this paper, we introduce the *Traveling Social Golfer Problem* (TSGP), as a generalization of the SGP; the TSGP allows us to take fairness, as measured by the difference in travel times between opposing teams, into account.

Another well-known problem related to the scheduling problem in the Volleyball Nations League is the Travelling Tournament Problem (TTP), see [6] for a precise description. In contrast to our problem, in the TTP pairs of teams meet in the venue of one of the two opposing teams. Moreover, the objective in the TTP is to minimize total travel distance; difference in travel time between opposing teams is not considered in the TTP. We refer to [7] and [5] for an overview concerning the TTP.

A number of studies has been devoted to the scheduling of national volleyball leagues where mainly for cost reasons, the objective is to minimize total travel time. We mention [2] who model the Argentine national volleyball league as an instance of the Traveling Tournament Problem, and [3] who investigate the Italian volleyball league. Further, [12] study the Norwegian Volleyball League; one of their models, motivated by a cost-objective, is devoted to minimizing total travel distance in that league. These leagues are organized in the format of a Double Round Robin, and as such differ from the VNL.

2   The Traveling Social Golfer Problem (TSGP)

2.1   Definition of the TSGP

As described in Sect. 1, the SOCIAL GOLFER PROBLEM is a well known combinatorial question, where the task is to schedule golfers in groups of size p over multiple rounds, such that no golfer plays with another golfer in the same group twice. In the TRAVELING SOCIAL GOLFER PROBLEM (TSGP), all groups have to play at (different) venues, where the objective is to create a schedule that minimizes the *unfairness* arising from golfers having different travel times between the venues.

In order to give a precise formulation of the TSGP, we use the following notation to describe the input:

- N: the number of teams,
- k: a poule size,
- V: the set of venues,
- $d(v, w)$: a distance between each pair of venues $v, w \in V$, and
- c_v: a multiplicity for each $v \in V$.

The multiplicities c_v indicate the exact number of times venue $v \in V$ must host a poule; indeed, in the practical situation of the VNL, it is not uncommon that a venue is host to different poules in different rounds. The multiplicities allow us to accommodate such situations.

Furthermore, we use the following notation to describe a solution:

- R: a set of rounds,
- P_i^r: the set of teams in poule i in round r, $1 \le i \le \frac{N}{k}$, $r \in R$,
- $v_r(t)$: the venue of the poule in which team $t \in \{1, \ldots, N\}$ plays in round $r \in R$.

Finally, we measure the value of a schedule S by its unfairness $u(S)$ as follows:

$$u(S) = \sum_{r \in R \setminus \{1\}} \sum_{i=1}^{\frac{N}{k}} \max_{s,t \in P_i^r} |d(v_r(s), v_{r-1}(s)) - d(v_r(t), v_{r-1}(t))|. \tag{1}$$

Thus, for every poule P_i^r in every round $r \in R \setminus \{1\}$, we consider the two teams whose difference in travel distance needed to arrive at the corresponding venue, is maximum over all pairs of teams in the poule; this quantity is summed over all poules, and all rounds (except the first round, as we assume that all teams have ample time to arrive at their first venue). Thus, a lower value of $u(S)$ indicates that the difference in travel times between opposing teams was less and thus the schedule was more fair. The measure u is applicable to any schedule for N teams that has a poule/round-structure.

Example 1. A tournament with $N = 4$, teams $1, \ldots, 4$, is organized over three rounds, and poules of size $k = 2$. There are four venues, $V = \{A, B, C, D\}$, with $c_A = c_D = 2$ and $c_B = c_C = 1$. Distances between venues are $d(A, B) = d(A, C) = d(B, D) = d(C, D) = 1$ and $d(A, D) = d(B, C) = 2$.

The following schedule S with poules P_i^r and venues v is used:

$$P_1^1 = \{1, 2\}, \quad v = A \qquad\qquad P_2^1 = \{3, 4\}, \quad v = D$$
$$P_1^2 = \{1, 3\}, \quad v = A \qquad\qquad P_2^2 = \{2, 4\}, \quad v = B$$
$$P_1^3 = \{1, 4\}, \quad v = D \qquad\qquad P_2^3 = \{2, 3\}, \quad v = C$$

Thus:

$$u(S) = |2 - 0| + |1 - 1| + |2 - 1| + |2 - 1| = 4$$

We state the following optimization problem that we call the (N, k)-Traveling Social Golfer Problem, or (N, k)-TSGP for short.

Problem 1. (N, k)-TSGP
Input: A number of teams $N \in \mathbb{N}$, a poule size $k \in \mathbb{N}$, a set of venues V each with multiplicity c_v ($v \in V$), and a distance function $d : V \times V \to \mathbb{R}$.
Output. A schedule S consisting of $|R|$ rounds minimizing $u(S)$ such that:

- there is an equi-partitioning of N teams in poules $P_1^r, \ldots P_{\frac{N}{k}}^r$ for each round $r \in R$, with for each pair of distinct teams s, t, $\exists!$ i, r with $s, t \in P_i^r$,
- an allocation of poules to venues that results in venues $v_r(t)$ ($r \in R$, $t = 1, \ldots, N$) such that venue $v \in V$ acts c_v times as a host for a poule.

It is clear that, depending on the input, a feasible schedule to (N, k)-TSGP need not exist; in fact, it is not difficult to find instances where there is no schedule S that satisfies all the constraints. Indeed, as the schedule asks for a partitioning of the N teams in poules of size k in each round, we immediately see that N should be a multiple of k, or $N \equiv_k 0$. In addition, as the schedule should correspond to a single round robin tournament, and as all teams play $k - 1$ matches per round, we conclude that $N - 1$ should be a multiple of $k - 1$, or $(N - 1) \equiv_{k-1} 0$. Thus, a solution of the (N, k)-TSGP can only exist if there is an integral ρ such that $N = k \cdot ((k - 1)\rho + 1)$.

The above are necessary conditions that need to be satisfied. In fact, the (N, k)-TSGP can only have a solution that satisfies the single round robin format, if the corresponding instance of the SGP is solvable. In general, solutions of the SGP are known to exist when $N = k^2$ and k is a prime power. Thus, for the Volleyball Nations League, the underlying $N = 16, k = 4$-SGP problem will be solvable.

Remark. For the rest of the paper, we will assume that $N = k^2$ and $|R| = k + 1$.

2.2 Decomposing the TSGP into Venue Assignment and Nation Assignment

The problem of solving an instance of $(N = k^2, k) - TSGP$ can be decomposed into two phases:

- Venue Assignment. In the first phase, we specify, for each round $r \in R$, which venues act as a host in each round $r \in R$. Let $U_r \subset V$, with $|U_r| = k$, $r = 1, \ldots, k + 1$ be the set of venues that act as hosts in round r.
- Nation Assignment. In the second phase, we decide upon the composition of the poules, i.e., we choose the sets P_i^r and allocate these poules to the venues in U_r, $r = 1, \ldots, k + 1$.

By going through these two phases, we find a schedule S. It is crucial to observe that the unfairness of S, i.e., $u(S)$, follows directly from the venue assignment when $N = k^2$. We record this observation formally.

Theorem 1. *For each schedule S of a given an instance of $(N = k^2, k)-TSGP$, $u(S)$ is determined only by the Venue Assignment, for each integer $k \geq 2$.*

Proof. We claim that for each schedule S:

$$u(S) = \sum_{r \in R \setminus \{1\}} \sum_{i=1}^{k} \max_{s,t \in P_i^r} |d(v_r(s), v_{r-1}(s)) - d(v_r(t), v_{r-1}(t))|$$

$$= \sum_{r \in [1,\ldots,k]} \sum_{u \in U_{r+1}} \max_{v,w \in U_r} |d(v, u) - d(w, u)|.$$

The latter equality follows from the fact that, independent of the composition of the poules, the k teams that play in a poule in some round, will not meet again in a next round, and hence these k teams will travel to each of the k distinct venues in the next round.   □

Theorem 1 allows us to compute the unfairness of a schedule S, $u(S)$, without specifying the schedule S. As a consequence, it becomes much easier in practice to find schedules for which $u(S)$ is minimum (see Sect. 5).

3   The Complexity of Venue Assignment

In this section, we formally define Venue Assignment, and establish its complexity. Given that feasible schedules to the $(N = k^2, k)$-TSGP exist, Theorem 1 implies that our task of finding an optimal solution to (N, k)-TSGP is reduced to finding an optimal venue assignment. Clearly, this is related to the differences in traveled distance between two opposing teams, which in turn follows from the venues that are selected in each round.

In an extreme case, if only a single venue v is given (with multiplicity $c_v = k(k + 1)$), then all matches in all poules in all rounds are played in the same venue, and there is no travel distance. However, in general, the set of venues V and their pairwise distances, are instrumental in finding good venue assignments. Of course, we assume that $\sum_{v \in V} c_v = k(k+1)$. We now give a formal description.

Problem 2. VENUE-ASSIGNMENT (VA)
Input. A value $k \in \mathbb{N}$, a set of venues V, an integral multiplicity c_v for $v \in V$, and a distance matrix $d(v, w)$ for each $v, w \in V$.

Output. For $r \in [1, \ldots, k+1]$, subsets $U_r \subset V$ with $|U_r| = k$, such that $\forall v \in V$, $c_v = |\{r : v \in U_r\}|$ that minimizes:

$$\Delta = \sum_{r \in [1, \ldots, k]} \sum_{u \in U_{r+1}} \max_{v, w \in U_r} |d(v, u) - d(w, u)|. \tag{2}$$

To establish the hardness of VENUE-ASSIGNMENT, we use the following decision problem.

Problem 3. LONGEST HAMILTONIAN PATH ON A COMPLETE GRAPH (LHP)
Input: A complete graph $G = (H, E)$, $|H| = n$ with nonnegative, symmetric weights $w(h_1, h_2)$ for each $h_1, h_2 \in H$, and an integer B.
Question: Does there exist a Hamiltonian Path $(h_{i_1}, \ldots, h_{i_n})$ in G such that $\sum_{j=1}^{n-1} d(h_{i_j}, h_{i_{j+1}}) \geq B$?

LHP is well-known to be NP-complete.

Theorem 2. VENUE-ASSIGNMENT *is NP-Hard.*

Proof. We prove this statement by a reduction from LONGEST HAMILTONIAN PATH ON A COMPLETE GRAPH.

Given an instance of LHP, with vertex set $H = \{h_1, \ldots, h_n\}$ and weights $w : L \times L \to \mathbb{R}$, we construct an instance of the decision problem corresponding to VA, using a parameter K, in the following way.

We choose $k = n - 1$. Further, the set of venues V consists of $V = V_1 \cup V_2$, where $V_1 = H$ and $|V_2| = k - 1$. For each $v \in V_1$, $c_v = 1$ and for each $v \in V_2$, $c_v = k + 1$. Let $D = \max_{h_1, h_2 \in H} w(h_1, h_2)$ and define a distance function d in the following way:

$$d(u, v) = \begin{cases} w(u, v) & u, v \in V_1 \\ 2D & u \in V_1, v \in V_2 \\ 0 & u, v \in V_2 \end{cases} \tag{3}$$

Notice that the resulting distances satisfy the triangle inequality when the instance of LHP does. Finally, we set $K = k^2 \cdot 2D - B$, and ask whether there exists a venue assignment with unfairness at most K. We have now specified an instance of the decision version of VA.

Let us argue that if there exists a solution to VA with unfairness at most K, LHP is a yes-instance, and vice versa.

To find a solution to any instance of VA, we need to find $U_r \subset V$ for $r \in [k+1]$ such that $\forall v \in V$, $c_v = |\{r : v \in U_r\}|$. As we know that for all $v \in V_2$, $c_v = k+1$, we see that any feasible solution must have $V_2 \subset U_r$ for each r, and as $c_v = 1$ for $v \in V_1$, we get that any feasible solution must schedule every venue $v \in V_1$ exactly once. Thus, any feasible solution to VA consists of $U_r = V_2 \cup v_{i_r}$ with $v_{i_r} \in V_1$ and $v_{i_r} = v_{i_{r'}} \iff r = r'$. In other words, any feasible solution to VA corresponds to an ordering $p = (v_{i_1}, \ldots, v_{i_{k+1}})$ of the venues in V_1. Given such an ordering p we get the following expression for the unfairness:

$$K = \sum_{r=1}^{k} \left((k-1) \cdot 2D + (2D - d(v_{i_{r+1}}, v_{i_r}))\right) \tag{4}$$

The first term in the summation results from the fact that there are $k-1$ venues from V_2 in every round and one from V_1, and since $d(v, w) - d(v, v') = 2D - 0$ for all $v, v' \in V_2$, $w \in V_1$, we get $k-1$ venues where the maximal travel difference is $2D$. The second term equals the difference in travel distance between the teams traveling from any of the $v \in V_2$ to the $v_{i_{r+1}} \in V_1$, and the team traveling from $v_{i_r} \in V_1$.

We find:

$$K = \sum_{r=1}^{k} \left((k-1) \cdot 2D + (2D - d(v_{i_{r+1}}, v_{i_r}))\right) \tag{5}$$

$$= k^2 \cdot 2D - \sum_{r=1}^{k} d(v_{i_{r+1}}, v_{i_r}) \tag{6}$$

$$= k^2 \cdot 2D - B. \tag{7}$$

Thus, solving this instance of the decision version of VA equals solving the corresponding instance of LHP, which implies that VA is NP-Hard.   □

4   An Integer Programming Formulation

In this section, we give an integer programming formulation of VENUE-ASSIGNMENT. Motivated by the current practice in the VNL, we incorporate the following issue in our formulation: each venue has a team that considers this venue as its home-venue. Next, in any schedule for the VNL it must be the case that when a venue is hosting a poule, the poule must contain the team for which this venue is the home-venue. And in case there are multiple venues that are the home-venue of the same team, it is a fact that those venues are never a host of a poule in the same round. In the context of the VNL, this property allows that each venue always hosts a poule that contains the national team; this team can be regarded as the home playing team, or host nation.

Let $x_{v,r}$ be the binary variables that indicate whether venue $v \in V$ hosts a poule in round $r \in \{1, \ldots, 5\} = R$. Further, we need real variables $s_{v,w,r}$ (capturing distances between venues v and w acting as host in rounds r and $r + 1$), $m_{v,r}$ (capturing the largest distance traveled to venue v in round r), and $K_{v,r}$ (capturing the difference in travel distance to venue v in round r). Let $\Delta = \max_{v,w} d(v,w)$, and let $W \subset V \times V$ be the set of pairs of venues that cannot both host a poule in the same round. The following IP minimizes u: the sum of the difference in travel distances per poule, over the poules.

$$\min \sum_{v \in V} \sum_{r \in R} K_{v,r} \tag{8}$$

$$\text{s.t.} \ \sum_{v \in V} x_{v,r} = k \qquad\qquad \forall r \in R, \tag{9}$$

$$\sum_{r \in R} x_{v,r} = c_v \qquad\qquad \forall v \in V, \tag{10}$$

$$x_{v,r} + x_{w,r} \leq 1 \qquad\qquad \forall r \in R, \ \forall (v,w) \in W, \tag{11}$$

$$s_{v,w,r} \geq d_{v,w}(x_{v,r} + x_{w,r-1} - 1) \qquad \forall v, w \in V, \ \forall r \in R \setminus 1, \tag{12}$$

$$s_{v,w,r} \leq \min(d_{v,w}x_{v,r}, d_{v,w}x_{w,r-1}) \qquad \forall v, w \in V, \ \forall r \in R \setminus 1, \tag{13}$$

$$m_{v,r} \geq s_{v,w,r} \qquad\qquad \forall v, w \in V, \ \forall r \in R \setminus 1, \tag{14}$$

$$K_{v,r} \geq m_{v,r} - s_{v,w,r} - D(1 - x_{w,r-1}) \qquad \forall v, w \in V, \ \forall r \in R \setminus 1, \tag{15}$$

$$x_{v,r} \in \{0, 1\}, K_{v,r} \geq 0 \qquad\qquad \forall v \in V, r \in R. \tag{16}$$

Constraints (9) ensure that in every round, k venues are host; constraints (10) ensure that every venue hosts as often as required; constraints (11) ensure that two venues that should not host simultaneously, will not host simultaneously. Auxiliary variables $s_{v,w,r}$ are at least as big as $d_{v,w}$, the distance traveled between w, v between round $r - 1$ and r if the venues host in the respective rounds, by (12), but never bigger than $d_{v,w}$ by (13). The variables $m_{v,r}$ equal the maximum distances traveled to venue v in round r (can equal zero 0 if v does not host in round r), as defined by (14), and $K_{v,r}$ resembles the difference in traveled distances towards v in round r compared to the maximum travel distance, where the terms $-D \cdot (1 - x_{v',r-1})$ in (15) nullify any influence by distances between a venue that does not host in round $r - 1$.

(8) is the objective function, minimizing $u = \sum_{v,r} K_{v,r}$.

5   Solving VNL in Practice

5.1   Do Feasible Schedules Exist?

Solving the VENUE-ASSIGNMENT with the IP from the previous section, does not automatically lead to a schedule for a practical instance of the VNL. As described in Sect. 4, in the Volleyball Nations League, the venues of a poule can be considered as *home* to one of the teams in the poule. A poule scheduled to play in China, will have the Chinese team in it as *home team* (as an aside, it is interesting to note here that [1] find the presence of (a significant) home advantage in volleyball matches played in Italian and Greek national leagues). Of course, each venue is a home venue to a single team; however, a team can have multiple home venues.

It is not true that, when given an assignment of venues to rounds, a schedule is guaranteed to exist such that every venue is a home venue. Then, in such a case, a venue hosts a poule of teams, none of which plays home. Example 2 shows how a feasible solution for the VENUE-ASSIGNMENT cannot be extended to a solution of the traveling social golfer problem, with all venues being a home venue.

Example 2. Let $N = 4$ be the number of teams with poule size $k = 2$, and let V be the set of home venues. All countries t have a venue $v_t \in V$, where for countries $t = 1, 2$, their venue has a multiplicity of 2 and the other venues have a multiplicity of 1. Solving the corresponding instance of VENUE-ASSIGNMENT could result in a solution as is given in Table 1.

Table 1. Infeasible venue-assignment

Poule	Round 1	Round 2	Round 3
1	v_1	v_1	v_3
2	v_2	v_2	v_4

The venue assignment in Table 1 clearly satisfies the given multiplicities. However, it is impossible to schedule match (t_1, t_2) in any round when restricting teams to play at their home venue whenever it is scheduled in a round. Moreover, venue assignments satisfying the given multiplicities yielding a feasible schedule do exist for the given example.

Thus, we see that solving the Venue-Assignment alone is not necessarily the same as solving the VNL-problem in practice. However, the following claim shows that for the particular dimensions of the VNL ($N = 16, k = 4$), a Venue-Assignment can *always* be extended to a solution for the Volleyball Nations League.

Claim. Let $N = 16$ be the number of teams and $k = 4$ the poule size. Let V be the set of venues, with multiplicity constraints such that each team has at least one home venue, and all venues need to be scheduled at least once - never with two venues of the same team hosting simultaneously. Then any solution of the corresponding VENUE-ASSIGNMENT instance, can be extended to a solution of the VNL-instance.

Although we do not formally prove this claim, we exhibit in Table 2 a 'blueprint' that can be extended to a feasible schedule for any venue assignment that follows from any set of multiplicities.

Table 2. Blueprint for finding a nation assignment

R_1	R_2	R_3	R_4	R_5
$1, 5, 9, 13$	$1, 6, 11, 16$	$1, 4, 10, 15$	$1, 3, 12, 14$	$1, 2, 7, 8$
$2, 6, 10, 14$	$2, 5, 12, 15$	$2, 3, 13, 16$	$2, 4, 9, 11$	$3, 4, 5, 6$
$3, 7, 11, 15$	$3, 8, 9, 10$	$6, 7, 9, 12$	$5, 7, 10, 16$	$10, 11, 12, 13$
$4, 8, 12, 16$	$4, 7, 13, 14$	$5, 8, 11, 14$	$6, 8, 13, 15$	$9, 14, 15, 16$

5.2   Results

As instances of VNL satisfy the conditions of the Claim, we can proceed applying the IP for the Venue-Assignment to the known instances of the Volleyball Nations League, and compare our solution to that of the schedules used in practice. The IP is implemented in Python 3 using Gurobi 9.0. All computations have been done on a laptop with an Intel Core i7-7700HQ CPU 2.8-GHz processor and 32 GB RAM. The distances between venues are obtained via https://www.distancecalculator.net/, and are divided by 100 and rounded down. The four instances that we analyse are the Women's and Men's tournaments of 2018 and 2019. All values resulting from the IP are given to be optimal by the solver and are found within approximately 2 h of computation time. In Table 3 we give the unfairness corresponding to the optimal venue assignment, $u(S_{opt})$, and we give the unfairness that corresponds to the venue assignments used in practice, $u(S_{real})$. Also we give the total travel distance for the two corresponding solutions, $d(S_{opt})$ and $d(S_{real})$, where the distance is given in units of 100 km. The final column gives the computation time in seconds.

Table 3. Unfairness of real life S_{real} and optimal S_{opt}, and their total travel distance.

Instance	$u(S_{opt})$	$u(S_{real})$	$d(S_{opt})$	$d(S_{real})$	Computation time (s)
M2018	233	1366	4272	4806	5105 s
W2018	381	1541	4956	4169	1036 s
M2019	347	1239	5237	4657	7230 s
W2019	491	1288	4214	3708	4650 s

As is imminent from Table 3, the fairness of the schedules used in the Volleyball Nations League can be much improved in comparison to the schedules that have been used. Moreover, these improvements in fairness do not come at the expense of the total travel distance; indeed, total travel distance is similar for our schedules when compared to the real life schedules.

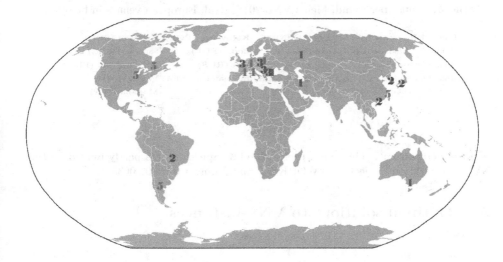

Fig. 1. Optimal venues per round, VNL Men 2018

We now discuss our schedules in more detail. In Table 4, the optimal venue selection for the 2018 Men's tournament, as given by the IP, is shown, and this venue assignment is also visualized in Fig. 1 - the output for the other instances is given in Appendix A. For comparison, the 'real' schedule is shown in Table 5.

In Fig. 1 we see that the optimal schedule creates two specific European rounds, where all poules are played within Europe, and two rounds without any poule in Europe. In contrast with that, the schedule that was used in practice had both European and non-European venues in every round - thus partially leading to a high amount of unfairness(Tables 6, 7 and 8).

As the unfairness in travel times as well as total traveled distances is completely determined by the venue assignment, any nation assignment is equally good with respect to the objectives. Up to satisfying the underlying SGP and assigning home nations to home venues, there is complete freedom to optimize the nations assignment to whatever other objectives the organizers see fit; this can be done without compromising on the original objectives.

Table 4. Venues per round, Men's VNL 2018, Optimal. European venues in boldface.

Round 1	Melbourne (AUS)	Tehran (IRA)	**Ufa (RUS)**	**Varna (BUL)**
Round 2	Goiânia (BRA)	Jiangmen (CHN)	Osaka (JPN)	Seoul (KOR)
Round 3	**Katowicze (POL)**	**Kraljevo (SRB)**	**Rouen (FRA)**	**Sofia (BUL)**
Round 4	**Aix-en-Prov. (FRA)**	**Lodz (POL)**	**Ludwigsb. (GER)**	**Modena (ITA)**
Round 5	Hoffman Est. (USA)	Ningbo (CHN)	Ottawa (CAN)	San Juan (ARG)

Table 5. Venues per round, Men's VNL 2018, Real. European venues in boldface.

Round 1	**Rouen (FRA)**	Ningbo (CHN)	**Katowicze (POL)**	**Kraljevo (SRB)**
Round 2	Goiânia (BRA)	**Sofia (BUL)**	**Lodz (POL)**	San Juan (ARG)
Round 3	Ottawa (CAN)	Osaka (JPN)	**Ufa (RUS)**	**Aix-en-Prov. (FRA)**
Round 4	Seoul (KOR)	**Ludwigsb. (GER)**	Hoffman Est. (USA)	**Varna (BUL)**
Round 5	Melbourne (AUS)	Jiagmen (CHN)	Tehran (IRA)	**Modena (ITA)**

Acknowledgement. The research of Frits C.R. Spieksma was partly funded by the NWO Gravitation Project NETWORKS, Grant Number 024.002.003.

A   Optimal solutions to VNL-instances

Table 6. Venues per round, Womens VNL 2018. Unfairness $u(S) = 381$

Round 1	Bangkok (THA)	Barneri (ITA)	Hong Kong (HKO/CHN)	Santa Fe (ARG)
Round 2	Eboli (ITA)	Rotterdam (NED)	Stuttgart (GER)	Walbrzyck (POL)
Round 3	Jiangmen (CHN)	Naklon (THA)	Suweo (KOR)	Toyota (JPN)
Round 4	Apeldoorn (NED)	Bydgozcz (POL)	Kraljevo (GER)	Ningbo (CHN)
Round 5	Ankara (TUR)	Yekaterinburg (RUS)	Lincoln (USA)	Macau (CHN)

Table 7. Venues per round, Men's VNL 2019. Unfairness $u(S) = 347$

Round 1	Katowicze (POL)	Novi Sad (SRB)	Plovdiv (BUL)	Urmia (IRN)
Round 2	Ardabi (IRN)	Brisbane (AUS)	Ufa (RUS)	Varna (BUL)
Round 3	Cuiaba (BRA)	Jiangmen (CHN)	Mendoza (ARG)	Tokyo (JPN)
Round 4	Cannes (FRA)	Gondomas (POR)	Leipzig (GER)	Milan (ITA)
Round 5	Brasilia (BRA)	Hofman Est. (USA)	Ningbo (CHN)	Ottawa (CAN)

Table 8. Venues per round, Women's VNL 2019. Unfairness $u(S) = 491$

Round 1	Ankara (TUR)	Macau (CHN)	Opole (POL)	Ruse (BUL)
Round 2	Boryeong (KOR)	Yekaterinburg (RUS)	Ningbo (CHN)	Tokyo (JPN)
Round 3	Bangkok (THA)	Brasilia (BRA)	Jiangmen (CHN)	Lincoln (USA)
Round 4	Apeldoorn (NED)	Conegliano (ITA)	Kortrijk (BEL)	Stuttgart (GER)
Round 5	Ankara (TUR)	Belgrade (SRB)	Hong Kong (HKO/CHN)	Perugia (ITA)

References

1. Alexandros, L., Panagiotis, K., Miltiades, K.: The existence of home advantage in volleyball. Int. J. Performance Anal. Sport **12**, 272–281 (2012)
2. Bonomo, F., Cardemil, A., Durán, G., Marenco, J., Sabán, D.: An application of the traveling tournament problem: the argentine volleyball league. Interfaces **42**, 245–259 (2012)
3. Cocchi, G., Galligari, A., Nicolino, F., Piccialli, V., Schoen, F., Sciandrone, M.: Scheduling the Italian national volleyball tournament. Interfaces **48**, 271–284 (2018)
4. Dotú, I., Van Hentenryck, P.: Scheduling social golfers locally. In: Barták, R., Milano, M. (eds.) CPAIOR 2005. LNCS, vol. 3524, pp. 155–167. Springer, Heidelberg (2005). https://doi.org/10.1007/11493853_13
5. Durán, G., Durán, S., Marenco, J., Mascialino, F., Rey, P.: Scheduling Argentina's professional basketball leagues: a variation on the travelling tournament problem. Eur. J. Oper. Res. **275**, 1126–1138 (2019)
6. Easton, K., Nemhauser, G., Trick, M.: Solving the travelling tournament problem: a combined integer programming and constraint programming approach. In: Burke, E., De Causmaecker, P. (eds.) PATAT 2002. LNCS, vol. 2740, pp. 100–109. Springer, Heidelberg (2003). https://doi.org/10.1007/978-3-540-45157-0_6
7. Goerigk, M., Westphal, S.: A combined local search and integer programming approach to the traveling tournament problem. Ann. Oper. Res. **239**(1), 343–354 (2014). https://doi.org/10.1007/s10479-014-1586-6
8. Harvey, W., Winterer, T.: Solving the MOLR and social golfers problems. In: van Beek, P. (ed.) CP 2005. LNCS, vol. 3709, pp. 286–300. Springer, Heidelberg (2005). https://doi.org/10.1007/11564751_23
9. Huyghe, T., Scanlan, A., Dalbo, V., Calleja-González, J.: The negative influence of air travel on health and performance in the national basketball association: a narrative review. Sports **6**, 89 (2018)
10. Liu, K., Löffler, S., Hofstedt, P.: Solving the social golfers problems by constraint programming in sequential and parallel. In: Proceedings of the 11th International Conference on Agents and Artificial Intelligence (ICAART 2019), pp. 29–39 (2019)

11. Lo, M., Aughey, R.J., Stewart, A.M., Gill, N., McDonald, B.: The road goes ever on and on-a socio-physiological analysis of travel-related issues in super rugby. J. Sports Sci. (2020)
12. Raknes, M., Pettersen, K.H.: Optimizing sports scheduling: mathematical and constraint programming to minimize traveled distance with benchmark from the norwegian professional volleyball league. Master Thesis, Norwegian Business School (2018)
13. Rothuizen, L.: A variation of the travelling tournament problem: fairness in the volleyball nations league. Bachelor Thesis, Eindhoven University of Technology (2020)
14. Samuels, C.: Jet lag and travel fatigue: a comprehensive management plan for sport medicine physicians and high-performance support teams. Clin. J. Sport Med. **22** (2012)
15. Song, A., Severini, T., Allada, R.: How jet lag impairs major baseball performance. Proc. Natl. Acad. Sci. **114**, 1407–1412 (2017)
16. Stevens, C., Thornton, H., Fowler, P., Esh, C., Taylor, L.: Long-haul northeast travel disrupts sleep and induces perceived fatigue in endurance athletes. Front. Physiol. **20** (2018)
17. Triska, M., Musliu, N.: An effective greedy heuristic for the social golfer problem. Ann. Oper. Res. **194**, 413–425 (2012)
18. Winter, W.C., Hammond, W.R., Green, N.H., Zhang, Z., Bliwise, D.L.: Measuring circadian advantage in major league baseball: a 10-year retrospective study. Int. J. Sports Physiol. Performance **4**, 394–401 (2009)

Towards a Compact SAT-Based Encoding of Itemset Mining Tasks

Ikram Nekkache[1,2]([✉]) [iD], Said Jabbour[1]([✉]) [iD], Lakhdar Sais[1] [iD],
and Nadjet Kamel[2] [iD]

[1] CRIL - CNRS UMR 8188, University of Artois, Lens, France
{nekkache,jabbour,sais}@cril.fr
[2] LRSD Laboratory, Department of Computer Science, Faculty of Sciences,
University Ferhat Abbas Sétif-1, Sétif, Algeria
{ikram.nekkache,nkamel}@univ-setif.dz

Abstract. Many pattern mining tasks have been modeled and solved using constraints programming (CP) and propositional satisfiability (SAT). In these two well-known declarative AI models, the problem is encoded as a constraints network or a propositional formula, whose associated models correspond to the patterns of interest. In this new declarative framework, new user-specified constraints can be easily integrated, while in traditional data mining, such additional constraints might require an implementation from scratch. Unfortunately, these declarative data mining approaches do not scale on large datasets, leading to huge size encodings. In this paper, we propose a compact SAT-based encoding for itemset mining tasks, by rewriting some key-constraints. We prove that this reformulation can be expressed as a Boolean matrix compression problem. To address this problem, we propose a greedy approach allowing us to reduce considerably the size of the encoding while improving the pattern enumeration step. Finally, we provide experimental evidence that our proposed approach achieves a significant reduction in the size of the encoding. These results show interesting improvements of this compact SAT-based itemset mining approach while reducing significantly the gap with the best state-of-the-art specialized algorithm.

Keywords: Data mining · Itemset mining · Satisfiability

1 Introduction

Frequent itemset mining problem is a fundamental task in data mining, knowledge discovery and data analysis. Initially proposed for the well-known market basket analysis application [1], it is now widely used in various fields and tasks that require the discovery of regularities between items or attributes. This real interest has been accompanied by numerous algorithm developments for enumerating interesting patterns. Different classes of patterns have been identified allowing to reduce the size of the output. Closed and maximal patterns are some

© Springer Nature Switzerland AG 2021
P. J. Stuckey (Ed.): CPAIOR 2021, LNCS 12735, pp. 163–178, 2021.
https://doi.org/10.1007/978-3-030-78230-6_11

of these traditional condensed representations. In addition, users can control the set of required patterns by enumerating those covering at least λ (called a minimum support threshold) transactions (see [8] for a survey).

Recently, new declarative approaches for data mining have been emerged. Initiated by De Raedt et al., this new research trend proposes to make use of constraint programming for modeling and solving data mining tasks including itemset mining (CP4IM) in [5]. The goal is two folds. First, in this declarative and flexible framework, new constraints can be easily integrated in contrast to specialized approaches where new implementations are often required. Secondly, data mining tasks might benefit from the continuous progress in the efficiency of CP solvers. In such CP framework, usual itemset mining constraints (e.g. frequency, maximality, monotonicity) can be elegantly formulated and easily integrated [11].

Encouraged by these promising results, several contributions used the two well-known AI models, CP and SAT, to solve other data mining problems. The problem of discovering frequent, closed and maximal patterns in a sequence of items and a sequence of itemsets has been formalized using propositional satisfiability [12]. In [4], the authors solve the frequent itemset mining problem by compiling the set of all itemsets into a binary decision diagram (BDD) (augmented with counts). Then frequent itemsets are extracted by querying the BDD. By considering the relationship between local constraint-based mining and constraint satisfaction problems, Khiari et al. [13] proposed a model for mining patterns that combines several local constraints, i.e., patterns defined by n-ary constraints. In addition, new constraint-based languages have been designed for modeling and solving data mining problems. We can mention the constraint-based language defined in [15], which enables the user to define queries in a declarative manner to handle pattern sets and global patterns. All primitive constraints of the language are modeled and solved using the SAT framework. More recently, Guns et al. [10], introduced a general-purpose declarative mining framework called MiningZinc. Compared with the CP4IM framework [11], MiningZinc supports a wide variety of different solvers (including DM algorithms and general-purpose solvers) and uses a significantly more expressive high-level modeling language.

CP and SAT-based approaches for data mining have real advantages in terms of "declarativeness" and genericity. Their major bottleneck rises in their lack of scalability. Recently, several attempts to solve such challenging issue have been initiated, using for example decomposition and parallel approaches as a mean to compete with specialized approaches [2].

Our goal in this paper, is to effectively compact and reduce the size of SAT-based encoding for itemset mining problem proposed in [6], on large real datasets. Our approach involves rewriting some important constraints compactly. We also show that the compression process can be expressed as a matrix compression problem.

The paper is structured as follows. Section 2 provides the preliminary knowledge about propositional logic, SAT problems and itemset mining. Section 3 reviews the SAT-based encoding of itemset mining. Section 4 presents our new a compact SAT encoding for the itemset mining problem. Section 5 focuses on the experimental results. Finally we conclude and propose some perspectives.

2   Technical Background

In this section, we provide a brief description of itemset mining, propositional logic and propositional satisfiability.

2.1   Propositional Logic and SAT Problem

A propositional language $\mathcal{L}$ defined from a finite set of propositional variables $Var = \{p, q, r, \ldots\}$, the logical constants $\perp, \top$, and the usual logical connectives (namely, $\neg$, $\wedge$, $\vee$, $\rightarrow$, and $\leftrightarrow$) is considered. Propositional formulas will be denoted by Greek letters Σ, Δ, etc. $Var(\Sigma)$ denotes the set of propositional variables appearing in the formula Σ. It is common for logical reasoning algorithms to operate on *normal form* representations instead of arbitrary formulas. A formula in *conjunctive normal form* (CNF) is a conjunction ($\wedge$) of clauses, where a *clause* is a disjunction ($\vee$) of literals. A *literal* is a propositional variable (p) or its negation ($\neg p$). In addition, a *Boolean interpretation* μ of a formula Σ is a total function from $Var(\Sigma)$ to $\{0, 1\}$ (0 corresponds to *false* and 1 to *true*). We denote by $\Sigma_{|x}$ the formula Σ where x assigned *true*, i.e., $\Sigma \wedge x$. μ is a *model* of Σ iff it makes it true in the usual truth-functional way. Then, Σ is *satisfiable* if there exists a model of Σ. $models(\Sigma)$ denotes the set of models of a formula Σ. Lastly, SAT is the NP-complete problem that consists in deciding whether a given CNF formula is satisfiable or not.

2.2   An Overview of Itemset Mining

We consider Ω a set of items. The elements of Ω are indicated by the letters a, b, c, etc. An *itemset* I over Ω is a subset of Ω, i.e., $I \subseteq \Omega$. 2^{Ω} denotes the set of all itemsets over Ω. Typically, a *transaction* T_i is a pair (i, I) with $1 \leq i \leq m$, called the *transaction identifier*, and I an itemset, i.e., $(i, I) \in \mathbb{N} \times 2^{\Omega}$. For $T_i = (i, I)$, the size of T_i, is defined as $|T_i| = |I|$. A *transaction database* $\mathcal{D}$ is a set of transactions ($\mathcal{D} \subseteq \mathbb{N} \times 2^{\Omega}$) where each transaction identifier refers to a unique itemset. The maximum size of the transactions of $\mathcal{D}$, is noted $|T| = max_{(i,I) \in \mathcal{D}} |I|$. Given a transaction database $\mathcal{D}$ and an itemset I, the *cover* of I in $\mathcal{D}$, denoted $\mathcal{C}(I, \mathcal{D})$, is defined as follows: $\{i \in \mathbb{N} \mid (i, J) \in \mathcal{D} \text{ and } I \subseteq J\}$. The *support* of I in the database $\mathcal{D}$, denoted as $Supp(I, \mathcal{D})$, is defined as the cardinality of $\mathcal{C}(I, \mathcal{D})$, i.e., $Supp(I, \mathcal{D}) = |\mathcal{C}(I, \mathcal{D})|$. An itemset $I \subseteq \Omega$ such that $Supp(I, \mathcal{D}) \geq 1$ is *closed* iff, for all itemsets J with $I \subset J$, $Supp(J, \mathcal{D}) < Supp(I, \mathcal{D})$.

In the transaction database depicted in Table 1, we have $Supp(\{b, d, e\}, \mathcal{D}) = |\{2, 10\}| = 2$.

Table 1. Transaction database $\mathcal{D}$.

Tid	Itemset							
1			c		e	f	g	
2		b		d	e	f	g	h
3	a	b		d				
4	a	b		d		f		h
5		b	c		e	f	g	h
6			c				g	
7	a	b		d				h
8			c		e		g	
9	a	b		d				
10		b	c	d	e	f	g	h

Let $\mathcal{D}$ be a transaction database over Ω and λ a minimum support threshold. The *frequent itemset mining problem* consists in computing the following set:

$$\mathcal{FIM}(\mathcal{D}, \lambda) = \{I \subseteq \Omega \mid Supp(I, \mathcal{D}) \geq \lambda\}$$

It is well known that the main problem in itemset mining lies in the size of the output, which could be exponential, even when considering condensed representation of patterns.

3 SAT-based Encoding of Itemset Mining

Here, we review the SAT encoding scheme of the problem of mining itemsets of a transaction database $\mathcal{D}$ as proposed in [6]. Basically, to encode the itemset mining problem into SAT, one must introduce a set of variables and a set of constraints on those variables. More precisely, different variables are used to represent the cover of an itemset X. These variables are used in 0/1 linear inequalities to ensure the support of X.

Given a transaction database $\mathcal{D} = \{(1, T_1), \ldots, (m, T_m)\}$, a minimum support threshold λ. To represent the candidate itemset X, a propositional variables p_a is associated to each item a in order to guarantee if a belongs to X (i.e., $p_a = true$). For the cover of X, new variables q_i are also introduced for each transaction identifier $i \in \{1 \ldots m\}$. In addition, a set of constraints are introduced on the variables to define a one-to-one mapping between the models of the obtained CNF formula, denoted as $\Sigma_{\mathcal{D}, \lambda}$, and the set of itemsets.

First, the constraint allowing to capture all the transactions where the candidate itemset does not appear:

$$\bigwedge_{i=1}^{m} (q_i \leftrightarrow \bigwedge_{a \notin T_i} \neg p_a) \tag{1}$$

This constraint expresses that q_i is false if and only if the candidate itemset is not in the transaction T_i i.e. there is at least one item a in the candidate itemset that does not belong to the transaction.

By the following constraint, the candidate itemset is forced to be closed:

$$\bigwedge_{a \in \Omega} (p_a \vee \bigvee_{i, \, a \notin T_i} q_i) \tag{2}$$

In fact, if the set of all transactions containing a are false, then the itemset candidate is mapped to a subset of transactions containing a. Consequently, a must be in the final itemset.

Finally, to allow only the itemsets respecting the minimum size threshold min can be expressed using the following cardinality constraint:

$$\sum_{a \in \mathcal{I}} p_a \geq min \tag{3}$$

Proposition 1. *The set of models of* (1)$\wedge$(2)$\wedge$(3) *corresponds to the set of closed itemsets of size at least* min.

Finally, the frequency constraint, can be simply expressed as follows:

$$\sum_{i=1}^{m} q_i \geq \lambda \tag{4}$$

The frequent itemset mining task corresponds to the conjunction of (1) and (4).

Example 1. Let us reconsider the transaction database of Table 1. The itemset mining problem is defined as:

$$q_1 \leftrightarrow (\neg p_a \wedge \neg p_b \wedge \neg p_d \wedge \neg p_h)$$
$$q_2 \leftrightarrow (\neg p_a \wedge \neg p_c)$$
$$q_3 \leftrightarrow (\neg p_c \wedge \neg p_e \wedge \neg p_f \wedge \neg p_g \wedge \neg p_h)$$
$$q_4 \leftrightarrow (\neg p_c \wedge \neg p_e \wedge \neg p_g)$$
$$q_5 \leftrightarrow (\neg p_a \wedge \neg p_d)$$
$$q_6 \leftrightarrow (\neg p_a \wedge \neg p_b \wedge \neg p_d \wedge \neg p_e \wedge \neg p_f \wedge \neg p_h)$$
$$q_7 \leftrightarrow (\neg p_d \wedge \neg p_e \wedge \neg p_f \wedge \neg p_g)$$
$$q_8 \leftrightarrow (\neg p_a \wedge \neg p_b \wedge \neg p_d \wedge \neg p_f \wedge \neg p_h)$$
$$q_9 \leftrightarrow (\neg p_c \wedge \neg p_e \wedge \neg p_f \wedge \neg p_g \wedge \neg p_h)$$
$$q_{10} \leftrightarrow (\neg p_a)$$
$$q_1 + q_2 + q_3 + q_4 + q_5 + q_6 + q_7 + q_8 + q_9 + q_{10} \geq \lambda$$

4   A Compact SAT-Based Encoding

In this section, we propose an enhancement of the SAT-based encoding of itemset mining tasks. Let us take a look to the closeness constraint (2). For large real data, often $|T| \ll |\Omega|$. I.e., the number of missed items in each transaction is huge compared to those appearing in it. Consequently, the number of clauses associated to the constraint (2) is nearly $|D| \times |\Omega|$. Clearly, for large transaction databases such number of clauses is the main limitation for the scalability of the SAT-based itemset mining approaches. Moreover, for the closeness constraint, the size of the derived clauses might be very large i.e., for an item a, we need to consider all the transactions not containing a.

To deal with this important bottleneck, we propose an enhancement of the encoding described in Sect. 3. Let us first remark that the formula (1) can be equivalently formulated, by eliminating $\leftrightarrow$, as the conjunction of the two following clausal formulas (5) and (6):

$$\bigwedge_{a \in \Omega} \bigwedge_{a \notin T_i} (\neg p_a \vee \neg q_i) \tag{5}$$

$$\bigwedge_{T_i \in D} ((\bigvee_{a \notin T_i} p_a) \vee q_i) \tag{6}$$

Constraint (5) links each item to transactions where it does not appear. This constraint can also be reformulated as a set of implications:

$$\bigwedge_{a \in \Omega} (p_a \rightarrow \bigwedge_{i \in 1...m \mid a \notin T_i} \neg q_i) \tag{7}$$

Constraint (7) expresses that if p_a is assigned to true, then the set of all Boolean variables associated to transactions T_i not containing a are propagated to false. A first approach to reduce the size of the encoding consists in detecting sub-terms in ($\bigwedge_{i \in 1...m \mid a \notin T_i} \neg q_i$), the right side of the implications in (7), and introducing auxiliary variables to represent such sub-terms. To illustrate such a method, let us consider again Example 1. The sub-term $(\neg q_3 \wedge \neg q_6 \wedge \neg q_9)$ appears four times in (7).

$$p_a \rightarrow (\neg q_1 \wedge \neg q_2 \wedge \neg q_5 \wedge \neg q_6 \wedge \neg q_8 \wedge \neg q_{10})$$
$$p_b \rightarrow (\neg q_1 \wedge \neg q_6 \wedge \neg q_8)$$
$$p_c \rightarrow (\neg q_2 \wedge \neg q_3 \wedge \neg q_4 \wedge \neg q_7 \wedge \neg q_9)$$
$$p_d \rightarrow (\neg q_1 \wedge \neg q_5 \wedge \neg q_6 \wedge \neg q_8)$$
$$p_e \rightarrow (\neg \mathbf{q_3} \wedge \neg q_4 \wedge \neg \mathbf{q_6} \wedge \neg q_7 \wedge \neg \mathbf{q_9})$$
$$p_f \rightarrow (\neg \mathbf{q_3} \wedge \neg \mathbf{q_6} \wedge \neg q_7 \wedge \neg q_8 \wedge \neg \mathbf{q_9})$$
$$p_g \rightarrow (\neg \mathbf{q_3} \wedge \neg q_4 \wedge \neg \mathbf{q_6} \wedge \neg \mathbf{q_9})$$
$$p_h \rightarrow (\neg q_1 \wedge \neg \mathbf{q_3} \wedge \neg q_4 \wedge \neg \mathbf{q_6} \wedge \neg q_8 \wedge \neg \mathbf{q_9})$$

Consequently, a new variable r can be used to represent such sub-term. As the sub-term $(\neg q_3 \wedge \neg q_6 \wedge \neg q_9)$ is of positive polarity, we only need to add the definition $(r \rightarrow (\neg q_3 \wedge \neg q_6 \wedge \neg q_9))$, and substitute such sub-term with r in each implication containing it. The resulting formula is equivalent with respect to satisfiability. Such process allows us to earn 5 clauses in total i.e., 12 clauses are replaced with 7 clauses: 4 clauses (implications involving r) and 3 binary clauses (associated the definition of r). Nevertheless, such process requires an algorithm looking for such frequent sub-terms.

In this paper, we propose a new approach to identify efficiently such sub-terms. It is based on transactions reordering and the introduction of new variables providing an original way to compact the encoding presented above.

In the sequel, we indicate by $g(a)$ (respectively $f(a)$) the identifier of the last (respectively first) transaction involving a in the database.

Definition 1. *For an item a, we denote by $f(a)$ (respectively $g(a)$) the inner (resp. outer) transaction containing a. i.e., $f(a) = \min\limits_{1 \leq i \leq m} \{i \mid a \in T_i\}$ and $g(a) = \max\limits_{1 \leq i \leq m} \{i \mid a \in T_i\}$*

Using $g(a)$, the formula (7) can be expressed as the conjunction of (8) and (9).

$$\bigwedge_{a \in \Omega} (p_a \rightarrow \bigwedge_{i < g(a) \mid a \notin T_i} \neg q_i) \tag{8}$$

$$\bigwedge_{a \in \Omega} (p_a \rightarrow \bigwedge_{g(a) < i \leq m} \neg q_i) \tag{9}$$

In fact, let us look to the formula (9), starting from $g(a)$, all the transactions with greater identifiers, their associated variables have to be propagated to false. In the sequel, we show how such propagation can be captured differently by reformulating compactly such formulas using additional variables.

By associating to each transaction T_i a new Boolean variable r_i, the formula (9) can be reformulated as the conjunction of the formulas (10) and (11):

$$\bigwedge_{a \in \Omega} (p_a \rightarrow r_{g(a)} \wedge \bigwedge_{i \mid a \notin T_i, \, i < g(a)} \neg q_i) \tag{10}$$

$$\bigwedge_{i \in 1 \ldots m-1} (r_i \rightarrow \neg q_i) \wedge (r_i \rightarrow r_{i+1}) \tag{11}$$

Formula (11) forms a propagation chain allowing, when r_i is set to true, to propagate all q_i to false from i to m. Such chain of propagated literals can be used for different items a allowing to reduce the encoding size. Then, when p_a is set to true, $r_{g(a)}$ is then assigned to true, allowing using the propagation chain (11), to assign all the variables associated to transactions greater than $g(a)$ to false. Let us remark that the encoding size reduction depends on the chosen transactions ordering.

For the closeness constraint (2), it can be rewritten through the new additional variables as follows:

$$\bigwedge_{a \in \Omega} (p_a \vee \neg r_{g(a)} \vee \bigvee_{i \mid a \notin T_i, \, i < g(a)} \neg q_i) \tag{12}$$

Indeed, the formula (12) is derived from the formula (2) by substituting all literals $\neg q_i$ with i greater than $g(a)$ with $\neg r_{g(a)}$.

To significantly reduce the size of the encoding, one need to find an ordering of the transactions that maximize both the distance between $g(a)$ and m and between 1 and $f(a)$. Equivalently, one need to minimize the distance between $f(a)$ and $g(a)$ the first and the last transactions containing a respectively.

Similarly, considering a chain of implications in the reverse order, the previous transformation can be formulated as follows:

$$\bigwedge_{i \in 1..m-1} (r_i \rightarrow \neg q_i \wedge r_{i+1}) \tag{13}$$

$$\bigwedge_{i \in 2..m} (s_i \rightarrow \neg q_i \wedge s_{i-1}) \tag{14}$$

$$\bigwedge_{a \in \Omega} (p_a \rightarrow r_{g(a)} \wedge s_{f(a)} \wedge \bigwedge_{i \in]f(a),g(a)[, a \notin T_i} \neg q_i) \tag{15}$$

$$\bigwedge_{a \in \Omega} (p_a \vee \neg r_{g(a)} \vee \neg s_{f(a)} \vee \bigvee_{i \in]f(a),g(a)[, a \notin T_i} \neg q_i) \tag{16}$$

As for r_i, the Boolean variables s_i are added to guarantee the propagation of the transaction variables to the top i.e., if s_i is true, then all transaction q_j such that $j \leq i$ are propagated to false. As explained above, to reduce the size of the encoding while enhancing the propagation process, we need to find the best possible transaction ordering.

For a given ordering over the transactions of $\mathcal{D}$, the number of clauses of our encoding is in the worst case bounded by:

$$\sum_{a \in \Omega} (g(a) - f(a) + 1) + 3 \times |\Omega| + 4 \times |\mathcal{D}|$$

Indeed, each of the formulas (13) and (14) admits $2 \times |D|$. The formula (16) is in clausal form with $|\Omega|$ clauses. Finally, the formula (15) is an implication, that leads to $2 \times |\Omega| + \sum_{a \in \Omega} (g(a) - f(a) + 1)$ binary clauses.

Remark 1. Let us remark that, the best ordering corresponds to those where the transactions containing each item are contiguous. In this case, the best encoding size is equal to $3 \times |\Omega| + 4 \times |D|$.

Our optimisation problem consists in minimizing $\sum_{a \in \Omega} (g(a) - f(a))$. The problem of finding the best transaction ordering can be seen as Boolean matrix compression. In fact, by considering a 0–1 formulation of the transaction database, our goal is to rearrange the matrix rows such that the 1 values of each row are as consecutive as possible i.e., make the interval $]f(a), g(a)[$ as tight as possible. Our problem generalize the well known optimal linear arrangement problem [9]. We note this problem GOLA for Generalized Optimal Linear Arrangement problem.

Let us formally define our Boolean compression matrix problem.

Definition 2 (Problem Definition). *Given a 0–1 matrix M. The optimal compression matrix problem consists in finding a permutation σ over rows of M such that $\sum_{1 \leq i \leq m} h(i)$ is minimized, where $h(i) = (g(i) - f(i))$.*

Example 2. Let us reconsider the dataset of Example 1. The 0–1 matrix representation of the table $\mathcal{D}$ is depicted in Table 2. By reordering the transactions of the table, we can leads to matrix represented in Table 3.

Table 2. Boolean matrix of transaction database $\mathcal{D}$.

	a	b	c	d	e	f	g	h
t_1	0	0	1	0	1	1	1	0
t_2	0	1	0	1	1	1	1	1
t_3	1	1	0	1	0	0	0	0
t_4	1	1	0	1	0	1	0	1
t_5	0	1	1	0	1	1	1	1
t_6	0	0	1	0	0	0	1	0
t_7	1	1	0	1	0	0	0	1
t_8	0	0	1	0	1	0	1	0
t_9	1	1	0	1	0	0	0	0
t_{10}	0	1	1	1	1	1	1	1

Table 3. Boolean matrix of transaction database $\mathcal{D}$ reordred.

	a	b	c	d	e	f	g	h
t_3	1	1	0	1	0	0	0	0
t_9	1	1	0	1	0	0	0	0
t_7	1	1	0	1	0	0	0	1
t_4	1	1	0	1	0	1	0	1
t_2	0	1	0	1	1	1	1	1
t_{10}	0	1	1	1	1	1	1	1
t_5	0	1	1	0	1	1	1	1
t_1	0	0	1	0	1	1	1	0
t_8	0	0	1	0	1	0	1	0
t_6	0	0	1	0	0	0	1	0

Proposition 2. *GOLA is a NP-hard problem.*

Proof. Let us note Garey et al. [9] have proven that the classical problem of *optimal linear arrangement* (OLA in short) is NP-Hard. The decision version of this problem is defined as: given a graph $G = (V, E)$ and a positive integer

k and the question is to answer if there exists a one-to-one mapping $f : V \rightarrow \{1, 2, \ldots, |V|\}$ such that $\sum_{(u,v) \in E} |f(u) - f(v)| \leq k$. It is easy to remark that our problem generalize the *optimal linear arrangement*. When modeled as a graph the solution of GOLA is exactly the one of optimal linear arrangement. Here edges represent rows and vertices represent columns.

Several works in the literature, addressed the problem of compressing a large Boolean matrix.

For example, Booth and Lueker [3] showed in 1976 that for a given matrix M, there is a linear time algorithm that determines whether M has the consecutive-ones property and produces the desired permutation if so. Thus, if the relation has the consecutive-ones property, we can reorder the columns on disk so that the elements of each row can be accessed in a single seek. However, this will in general not be possible and minimizing the number of runs when a matrix does not have the consecutive-ones property is hard.

4.1 Solving Generalized Optimal Linear Arrangement Problem

In this section, we discuss the GOLA problem using complete and greedy approaches. Let us note that as for OLA problem, it is possible to consider some dedicated heuristics allowing high compression rate by searching the best possible values of $\sum_{a \in \Omega} (g(a) - f(a))$. To obtain the optimal solution, a possible approach consists in encoding the problem into Partial MaxSAT in order to benefit from the state-of-the-art solvers continuously improved these recent years. To do so, we consider a mapping between the set of transactions $\{T_1, \ldots, T_m\}$, the rows of the matrix, and the required positions. For each transaction T_i, m new Boolean variables are introduced to catch the possible positions in the requested optimal solution of T_i i.e., t_{ij} is true iff T_i is ranked in position j.

The combination of constraints (17) and (18) express a one-to-one mapping between the set of the rows and the new positions i.e., each row must be set exactly in one position from 1 to m. (17) requires that each transaction have to be set in one position and (18) expresses that two transactions could not be put in the same position.

$$\sum_{j \in 1 \ldots m} t_{ij} = 1 \quad \text{for all } i \in [1..m] \tag{17}$$

$$\sum_{j \in 1 \ldots m} t_{ji} \leq 1 \quad \text{for all } i \in [1..m] \tag{18}$$

To capture the bounds of each item a, we associate to each item a two sets of variables namely $x_{a,1}, \ldots, x_{a,m}$. $x_{a,i}$ indicates that transaction i contains item a. Constraint (19) allows to link $x_{a,i}$ to t_{ij}.

$$\bigwedge_{a \in \Omega} \bigwedge_{j=1}^{m} \left(\bigvee_{T_i \mid a \in T_i} t_{ij} \rightarrow x_{a,j} \right) \tag{19}$$

$f(a)$ and $g(a)$ can be expressed as follows:

$$f(a) = \sum_{i=1}^{m} i \times (x_{a,i} \wedge \bigwedge_{1 \leq j < i} \neg x_{a,j})$$

$$g(a) = \sum_{i=1}^{m} i \times (x_{a,i} \wedge \bigwedge_{i < j \leq m} \neg x_{a,j})$$

Finally, the objective function can be expressed as the follows:

$$\text{minimize} \quad \sum_{a \in \Omega} (g(a) - f(a)) \tag{20}$$

A solution of (17) $\wedge$ (18) $\wedge$ (19) subject to the objective function (20) corresponds to a best rearrangement for matrix compression allowing to compress efficiently our encoding. Nevertheless, the NP-Hardness of this problem is clearly an issue in practice. Thus, it will be convenient to design an alternative approach providing high reduction factor while maintaining a reasonable amount of time.

To do so, a greedy approach can be used instead. It proceeds as follows. It recursively picks up the less frequent item a in the current database. Then, all transactions containing a are placed on the top of $\mathcal{D}$. The goal is to minimize $g(a)$. This operation is repeated by considering the less frequent item in the remaining transactions i.e., $\mathcal{D} \setminus \{T_i, \ a \in T_i\}$. Let us remark that two strategies can be considered: either to choose the less frequent item in $\mathcal{D} \setminus \{T_i, \ a \in T_i\}$ or in the whole database $\mathcal{D}$.

Example 3. Let us reconsider the transaction database $\mathcal{D}$ depicted in Table 1. By applying, the greedy approach, the obtained transaction database is depicted in Fig. 1. The associated chain of implications, are illustrated on the left side of the figure.

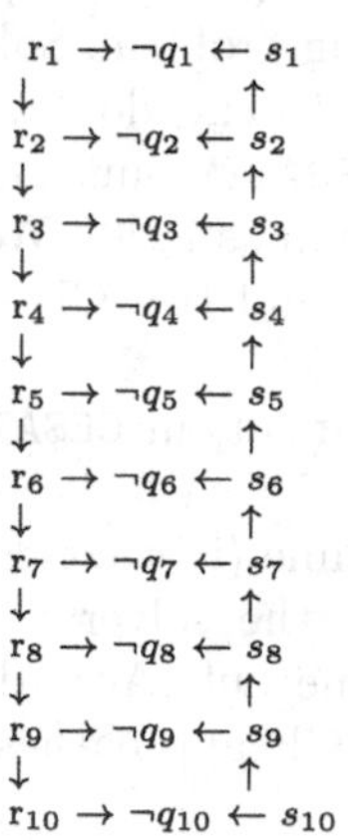

Tid	Itemset							
1	a	b		d				
2	a	b		d				
3	a	b		d				h
4	a	b		d		f		h
5		b		d	e	f	g	h
6		b	c	d	e	f	g	h
7		b	c		e	f	g	h
8			c		e	f	g	
9			c		e		g	
10			c				g	

Chain of implications (left side of figure):

$r_1 \to \neg q_1 \leftarrow s_1$
$\downarrow \qquad \uparrow$
$r_2 \to \neg q_2 \leftarrow s_2$
$\downarrow \qquad \uparrow$
$r_3 \to \neg q_3 \leftarrow s_3$
$\downarrow \qquad \uparrow$
$r_4 \to \neg q_4 \leftarrow s_4$
$\downarrow \qquad \uparrow$
$r_5 \to \neg q_5 \leftarrow s_5$
$\downarrow \qquad \uparrow$
$r_6 \to \neg q_6 \leftarrow s_6$
$\downarrow \qquad \uparrow$
$r_7 \to \neg q_7 \leftarrow s_7$
$\downarrow \qquad \uparrow$
$r_8 \to \neg q_8 \leftarrow s_8$
$\downarrow \qquad \uparrow$
$r_9 \to \neg q_9 \leftarrow s_9$
$\downarrow \qquad \uparrow$
$r_{10} \to \neg q_{10} \leftarrow s_{10}$

Fig. 1. Transactions reordering: a greedy approach.

Finally, the complete encoding associated to the transaction database reordered greedily (Fig. 1) is given by the following Boolean formula:

$$(p_a \rightarrow r_5) \wedge (p_b \rightarrow r_8) \wedge (p_c \rightarrow s_5) \wedge (p_d \rightarrow r_7) \wedge$$
$$(p_e \rightarrow \neg q_{10} \wedge s_4) \wedge (p_f \rightarrow r_9 \wedge s_3) \wedge (p_g \rightarrow s_4) \wedge (p_h \rightarrow r_8 \wedge s_2) \wedge$$
$$\bigwedge_{5 \leq i \leq 9} (r_i \rightarrow r_{i+1} \wedge \neg q_i) \wedge \bigwedge_{2 \leq i \leq 7} (s_i \rightarrow s_{i-1} \wedge \neg q_i)$$

5   Experimental Evaluation

We now present the experiments carried out to evaluate the performance of our approach. In particular, we study the compression rate and the time needed to enumerate models of the obtained formula and then of the pattern of interest. For comparison, we consider the solver MiniSAT [7], adapted to enumerate all models by performing a backtrack search without restart as depicted in DPLL-type Algorithm 1. More precisely, the algorithm branch first on items variables. In fact, each assignment over items variables allows to propagate all the q_i variables. The items variables are chosen according to the frequency of the items in the database i.e., the solver branch on the less frequent first. No conflict analysis is performed. A simple backtrack is performed after each conflict or when a model is found.

Our experiments were performed on a machine with Intel Xeon quad-core processors with 32 GB of RAM running at 2.66 GHz on Linux CentOS. Time-out was set to 1800 s and memory-out to 10 GB in all runs. We consider the datasets coming from FIMI[1] and CP4IM[2]. The characteristics of the considered datasets are given in Table 4. The sources of the model enumeration are available at : https://github.com/ikramnekkache/CESATIM.

We follow the experimental schema of Dlala et al. [6]: The enumeration problem is decomposed into the enumeration of a sequence of sub-problems $\Sigma_i = \Sigma \wedge \neg p_{a_1} \wedge \ldots \wedge \neg p_{a_{i-1}} \wedge p_{a_i}$ where $\Omega = \{a_1, \ldots, a_n\}$ and Σ is the formula of SAT-Based encoding of itemsets mining problem. Solving Σ_i consists in considering transactions containing a_i. Then, to encode Σ_i, our compact approach is used. In the sequel, we denote by CESATIM our compact encoding for SAT-Based itemsets mining using decomposition as in [6]. We compare the performances of CESATIM against ParaSATMiner and two CP approaches namely ClosedPattern [14] and CoverSize [16].

In Table 5, we report the comparative results of our CESATIM approach using different minimum support threshold values. For each dataset, the number of models (closed patterns) and the total CPU time (in seconds) are reported.

The symbol (-) is used to mention that the solver is not able to finish the enumeration process within the fixed time out. According to the results, CESATIM outperform ParaSATMiner and the CP approaches on almost all the

---

[1] http://fimi.ua.ac.be/data/.
[2] http://dtai.cs.kuleuven.be/CP4IM/datasets/.

Algorithm 1: DPLL for all model enumeration

Input: a CNF formula Σ
Output: SAT or UNSAT
$\mathcal{I} = \emptyset$;   /* interpretation */
$dl = 0$;   /* decision level */
while *(true)* **do**
   $c = \text{unitPropagation}(\Sigma, \mathcal{I})$;
   if *(c != null)* **then**
      $dl \leftarrow dl - 1$;
      if *(dl < 0)* **then**
         return UNSAT;
      else
         $\text{backtrack\_until}(dl - 1)$;
   else
      if *($\mathcal{I} \models \Sigma$)* **then**
         $S \leftarrow S \cup \{\mathcal{I}\}$;
         $dl \leftarrow dl - 1$;
         if *(dl < 0)* **then return** UNSAT;
         $\text{backtrack\_until}(dl)$
      else
         $x = \text{selectDecisionVariable}(\Sigma)$;
         $dl = dl + 1$;
         $\mathcal{I} = \mathcal{I} \cup \{x\}$;

Table 4. Data characteristics.

Instance	#Transactions	#Items	Density	Size
Chess	3196	75	49.0%	340K
Mushroom	8124	119	19.0%	516K
T10I4D100K	100000	870	1.0%	3.9M
Retail	88162	16470	0.06%	4,2M
Connect	67558	129	35.62%	8.9M
T40I10D100K	100000	942	4.31%	15M
Pumsb	49046	2113	3.0%	16,7M
Kosarak	990002	41267	0.01%	32M
Accidents	340183	468	7.0%	34M

datasets. Particularly, our approach enumerate the set of all closed itemsets of pumbs in only 194.70 s while ParaSATMiner needs 643.11 s. ClosedPattern and CoverSize are not able to finish the enumeration of all patterns under the fixed time out. Notably, except for retail where ParaSATMiner outperfoms CESATIM, on all the remaining datasets, CESATIM achieves the best performances.

Table 5. ParaSATMiner vs CESATIM vs ClosedPattern vs CoverSize.

Instance	Θ	Closed pattern	Cover size	ParaSat Miner	#Clauses	CESATIM	#Clauses	#Models
Retail	80	–	265.10	**12.51**	1617596	13.84	964643	$> 8.10^3$
	60	–	295.47	**16.48**	2091800	18.50	1179210	$> 41.10^4$
	40	–	334.23	**23.53**	3077550	27.45	1591962	$> 2.10^4$
	20	–	439.94	**39.95**	6520011	48.13	2949890	$> 5.10^4$
	10	–	586.16	**72.23**	15415367	88.36	6934878	1.10^5
Connect	40000	7.54	14.95	6.32	1518140	**3.39**	1065923	$> 7.10^4$
	20000	50.22	75.48	21.51	3205501	**8.45**	2459758	$> 5.10^5$
	10000	526.43	431.19	64.11	5448843	**21.96**	4428425	$> 3.10^6$
	5000	–	–	166.43	7549685	**51.23**	6310490	$> 1.10^7$
Chess	2000	1.51	1.22	0.21	55083	**0.12**	38584	$\simeq 7.10^4$
	1500	6.30	4.09	0.79	128897	**0.38**	96691	$> 5.10^5$
	1000	51.35	28.62	5.04	214078	**2.21**	171067	$> 4.10^6$
	500	577.29	311.47	46.07	339787	**20.37**	285651	$> 45.10^6$
	250	–	–	173.44	414786	**81.92**	352443	$\simeq 2.10^8$
	100	–	–	463.07	465613	**230.13**	395220	$> 5.10^8$
Accidents	100000	101.68	145.96	73.03	14765606	**22.12**	11654191	$\simeq 1.10^5$
	80000	319.25	283.98	142.73	17435639	**35.76**	14024172	$\simeq 4.10^5$
	60000	–	866.21	294.79	22051341	**66.49**	17995214	$> 1.10^6$
	40000	–	–	723.79	31196169	**159.74**	26000938	$\simeq 6.10^6$
Pumbs	40000	20.99	389.43	4.12	428574	**2.30**	288083	$> 2.10^4$
	35000	103.20	325.42	9.76	727145	**4.25**	500954	$\simeq 2.10^5$
	30000	434.25	404.26	29.20	1341404	**10.67**	974194	$\simeq 9.10^5$
	25000	–	994.35	132.86	2111763	**39.65**	1569301	$\simeq 6.10^6$
	20000	–	–	643.11	3801324	**194.70**	2821021	$> 3.10^7$
T40I10D100K	10000	4.16	51.43	2.69	0	**1.48**	0	$\simeq 1.10^2$
	8000	5.08	51.13	4.11	0	**1.88**	0	$> 1.10^2$
	6000	10.38	52.39	7.28	138699	**2.65**	138704	$> 2.10^2$
	4000	30.51	53.26	9.98	1198000	**3.80**	893000	$> 4.10^2$
	2000	144.89	58.22	23.63	16880498	**10.59**	11951802	$> 1.10^3$
T10I4D100K	500	106.87	24.04	3.00	916075	**1.73**	695817	$> 1.10^3$
	400	147.14	25.56	3.60	1709425	**2.21**	1108283	$> 1.10^3$
	300	217.40	27.73	4.61	3345488	**3.02**	1824098	$> 4.10^3$
	200	314.17	29.12	6.51	6226063	**4.39**	3021183	$> 1.10^4$
	100	497.10	32.40	15.24	14354996	**10.79**	6642109	$> 2.10^4$
	50	–	45.05	85.52	34040358	**67.90**	17041570	$> 4.10^4$
Kosarak	4000	–	–	26.81	8298064	**24.22**	5125364	$> 2.10^3$
	3000	–	–	37.35	10867305	**35.41**	6523942	$> 4.10^3$
	2000	–	–	62.27	15819901	**60.15**	9163330	$> 3.10^4$
	1000	–	–	141.90	35068843	**136.87**	20604547	$\simeq 5.10^5$
Mushroom	250	1.14	2.54	1.23	956487	**0.90**	742560	$> 1.10^4$
	100	1.64	1.91	2.16	1149719	**1.65**	868463	$> 3.10^4$
	50	2.70	2.47	2.37	1209497	**1.86**	893076	$> 5.10^4$
	25	2.82	3.16	2.82	1251890	**2.30**	906841	$\simeq 8.10^4$
	5	5.57	4.20	4.31	1300095	**3.74**	917710	$> 1.10^5$

To explain the performances of our compact encoding we report in columns 6 and 8 the number of total clauses generated by ParaSATMiner and CESATIM respectively. The number of clauses generated by CESATIM corresponds to the total number of clauses of all the sub-problems generated during the decomposition steps. As we can observe, using our approach the total clauses is clearly lower. For instance, for `T10I4D100K` data, `ParaSATMiner` generates *34040358* clauses where `CESATIM` needs only *17041570* clauses.

6 Conclusion

In this paper, we presented an efficient approach for compacting the SAT-based encoding of the itemset mining task. We proved that this compact reformulation can be expressed as a Boolean matrix compression problem. We proposed a greedy approach that allows us to significantly reduce the size of the encoding. Experimental results show significant compression rate with respect to the original encoding size. Interestingly, our approach leads to better performances improvements compared to both the original `ParaSATMiner` and the two well known CP approaches, namely `ClosedPattern` and `CoverSize`.

As a future work, it may be interesting to extend our approach to SAT-based encoding of other data mining problems such as sequential pattern mining. The second issue is to improve our greedy approach, by finding more better ordering heuristics.

References

1. Agrawal, R., Imielinski, T., Swami, A.N.: Mining association rules between sets of items in large databases. In: ACM SIGMOD International Conference on Management of Data, pp. 207–216. ACM Press, Baltimore (1993)
2. Belaid, M., Bessiere, C., Lazaar, N.: Constraint programming for mining borders of frequent itemsets. In: Proceedings of the Twenty-Eighth International Joint Conference on Artificial Intelligence, IJCAI 2019, Macao, China, 10–16 August 2019, pp. 1064–1070 (2019)
3. Booth, K.S., Lueker, G.S.: Testing for the consecutive ones property, interval graphs, and graph planarity using PQ-tree algorithms. J. Comput. Syst. Sci. **13**(3), 335–379 (1976)
4. Cambazard, H., Hadzic, T., O'Sullivan, B.: Knowledge compilation for itemset mining. In: ECAI 2010, pp. 1109–1110 (2010)
5. De Raedt, L., Guns, T., Nijssen, S.: Constraint programming for itemset mining. In: ACM SIGKDD, pp. 204–212 (2008)
6. Dlala, I.O., Jabbour, S., Raddaoui, B., Sais, L.: A parallel sat-based framework for closed frequent itemsets mining. In: Principles and Practice of Constraint Programming - 24th International Conference, CP 2018, Proceedings, Lille, France, 27–31 August 2018, pp. 570–587 (2018)
7. Eén, N., Sörensson, N.: An extensible sat-solver. In: Proceedings of the Sixth International Conference on Theory and Applications of Satisfiability Testing (SAT 2003), pp. 502–518 (2002)
8. Fournier-Viger, P., Lin, J.C., Vo, B., Truong, T.C., Zhang, J., Le, H.B.: A survey of itemset mining. Wiley Interdiscip. Rev. Data Min. Knowl. Discov. **7**(4) (2017)
9. Garey, M.R., Johnson, D.S., Stockmeyer, L.J.: Some simplified np-complete graph problems. Theor. Comput. Sci. **1**(3), 237–267 (1976)

10. Guns, T., Dries, A., Tack, G., Nijssen, S., De Raedt, L.: Miningzinc: a modeling language for constraint-based mining. In: Proceedings of the Twenty-Third International Joint Conference on Artificial Intelligence, IJCAI 2013, pp. 1365–1372 (2013)
11. Guns, T., Nijssen, S., De Raedt, L.: Itemset mining: a constraint programming perspective. Artif. Intell. **175**(12–13), 1951–1983 (2011)
12. Jabbour, S., Sais, L., Salhi, Y.: Boolean satisfiability for sequence mining. In: CIKM, pp. 649–658 (2013)
13. Khiari, M., Boizumault, P., Crémilleux, B.: Combining CSP and constraint-based mining for pattern discovery. In: Taniar, D., Gervasi, O., Murgante, B., Pardede, E., Apduhan, B.O. (eds.) ICCSA 2010. LNCS, vol. 6017, pp. 432–447. Springer, Heidelberg (2010). https://doi.org/10.1007/978-3-642-12165-4_35
14. Lazaar, N., Lebbah, Y., Loudni, S., Maamar, M., Lemière, V., Bessiere, C., Boizumault, P.: A global constraint for closed frequent pattern mining. In: International Conference on Principles and Practice of Constraint Programming, pp. 333–349 (2016)
15. Metivier, J.P., Boizumault, P., Crémilleux, B., Khiari, M., Loudni, S.: A constraint-based language for declarative pattern discovery. In: 2011 IEEE 11th International Conference on Data Mining Workshops (ICDMW), pp. 1112–1119. Vancouver, Canada (2011)
16. Schaus, P., Aoga, J.O.R., Guns, T.: Coversize: A global constraint for frequency-based itemset mining. In: International Conference on Principles and Practice of Constraint Programming, pp. 529–546 (2017)

A Pipe Routing Hybrid Approach Based on A-Star Search and Linear Programming

Marvin Stanczak[1,2]([⊠]), Cédric Pralet[1]([⊠]), Vincent Vidal[1]([⊠]),
and Vincent Baudoui[2]([⊠])

[1] ONERA/DTIS, Université de Toulouse, 31055 Toulouse, France
{marvin.stanczak,cedric.pralet,vincent.vidal}@onera.fr
[2] Airbus Defence and Space, Toulouse, France
{marvin.stanczak,vincent.baudoui}@airbus.com

Abstract. In this work, we consider a pipe routing problem encountered in the space industry to design waveguides used in telecommunication satellites. The goal is to connect an input configuration to an output configuration by using a pipe composed of a succession of straight sections and bends. The pipe is routed within a 3D continuous space divided into non-regular convex cells in order to take obstacles into account. Our objective is to consider several non-standard features such as dealing with a set of bends restricted to a bend catalog that can contain both orthogonal and non-orthogonal bends, or with pipes that have a rectangular section, which makes the pipe orientation important. An approach is introduced to automate the design of such pipe systems and help designers to manage the high number of constraints involved. Basically, this approach formulates the pipe routing problem as a shortest path problem in the space of routing plans, where a routing plan is specified by the parts composing a pipe and by geometrical constraints imposed on these parts. The problem is then solved using weighted A* search combined with a linear program that quickly evaluates the feasibility and the cost of a routing plan. The paper shows that the approach obtained has the capacity to solve realistic instances inspired by industrial problems.

Keywords: Pipe routing · Linear programming · Weighted A*

1 Introduction

Waveguide routing plays an important role during the design phase of a telecommunication satellite. A waveguide can be seen as a pipe with a rectangular section which carries an electromagnetic signal between two components of the satellite payload. Traditionally, waveguide designers define routes manually and build a route as a succession of rigid straight sections and rigid bends using their expertise to select a routing configuration that saves length and bends. The bends that can be used are defined in catalogs provided by waveguide manufacturers. These catalogs can include orthogonal bends (90° bends, when the direction of the pipe before the bend is orthogonal to the direction of the pipe after the bend)

© Springer Nature Switzerland AG 2021
P. J. Stuckey (Ed.): CPAIOR 2021, LNCS 12735, pp. 179–195, 2021.
https://doi.org/10.1007/978-3-030-78230-6_12

but also non orthogonal ones (30° bends, 45° bends, 60° bends, etc.) which are useful to save length in the very constrained routable space inside a satellite.

Due to the large number of waveguides to be routed and to the large number of constraints to take into account, waveguide routing is usually a highly time-consuming task. In other industries, research has been conducted to automate the pipe routing process, to save time and money during the design phase.

Existing Methods. Historically, the classic pipe routing algorithm is the *Maze algorithm* introduced by Lee [23]. It considers a discrete routing environment represented as a regular grid where the cells occupied by obstacles are labeled. Then, starting from the source cell, the best route is computed by exploring the cells' neighborhood until the target is reached. Other algorithms were proposed to consume less memory space, based on Particle Swarm Optimization [3], Genetic Algorithm [18,21], or Ant Colony Optimization [9,19]. Recently, Belov also introduced a Constraint Programming formulation [4,5]. With regard to non orthogonal bends, Ando proposed to introduce additional edges in the cell adjacency graph [2], but this approach does not guarantee to find a feasible solution even if one exists in the continuous space.

Other methods called *Skeleton approaches* build a graph of possible route candidates using rules inspired by experience. The pipe routing problem is still modeled as a shortest path problem in a discrete graph, but this time the nodes of the graph are associated with continuous positions that are not forced to be located on a regular grid. The skeleton approach was tackled using the Dikjstra algorithm [20] or Mixed Linear Integer Programming [13]. A more recent paper also explored Evolutionary Algorithms considering multi-objective routing in a skeleton graph [24]. Similarly to the cell decomposition strategy, the skeleton approach can lead to sub-optimal solutions in the continuous space.

To improve the quality of the solutions, some authors proposed a two-stage approach that first searches for a global routing channel, and then details the routes inside each cell of this channel [31], sometimes using patterns [26].

To avoid discretizing the routing space, other methods discretize the set of possible physical pipe components and build pipes by adding components one by one from a catalog containing bends and straight sections of fixed length [12]. Again, this method can fail to find feasible solutions due to the discretization.

Another way to route pipes in a continuous environment is the *Escape Algorithm* or *Line Search Algorithm*. This technique introduced by Hightower [15] extends lines from the origin point using a set of directions and repeats this extension each time a line intersects an obstacle, until the destination is reached. This approach allows dealing with non orthogonal bends but it does not take into account the orientation of the pipe sections, which is required in the case of waveguide routing. Indeed, the waveguide must reach the destination with the right orientation, that is to say with the right angular position of the section around the *neutral fibre* followed by the barycentre of the section along the pipe.

Last, parametric models were also proposed using adaptable pipe patterns with parameters such as the length of straight sections. The pipe route can then be optimized using Mathematical Programming [25,28] or Genetic Algorithms [17], but existing models do not consider non orthogonal bends.

Contribution. Most of the previous methods rely on the classical assumptions that the pipe has a circular section, has axis-parallel segments and/or uses orthogonal bends only, but these hypotheses are not acceptable for waveguides. For these reasons, no method from the literature can be reused to solve the waveguide routing problem. This article introduces a new pipe routing technique that addresses the three following challenges: optimize pipe cost in a 3D continuous routing space, use non orthogonal bends from a catalog and take into account unsymmetrical pipe sections. We consider the routing of a single pipe only. In an industrial context, such a routing problem must be solved in few seconds in order to ensure quick iterations during the design phase. The extended problem which deals with several pipes is left for future work.

The rest of the article is organized as follows. Section 2 introduces a formal definition of the pipe routing problem considered. Section 3 describes the notion of routing plan and presents a linear program that evaluates the feasibility and cost of a plan. Section 4 formulates the pipe routing problem as a shortest path problem in the space of routing plans and defines heuristics for estimating the cost required to reach the goal configuration from the current routing plan. Section 5 provides experimental results obtained with the weighted A* search, and Sect. 6 gives future work perspectives.

2 Problem Definition

2.1 Routing Space

The routing space $W \subseteq \mathbb{R}^3$ describes the physical space that the pipe neutral fibre can cross. The structure of a telecommunication satellite is made of several panels on which waveguides must be fixed. Thus, the routable space is naturally split into several cells that correspond to the close environment of each panel. As about one hundred obstacles can be fixed on a panel, these cells are divided into sub-cells that avoid obstacles, using a Delaunay constrained triangulation. This way, we model the routing space W as a set of cells C where each cell $c \in C$ is a convex polyhedron P_c (see Fig. 1).

Two cells $c, c' \in C$ that overlap are called adjacent, and their *interface* i is the non-empty polyhedron $P_i = P_c \cap P_{c'}$. We denote by I the set of routing interfaces between cells, and by $I(c)$ the set of interfaces involving cell c. We assume here that each interface is a surface, even if the techniques defined can be extended to the case where interfaces are volumes. From these definitions, the routing space can be represented as a graph containing one node per routing cell and one edge per routing interface between two cells. In the following, this graph is assumed to be connected.

2.2 Input and Output Configurations

In pipe routing, the goal is to connect an input configuration with an output configuration. Mathematically speaking, a *configuration* θ is a pair (P_θ, o_θ) where

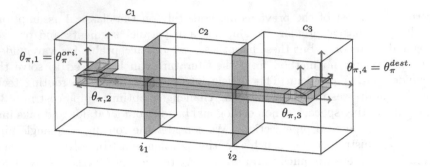

Fig. 1. Pipe in a simple routing space containing three cells (interfaces between cells are depicted in blue, straight sections in orange and bends in green) (Color figure online)

P_θ is a point in $\mathbb{R}^3$ and o_θ is an orientation defined by three vectors $(\overrightarrow{e_{o,1}}, ..., \overrightarrow{e_{o,3}})$ that form an orthonormal basis of $\mathbb{R}^3$ with $\overrightarrow{e_{o,3}}$ normal to the pipe section.

The origin point is located in a convex polyhedron $\mathcal{P}^{ori.}$ included in an origin cell $c^{ori.} \in \mathcal{C}$ and the destination point is located in a convex polyhedron $\mathcal{P}^{dest.}$ included in a destination cell $c^{dest.} \in \mathcal{C}$. Using polyhedrons $\mathcal{P}^{ori.}$ and $\mathcal{P}^{dest.}$ instead of fixed origin and destination points allows keeping some flexibility in early design phases. For instance, if a straight pipe section can connect an origin and a destination configuration up to a small position error, it can be relevant to slightly move the origin or destination point instead of defining a pipe that contains a loop to compensate for the position error.

Additionally, the (unique) origin orientation $o^{ori.}$ can be set as the reference orientation, and there exists a set of possible destination orientations $\mathcal{O}^{dest.}$, either because the destination orientation is still flexible, or because some orientations are equivalent from the pipe section point of view.

2.3   Straight Sections and Bends

To connect input and ouput configurations, engineers use catalogs of parts referred to as *straight sections* and *bends*. In our modeling, these parts are approximated as functions that transform a configuration $\theta = (P_\theta, o_\theta)$ into another configuration $\theta' = (P_{\theta'}, o_{\theta'})$.

A straight section u of length $L_u \in \mathbb{R}^+$ is modeled as an application that transforms a configuration $\theta = (P_\theta, o_\theta)$ into a configuration $\theta' = u(\theta) = (P_{\theta'}, o_{\theta'})$ where point $P_{\theta'}$ is obtained by a translation of P_θ along direction $\overrightarrow{e_{o_\theta,3}}$ ($P_{\theta'} = P_\theta + L_u \overrightarrow{e_{o_\theta,3}}$) and where the orientation is unchanged ($o_{\theta'} = o_\theta$). In the following, a translation of length L is referred to as t_L, therefore $u(\theta) = t_{L_u}(\theta)$.

The modeling of bends is a bit more complex. Basically, a bend b has a certain length and changes the direction of the pipe according to a given bend radius. It is approximated as a pipe section composed of a straight section of length $L_b \in \mathbb{R}^+$ called the *half-length* of the bend, an orientation change defined by a rotation matrix M_b, and another straight section of length L_b again (see Fig. 1). More formally, a bend b is modeled as a function that transforms a configuration $\theta = (P_\theta, o_\theta)$ into a configuration $t_{L_b} \circ r_{M_b} \circ t_{L_b} (\theta)$ obtained by applying (1) a translation t_{L_b} of length L_b, (2) a rotation r_{M_b} defined by matrix M_b, and (3) a second translation t_{L_b} of length L_b. For a given bend, the intermediate configuration $r_{M_b} \circ t_{L_b} (\theta)$ is called the *transition configuration*. It corresponds to the configuration obtained just after the break point of the bend.

In the following, we consider a restricted set of bends B called the *catalog of bends*. It can contain only 90° bends for instance, or 90° and 45° bends, or any other combination of bend types and angles. Using a catalog of bends can seem counter-intuitive at first since bend suppliers often propose a continuous choice of bend angle radius, and since complex bends could be obtained through 3D-printing. However, using a catalog of bends allows reusing bends and ordering numerous bends of the same type, which reduces the costs.

2.4 Pipe and Polyline Approximation

Given the catalog of bends B, a pipe π of length $N_\pi > 0$ is then a pair $\pi = (\theta_\pi^{ori.}, \sigma_\pi)$ composed of an initial configuration $\theta_\pi^{ori.}$ and a composition

$$\sigma_\pi = u_{\pi,N_\pi} \circ b_{\pi,N_\pi - 1} \circ u_{\pi,N_\pi - 1} \circ ... \circ u_{\pi,2} \circ b_{\pi,1} \circ u_{\pi,1}$$

alternating between straight sections $u_{\pi,i}$, for $i \in [\![1, N_\pi]\!]$, and bends $b_{\pi,i} \in B$, for $i \in [\![1, N_\pi - 1]\!]$ (see Fig. 1).

As straight sections and bends apply translation and rotation operations on the initial configuration, the neutral fibre of pipe π describes a polyline $[P_1, ..., P_{N_\pi+1}]$ in $\mathbb{R}^3$ whose points correspond to the transition configurations of the bends (except for the first and last points). The ith point of this polyline is $P_{\pi,i}$, and the ith segment is $[P_{\pi,i}, P_{\pi,i+1}]$. We also denote by $\ell_{\pi,i}$ the length of the ith segment of pipe π, and by $\theta_{\pi,i} = (P_{\pi,i}, o_{\pi,i})$ the ith transition configuration of π, for $i \in [\![1, N_\pi + 1]\!]$ (with the convention that $\theta_{\pi,1} = \theta_\pi^{ori.}$, and that $\theta_{\pi,N_\pi+1}$ is the configuration obtained at the end of the pipe).

2.5 Constraints

We now list the constraints that must be satisfied by a pipe π.

- The initial and final configurations of π must be consistent with the required input and output configurations, that means respectively $P_{\pi,1} \in \mathcal{P}^{ori.}$ and $P_{\pi,N_\pi+1} \in \mathcal{P}^{dest.}$, but also $o_{\pi,1} = o^{ori.}$ and $o_{\pi,N_\pi+1} \in \mathcal{O}^{dest.}$.
- The polyline defined by the pipe must contain at most N^{max} segments. This limits the number of break points, which is particularly useful when the pipe conveys a flow whose quality is impacted by such break points.
- Every bend used in π must belong to catalog B.
- All segments of the polyline traversed by the neutral fibre of the pipe must be contained within the routing space, that is $[P_{\pi,i}, P_{\pi,i+1}] \subset \mathcal{W}$ for every

$i \in [\![1, N_\pi]\!]$. For computational reasons, the constraints enforcing that a pipe section must not cross other pipe sections are omitted. They can be checked afterwards, and in case of violation, a manual modification of the pipe design is required.

- For stability reasons, the pipe must be attached to floors or walls of the structure within which it is routed. As a result, there is a set of orientations $\mathcal{O}_c$ allowed within each cell $c \in \mathcal{C}$. Typically, for a rectangular pipe section we impose that the orientation o of each pipe segment passing through a cell c must be orthogonal to a facet of c called a *wall*. If $\overrightarrow{u}$ denotes the normal of this wall, we must ensure that either $\overrightarrow{e_{o,1}}.\overrightarrow{u} = 0$ or $\overrightarrow{e_{o,2}}.\overrightarrow{u} = 0$ (possibility to attach a bracket to one of the sides of the rectangular section).
- Every straight section of the pipe must have a minimum length $L^{min} \in \mathbb{R}^+$. This specification comes from constraints imposed by suppliers.

2.6 Objective Function

The objective in the pipe routing problem is to minimize the overall cost of the pipe, given that each bend b has a cost $C_b \in \mathbb{R}^+$ and each straight section of length ℓ has a cost $\ell \mu$ where $\mu \in \mathbb{R}^+$ is a linear cost. The objective is then to minimize the global cost C_π of the pipe defined by:

$$C_\pi = \mu \sum_{i=1}^{N_\pi} \ell_{\pi,i} + \sum_{i=1}^{N_\pi - 1} C_{b_{\pi,i}}$$

3 Routing Plan

In order to route a pipe, we propose to iteratively build its neutral fibre starting from the origin configuration, by making at each step decisions such as the addition of a new bend at the end of the pipe or the crossing of an interface between two adjacent cells. To formalize the approach, the concept of *routing plan* is introduced to represent the decisions made so far on the pipe components.

3.1 Definition

A routing plan s describes, in an abstract way, a neutral fibre composed of N_s successive segments with for each segment i:

- the bend $b_{s,i} \in \mathcal{B}$ applied at the end point of segment i, for $i \in [\![1, N_s - 1]\!]$;
- the sequence of interfaces $\mathcal{I}_{s,i} \subseteq \mathcal{I}$ crossed by segment i, for $i \in [\![1, N_s]\!]$.

Moreover, a routing plan can be terminated or not. When a routing plan is terminated, the neutral fibre has to reach the pipe destination. The set of routing plans is denoted by $\mathcal{S}$. Several data can be derived from the basic definition of a routing plan, including:

- the orientation $o_{s,i} \in \mathcal{O}$ of segment i, for $i \in [\![1, N_s]\!]$; this orientation can be computed from the origin orientation and from the list of bends applied;
- the cell $c_{s,i} \in \mathcal{C}$ to which the end point of segment i belongs, for $i \in [\![1, N_s]\!]$; this cell can be computed from the last interface crossed by segment i.

3.2   Feasibility and Cost

A routing plan $s \in \mathcal{S}$ is *feasible* if it is possible to create a neutral fibre following the choices made in s and satisfying the pipe routing constraints introduced in Sect. 2.5. This feasibility problem can be formulated as a linear program, referred to as LP_s, that contains four kinds of variables:

- *length variables* ℓ_i^s such that, for $i \in [\![1, N_s]\!]$, the real variable ℓ_i^s is the length of the ith segment of the neutral fibre;
- *position variables* $p_i^s = (p_{i,x}^s, p_{i,y}^s, p_{i,z}^s)$ such that, for $i \in [\![1, N_s + 1]\!]$, the real variable $p_{i,x}^s$ (respectively $p_{i,y}^s$ and $p_{i,z}^s$) is the x-coordinate (respectively y-coordinate and z-coordinate) of the ith point of the neutral fibre;
- *interface variables* $q_{i,j}^s = (q_{i,j,x}^s, q_{i,j,y}^s, q_{i,j,z}^s)$ such that, for $i \in [\![1, N_s]\!]$ and $j \in \mathcal{I}_{s,i}$, the real variable $q_{i,j,x}^s$ (respectively $q_{i,j,y}^s$ and $q_{i,j,z}^s$) is the x-coordinate (respectively y-coordinate and z-coordinate) of the intersection between the ith segment of the neutral fibre and an interface j it has to cross;
- *interface distance variables* $\alpha_{i,j}^s$ such that, for $i \in [\![1, N_s]\!]$ and $j \in \mathcal{I}_{s,i}$, the real variable $\alpha_{i,j}^s$ is the distance between the ith point of the neutral fibre and the intersection $q_{i,j}^s$ with the interface j it has to cross.

Linear program LP_s can be formulated as follows:

$$\text{minimize } \mu \sum_{i=1}^{N_s} \ell_i^s + \sum_{i=1}^{N_s - 1} C_{b_{s,i}} \tag{1}$$

subject to :

$$p_1^s \in \mathcal{P}^{ori.} \tag{2}$$

$$p_{N_s+1}^s \in \mathcal{P}^{dest.} \qquad \text{if } s \text{ is terminated} \tag{3}$$

$$p_{i+1}^s \in \mathcal{P}_{c_{s,i}} \qquad \forall i \in [\![1, N_s]\!] \tag{4}$$

$$\ell_i^s \geq L_{b_{s,i-1}} + L^{min} + L_{b_{s,i}} \qquad \forall i \in [\![1, N_s]\!] \tag{5}$$

$$\overrightarrow{p_i^s p_{i+1}^s} = \ell_i^s \overrightarrow{e_{o_{s,i},3}} \qquad \forall i \in [\![1, N_s]\!] \tag{6}$$

$$q_{i,j}^s \in \mathcal{P}_j \qquad \forall i \in [\![1, N_s]\!] \quad \forall j \in \mathcal{I}_{s,i} \tag{7}$$

$$\alpha_{i,j}^s \leq \ell_i^s \qquad \forall i \in [\![1, N_s]\!] \quad \forall j \in \mathcal{I}_{s,i} \tag{8}$$

$$\overrightarrow{p_i^s q_{i,j}^s} = \alpha_{i,j}^s \overrightarrow{e_{o_{s,i},3}} \qquad \forall i \in [\![1, N_s]\!] \quad \forall j \in \mathcal{I}_{s,i} \tag{9}$$

$$\ell_i^s \in \mathbb{R}^+ \qquad \forall i \in [\![1, N_s]\!] \tag{10}$$

$$p_i^s \in \mathbb{R}^3 \qquad \forall i \in [\![1, N_s + 1]\!] \tag{11}$$

$$q_{i,j}^s \in \mathbb{R}^3 \quad \alpha_{i,j}^s \in \mathbb{R}^+ \qquad \forall i \in [\![1, N_s]\!] \quad \forall j \in \mathcal{I}_{s,i} \tag{12}$$

Constraint 2 states that the neutral fibre must start from the origin cell. Such an inclusion constraint within a convex polyhedron can be expressed as a set of linear constraints. Constraint 3 imposes that if plan s is terminated, the neutral fibre must reach the destination cell. In the same way, Constraints 4 force the successive break points of the neutral fibre to belong to the cell to which they are allocated given plan s. Constraints 5 impose a minimal length on the segments. For the ith segment, this minimum length is obtained from the minimal length L^{min} of straight sections and from the respective contributions

$L_{b_s,i-1}$ and $L_{b_s,i}$ of the previous and next bends (see Fig. 2). By convention, we assume that $L_{b_s,0} = 0$ and $L_{b_s,N_s} = 0$, since the first and last segments do not have a previous or a next bend respectively. Constraints 6 define the coordinates of the $(i+1)^{\text{th}}$ point of the neutral fibre from the coordinates of the ith point, the orientation of the ith segment as specified by routing plan s, and the length of this segment. Constraints 7–9 impose that the intersection between the ith segment and an interface j it has to cross must belong both to the interface and to the segment.

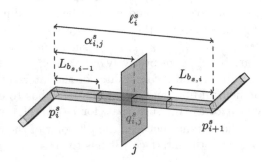

Fig. 2. Portion of a pipe between two successive break points of the neutral fibre (the straight section with variable length is depicted in orange). (Color figure online)

If LP_s does not have a solution, then routing plan s is not feasible and, by convention, its cost is infinite. On the contrary, when there is a solution, routing plan s is *feasible*. In this case, the optimum value $g(s)$ of LP_s is a lower bound on the cost of any pipe that satisfies the constraints defined by routing plan s. Furthermore, the straight sections of the pipe can be rebuilt from the optimal lengths found for the pipe segments, while the bends are already defined by s.

4   Shortest Path Problem Formulation

Based on the routing plan presented above, the pipe routing problem can be seen as a shortest path problem in a graph whose nodes correspond to routing plans and whose arcs represent the basic updates that can be made on routing plans: bend addition, interface crossing, or pipe termination. The goal is then to explore the space of routing plans $\mathcal{S}$ from an empty routing plan starting at the origin cell $c^{ori.}$ with the origin orientation $o^{ori.}$, in order to reach a feasible and terminated plan, ending at destination cell $c^{dest.}$ with a destination orientation in $\mathcal{O}^{dest.}$. Thus, the pipe routing problem can be solved using an A*-like algorithm after formally defining the possible successors of a routing plan in the search space (see Sect. 4.2) and a path-finding heuristic (see Sect. 4.3).

4.1   Search Algorithms

Various heuristic search schemes can be considered to solve shortest path problems. Several of them were compared in [29], showing that hill climbing methods can provide fast results [16,22], while best-first search [1,8,11] and beam search [6,10,30] often offer a better trade-off between solution quality and computation time. Also, compared to beam search, best-first search is more suitable for state spaces in which destination states cannot be reached from some states. We focus here on best-first search techniques, that develop at each step the search node that seems the more promising among all search nodes created so far.

Among the best-first search algorithms, the Weighted A* approach [27] provides competitive results in many fields. Basically, in the A* algorithm [14], the evaluation of a state s is given as $f(s) = g(s) + h(s)$ where $g(s)$ stands for the minimal cost to reach s from the initial state and $h(s)$ stands for the heuristic evaluation of the minimum cost to reach a goal state starting from s. In Weighted A* (denoted WA*), this evaluation is replaced by $f(s) = g(s) + \epsilon \cdot h(s)$ where $\epsilon > 1$ gives a higher weight to the heuristic evaluation, so as to favor the expansion of search nodes corresponding to configurations that seem to be close to the destination. In our case, $g(s)$ is obtained from the value of an optimal solution to LP_s, while three possible versions of $h(s)$ are presented in Sect. 4.3.

4.2   Neighborhood

By definition, a terminated routing plan cannot be expanded. On the contrary, if routing plan $s \in S$ is not terminated, it can be expanded by adding new routing decisions, as defined below. During plan expansions, only the feasible plans are considered as successors. Indeed, if LP_s is not feasible and if plan $s' \in S$ extends s, then $LP_{s'}$ is not feasible either because LP_s is a subproblem of $LP_{s'}$.

Bend Addition. Let b be a bend of catalog B. If $N_s < N^{max}$ and if the orientation $r_{M_b}(o_{s,N_s})$ obtained after applying bend b from the current last orientation o_{s,N_s} is acceptable in the current last cell c_{s,N_s} ($r_{M_b}(o_{s,N_s}) \in \mathcal{O}_{c_{s,N_s}}$), then bend b can be added to routing plan s. This adds a new segment to the neutral fibre and definitively allocates the end point of the N_s^{th} segment to cell c_{s,N_s}. Formally, the resulting routing plan s' is defined as follows:

$$N_{s'} = N_s + 1 \tag{13}$$

$$b_{s',i} = b_{s,i} \qquad \forall i \in [\![1, N_{s'} - 2]\!] \tag{14}$$

$$b_{s',N_{s'}-1} = b \tag{15}$$

$$\mathcal{I}_{s',i} = \mathcal{I}_{s,i} \qquad \forall i \in [\![1, N_{s'} - 1]\!] \tag{16}$$

$$\mathcal{I}_{s',N_{s'}} = [\,] \tag{17}$$

Interface Crossing. Let j be a neighbor interface of the current cell, that is $j \in \mathcal{I}(c_{s,N_s})$. Let n_j be the normal to the interface that is oriented towards the current cell. If the scalar product between the current orientation o_{s,N_s} and n_j is negative and if o_{s,N_s} satisfies the orientation constraints of the cell c_j located on the other side of the interface ($o_{s,N_s} \in \mathcal{O}_{c_j}$), then interface j can be crossed. We also limit ourselves to the destination cells that have not been visited before, meaning that $c_j \neq c_{s,i}$ for $i \in [\![1, N_s]\!]$. Formally, the resulting plan s' is defined as follows:

$$N_{s'} = N_s \tag{18}$$
$$b_{s',i} = b_{s,i} \qquad \mathcal{I}_{s',i} = \mathcal{I}_{s,i} \qquad \forall i \in [\![1, N_{s'} - 1]\!] \tag{19}$$
$$\mathcal{I}_{s',N_{s'}} = \mathcal{I}_{s,N_{s'}} \cup [j] \tag{20}$$

Pipe Termination. Last, if the destination cell and the destination orientation have been reached, meaning that $c_{s,N_s} = c^{dest.}$ and $o_{s,N_s} \in \mathcal{O}^{dest.}$, then routing plan s can be terminated. The resulting plan s' is the same as plan s but it is terminated and it has to reach the destination position (see Constraint 3).

4.3   Trail Heuristic

In order to choose a routing plan s to expand at each step, one key component is the heuristic evaluation of s, that must give an estimation of the minimum cost required to extend s and reach a feasible and terminated routing plan. This section proposes three heuristic evaluations.

Distance as the Crow Flies. A first naive heuristic evaluation is the distance as the crow flies from the end point of the routing plan to the destination position, referred to as $h_{a.c.f.}$. However, this heuristic does not take into account the routing space constraints. The introduced method with this heuristic and $\epsilon = 1$ is used as baseline because of the lack of a usable approach from the literature.

Trail Space. To get more accurate estimations of the distance to the destination, we introduce two other heuristics based on a graph $G(\mathcal{M}, \mathcal{D})$ defining so-called *candidate trails* between some specific points of the routing space. Formally, a *trail* for a routing plan $s \in \mathcal{S}$ is a polyline inside the routing space that connects the end point $p^s_{N_s+1}$ of s as provided by LP_s to the destination polyhedron $\mathcal{P}^{dest.}$. In a trail, the constraints expressing that each orientation change of the polyline must correspond to a bend of the catalog are ignored, the goal being only to estimate the quality of a routing plan. Furthermore, the sequence of interfaces crossed by a trail is called a *channel*.

   We propose to discretize the trail space during a preprocessing step by sampling a set of points $\mathcal{M}(i)$ on each interface $i \in \mathcal{I}$ using Bridson's algorithm [7]. It is a maximum Poisson disk sampling method which ensures that sampled points are separated by a *sampling radius* $\rho \in \mathbb{R}^+$, such that when ρ decreases, the density of sampled points increases. Then, for two distinct interfaces i, i' involving a common cell c ($i, i' \in \mathcal{I}(c)$), each point in $\mathcal{M}(i)$ is connected to all the points in $\mathcal{M}(i')$. Last, a set of points $\mathcal{M}(\mathcal{P}^{dest.})$ is generated in the destination polyhedron $\mathcal{P}^{dest.}$. The points in $\mathcal{M}(\mathcal{P}^{dest.})$ are connected to points on the interfaces of $c^{dest.}$. The graph obtained is denoted $G(\mathcal{M}, \mathcal{D})$.

Trail Length Heuristic. During a preprocessing step, the distance L_M from each point $M \in \mathcal{M}$ to the destination cell as well as the corresponding shortest trail is computed using a Dijkstra procedure. Then, the remaining cost to the destination for a routing plan $s \in \mathcal{S}$ can be evaluated based on the length of the shortest trails in $G(\mathcal{M}, \mathcal{D})$ (see Fig. 3). The estimation is further improved by using the shortest trail for an *extended routing plan* $\tilde{s} \in \mathcal{S}$ that can be computed by crossing as many interfaces as possible in the channel of the shortest trail from plan s without adding any bends. The feasibility of the extended plans is evaluated using the linear programs, whose solutions can be reused when the extended plans are expanded in turn. Finally, this heuristic is defined by:

$$h_{length}(s) = \begin{cases} \mu \cdot \min_{M \in \mathcal{M}(\mathcal{P}^{dest.})} \|\overrightarrow{p_{N_{\tilde{s}}+1}^{\tilde{s}} M}\| & \text{if } c_{\tilde{s}, N_{\tilde{s}}} = c^{dest.}, \\ \mu \cdot \min_{M \in \cup_{i \in \mathcal{I}\left(c_{\tilde{s}, N_{\tilde{s}}}\right)} \mathcal{M}(i)} \left(\|\overrightarrow{p_{N_{\tilde{s}}+1}^{\tilde{s}} M}\| + L_M \right) & \text{otherwise.} \end{cases}$$

(21)

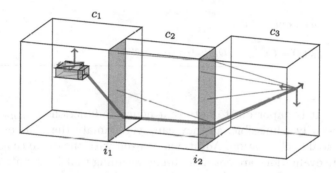

Fig. 3. Trail heuristic on a simple routing case (trails are depicted in green). (Color figure online)

Trail Cost Heuristic. Previous heuristics only estimate the lineic contribution to the cost of a pipe. However, the polyline $[M_1, ..., M_K]$ of the shortest trail for extended plan $\tilde{s}$ can also be used to estimate the cost of the remaining bends that will be added to the pipe to reach the destination. This approach is described in Algorithm 1, that returns a heuristic evaluation referred to as h_{cost}. This algorithm analyzes the successive orientation changes on the trail and tries to reproduce these changes with the bends of the catalog. It estimates the bend cost using a function σ that takes as an input the orientation o of the pipe and a vector $\overrightarrow{u}$, and that returns as an output a quantity $\sigma(o, \overrightarrow{u}) = \overrightarrow{e_{o,3}} \cdot \frac{\overrightarrow{u}}{\|\overrightarrow{u}\|}$ estimating through a scalar product the angular proximity between the main direction $\overrightarrow{e_{o,3}}$ of orientation o and the direction defined by $\overrightarrow{u}$. The closer $\sigma(o, \overrightarrow{u})$ is to 1, the closer o is from direction $\overrightarrow{u}$. Note that with such an approach, we ignore the orientation of the pipe section around the neutral fibre. Last, from the last orientation o reached when following the trail, Algorithm 1 also adds the cost $Dijkstra\left(o, \mathcal{O}^{dest.}\right)$ of the best bend combination that reaches a destination orientation from o. Such a cost can be precomputed for any reachable orientation o, before using Algorithm 1.

Algorithm 1. Evaluate $h_{cost}(s)$

Require: $s \in \mathcal{S}$, shortest trail polyline $[M_1, ..., M_K]$ from extended plan $\tilde{s}$.

$o \leftarrow o_{s,N_s}$

$C_{bends} \leftarrow 0$

for $k \in [\![1, K-1]\!]$ **do**

    repeat

        $improvement \leftarrow false$

        $b^* \leftarrow \mathrm{argmax}_{b \in B}\left(\sigma\left(r_{M_b}(o), \overrightarrow{M_k M_{k+1}}\right)\right)$

        if $\sigma\left(r_{M_{b^*}}(o), \overrightarrow{M_k M_{k+1}}\right) > \sigma\left(o, \overrightarrow{M_k M_{k+1}}\right)$ **then**

            $improvement \leftarrow true$

            $C_{bends} \leftarrow C_{bends} + C_{b^*}$

            $o \leftarrow r_{M_{b^*}}(o)$

        end if

    until $improvement = false$

end for

return $h_{length}(s) + C_{bends} + Dijkstra\left(o, \mathcal{O}^{dest.}\right)$

Admissibility. It is important to note that heuristic functions h_{length} and h_{cost} are not necessarily admissible, as they can overestimate the real cost required to reach the goal configuration. Additionally, even with the distance as the crow flies, the state evaluations are not monotonic, meaning that for a state s' extending s, we might have $f(s') < f(s)$ (proof omitted for space reasons, but basically it is possible to define an example in which the intermediate end point implicitly chosen when solving LP_s does not belong to an optimal path). These remarks imply that the first feasible and terminated routing plan reached by A*-like algorithms with these heuristics is not necessarily optimal in our case.

5   Experiments

The shortest path formulation presented in Sect. 4 for the pipe routing problem can be solved using a Weighted A* (WA*) search hybridized with a linear program which evaluates the cost of routing plans. This hybrid approach has been implemented using the Java language and the simplex solver from `Apache Commons Math 3.6.1` (but other LP solvers might be faster). It has been tested on four routing problems of increasing complexity, with the three proposed heuristics $h_{a.c.f.}$, h_{length} and h_{cost}. Experiments were conducted on an Intel 2.70 GHz processor with 16 GB of RAM. In absence of a reusable approach from the literature, the baseline is the proposed method with $h_{a.c.f.}$ and $\epsilon = 1$.

5.1   Test Cases

In practice, the proposed method is used by adding conflicting obstacles iteratively into the cell decomposition. Thus, it is possible to reduce industrial cases
to simple setups like the one presented on Fig. 4. The dimensions of this routing
space are 380 by 120 by 120. Instances 1, 2, 3 and 4 split this routing space
into respectively 8, 34, 62 and 78 cells after adding respectively 0, 18, 42 and
58 obstacles into the cell decomposition. All instances aim at connecting configuration θ_s to configuration θ_d using a maximum number of 100 bends (that
defines the maximum number of segments). The bend catalog contains 90° and
45° bends that have a common cost $C_b = 100$. The linear cost is set to 1. In
the industry, such instances have to be solved many times, so a solution must
be found quickly. For this purpose, the runtime has been limited to 3 min here,
and the search is stopped when the first solution is found.

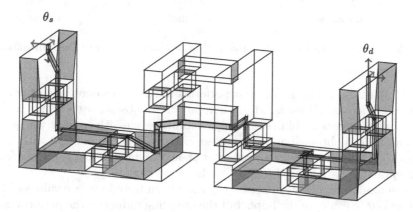

Fig. 4. Configurations of the instances, with the routing space and the solution of
instance 2 found using WA* search and heuristic h_{cost} with $\epsilon = 1.2$ and $\rho = 5$.

5.2   Results and Discussion

Our approach has been tested with the WA* search using ϵ values set to 1
(corresponding to A*), 1.2, 1.5, 2, 3 and 5. For the trail heuristics, the sampling
radius ρ is set to 5. Figure 5 shows the performances and the number of visited
plans for the different heuristics. For clarity reasons, only the best values of ϵ
are shown for each heuristic.

  As expected, heuristic $h_{a.c.f.}$ which uses the distance as the crow flies does
not succeed to solve difficult instances and visits many more routing plans than
other heuristics. In the same way, the trail length heuristic h_{length} requires a
high value of $\epsilon = 5$ to be able to solve more complex instances, and it does not
manage to solve Instance 4 within an acceptable runtime. Nevertheless, when
heuristics $h_{a.c.f.}$ and h_{length} succeed, they provide good solutions.

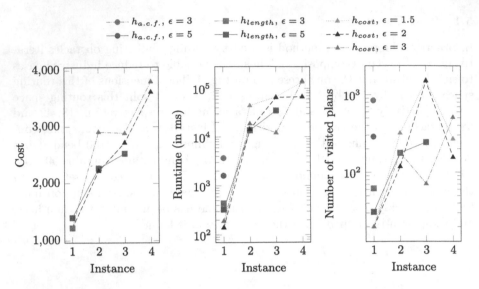

Fig. 5. Results of the three heuristics with $\rho = 5$ on the four test case instances.

In comparison, heuristic h_{cost} outperforms the previous ones when the number of cells increases, that is when there are more obstacles to avoid. Indeed, it solves all instances within reasonable runtimes, even with smaller values of ϵ, and the solutions obtained on the smallest instances are competitive with $h_{a.c.f.}$ and h_{length}. Difficult cases are solved after about one minute and the pipes found are acceptable by a human designer, as shown on Fig. 4.

In a second experiment, heuristic h_{cost} has been tested using ρ values of 1, 5, 10, 25 and 50 to evaluate the impact of the sampling radius on the performances. The results are shown on Fig. 6. WA*-search tends to visit more routing plans when the sampling radius ρ decreases, meaning that more points are sampled on each interface. Consequently, the resolution of the routing instances takes more time. However, increasing the sampling density seems to improve the cost of the solutions. This trend must be confirmed on more test cases, but the adjustment of the sampling radius would be a trade-off between the runtime and the quality of the solutions. On our instances, the best complete tuning uses $\epsilon = 2$ and $\rho = 5$, which provides the solution given in Fig. 4.

Obviously, the regularity of the solutions depends on the bend cost. The instances used in this paper strongly penalize the bends in comparison with the linear cost, which favors the regularity of the pipes. Bend costs closer to the lineic cost would lead to less regular optimal pipes and the approach would provide solutions that might be rejected by designers because of a high number of bends.

Last, real test cases have confirmed that linear program LP_s does not prevent a pipe from colliding with itself. It particularly happens to correct an infinitesimal gap on one axis between the origin and destination positions. This case can be fixed using a tolerance on positions, but the phenomenon also occurs when

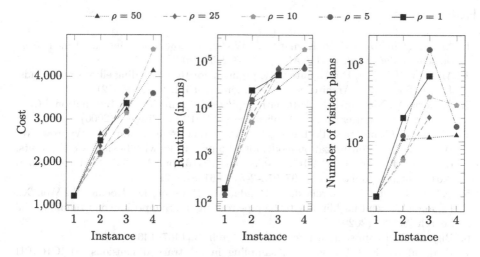

Fig. 6. Impact of the sampling radius ρ using WA*-search with $\epsilon = 2$ and h_{cost}.

the pipe orientation has to change in a small space. The automated management of self interfering constraints is a future work direction.

6   Conclusion

This paper presented an algorithm for routing a single pipe in a continuous 3D routing space divided into convex cells, using both orthogonal and non-orthogonal bends. The approach is able to take into account the orientation of the pipe section, which is particularly useful for rectangular pipes. It is based on a combination between weighted A* search in the discrete space of routing plans and linear programming for evaluating the feasibility and the cost of a routing plan in the continuous routing space. Several heuristics were proposed and compared on realistic instances. The best one, based on a quick estimation of the remaining trail to reach the destination, provides acceptable solutions within reasonable runtimes and can be used on industrial cases.

For future work, the performance of the proposed approach could be improved by adding to the routing plan the definition of the next interface that should be crossed. The underlying idea is that the linear program could exploit such a specification to generate a better intermediate end point. Furthermore, the linear program LP_s is solved from scratch at each iteration, but it could be warm-started using the solution formed for the prior routing plan. Additionally, instead of A* or weighted A*, a full Mixed Integer Linear Programming (MILP) model could be tested, even if experiments performed on obstacle-free routing spaces showed that getting results within a few seconds can be challenging for MILP. Last, the method must be extended to the multiple pipe routing problem using an high level approach like in [5].

References

1. Aine, S., Chakrabarti, P., Kumar, R.: AWA*-a window constrained anytime heuristic search algorithm. In: IJCAI, pp. 2250–2255 (2007)
2. Ando, Y., Kimura, H.: An automatic piping algorithm including elbows and bends. J. Jpn. Soc. Naval Architects Ocean Engineers **15**, 219–226 (2012)
3. Asmara, A., Nienhuis, U.: Automatic piping system in ship. In: International Conference on Computer and IT Application (COMPIT). Citeseer (2006)
4. Belov, G., Czauderna, T., Dzaferovic, A., Garcia de la Banda, M., Wybrow, M., Wallace, M.: An optimization model for 3d pipe routing with flexibility constraints. In: Beck, J.C. (ed.) CP 2017. LNCS, vol. 10416, pp. 321–337. Springer, Cham (2017). https://doi.org/10.1007/978-3-319-66158-2_21
5. Belov, G., Du, W., Garcia de al Banda, M., Harabor, D., Koenig, S., Wei, X.: From multi-agent pathfinding to 3D pipe routing. In: Symposium on Combinatorial Search (SoCS) (2020)
6. Bisiani, R.: Beam search. Encycl. Artif. Intell. **2**, 1467–1468 (1992)
7. Bridson, R.: Fast Poisson disk sampling in arbitrary dimensions. SIGGRAPH Sketches **10**, 1 (2007)
8. Ebendt, R., Drechsler, R.: Weighted A* search-unifying view and application. Artif. Intell. **173**(14), 1310–1342 (2009)
9. Fan, X., Lin, Y., Ji, Z.: The ant colony optimization for ship pipe route design in 3D space. In: 2006 6th World Congress on Intelligent Control and Automation, vol. 1, pp. 3103–3108. IEEE (2006)
10. Furcy, D., Koenig, S.: Limited discrepancy beam search. In: IJCAI (2005)
11. Furcy, D., Koenig, S.: Scaling up WA* with commitment and diversity. In: IJCAI, pp. 1521–1522 (2005)
12. Furuholmen, M., Glette, K., Hovin, M., Torresen, J.: Evolutionary approaches to the three-dimensional multi-pipe routing problem: a comparative study using direct encodings. In: Cowling, P., Merz, P. (eds.) EvoCOP 2010. LNCS, vol. 6022, pp. 71–82. Springer, Heidelberg (2010). https://doi.org/10.1007/978-3-642-12139-5_7
13. Guirardello, R., Swaney, R.E.: Optimization of process plant layout with pipe routing. Comput. Chem. Eng. **30**(1), 99–114 (2005)
14. Hart, P.E., Nilsson, N.J., Raphael, B.: A formal basis for the heuristic determination of minimum cost paths. IEEE Trans. Syst. Sci. Cybern. **4**(2), 100–107 (1968)
15. Hightower, D.W.: A solution to line-routing problems on the continuous plane. In: Proceedings of the 6th Annual Design Automation Conference, pp. 1–24 (1969)
16. Hoffmann, J.: Extending FF to numerical state variables. In: ECAI pp. 571–575. Citeseer (2002)
17. Ikehira, S., Kimura, H., Ikezaki, E., Kajiwara, H.: Automatic design for pipe arrangement using multi-objective genetic algorithms. J. Jpn Soc. Naval Architects Ocean Engineers **2**, 155–160 (2005)
18. Ito, T.: A genetic algorithm approach to piping route path planning. J. Intell. Manuf **10**(1), 103–114 (1999)
19. Jiang, W.Y., Lin, Y., Chen, M., Yu, Y.Y.: A co-evolutionary improved multi-ant colony optimization for ship multiple and branch pipe route design. Ocean Eng. **102**, 63–70 (2015)
20. Kim, S.H., Ruy, W.S., Jang, B.S.: The development of a practical pipe auto-routing system in a shipbuilding CAD environment using network optimization. Int. J. Naval Architecture Ocean Eng. **5**(3), 468–477 (2013)

21. Kimura, H.: Automatic designing system for piping and instruments arrangement including branches of pipes. In: International Conference on Computer Applications in Shipbuilding (ICCAS), pp. 93–99 (2011)

22. Koenig, S., Sun, X.: Comparing real-time and incremental heuristic search for real-time situated agents. Auton. Agents Multi-Agent Syst. **18**(3), 313–341 (2009)

23. Lee, C.Y.: An algorithm for path connections and its applications. IRE Trans. Electron. Comput. **3**, 346–365 (1961)

24. Liu, L., Liu, Q.: Multi-objective routing of multi-terminal rectilinear pipe in 3D space by MOEA/D and RSMT. In: 2018 3rd International Conference on Advanced Robotics and Mechatronics (ICARM), pp. 462–467. IEEE (2018)

25. Medjdoub, B., Bi, G.: Parametric-based distribution duct routing generation using constraint-based design approach. Autom. Constr. **90**, 104–116 (2018)

26. Park, J.H., Storch, R.L.: Pipe-routing algorithm development: case study of a ship engine room design. Exp. Syst. Appl. **23**(3), 299–309 (2002)

27. Pohl, I.: First results on the effect of error in heuristic search. Mach. Intell. **5**, 219–236 (1970)

28. Sakti, A., Zeidner, L., Hadzic, T., Rock, B.S., Quartarone, G.: Constraint programming approach for spatial packaging problem. In: Quimper, C.-G. (ed.) CPAIOR 2016. LNCS, vol. 9676, pp. 319–328. Springer, Cham (2016). https://doi.org/10. 1007/978-3-319-33954-2_23

29. Wilt, C.M., Thayer, J.T., Ruml, W.: A comparison of greedy search algorithms. In: 3rd Annual Symposium on Combinatorial Search (2010)

30. Zhou, R., Hansen, E.A.: Beam-stack search: integrating backtracking with beam search. In: ICAPS, pp. 90–98 (2005)

31. Zhu, D., Latombe, J.C.: Pipe routing-path planning (with many constraints). In: Proceedings. 1991 IEEE International Conference on Robotics and Automation, pp. 1940–1941. IEEE Computer Society (1991)

MDDs Boost Equation Solving
on Discrete Dynamical Systems

Enrico Formenti⬡, Jean-Charles Régin⬡, and Sara Riva$^{(\boxtimes)}$⬡

Université Côte d'Azur, CNRS, I3S, Sophia Antipolis, France
{enrico.formenti,jean-charles.regin,sara.riva}@univ-cotedazur.fr

Abstract. Discrete dynamical systems (DDS) are a model to represent complex phenomena appearing in many different domains. In the finite case, they can be identified with a particular class of graphs called dynamics graphs. In [9] polynomial equations over dynamics graphs have been introduced. A polynomial equation represents a hypothesis on the fine structure of the system. Finding the solutions of such equations validate or invalidate the hypothesis.

This paper proposes new algorithms that enumerate all the solutions of polynomial equations with constant right-hand term outperforming the current state-of-art methods [10]. The boost in performance of our algorithms comes essentially from a clever usage of Multi-valued decision diagrams.

These results are an important step forward in the analysis of complex dynamics graphs as those appearing, for instance, in biological regulatory networks or in systems biology.

Keywords: Multi-valued decision diagrams · Discrete dynamical systems · Graphs semiring

1   Introduction

Multi-valued Decision Diagrams (MDD) are a generalization of Binary Decision Diagrams (BDD) [1,5] used to obtain efficient representations of functions (with finite domains) or (finite) sets of tuples. An MDD is a Directed Acyclic Graph (DAG) created from a finite set of variables with specific (finite) domains. Associating each variable with a level of the structure, the MDD represents a set of feasible assignments as a path from the *root* to the *final* node. A crucial aspect of MDDs is the exponential compression power of the reduction operation and the fact that many classical operations (intersection, union, *etc.*.) can be performed without decompression. In the last years, MDDs have been applied in many disparate research domains proving the potential of this structure. MDDs are used, for instance, to improve random forest algorithms replacing the classic binary decision trees [12], to represent and analyze automotive product data of valid/invalid product configurations [6], and to perform trust analysis in social networks [16]. There are also applications related to mathematical models like

ⓒ Springer Nature Switzerland AG 2021
P. J. Stuckey (Ed.): CPAIOR 2021, LNCS 12735, pp. 196–213, 2021.
https://doi.org/10.1007/978-3-030-78230-6_13

Multi-State Systems (MSS) [15]. In this case, MDDs represent the MSS structure in terms of Multi-Valued Logic to compute some measures. MDDs find applications also in the analysis of the discrete dynamical systems. Indeed, in [13], MDD represents logic rules to analyze some properties and dynamics aspects in the case of regulatory networks (for example, to perform a stable states identification).

In this work, we propose to apply MDDs to equations over Discrete Dynamical Systems (DDS). DDS are a model to represent complex phenomena which evolve in (discrete) time coming from different domains such as genetic regulatory networks, boolean automata networks, population dynamics, and many others. DDS consist in a finite set of states and a next-state function. In the finite case, DDS correspond to a particular class of graphs, called dynamic graphs (*i.e.* graphs with outgoing degree one). Therefore, DDS are simple structures that can be applied to any phenomena that evolve according to a function. Often, when representing phenomena with DDS, one needs to validate "macro" dynamics coming from experimental results, or to identify the set of "micro" dynamics which generate a given "macro" behavior observed.

In [9], operations of sum and multiplication have been introduced to explain how DDS can be combined to generate new dynamical behaviors. Equipping the set of DDS with these operations provides a commutative semiring, and this naturally leads to write polynomial equations over DDS. This is interesting because these equations (with a constant right-hand term) can model hypotheses over the dynamics, and if one can prove properties for DDS then they can be automatically lifted to the application models. The idea is to solve these equations to validate the corresponding hypotheses. This paper introduces a pipeline to solve equations over the cycles of dynamics graphs (*i.e.* equations/hypotheses over the long term behavior of a phenomenon), formally introduced in [10]. The first part of this paper proposes a new algorithm which outperforms the state-of-art technique (Colored-Tree method) using MDDs to enumerate the solutions set of simple equations. The second part introduces a pipeline to solve general equations using MDDs to solve basic cases and to limit the exploration of the solutions space. The overall purpose is to introduce the first complete pipeline, based on MDD, to solve the equations introduced in [9,10] to study the hypothesis over the long term behavior of a phenomenon modeled by DDS.

2   Preliminaries

2.1   Multi-valued Decision Diagrams (MDDs)

Multi-valued decision diagrams are the extension of Binary decision diagrams (BDD) in which the diagram consists of a rooted acyclic graph able to represent a multi-valued function $f : \{0...d-1\}^r \to \{true, false\}$. Considering a generic MDD, each level represents a variable and a final layer contains the true terminal node (tt). Therefore, the data-structure contains $r + 1$ layers and a path from the *root* to the tt node represents a valid set of assignments. Each node is characterised by the variable represented in the layer and at most d outgoing edges. In general, the edges are directed from an upper layer to a lower one. An edge corresponds to an assignment of the variable to a certain value specified over the edge. This data-structure presents only one root and one leaf. The false node, and the corresponding paths to reach this node, are omitted.

According to [2,8] , an MDD is **deterministic** if all the nodes have pairwise distinct labels on outgoing edges. Moreover, an MDD is **ordered** if given two nodes A and B such that A is the parent of B, we have that the variable represented into the node A is the smallest (*w.r.t.* a given total order over the set of variables).

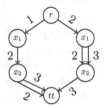

Fig. 1. The MDD corresponding to the valid assignments $\{(1,2,3),(2,2,3),(2,3,3),(1,2,2)\}$ for the variables x_0, x_1 and x_2. The structure presents 4 layers. The *root* and its edges represent the variable x_0 and its possible assignments.

One of the most interesting aspects of MDDs is their **reduction**. According to this procedure, they can gain an exponential factor in representation space. The reduction of MDDs aims at merging equivalent nodes *i.e.* that have equivalent outgoing paths. In particular, two nodes are equivalent if they have the same label and destination for each edge. The idea is to identify and merge equivalent nodes from the bottom to the top of the MDD, see Fig. 2 for an example.

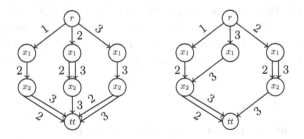

Fig. 2. An MDD before reduction (left) and its reduced version (right).

Several reduction algorithms for MDDs have been introduced [3,7]. In our software we use the **pReduce algorithm** [14] since the time complexity per node is bounded by its number of outgoing arcs and the complexity is linear on the size of the MDD.

Another big advantage of using MDDs is that classical set operations such as Cartesian product, complement, intersection, union, difference, and many others can be performed without decompressing the structure. For instance, the cartesian product over MDDs can be performed just by transforming the *root* of an MDD into the *tt* node of another one (see Fig. 3 for an example).

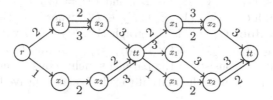

Fig. 3. The MDD representing the Cartesian product of the MDD in Fig. 1 and the MDD in Fig. 2 (right). The diagram is drawn horizontally for lack of room.

Given two MDDs, the intersection algorithm creates a new MDD to represent the solutions contained in both the original ones (see Fig. 4). To achieve this goal, the process starts with the creation of a new root, and only the outgoing edges that are common between the two structures are recreated. In this way, each new node corresponds to two original nodes and the process is iterated. It is important to remember that in the end, it is necessary to verify if there are nodes without children; if any, they must be deleted. The drawback of this algorithm is that the resulting MDD can be larger than the original ones. For more details, we refer the reader to [14] and [4].

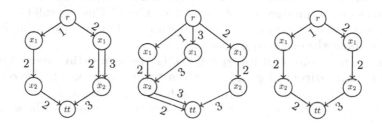

Fig. 4. Two MDDs (left and center) and their intersection (right).

2.2   Discrete Dynamical Systems and Dynamics Graphs

A Discrete Dynamical System (DDS) is a structure $\langle \chi, f \rangle$ where χ is a finite **set of states** and $f : \chi \to \chi$ is a function called the **next state map**. When modelling a phenomenon by a DDS $\langle \chi, f \rangle$, χ is its set of states and f is the law which brings from state $\alpha \in \chi$ at time t to the state $f(\alpha)$ at time $t + 1$.

When χ is finite, any DDS $\langle \chi, f \rangle$ can be identified with its **dynamics graph** $G \equiv \langle V, E \rangle$ where $V = \chi$ and $E = \{(\alpha, \beta) \in V \times V, f(\alpha) = \beta\}$. Therefore, all the properties of the DDS can be deduced from the properties of its dynamics graph. As a first property, one can remark that they are graphs with outgoing degree one and hence each strongly connected components of such graphs is made by a single cycle (or loop). From now on, we will turn all the discussion about DDS in terms of dynamics graphs. Call $\mathcal{DG}$ the set of all dynamics graphs up to (graph) isomorphism. One can define on $\mathcal{DG}$ two operations: sum and product as follows. Given two graphs, $G_1 = \langle V_1, E_1 \rangle \in \mathcal{DG}$ and $G_2 = \langle V_2, E_2 \rangle \in \mathcal{DG}$, the **sum** $G_1 + G_2$ is the graph $G = \langle V_1 \sqcup V_2, E_1 \sqcup E_2 \rangle \in \mathcal{DG}$, where $\sqcup$ is the disjoint union operator. The **product** $G_1 \cdot G_2$ is the graph $\langle V', E' \rangle \in \mathcal{DG}$ with $V' = V_1 \times V_2$ and $E' = \{((\alpha_1, \alpha_2), (\beta_1, \beta_2)) \in V' \times V', (\alpha_1, \beta_1) \in E_1 \text{ and } (\alpha_2, \beta_2) \in E_2\}$. The product defined above consists in the parallel synchronous execution of the two dynamics graphs. The sum is the mutually exclusive alternative between the two behaviours.

$R := \langle \mathcal{DG}, +, \cdot \rangle$ is a commutative semiring in which $\langle \emptyset, \emptyset \rangle$ is the neutral element *w.r.t.* $+$ and $\langle \{\alpha\}, \{(\alpha, \alpha)\} \rangle$ is the neutral element *w.r.t.* multiplication.

Now, consider the semiring $R[x_1, x_2, \ldots, x_s]$ of polynomials over R in the variables x_i, naturally induced by R. Polynomial equations of the form (1) model hypotheses about a certain dynamics deduced from experimental data.

$$a_1 \cdot x_1^{w_1} + a_2 \cdot x_2^{w_2} + \ldots + a_k \cdot x_s^{w_s} = C \tag{1}$$

Equation (1) can be interpreted as follows. The constant term C on the right-hand side is the dynamical system deduced from experimental data. The point is that C comes just from experimental data and hence it might be the "macro" result of many cooperating hidden variables at a "micro" level. On the left hand side of (1), we have a hypothesis on the "micro" structure based on partial information (the coefficients) and unknown information (the variables). In other words, the coefficients a_i are hypothetical sub-dynamical systems that should cooperate to produce the observed dynamics C. Finding valid values for the unknown variables provides a finer structure for C which can bring further knowledge about the observed phenomenon.

More generally, **one can interpret Eq. (1) as a question over dynamics graphs (*i.e.* directed graphs with outgoing degree 1)**. The constant right-hand side represents the current graph and the left-hand side is a question (hypothesis) about a possible decomposition (according to the semi-ring operations).

In [9], it has been proved that finding solutions to generic polynomial equations (*i.e.* in which both left and right-hand side of the equation are made by polynomials) over DDS is undecidable, while the problem is decidable when the right-hand side of the equation is constant. However, even in the decidable case, the complexity of the problem is beyond NP, except for very particular cases.

Some abstractions are introduced to progressively filter the solutions space according to features of the real solutions. Therefore, the general solutions are found in the intersection of the solutions of these abstractions. For this reason, the solutions enumeration of each abstraction is fundamental.

At least three abstractions can be devised on dynamics graphs, namely, abstraction on cycles, on cardinality (of the vertex set) and on paths. Studying each abstraction separately allows to study different aspects of a dynamics. In particular, solving equations over cycles leads to the validation of hypotheses over the long term behavior of a phenomenon.

3   The Abstraction on Cycles

In this paper we analyze equations over the cyclic part of the dynamics. Considering a generic Eq. (1), we introduce a pipeline to solve the abstraction over the long term behavior but before we need to recall some notation and some concepts that will be useful in the sequel.

For a dynamics graph G, let $\mathring{G}$ be the subgraph of G which contains only the cycles and the loops. Denote $\mathcal{R}$ the restriction of R such that for any $G \in R$, $\mathring{G} \in \mathcal{R}$. It is not difficult to see that $\langle \mathcal{R}, +, \cdot \rangle$ is a sub-semiring of $\langle R, +, \cdot \rangle$. Thus, solving Eq. (1) over $\mathcal{R}$ is a necessary step for solving it over R. This also implies that we have to enumerate all solutions in $\mathcal{R}$ first and then filter them out using the other abstractions to find the general solutions.

Notation 1. *A cycle $\{\alpha_1, \alpha_2, \dots, \alpha_p\}$ of length p can be conveniently denoted by C_p^1. A subgraph, with K different lengths of cycles is denoted $\bigoplus_{i=1}^{K} C_{p_i}^{n_i}$ (Fig. 5).*

Fig. 5. A dynamics graph G, in which the subgraph $\mathring{G}$ is the red part and it is denoted $C_2^2 \oplus C_3^1 \oplus C_5^1$ according to our notation.

The operations of sum and product over $\mathcal{R}$ can be conveniently applied to the new notation. Graphically, we can consider the result of a sum as a new system composed by all the cycles of the input systems, and the product one as a new system generated with the Cartesian product of the cycles of the input systems.

Definition 1 (Sum). *Consider two dynamics graphs $\mathring{A} \equiv \bigoplus_{i=1}^{K_A} C_{p_{Ai}}^{n_{Ai}}$ and $\mathring{B} \equiv \bigoplus_{j=1}^{K_B} C_{p_{Bj}}^{n_{Bj}}$, $\mathring{A} \oplus \mathring{B}$ is $\bigoplus_{i=1}^{K_A} C_{p_{Ai}}^{n_{Ai}} \oplus \bigoplus_{j=1}^{K_B} C_{p_{Bj}}^{n_{Bj}} = C_{p_{A1}}^{n_{A1}} \oplus \dots \oplus C_{p_{AK_A}}^{n_{AK_A}} \oplus C_{p_{B1}}^{n_{B1}} \oplus \dots \oplus C_{p_{BK_B}}^{n_{BK_B}}$.*

Considering two sets of cycles $C_{p_{Ai}}^{n_{Ai}}$ and $C_{p_{Bj}}^{n_{Bj}}$, if they have the same cycles length ($p_{Ai} = p_{Bj}$), then they can be rewritten like $C_{p_{Ai}}^{n_{Ai}+n_{Bj}}$.

Definition 2 (Product). *Consider* $\overset{\circ}{A} \equiv \overset{K_A}{\underset{i=1}{\bigoplus}} C_{p_{Ai}}^{n_{Ai}}$ *and* $\overset{\circ}{B} \equiv \overset{K_B}{\underset{j=1}{\bigoplus}} C_{p_{Bj}}^{n_{Bj}}$, $\overset{\circ}{A} \odot \overset{\circ}{B}$ *is*

$$\overset{K_A}{\underset{i=1}{\bigoplus}} C_{p_{Ai}}^{n_{Ai}} \odot \overset{K_B}{\underset{j=1}{\bigoplus}} C_{p_{Bj}}^{n_{Bj}} = \overset{K_A}{\underset{i=1}{\bigoplus}} \overset{K_B}{\underset{j=1}{\bigoplus}} C_{p_{Ai}}^{n_{Ai}} \odot C_{p_{Bj}}^{n_{Bj}} = \overset{K_A}{\underset{i=1}{\bigoplus}} \overset{K_B}{\underset{j=1}{\bigoplus}} C_{\mathrm{lcm}(p_{Ai},p_{Bj})}^{n_{Ai} \cdot n_{Bj} \cdot \gcd(p_{Ai},p_{Bj})} \quad \textit{where}$$

gcd *is the greatest common divisor and* lcm *is the least common multiple.*

With the help of the previous notation, Eq. (1) can be rewritten as

$$\left(\overset{K_1}{\underset{i=1}{\bigoplus}} C_{p_{1i}}^{n_{1i}} \odot \overset{\circ}{x_1}^{w_1}\right) \oplus \left(\overset{K_2}{\underset{i=1}{\bigoplus}} C_{p_{2i}}^{n_{2i}} \odot \overset{\circ}{x_2}^{w_2}\right) \oplus \ldots \oplus \left(\overset{K_s}{\underset{i=1}{\bigoplus}} C_{p_{si}}^{n_{si}} \odot \overset{\circ}{x_s}^{w_s}\right) = \overset{m}{\underset{j=1}{\bigoplus}} C_{p_j}^{n_j} \quad (2)$$

where K_z is the number of distinct cycles size in the system a_z with $z \in \{1, \ldots, s\}$ (*cf.* Eq. (1)) and n_{zi} is the number of cycles of length p_{zi} of a_z. In the right term C, there are m different periods, where for the jth different period there are n_j cycles of period p_j. However, Eq. (2) is still hard to solve in the present form. We can simplify it further by performing a **contraction step** which consists in cutting Eq. (2) into two simpler equations

$$\begin{cases} C_{p_{11}}^{n_{11}} \odot \overset{\circ}{x_1}^{w_1} = \overset{m}{\underset{j=1}{\bigoplus}} C_{p_j}^{u_j} & (3a) \\[4mm] C_1^1 \odot \overset{\circ}{y} = \overset{m}{\underset{j=1}{\bigoplus}} C_{p_j}^{v_j} & (3b) \end{cases}$$

with $\overset{\circ}{y} = \left(\overset{K_1}{\underset{i=2}{\bigoplus}} C_{p_{1i}}^{n_{1i}} \odot \overset{\circ}{x_1}^{w_1}\right) \oplus \left(\overset{K_2}{\underset{i=1}{\bigoplus}} C_{p_{2i}}^{n_{2i}} \odot \overset{\circ}{x_2}^{w_2}\right) \oplus \ldots \oplus \left(\overset{K_s}{\underset{i=1}{\bigoplus}} C_{p_{si}}^{n_{si}} \odot \overset{\circ}{x_s}^{w_s}\right)$ and $n_j = u_j + v_j$ for $j \in \{1, \ldots, m\}$. By recursively applying contraction steps, and for all possible u_j, v_j values in Equations (3a) and (3b), solving Eq. (2) boils down to solve multiple times **simple equations** *i.e.* equations of the form

$$C_p^1 \odot X = C_q^n \quad (4)$$

where $p \in \{p_{11}, p_{12}, \ldots, p_{sK_s}\}$, $q \in \{p_1, p_2, \ldots, p_m\}$, and n is smaller than the number of cycles of length q in the right part.

Assume that X is $\overset{\circ}{x_z}^{w_z}$ such that $z \in \{1, \ldots, s\}$. It is necessary to find $\overset{\circ}{x_z}$ from X *i.e.* we need to compute the w_z root of X. Given $2M$ integers $p_i, k_i \in \mathbb{N}$, with $p_i > 0$ for all $i \in \{1, \ldots, M\}$, let $l(p_1, p_2, \ldots, p_t, k_1, k_2, \ldots, k_t)$ be the lcm between the p_i for which $k_i \neq 0$ and $t \leq M$ (with $l(p_1, p_2, \ldots, p_t, k_1, k_2, \ldots, k_t) = 1$ iff $\forall 1 \leq i \leq t, k_i = 0$). Assume $\overset{\circ}{A} \equiv C_{p_1}^1 \oplus C_{p_2}^1 \oplus \ldots \oplus C_{p_M}^1$. Then,

$$\left(\overset{\circ}{A}\right)^w \equiv \overset{M}{\underset{i=1}{\bigoplus}} C_{p_i}^{p_i^{w-1}} \oplus \underset{\substack{k_1+k_2+\ldots+k_M=w \\ 0 \leq k_1, k_2, \ldots, k_M < w}}{\bigoplus} \binom{w}{k_1, k_2, \ldots, k_M} C_{l(p_1, p_2, \ldots, p_M, k_1, k_2, \ldots, k_M)}^{\prod_{t=1}^{M} p_t^{k_t-1} \cdot \prod_{t=2}^{M} \gcd(l(p_1, \ldots, p_{t-1}, k_1, \ldots, k_{t-1}), p_t)} \quad (5)$$

3.1   The State-of-art Method

In [10], two computational problems are devised concerning simple equations. The SOBFID (*SOlve equation on BIjective Finite DDS*) problem is a decision problem which takes in input $p, n, q \in \mathbb{N} \setminus \{0\}$ and returns true iff $C_p^1 \odot X = C_q^n$ admits a solution. EnumSOBFID is the problem which takes the same input as SOBFID and outputs the list of all solutions of $C_p^1 \odot X = C_q^n$.

To the best of our knowledge, the Colored-Tree Method (CTM) proposed in [10] is the best current technique to solve this problem. The method exploits a connection between EnumSOBFID and the well-known Change-making problem [11] coupled with a completeness-check (running in exponential time) to explore the feasible solutions space. Essentially, the right term n is decomposed in every possible way and these possibilities are represented in a tree. The method comprises two main phases: tree building and solutions aggregation. In the first one, the algorithm uses a tree to decompose the right part of the equation in a certain number of subgraphs, searching for each node of the tree (each subgraph) the minimum number of product operations (minimum number of cycles in the variable or minimum number of children) which are necessary to produce it. The subgraphs found are then divided into subsets of children. Iterating this idea on each subset produced, the method arrives to enumerate all the possible ways to generate the cycles involved in the right part. During the second phase, the method computes (bottom-up) the real solutions of the equation represented in the tree.

4   Boosting Everything up with MDDs

The enumeration of solutions for Eq. (2) is one of the main objective of this paper. In order to achieve this, we solve the EnumSOBFID problem and show that MDDs boost up the enumeration. In other words, we are interested to find all the possible ways to generate C_q^n cycles starting with one cycle of length p. This last problem is similar to the well-known Change-making problem (in its enumeration version) in which one aims at finding all the possible ways to change a total amount with a given set of coins. In our case, the total amounts are the C_q^n cycles but to complete the similarity we need to find the coins which can be part of a solution. Remark that a cycle C_u^1 generates C_q^r cycles in the right part iff r divides q, $u = \frac{q}{p} \cdot r$, $\gcd(p, \frac{q}{p} \cdot r) = r$ and $\mathrm{lcm}(p, \frac{q}{p} \cdot r) = q$. A cycle C_u^1 with the previous properties is called **feasible** and r is a **feasible divisor** of q.

Let $D_{p,q} = \{d_1, \ldots, d_l\}$ be the set of feasible divisors (*w.r.t.* Eq. (4) of course). There is a solution to (4) iff there exist $\overline{x_1}, \ldots, \overline{x_l}$ such that $\sum_{i=1}^{l} d_i \overline{x_i} = n$ (*i.e.* there is a solution to the Change-making problem for a total amount n and a coins system $D_{p,q}$). To solve EnumSOBFID we need to enumerate all solutions of the previous Change-making problem. At this point MDDs come into play. We are going to use them to have a compact and handful representation of the set of all possible solution to Eq. (4).

SB-MDD. The MDD $M_{p,q,n}$ containing all solutions to Eq. (4) is a labelled digraph $\langle V, E, \ell \rangle$ where $V = \bigcup_{i=1}^{Z} V_i$ with $Z = \lfloor \frac{n}{\min D_{p,q}} \rfloor + 1$ and $V_1 = \{root\}$, V_i is a multiset of $\{1, \ldots, n-1\}$, and, finally, $V_Z = \{tt\}$. For any node $\alpha \in V$, let $val(\alpha) = \alpha$ if $\alpha \neq root$ and $\alpha \neq tt$, $val(root) = 0$ and $val(tt) = n$. For any $i \in \{1, \ldots, Z-1\}$ and for any $\alpha \in V_i$ and $\beta \in V_{i+1} \cup \{tt\}$, $(\alpha, \beta) \in E$ iff $val(\beta) - val(\alpha) \in D_{p,q}$ and $val(\beta) \leq val(tt)$. The labelling function $\ell \colon E \to D_{p,q}$ is such that for any $(\alpha, \beta) \in E$, $\ell((\alpha, \beta)) = val(\beta) - val(\alpha) \in D_{p,q}$.

Graphically, $M_{p,q,n}$ can be represented by layers as usual, interpreting each layer as the usage of the i-th coin. Each node represents the sum of coins from the *root* to the node. Moreover, the longest path in $M_{p,q,n}$ is $Z = \lfloor \frac{n}{\min D_{p,q}} \rfloor + 1$. Remark that $M_{p,q,n}$ contains duplicated solutions, indeed, it contains all the permutations of a solution but according to Eq. (4) different permutations lead to the same solution. For this reason, we impose a **symmetry breaking constraint**: for any node α (different from tt), let e be the label of the incoming edge; the only allowed outgoing edges of α are those with label $\ell \leq e$. In this way all the paths of the MDD will be ordered and the size of the MDD will be smaller. An **SB-MDD** is an MDD which satisfies the symmetry breaking constraint. The building of $M_{p,q,n}$ ends with a pReduction which merges equivalent nodes and deletes all nodes (and the corresponding edges) which are not on a path from *root* to *tt*. Let us illustrate the construction with an example.

Example 1. Consider the simple equation $C_2^1 \odot X = C_6^6$. The set of divisors of q (smaller or equal to n) is $\{6, 3, 2, 1\}$. However, $D_{p,q} = \{1, 2\}$. Indeed, we have

$$r = 6 \wedge u = 18 \to gcd(2, 18) \neq 6 \wedge lcm(2, 18) \neq 6$$

$$r = 3 \wedge u = 9 \to gcd(2, 9) \neq 3 \wedge lcm(2, 9) \neq 6$$

$$r = 2 \wedge u = 6 \to gcd(2, 6) = 2 \wedge lcm(2, 6) = 6$$

$$r = 1 \wedge u = 3 \to gcd(2, 3) = 1 \wedge lcm(2, 3) = 6$$

Figure 6 shows $M_{2,6,6}$ in its classic form (left) and in its SB-MDD form (right). Remark that in Fig. 6 (left) many solutions are duplicated. For example, the solution $[2, 2, 1, 1]$ (in red) is represented (more than) twice. Once $M_{2,6,6}$ is built and reduced, reading solutions correspond to paths labels from *root* to *tt*: $\{[2, 2, 2], [2, 2, 1, 1], [2, 1, 1, 1, 1], [1, 1, 1, 1, 1, 1]\}$.

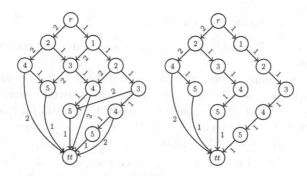

Fig. 6. The reduced MDD representing all the solutions of $C_2^1 \odot X = C_6^6$ in its classic form (left) and in its SB-MDD form (right).

The solution $[2, 2, 2]$ says that 6 is changed with 3 coins of value $r = 2$. Recalling that $u = \frac{q}{p} \cdot r$ we can express the solution in term of dynamics graphs as C_6^3. Operating similarly for all the other solutions we find $\{C_6^3, C_6^2 \oplus C_3^2, C_6^1 \oplus C_3^4, C_3^6\}$.

4.1 Equations over Dynamics Graphs

We have shown how MDDs can be used to compute the solutions set of simple equations. In this section, we introduce a pipeline to solve Equations of form (2) to validate hypotheses over discrete systems represented by DDS. The goal is the enumeration of solutions. The pipeline consists in the following steps:

- identification and resolution of the necessary equations (each necessary equation corresponds to an SB-MDD);
- enumeration of the contractions steps by an MDD structure;
- computation of the solutions of each contraction step with a particular technique to compute the intersection between SB-MDDs.

Necessary Equations. As we have already seen, to solve a generic equation over dynamics graphs we need to find the solutions of a certain number of contraction steps (Equations (3a) and (3b)). Each of these contraction steps ends up with a system of equations of the form

$$C_{p_{zi}}^{n_{zi}} \odot X_z = \bigoplus_{j=1}^{m} C_{p_j}^{v_j} \tag{6}$$

where $z \in \{1, \ldots, s\}$, $i \in \{1, \ldots, K_z\}$, and $v_j \leq n_j$. This means that each monomial is responsible for the generation of a subgraph of the right term. In other words, the solution of (6) is the Cartesian product of the solutions of a certain number of simple equations. Remark that a single simple equation might occur several times while searching for a solution of a single contraction step or in multiple contraction steps.

According to the product rules, each equation $C_{p_{zi}}^{n_{zi}} \odot X_z = C_{p_j}^{v_j}$ is equivalent to $C_{p_{zi}}^1 \odot X_z = C_{p_j}^{\frac{v_j}{n_{zi}}}$. Remark that if v_j/n_{zi} is not an integer, then the equation has no solutions.

Since there are many contraction steps that need to be explored, we aim at limiting the exploration by ignoring those which involve simple equations without solutions. Therefore, the method starts with the computation of the set of simple equations that can be involved in the result of a contractions step. This amounts to compute all the SB-MDDs $M_{p_{zi},p_j,\frac{n}{n_{zi}}}$ with $p_{zi} \in \{p_{11}, \ldots, p_{sK_s}\}$, $p_j \in \{p_1, \ldots, p_m\}$, for all $n \in \{1, \ldots, n_j\}$. In this way, even if a simple equation is involved in many contraction steps, it is solved only once. A simple equation with at least one solution is called **necessary equation**. It is important to notice that it is not necessary to explore an SB-MDD to decide if a solution exists. Indeed, by construction, an SB-MDD is generated iff it contains at least a solution.

Contractions Steps. Once the set of necessary equations has been computed, we create an MDD CS to enumerate all the contractions steps that must be taken into account to enumerate the solutions of Eq. (2). This MDD will be a Cartesian product of other MDDs, one for each p_j of the right term, $i.e.$ $CS = \bigtimes_{j=1}^m CS_j$.

Consider a cycle length p_j in the constant term. The MDD CS_j accounts for all the feasible ways (according to the set of necessary equations) to generate n_j cycles of length p_j using the monomials of the equation. Therefore, CS_j is a labelled digraph $\langle V_j, E_j, \ell_j \rangle$ where $V_j = \bigcup_{z=1}^{s+1} \bigcup_{i=1}^{K_z} V_{j,zi}$ with $K_{s+1} = 1$ (see Eq. (2) for the meaning of K_i) and $V_{j,11} = \{root\}$, $V_{j,zi} \subseteq \{0, \ldots, n_j\}$, and, finally, $V_{j,(s+1)1} = \{tt\}$. For any node $\alpha \in V_j$, let $val(\alpha) = \alpha$ if $\alpha \neq root$ and $\alpha \neq tt$, $val(root) = 0$ and $val(tt) = n_j$. The set of possible outgoing edges of level $V_{j,zi}$ is $D_{p_{zi},p_j} = \{g \in \mathbb{N} \mid 1 \leq g \leq n_j \text{ and } M_{p_{zi},p_j,\frac{g}{n_{zi}}} \in \text{necessary equations}\} \cup \{0\}$. For any $\alpha, \beta \in V_j$, $(\alpha, \beta) \in E_j$ iff

1. $\alpha \in V_{j,zi}$ and either $\beta \in V_{j,z(i+1)}$ for some $i < K_z$ or $\beta \in V_{j,(z+1)1}$;
2. $val(\beta) - val(\alpha) \in D_{p_{zi},p_j}$ and $val(\beta) \leq val(tt)$.

In this MDD the outgoing edges of a level $V_{j,zi}$ represent the cycles of length p_j generated by the monomial $C_{p_{zi}}^{n_{zi}} \odot X_z$ of the left part. The labelling function $\ell_j \colon E_j \to \bigcup_{z=1}^s \bigcup_{i=1}^{K_z} D_{p_{zi},p_j}$ is such that for any $(\alpha, \beta) \in E_j$, $\ell_j((\alpha, \beta)) = val(\beta) - val(\alpha) \in D_{p_{zi},p_j}$ with $\alpha \in V_{j,zi}$. Starting from each node, a label 0 means that the monomial is not involved in the generation of the cycles of length p_j. The sum of the labels of each path from the $root$ to the tt node will be equal to n_j, because n_j cycles of length p_j must be generated.

For each path of CS, it is necessary to perform some some additional steps to understand if it leads to feasible solutions of the equation as explained in the next section.

Solve a System. Each contraction step corresponds to a system of Equations of type (6). To compute the solutions set of these equations one needs to compute the Cartesian product between the different solutions of the corresponding simple equations. In its turn, a solution to a simple equation is represented by a SB-MDD. Therefore, their Cartesian product is computed in linear time by placing the SB-MDDs one on top of the other to form a new MDD. We call **SB-Cartesian MDD** an MDD build in such a way. Remark that an SB-Cartesian MDD is not a SB-MDD.

Recall that X_z in Eq. (6) represents a variable $\mathring{x}_z^{w_z}$ and if $\mathring{x}_z$ is involved in different monomials, then, an intersection operation is required. To compute the solutions set of the variable, we start considering equations involving $\mathring{x}_z$ with the same power w_z by computing the intersection over the corresponding MDDs. Moreover, remark that each $\mathring{x}_z^{w_z}$ corresponds to a SB-Cartesian MDD (except for the case in which a monomial is responsible for the generation of only one cycle length). However, notice that the classic algorithm to perform the intersection over MDDs cannot be used if the goal is the intersection between MDDs issued by a Cartesian product (*i.e.* SB-Cartesian MDDs) because the result depends on the order of the MDDs. In the next section, we propose a new algorithm to perform this task independently from the order.

Once $\mathring{x}_z^{w_z}$ is assigned with a set of solutions, we need to compute the w_z-th root for each value of $\mathring{x}_z^{w_z}$ and finally the intersection between all the roots found so far.

Finally, we stress that the root procedure is not a trivial step. Indeed, the inverse operations of sum and product are not definable in the commutative semiring of DDS. Therefore, we need an algorithmic technique to compute the result of the w-th root of $\mathring{x}$. Considering Formula (5), the root can be computed combinatorially or through a finite number of polynomials equations over real numbers of increasing degree. We combine both approaches in order to speedup the computations. Consider a generic system $\mathring{x} = C_{p_1'}^{n_1'} \oplus C_{p_2'}^{n_2'} \oplus \ldots \oplus C_{p_l'}^{n_l'}$ such that $\mathring{x}^w = C_{p_1}^{n_1} \oplus C_{p_2}^{n_2} \oplus \ldots \oplus C_{p_h}^{n_h}$, the number of cycles n_1', of the minimum length p_1, involved into the root's solution is computed with the polynomial equation $(p_1)^{w-1} \cdot (n_1')^w = n_1$ with $p_1 = p_1'$. The remaining part of the solution is combinatorially computed knowing that a length of cycle p_i may be involved in the root solution iff $n_i \geq (p_i)^{w-1}$, and we will have a maximum number of cycles of this length equals to $\frac{n_i}{(p_i)^{w-1}}$.

SB-Cartesian Intersection. Consider a set $\overline{M}$ of MDDs in which some are SB-Cartesian MDDs and others are not. We propose an algorithm (Algorithm 1) which computes the intersection of all the elements in $\overline{M}$. Our algorithm needs to be started with an initial set of candidate solutions $\overline{S}$ (*initial guess*). If $\overline{M}$ contains at least a simple SB-MDD (*i.e.* one which is not a SB-Cartesian MDD), then $\overline{S}$ is the set of solutions read in the MDD resulting from classical intersection of the SB-MDDs in $\overline{M}$; otherwise $\overline{S}$ is the set of solutions read in an arbitrarily chosen SB-Cartesian of $\overline{M}$. Using $\overline{S}$, we will compute the intersection between the remaining SB-Cartesians. The idea is to search the solutions into each SB-Cartesian MDD and update each time the remaining solutions.

Each candidate solution is ordered and recursively searched in a SB-Cartesian MDD in $\overline{M}$ (Algorithms 2 and 3). If it is not found, then it is removed from the set of candidate solutions. Given a SB-Cartesian element M of $\overline{M}$, a candidate solution s is validated if it possible to visit each SB-MDD involved in M using a subset of the elements of the solution. Starting from the biggest elements of s, we try to find a path from the *root* to the first *tt* node (recall that we are visiting a SB-Cartesian MDD). We proceed in this way because the paths of each SB-MDD are ordered. After having gone through the first SB-MDD, we need to recursively repeat the procedure with the remaining elements of s.

Two special cases need our attention:

- a generic node in which it is not possible to find a common element between the remaining elements of a solution and the outgoing edges;
- a node (different from the final *tt* node) in which there are no more elements of s to compare with the outgoing edges.

In both cases, it is necessary to go back in the visited nodes until there is another possible outgoing edge that can be taken into consideration. Once the validation procedure arrives at the last SB-MDD of M, a linear search is performed with the remaining elements. Finally, $s \in M$ if there exists a path from the last *root* to the final *tt* node following the remaining elements (Algorithm 4). If the linear search fails, then we return to the first node with a different feasible outgoing edge. If no different feasible edges exist, then s is not a solution. The previous procedure is performed over each candidate solution of the initial guess.

We stress that in the worst case, for a given candidate solution, our algorithm explores only the subgraph of a SB-Cartesian MDD made of feasible edges.

Algorithm 1: SB-Cartesian Intersection

Input : $\overline{M}$, set of MDDs
Output: $\overline{S}$, solutions of the intersection in $\overline{M}$
Cartesian$\leftarrow \emptyset$;
Traditional$\leftarrow \emptyset$;
forall $m \in \overline{M}$ **do**
    if m *is a SB-Cartesian MDD* **then**   Cartesian.add(m) ;
    else   Traditional.add(m) ;

$\overline{S} \leftarrow \emptyset$;
if $|Traditional| > 0$ **then**
    if $|Traditional| = 1$ **then**
        $\overline{S} \leftarrow$ Traditional[0].readSolutions();
    else
        MddIntersected$\leftarrow$ClassicIntersection(Traditional[0],Traditional[1]);
        forall $m \in Traditional \setminus \{Traditional[0], Traditional[1]\}$ **do**
            MddIntersected$\leftarrow$ClassicIntersection(MddIntersected,m);
        $\overline{S} \leftarrow$ MddIntersected.readSolutions();

else
    $\overline{S} \leftarrow$ Cartesian[0].readSolutions();
    Cartesian.remove(0);

if $|\overline{S}| \neq 0$ **then**
    forall $m \in Cartesian$ **do**
        CartesianSearch($\overline{S}$,m);

return $\overline{S}$

The approach introduced to compute the intersection of SB-Cartesian MDDs is suitable for our needs, its improvement constitutes an interesting future research direction. We need to precise an additional aspect of our application. As explained above, an SB-MDD corresponding to a simple equation has labels based on the feasible divisors set (or feasible coins) and each divisor/coin corresponds to a certain cycle C_u^1. However, two different coins of two different simple equations can correspond to the same C_u^1 as well as the same coin may correspond to different cycles lengths for different simple equations. Therefore, when searching for a solution in a SB-Cartesian MDD, we must take into account these cases. In our application, two divisors/coins are considered equivalent if they correspond to the same C_u^1.

Algorithm 2: SB-Cartesian Search

Input : S set of solutions, M a SB-Cartesian MDD
Output: $\overline{S}$ solutions involved in M
$\overline{S} \leftarrow \emptyset$;
forall $s \in S$ **do**
 s.order();
 find←FindSolution(s,0,M);
 if *find* **then** $\overline{S}$.add(s) ;
return $\overline{S}$

Algorithm 3: FindSolution

Input : s solution, $0 \leq i < |M|$, M a SB-Cartesian MDD
Output: true if $s \in M$, false otherwise
return *FindSolutionNode(s,M[i].root,i)*

Algorithm 4: FindSolutionNode

Data: S sub-solution, N node, $0 \leq i < |M|$, M a SB-Cartesian MDD
Result: true if $S \in M$, false otherwise
find←FALSE;
if N *is not a tt* **then**
 forall $e \in S$.*removeDuplicates()* $\wedge$ *find=FALSE* **do**
  if *N.edge.contains(e)* **then**
   newSubSolution← $S \setminus \{e\}$;
   find←findSolutionNode(newSubSolution,N.children(e),i);
else
 $i \leftarrow i + 1$;
 if $i = |M|-1$ **then**
  valid←LinearSearch(S,M[i].root);
  if *valid* **then return** *TRUE* ;
 else
  return *FindSolution(S,i)*
return *find*

---

5 Experiments

The experimental evaluation is divided into two parts: one concerns simple equations and the other one is devoted to the complete method which solves generic equations.

Concerning equations of the form $C_p^1 \odot X = C_q^n$, in our experiments, we set $p = q$ since this grants the existence of at least a solution. Using the MDDs it is possible to outperform the Colored-Tree method (CTM) *w.r.t.* both memory and time. If we compare the dimension (in terms of nodes) of a colored-tree with the corresponding SB-MDD for a given equation, the second one is smaller. CTM presents some out of memory cases even for equations with n, q, and p smaller than 30 and memory limit of 30GB (see Fig. 7 (left)). Using MDDs, we solved equations with n, q, and p up to 100 without any out of memory case and only 6GB RAM limit (see Fig. 7 (right)).

Analysing the time to solve equations with parameters up to 30, it turns out that the new technique is faster than the previous one (see Table 1). The reason is that CTM requires a time consuming check procedure to ensure the completeness of the solutions which is not necessary in the MDD case. Due to too high memory and time costs, CTM is unsuitable to solve simple equations coming from contractions steps. The new method fixes these issues allowing to solve generic polynomial equations.

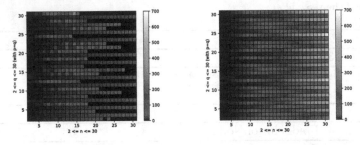

Fig. 7. The number of nodes for the colored-tree with memory limit of 30GB (left) and for the SB-MDD with memory limit of 4GB (right) in the case of equation for type $C_q^1 \odot X = C_q^n$. Remark that the black square in the right part of the left diagram are out of memory cases.

Turning to generic equations, we use the proposed pipeline to find the solutions (if any). If roots computation is not performed then everything depends on the number of monomials, the number of distinct sizes of cycles in the right side of the equations. If these quantities grow also the number of contraction steps to consider grows. Solutions spaces are limited by considering only contraction steps that are feasible according to the necessary equations. For example, consider the equation $C_5^1 \odot \mathring{y} \oplus C_4^1 \odot \mathring{y}^2 \oplus C_3^1 \odot \mathring{x}^2 \oplus C_2^1 \odot \mathring{x}^3 = C_{10}^3 \oplus C_4^{68} \oplus C_3^9 \oplus C_6^9 \oplus C_{12}^{136}$, the number of contraction steps is $\approx 2,42 \cdot 10^{16}$, but only 6665400 are considered.

The computation of roots is the most time consuming operation. Indeed, consider the equation $C_3^2 \odot \mathring{x} \oplus C_3^4 \odot \mathring{y} = C_3^{162} \oplus C_6^{20} \oplus C_{12}^{104}$, we can find 49329000 solutions considering the 6642 feasible contraction steps in only 58 s. If we consider the same equation but with $\mathring{x}^2$ in place of $\mathring{x}$, then the number of contraction steps explored is the same, but we found 4510 solutions in 5.6 hours (therefore the average computational time per contraction is 3.04 s).

Roots computation is expensive and this is reasonable in a sense. Since the division and subtraction operations cannot be performed directly, one needs sophisticated techniques. The pipeline introduced above leads to the first complete technique to solve equations over the asymptotic behavior of DDS. In the end, the pipeline is already proving itself suitable for applications over real experimental data.

Table 1. Computation times (millisec) of CTM (left) and MDD (right) over different input parameters. Symbols '-' represent out of memory cases.

q,p \ n	2	3	4	5	6	7	8	9	10
2	56|0	56|0	54|0	54|0	53|0	54|0	54|0	54|0	53|0
3	54|1	54|0	55|1	54|0	55|1	54|0	55|1	54|0	55|1
4	55|2	54|1	56|2	53|0	57|2	54|0	55|2	55|1	55|2
5	55|2	56|2	59|2	54|0	61|30	54|0	58|2	55|3	56|2
6	58|2	56|2	60|2	58|1	63|9	55|0	60|2	56|2	62|2
7	60|3	58|2	63|19	56|1	78|21	54|0	63|29	56|2	61|2
8	63|2	59|2	96|10	60|2	107|20	56|1	97|9	61|2	65|2
9	66|3	60|3	106|21	57|2	153|22	57|1	168|21	60|3	76|20
10	84|3	62|2	140|11	58|2	185|21	57|2	369|11	70|2	120|17

q,p \ n	11	12	13	14	15	16	17	18	19	20
11	55|0	824|25	55|0	116|3	74|21	406|22	55|0	1071|22	57|0	1334|23
12	58|1	17678|26	57|0	132|4	88|21	4105|12	56|0	6022|22	56|0	3672|22
13	61|2	177894|27	56|0	246|21	92|21	4163|27	55|0	5967|24	56|0	3332|24
14	59|2	1277979|26	61|1	900|11	116|22	19895|24	56|0	27381|27	56|0	96697|26
15	60|2	-|28	60|2	3721|22	169|22	19711|25	56|0	637457|26	59|0	419000|25
16	62|2	-|29	61|2	19900|12	502|20	-|13	57|0	1185947|26	60|0	759365|26
17	62|2	-|30	62|2	25908|24	554|23	-|26	57|0	-|27	57|0	-|27
18	64|2	-|46	62|2	164167|13	1102|22	-|26	61|2	-|28	61|0	-|30
19	66|2	-|32	63|2	226315|25	950|24	-|27	62|2	-|34	57|0	-|39
20	68|2	-|32	65|2	1707299|25	2749|24	-|16	63|2	-|31	62|2	-|29

q,p \ n	21	22	23	24	25	26	27	28	29	30
21	2712|23	343542|26	60|0	-|35	95|4	389971|7	2034|12	-|28	58|0	-|37
22	23339|24	-|14	62|0	-|36	103|4	381929|7	2711|24	-|30	59|0	-|35
23	27430|24	-|27	59|0	-|38	134|4	-|7	2712|24	-|31	59|0	-|37
24	149296|25	-|16	64|2	-|39	149|4	-|8	20641|14	-|42	59|0	-|41
25	162413|25	-|28	65|2	-|40	160|4	-|26	24632|24	-|34	59|0	-|39
26	212277|25	-|16	66|2	-|42	403|5	-|15	24177|25	-|33	60|0	-|40
27	-|25	-|27	69|2	-|45	454|4	-|33	-|15	-|34	59|0	-|45
28	-|26	-|17	70|2	-|47	488|5	-|16	-|25	-|35	60|0	-|44
29	-|28	-|28	69|2	-|66	506|5	-|28	-|27	-|36	61|1	-|46
30	-|26	-|27	69|3	-|51	838|6	-|17	-|17	-|38	65|2	-|48

6   Conclusions and Perspectives

Equations on DDS are useful to analyze dynamics of phenomena. They allow to model hypotheses on the dynamical behavior and solving them leads to their validation or invalidation. In particular, Eq. (2) allow studying the long-term aspects of the dynamics.

This paper introduces new algorithms to solve these equations over cycles (*i.e.* long term behavior). In the case of simple equations, our technique outperforms the CTM. This is an important breakthrough because CTM was used to solve simple equations (generated by contractions steps) and it was practically unusable due to huge memory consumption and time-consuming check procedures. Moreover, this paper proposes a pipeline to solve general equations. The pipeline computes the necessary equations to limit the exploration of the contractions steps and, in the end, it solves each system of equations corresponding to a contractions step. This allows the treatment of much larger dynamics graphs and much more complicated hypotheses.

Future perspectives include the improvement of the pipeline by parallelising the identification of solutions of different feasible contractions steps. Another research direction would try to speedup roots computation by increasing the number of coefficients computed through polynomial equations to reduce the combinatorics. In the end, this work aims to call for further research in MDDs. In fact, studying different ways to perform the intersection between SB-Cartesian MDDs is surely another subject that is worth exploring.

This research work leads to a complete and performing pipeline to validate hypotheses over the long term behavior of dynamics graphs adding one more item to the growing list of successful applications of MDDs.

Acknowledgments. This work has been supported by the French government, through the 3IA Côte d'Azur Investments in the Future project managed by the National Research Agency (ANR) with the reference number ANR-19-P3IA-0002.

References

1. Akers, S.B.: Binary decision diagrams. IEEE Trans. Comput. **27**(06) ,509–516 (1978)
2. Amilhastre, J., Fargier, H., Niveau, A., Pralet, C.: Compiling CSPs: a complexity map of (non-deterministic) multivalued decision diagrams. Int. J. Artif. Intell. Tools **23**(04), 1460015 (2014)
3. Andersen, H.R.: An introduction to binary decision diagrams. Lecture notes, available online, IT University of Copenhagen, p. 5 (1997)
4. Bergman, D., Cire, A.A., van Hoeve, W.: MDD propagation for sequence constraints. J. Artif. Intell. Res. **50**, 697–722 (2014)
5. Bergman, D., Cire, A.A., Van Hoeve, W.J., Hooker, J.: Decision diagrams for optimization, vol. 1. Springer, Berlin (2016). https://doi.org/10.1007/978-3-319-42849-9
6. Berndt, R., Bazan, P., Hielscher, K.S., German, R., Lukasiewycz, M.: Multi-valued decision diagrams for the verification of consistency in automotive product data. In: 2012 12th International Conference on Quality Software, pp. 189–192. IEEE (2012)
7. Cheng, K.C., Yap, R.H.: An MDD-based generalized arc consistency algorithm for positive and negative table constraints and some global constraints. Constraints **15**(2), 265–304 (2010)
8. Darwiche, A., Marquis, P.: A knowledge compilation map. J. Artif. Intell. Res. **17**, 229–264 (2002)
9. Dennunzio, A., Dorigatti, V., Formenti, E., Manzoni, L., Porreca, A.E.: Polynomial equations over finite, discrete-time dynamical systems. In: Mauri, G., El Yacoubi, S., Dennunzio, A., Nishinari, K., Manzoni, L. (eds.) ACRI 2018. LNCS, vol. 11115, pp. 298–306. Springer, Cham (2018). https://doi.org/10.1007/978-3-319-99813-8_27
10. Dennunzio, A., Formenti, E., Margara, L., Montmirail, V., Riva, S.: Solving equations on discrete dynamical systems. In: Cazzaniga, P., Besozzi, D., Merelli, I., Manzoni, L. (eds.) Computational Intelligence Methods for Bioinformatics and Biostatistics, pp. 119–132. Springer International Publishing, Cham (2020)

11. Martello, S., Toth, P.: Knapsack Problems: Algorithms and Computer Implementations. Wiley, New York, NY, USA (1990)
12. Nakahara, H., Jinguji, A., Sato, S., Sasao, T.: A random forest using a multi-valued decision diagram on an FPGA. In: 2017 IEEE 47th International Symposium on Multiple-Valued Logic (ISMVL), pp. 266–271. IEEE (2017)
13. Naldi, A., Thieffry, D., Chaouiya, C.: Decision diagrams for the representation and analysis of logical models of genetic networks. In: Calder, M., Gilmore, S. (eds.) CMSB 2007. LNCS, vol. 4695, pp. 233–247. Springer, Heidelberg (2007). https://doi.org/10.1007/978-3-540-75140-3_16
14. Perez, G., Régin, J.C.: Efficient operations on MDDs for building constraint programming models. In: IJCAI (2015)
15. Zaitseva, E., Levashenko, V., Kostolny, J., Kvassay, M.: A multi-valued decision diagram for estimation of multi-state system. In: Eurocon 2013, pp. 645–650. IEEE (2013)
16. Zhang, L., Xing, L., Liu, A., Mao, K.: Multivalued decision diagrams-based trust level analysis for social networks. IEEE Access **7**, 180620–180629 (2019)

Two Deadline Reduction Algorithms for Scheduling Dependent Tasks on Parallel Processors

Claire Hanen[1,2]([✉]) [iD], Alix Munier Kordon[1] [iD], and Theo Pedersen[1] [iD]

[1] Sorbonne Université, CNRS, LIP6, 75005 Paris, France
{Claire.Hanen,Alix.Munier}@lip6.fr
[2] UPL, Université Paris Nanterre, 92000 Nanterre, France

Abstract. This paper proposes two deadline adjustment techniques for scheduling non preemptive tasks subject to precedence relations, release dates and deadlines on a limited number of processors. This decision problem is denoted by $P|prec, r_i, d_i|\star$ in standard notations. The first technique is an extension of the Garey and Johnson algorithm that integrates precedence relations in energetic reasoning. The second one is an extension of the Leung, Palem and Pnueli algorithm that builds iteratively relaxed preemptive schedules to adjust deadlines.

The implementation of the two classes of algorithms is discussed and compared on randomly generated instances. We show that the adjustments obtained are slightly different but equivalent using several metrics. However, the time performance of the extended Leung, Palem and Pnueli algorithm is much better than that of the extended Garey and Johnson ones.

Keywords: Scheduling problem · Precedence constraints · Energetic reasoning · Preemptive relaxation

1 Introduction

This paper addresses the decision scheduling problem described in standard notations introduced in [14] as $P|prec, r_i, d_i|\star$. A set of tasks $\mathcal{T}$ and a precedence graph $\mathcal{G}$ are given. Each task $i \in \mathcal{T}$ has a deadline d_i, a release date r_i and a duration p_i. Tasks are performed on m identical processors. We address the existence of a feasible schedule. Notice that the problem is NP-hard in the strong sense, even in the special cases where no precedence constraints exists and one machine is considered $1|r_i, d_i|\star$ [11] or with unit execution times of tasks and common deadline $P|prec, r_i, d_i = D, p_i = 1|\star$ [29].

However, defining efficient polynomial algorithms providing necessary existence conditions is a challenging question since they might be used to improve the efficiency of constraint programming or branch and bound algorithms for the related optimization problems. Indeed, such necessary existence conditions combined to a binary search can provide a lower bound of the makespan ($C_{\max}$) or

© Springer Nature Switzerland AG 2021
P. J. Stuckey (Ed.): CPAIOR 2021, LNCS 12735, pp. 214–230, 2021.
https://doi.org/10.1007/978-3-030-78230-6_14

the maximum lateness ($L_{\max}$). Exact algorithms make use of these bounds [24], which appear particularly interesting when the branching scheme is based on splitting or reducing the tasks' intervals [2,5].

Such necessary conditions have been investigated by many authors since the early eighties, thoroughly improving the efficiency of exact algorithms. Several of them use interval adjustment techniques (ie. reducing deadlines and increasing release times) by relaxing precedence constraints. The special case of unit execution times has been also investigated with adjustment techniques considering both precedence and resource constraints.

Interval adjustment techniques based on energetic reasoning, ie. on the measure of the mandatory workload of time intervals, have been the subject of much attention when no precedence constraint is considered. These techniques developed for the problem $P|r_i, d_i|\star$ were extended to handle the cumulative scheduling problem (CuSP). In this case, each task i requires c_i resources for its execution, and the total number of resources is bounded. Baptiste et al. [1] and Derrien and Petit [9] have developed low time complexity algorithms by reducing the number of considered intervals. Ouellet and Quimper [27] and Carlier et al. [6] have improved the data structures used in the algorithms.

Tesch in [28] analyzed the time needed to reach a fixed point for the technique called energetic edge finding.

Most recent studies taking into account both precedence and resource constraints are devoted to the resource constrained scheduling problem (RCPSP) and extend earlier work on 1-machine and job-shop scheduling problems. According to Laborie and Nuijten [22], energetic constraints are either propagated on precedence relations, or precedence constraints are considered independently from the job's release times and deadlines in "energy precedence constraints". In order to compute a lower bound on the makespan, Haouari et al. proposed to integrate RCPSP resources and precedence constraints through linear programming via relevant relaxations of precedence and valid inequalities, either using energetic [18] or preemptive bounds [19].

In addition, the special case with unit processing times ($p_i = 1$ for each task $i \in \mathcal{T}$) has been investigated by several authors, most of them using reduction of deadline techniques embedding precedence and resource constraints. Garey and Johnson in [11] derived a polynomial algorithm (GJ algorithm in short) based on energetic reasoning that solves the decision problem for $m = 2$ processors. This algorithm was extended by Hanen and Zinder [17] to get an approximation algorithm for the $L_{\max}$ criteria for general parallel machines case when tasks have unit processing times. The Leung, Palem and Pnueli algorithm [23] (abbreviated to LPP algorithm) expresses necessary conditions on deadlines based on the iterative construction of schedules for relaxed sub-problems without precedence constraints. They also proved that this algorithm optimally solves several problems with particular precedence graphs. Hanen and Munier [4] showed that these two algorithms reach the same fixed point deadlines and an experimental study confirmed that the LPP algorithm is faster than the GJ one.

Our contribution in this paper is to extend both the GJ and LPP algorithms to handle tasks with any duration, and to compare the two approaches. Notice that, due to the inherent symmetry of the problem, all the algorithms considered here can be used to modify release dates as well, by simply reversing the orientation of the precedence arcs and swapping release times and deadlines. However, our experiments only considered deadline modifications.

The extension of the GJ algorithm theoretically dominates usual energetic reasoning due to the addition of precedence constraints. Our approach also considers stronger conditions than Laborie [21] who only considers the successors of a task to adjust the deadlines. The extension of LPP algorithm is based on the iterative construction of preemptive schedules. These two approaches were experimentally compared on randomly generated instances: we first proposed several measures of the effective deadline reduction, and the variation of the intrinsic parallelism of the instances. We observed that the reductions of the deadlines are roughly similar for the two extensions, even if the deadlines obtained are not necessarily equal. However, we also observed that according to the theoretical time complexity evaluation, the LPP algorithm has a much lower complexity than GJ.

The paper has six sections. In Sect. 2, we present the problem and the main notations. Section 3 is devoted to the extension of the GJ algorithm and the energetic reasoning. Section 4 presents the extension of the LPP algorithm. Section 5 presents our experiments. Finally, we conclude in Sect. 6.

2   Notations

An instance $\mathcal{I}$ of our scheduling problem is given by a set of n tasks $\mathcal{T}$, a precedence graph $\mathcal{G} = (\mathcal{T}, \mathcal{A})$ and m identical processors. For every task $i \in \mathcal{T}$, we denote by p_i the execution time of i. We suppose that the release time r_i and the deadline d_i of each task i are given, and satisfy

$$r_i + p_i \leq d_i. \tag{1}$$

A feasible schedule assigns a starting time t_i, such that $r_i \leq t_i \leq d_i - p_i$, and a processor among the m available ones, so that two tasks assigned to the same processor do not overlap.

We consider the decision problem of the existence of a feasible schedule denoted by $P|prec, r_i, d_i|\star$.

For any pair of tasks $(i, j) \in \mathcal{T}^2$, we note $i \rightarrow j$ if there exists a path in $\mathcal{G}$ from i to j. Then, $\Gamma^{+\star}(i)$ (resp. $\Gamma^{-\star}(i)$) is the set of descendants (resp. ancestors) of i, which are tasks j such that $i \rightarrow j$ (resp. $j \rightarrow i$). For any pair of tasks (i, j) with $i \rightarrow j$, we denote by $\ell_{ij}^\star$ the maximum value $\sum_{k \in \nu, k \neq j} p_k$ of a path ν of $\mathcal{G}$ from i to j. We assume that these values are pre-processed. This can be done in time complexity $\mathcal{O}(n^3)$ by using the Floyd-Warshall algorithm [8].

We assume that release times and deadlines are consistent with the precedence constraints:

$$\forall (i, j) \in \mathcal{A}, r_i + p_i \leq r_j \text{ and } d_j - p_j \geq d_i. \tag{2}$$

In the sections below, we introduce deadline modification algorithms.

We consider the algorithm PROPAGATE(i, d) that computes a consistent deadline vector assuming that all values of the input deadline vector d, except maybe the modified deadline d_i, are consistent ie. follow conditions (1) and (2). PROPAGATE(i, d) returns false if $d_i - p_i < r_i$, otherwise it adjusts all the ancestors j of i by setting $d_j = \min(d_j, d_i - \ell_{ji}^* + p_j - p_i)$ and returns true. Notice that then for any ancestor j of i, $r_j + p_j \leq d_j$. The time complexity of PROPAGATE(i, d) is $\mathcal{O}(n)$ provided that the values ℓ_{ij}^* are preprocessed.

3   Extension of the Garey and Johnson Algorithm

In this section we first explain the deadline reduction principle on which the Garey and Johnson algorithm [11] is based. Then we present the extended Garey and Johnson algorithm (eGJ in short) in its weak form, and analyze its time complexity. Finally we present the strong form of eGJ.

3.1   Principles of Deadline Reductions

The idea of the original Garey and Johnson algorithm [11], which was designed to solve the problem for two processors and tasks with unit processing times, is to reduce the deadline of a job i based on the measure of the number of tasks that must be executed in an interval $[s, t]$ assuming i ends at its deadline. This idea is extended here for tasks with any processing time by considering energetic reasoning [1] on time intervals.

Let i be a task and let us consider two values $s \leq t$ such that i may end between s and t:

$$r_i \leq s \leq d_i \leq t. \tag{3}$$

Figure 1 presents the three subsets of tasks $I(i, s, t)$, $S(i, s, t)$ and $T(i, s, t)$ that should have a part processed between s and t.

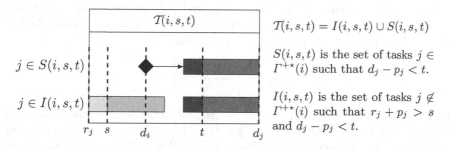

$T(i, s, t) = I(i, s, t) \cup S(i, s, t)$

$S(i, s, t)$ is the set of tasks $j \in \Gamma^{+*}(i)$ such that $d_j - p_j < t$.

$I(i, s, t)$ is the set of tasks $j \notin \Gamma^{+*}(i)$ such that $r_j + p_j > s$ and $d_j - p_j < t$.

Fig. 1. Sets of tasks $I(i, s, t)$, $S(i, s, t)$ and $T(i, s, t)$ with some mandatory part in $[s, t]$

Any task $j \in I(i, s, t)$ has no precedence relation with i; we set $w_j(i, s, t)$ as the minimum part of the task j that must be performed between s and t in any feasible schedule, i.e. when j is left shifted and right shifted as illustrated by the blue and red parts in Fig. 1. Clearly, for any task $j \in I(i, s, t)$,

$$w_j(i, s, t) = \min(t - s, p_j, \max(0, r_j + p_j - s), \max(0, t - (d_j - p_j))).$$

Similarly, tasks j from $S(i, s, t)$ are descendants of i with a minimum part $w_j(i, s, t)$ that must be performed between s and t **assuming i may end between s and d_i**. Notice that the task j starts after s and we just have to consider the contribution of j when it is right shifted.

$$\forall j \in S(i, s, t), \quad w_j(i, s, t) = \min(p_j, \max(0, t - (d_j - p_j))).$$

The total amount of work that is to be performed between s and t considering the minimum contribution of i in the interval $[s, t]$ is then:

$$W(i, s, t) = \max(0, r_i + p_i - s) + \sum_{j \in T(i, s, t)} w_j(i, s, t).$$

We define the associated slack $\Delta(i, s, t) = W(i, s, t) - m(t - s)$. If $\Delta(i, s, t) > 0$, there is not enough room in the time interval $[s, t]$ to execute the energy $W(i, s, t)$. This situation will fall into one of two cases, as in the following properties:

Property 1. Let us consider a task $i \in T$ and two values s and t such that $r_i \leq s \leq d_i \leq t$. If $\Delta(i, s, t) > 0$ and $S(i, s, t) = \emptyset$ then no feasible schedule exists.

Proof. If $S(i, s, t) = \emptyset$, then $T(i, s, t) = I(i, s, t)$; any task $j \in I(i, s, t)$ has no precedence relation with i and thus $w_j(i, s, t)$ part of j must be executed in the time interval $[s, t]$. The total energy $W(i, s, t)$ must then be executed in $[s, t]$ in any feasible schedule. Since $\Delta(i, s, t) > 0$, no feasible schedule exists, the result. □

Property 2. Let us consider a task $i \in T$ and two values s and t such that $r_i \leq s \leq d_i \leq t$. If $\Delta(i, s, t) > 0$ and $S(i, s, t) \neq \emptyset$, then in any feasible schedule, the completion time C_i of i verifies the inequality

$$C_i \leq t - \left\lceil \frac{W(i, s, t)}{m} \right\rceil. \tag{4}$$

Proof. The part $w_j(i, s, t)$ of any task $j \in S(i, s, t)$ must be executed before time t and after the end of i. Otherwise, if $j \in I(i, s, t)$, $w_j(i, s, t)$ is the part of j that must be executed between s and t. Since $W(i, s, t) > 0$, the only way to execute these tasks is to decrease the completion time C_i of i in order to fit tasks from $T(i, s, t)$ between C_i and t, thus the completion time C_i of i must verify Eq. (4). □

3.2   Description of the eGJ Algorithm

The eGJ algorithm takes as input an instance (T, G, m, r, d) of the problem and outputs either a set of modified deadlines $d^* = (d_i^*)_{i \in T}$ that should be fulfilled by any feasible schedule, or indicates that no feasible schedule exists, based on the conditions expressed in Properties 1 and 2.

Triples (i, s, t) are enumerated in a way that will be described below, and at each step if $\Delta(i, s, t) > 0$ then it either results in an infeasibility or defines a modification of the deadline of i based on the inequality (4)

$$d_i \leftarrow t - \left\lceil \frac{W(i, s, t)}{m} \right\rceil.$$

Once a modification of d_i occurs, the algorithm propagates the modification to the nodes of $\Gamma^{-*}(i)$ using PROPAGATE(i, d).

Not all possible triples (i, s, t) following inequality (3) need to be considered; Hanen and Munier [15], inspired by Carlier et al. [7] show that the slack $\Delta(i, s, t)$ has a local maxima only at the dominant triples with the forms (i, r_j, d_k), (i, d_i, d_k), $(i, r_k + d_k - d_j, d_k)$, $(i, d_i, r_k + d_k - d_i)$ with $d_i > r_k$ or $(i, r_j, r_k + d_k - r_j)$ with $r_j > r_k$.

For a current deadline vector d, we denote by $R(d)$ the set of values t such that there exists a dominant triple (i, s, t). If $t \in R(d)$, we denote by $X_t(d)$ the set of tasks i such that there exists a dominant triple (i, s, t). Finally, for $t \in R(d)$, and $i \in X_t(d)$ we denote by $L_{i,t}(d)$ the set of values s, such that (i, s, t) is a dominant triple.

Algorithm 1 enumerates the dominant triples in three nested loops. The outer loop enumerates t in decreasing order by maintaining a sorted list R corresponding to the set $\{\tau \in R(d), \tau < t\}$. The intermediate loop browses elements i of a list X corresponding to the set $X_t(d)$ in decreasing order of deadlines, and the inner loop browses elements s of a list L corresponding to the set $L_{i,t}(d)$ in increasing order. At each iteration, $\Delta(i, s, t)$ is computed and either a contradiction is found, or d_i is updated and propagated. Ordered lists X and R are then updated.

3.3   Complexity Analysis of eGJ

The next lemma will be later used to bound the number of iterations of the outer loop.

Lemma 1. *Consider an iteration t of the outer loop for which at the beginning of the iteration at least one task k satisfies $t = d_k$. At the end of this iteration, either infeasibility is detected or there is at least one task removed from X in line 16.*

Proof. Let us suppose by contradiction that at iteration $t = d_i$ for $i \in X$, no infeasibility is detected and no task is removed from X. If d_i is modified with an interval $[s, t]$, then $\Delta(i, s, t) > 0$ and according to the propagation at each step, any task $j \in \Gamma^{+*}(i)$ satisfies $d_j - p_j \geq d_i = t$, so $S(i, s, t) = \emptyset$. Following Property 1, the algorithm returns infeasibility (the contradiction). □

Algorithm 1. eGJ algorithm

Require: A precedence graph $\mathcal{G}$, release dates vector r, deadline vector d, processing
times vector p and m identical processors
Ensure: Modified deadlines vector d^* or infeasibility
1: $d_{\max} = \max_{i \in T} d_i$,
2: $R = R(d_{\max})$ in decreasing order, $X = X_{d_{max}}(d)$ in decreasing order
3: **while** $R \neq \emptyset$ and $X \neq \emptyset$ **do**
4:     $t =$ first element of R, remove t from R
5:     **for all** $i \in X$ **do**
6:         $L = L_{i,t}(d)$ in increasing order
7:         **repeat**
8:             $s =$ first element of L, remove s from L
9:             **if** $\Delta(i, s, t) > 0$ **then**
10:                Update d_i or return false (infeasibility) per Properties 1, 2
11:                $d =$ PROPAGATE(i, d), return false if inconsistency with release times.
12:                Update R (sorted list of $\{\tau < t, \tau \in R(d)\}$)
13:            **end if**
14:         **until** $s \geq d_i$
15:     **end for**
16:     Remove from X the tasks j for which $d_j = t$
17: **end while**
18: **return** $d^* = (d_i)_{i \in T}$

The next lemma is an outcome of Lemma 1 that bounds the number of iterations
of the outer loop of Algorithm 1.

Lemma 2. *The successive values of t are strictly decreasing. Moreover, the total
number of these successive values belongs to $\mathcal{O}(n^2)$.*

Proof. At the initialization step, all the values of R are different. At line 12, R
is updated with a list of values strictly less than t, thus the successive values of
t are strictly decreasing.

Let us consider now all the possible successive values for t:

- If $t = d_k$, then following Lemma 1, at least one task i is removed from X.
  Thus, there are at most n iterations in this case.
- Now, if $t = r_k + d_k - d_i$ with $d_i > r_k$, then $d_k > t$ and thus will not be later
  modified by the algorithm. Moreover, if d_i is decreased to d_i', $t' = r_k + d_k - d_i' >$
  t and thus will not be considered after t. There are then at most n^2 iterations
  in this case.
- Lastly, if $t = r_k + d_k - r_j$ with $r_j > r_k$, then $d_k > t$ and will also not be
  decreased further. There are also n^2 iterations in this case.

We deduce that the number of the successive values of t belongs to $\mathcal{O}(n^2)$, which
concludes the lemma.

$\square$

Theorem 1. *Algorithm eGJ is in time $\mathcal{O}(n^5 \log(n))$.*

Proof. For a fixed task $i \in X$, the execution time of the inner loop belongs to $\mathcal{O}(n^2 \log n)$. Indeed, there are $\mathcal{O}(n^2)$ values of s that must be sorted (in time $\mathcal{O}(n^2 \log(n))$). The time of the computation of the slack is in $\mathcal{O}(n)$. The modification of d_i and the propagation happen only once per iteration on i; the next value of d_i is less than s and thus the inner loop on s ends. So the time complexity of this modification and the propagation is $\mathcal{O}(n)$. Lastly, updating of the sorted list R is in time $\mathcal{O}(n^2 \log(n))$.

Now, the size of X is bounded by n, while by Lemma 2, the total number of iterations of the outer loop belongs to $\mathcal{O}(n^2)$, thus the theorem is proved.   □

The tightness of this bound is not proved. We will show in Sect. 5 that the experimental complexity of this algorithm is much smaller.

3.4 Strong Form of eGJ

The deadline reduction can be strengthened by considering for any valid triple (i, s, t) a new slack $\overline{\Delta}(i, s, t)$ assuming i is right shifted i.e. i ends exactly at its deadline:

$$\overline{\Delta}(i, s, t) = \min(p_i, d_i - s) + \sum_{j \in \mathcal{T}(i,s,t)} w_j(i, s, t) - m(t - s) \tag{5}$$

Now assume that $\Delta(i, s, t) \leq 0$ and $\overline{\Delta}(i, s, t) > 0$. In any feasible schedule the completion time of i satisfies $C_i - s + \sum_{j \in \mathcal{T}(i,s,t)} w_j(i, s, t) \leq m(t - s)$ so that

$$C_i \leq s + \min(p_i, d_i - s) - \overline{\Delta}(i, s, t). \tag{6}$$

The deadline of i can thus be reduced to the right term of Eq. (6). This condition can be inserted in Algorithm 1 adding to the inner loop (line 13) the case where $\Delta(i, s, t) \leq 0$ and $\overline{\Delta}(i, s, t) > 0$ in which modification is done according to (6). The modification is then propagated with PROPAGATE(i, d), and lists L and R are updated.

The arguments stated in Lemma 1 do not apply in this case; instead, all possible values of $0 \leq t \leq d_{\max}$ might be considered without updating R., which would lead to a pseudo-polynomial complexity detailed in [16].

4 Extension of the Leung Palem and Pnueli Algorithm

This section is devoted to the description of two extended forms of the Leung, Palem and Pnueli algorithm [23] for tasks with different execution times. Subsection 4.1 presents a general possible extension of this algorithm (eLPP in short), based on an optimization scheduling problem BACKWARDSCHEDULE. Two implementations are then discussed. In Subsect. 4.2, this problem is relaxed to obtain a polynomial time algorithm while an exact pseudo-polynomial time algorithm is presented is Subsect. 4.3.

4.1 Description of the eLPP Algorithm

For any task $i \in \mathcal{T}$, we note $Indep(i)$ the set of tasks $j \in \mathcal{T}$ such that $i \not\rightarrow j$ and $j \not\rightarrow i$. We set also $\mathcal{T}_i = \Gamma^{+\star}(i) \cup Indep(i)$.

Consider release and deadline vectors r and d and a task $i \in \mathcal{T}$. For any value $t \in \{r_i, \ldots, d_i - p_i\}$ corresponding to a possible starting time of i, we define t-dependent temporary release dates and deadlines for tasks in $\mathcal{T}_i \cup \{i\}$ as:

$$\hat{r}_j(t) = \begin{cases} \max\{r_j, t + \ell_{ij}^\star\} & \text{if } j \in \Gamma^{+\star}(i) \\ t & \text{if } i = j \\ r_j & \text{if } j \in Indep(i) \end{cases} \quad \text{and} \quad \hat{d}_j(t) = \begin{cases} d_j & \text{if } j \in \mathcal{T}_i \\ t + p_i & \text{if } j = i. \end{cases}$$

Consider a function $\text{EXISTENCE}(i, t, r, d)$ which checks the feasibility of the preemptive relaxation of the problem for tasks in $\mathcal{T}_i \cup \{i\}$ with release dates $\hat{r}(t)$, due dates $\hat{d}(t)$ and the m machine constraint. In a preemptive schedule each task might be interrupted and resumed on different machines.

$\text{EXISTENCE}(i, t, r, d)$

Input: A task $i \in \mathcal{T}$, release dates and deadlines vectors r and d, and $t \in \{r_i, \ldots, d_i - p_i\}$.

Question: Is there a feasible preemptive schedule of tasks from $\mathcal{T}_i \cup \{i\}$ meeting the release dates $\hat{r}(t)$ and the deadlines $\hat{d}(t)$?

This decision problem belongs to the class $P|r_i, d_i, pmtn|\star$. As shown by Martel [25], it can be transformed polynomially into a network flow problem and thus polynomially solved using a classical maximum-flow algorithm [13].

Let us now define the function $\text{BACKWARDSCHEDULE}(i, r, d)$ that returns the maximum value $t^\star \in \{r_i, \ldots, d_i - p_i\}$ such that $\text{EXISTENCE}(i, t^\star, r, d)$ is true. If such a value exists, $t^\star + p_i$ is an upper bound of the completion time of i in any feasible schedule of the initial scheduling problem. Otherwise, no feasible schedule exists and the function returns false. We will discuss in the following several implementations of this function.

Algorithm 2 presents the extended version of the LPP algorithm. Tasks are first sorted by decreasing release date. Deadlines of the tasks are then improved iteratively in this order using the previous function BACKWARDSCHEDULE. The calls to PROPAGATE maintain consistent deadlines considering precedence constraints.

The main problem addressed below is that no usual binary search on t can be considered to solve BACKWARDSCHEDULE because of the resource constraint for the task i. Indeed, a binary search can be considered to compute $t^\star$ if $\text{EXISTENCE}(i, t, r, d) = \text{true}$ for each value $t \in \{r_i, \ldots t^\star\}$, and $\text{EXISTENCE}(i, t, r, d) = \text{false}$ for each value $t \in \{t^\star + 1, \ldots d_i - p_i\}$. This property on $t^\star$ is not verified.

Then, a simple approach to solve the optimization problem BACKWARDSCHEDULE would be to start with $t = d_i - p_i$ and check each integer value in decreasing order until $\text{EXISTENCE}(i, t, r, d)$ is true. The number of steps would

Algorithm 2. eLPP algorithm

Require: A precedence graph $\mathcal{G}$, release dates r, deadlines d, processing times p and m identical processors

Ensure: Modified deadlines $d^\star$ or infeasibility

1: Adjust all r_i in topological order to reflect precedence
2: Adjust all d_i in reverse topological order to reflect precedence
3: Renumber tasks such that $r_1 \geq r_2 \geq \ldots \geq r_n$
4: **for** $i = 1$ to n **do**
5:    resultB=BACKWARDSCHEDULE(i, r, d)
6:    **if** not resultB **then**
7:      **return** false
8:    **end if**
9:    d_i =resultB
10:   d =PROPAGATE(i, d), return false if inconsistency with release times
11: **end for**
12: **return** $d^\star = (d_i)_{i \in \mathcal{T}}$

---

then be not polynomially bounded. Two implementations of BACKWARDSCHEDULE were developed in the following to cope with this problem.

4.2 Weak eLPP Algorithm

The simplest way to speed-up the time complexity of the eLPP algorithm is to limit the function EXISTENCE to tasks from $\mathcal{T}_i$ instead of $\mathcal{T}_i \cup \{i\}$. Indeed, if the task i is removed, the problem EXISTENCE(i, t, r, d) is more constrained when t increases, and thus a binary search on t can be considered to implement BACKWARDSCHEDULE. The complexity of the deadlines reduction algorithm is in polynomial time in this case as proven in Theorem 2. However, the deadlines obtained might be greater than the ones given by Algorithm 2.

Theorem 2. *The weak eLPP algorithm is in time $\mathcal{O}(n^4 \times \max_{i \in \mathcal{T}} \log(d_i - p_i - r_i))$.*

Proof. For each task $i \in \mathcal{T}$ and any value $t \in \{r_i, \ldots, d_i - p_i\}$, the time complexity for the computations of the vectors $\hat{r}$ and $\hat{d}$ is $\mathcal{O}(n^2)$. The number of nodes (resp. arcs) of the graph associated with the flow problem belongs to $\mathcal{O}(n)$ (resp. $\mathcal{O}(n^2)$) [25]. The time complexity of the flow algorithm is in $\mathcal{O}(n^3)$ using a push-relabel algorithm with a FIFO vertex selection rule [13]. As $t^\star$ is computed using a binary search in the time interval $\{r_i, \ldots, d_i - p_i\}$, we conclude that the overall time complexity of weak eLPP algorithm is $\mathcal{O}(n^4 \times \max_{i \in \mathcal{T}} \log(d_i - p_i - r_i))$, proving the theorem. $\square$

4.3 Strong eLPP Algorithm

The purpose of the strong version of eLPP is to develop an implementation of Algorithm 2 that is faster but remains exact.

Let us consider the relaxed decision problem $\text{EXISTENCER}(i, u, v, r, d)$ defined as follows which dissociates the parameters for the computation of the release dates and deadlines:

$\text{EXISTENCER}(i, u, v, r, d)$

Input: A task $i \in \mathcal{T}$, release dates and deadlines vectors r and d, and a pair of values $(u, v) \in \{r_i, \ldots, d_i - p_i\}^2$ with $u \geq v$.

Question: Is there a feasible preemptive schedule of tasks from $\mathcal{T}_i \cup \{i\}$ meeting the release dates $\hat{r}(v)$ and the deadlines $\hat{d}(u)$?

This decision problem also belongs to the class $P|r_i, d_i, pmtn|\star$. Thus, as for EXISTENCE it can be solved using Martel's transformation [25] to a network flow problem coupled with a classical maximum-flow algorithm [13]. We can also note that EXISTENCE is a special case of EXISTENCER for which $t = u = v$.

Now, let us define the function $\text{BACKWARDSCHEDULER}(i, u, r, d)$ with $u \in \{r_i, \ldots, d_i - p_i\}$ that returns the maximum value $v^\star \in \{r_i, \ldots, u\}$ such that $\text{EXISTENCER}(i, u, v^\star, r, d)$ is true if any, and false otherwise. Observe that, for any $(v, v') \in \{r_i, \ldots, d_i - p_i\}^2$ with $v < v'$, $\text{EXISTENCER}(i, u, v, r, d)$ is less constrained than $\text{EXISTENCER}(i, u, v', r, d)$. Thus, if $\text{EXISTENCER}(i, u, v', r, d) =$ true, then so is $\text{EXISTENCER}(i, u, v, r, d)$ and a binary search can be considered to solve BACKWARDSCHEDULER.

The remaining problem is then to find the maximal fixed point of BACKWARDSCHEDULER, that is, $u^\star$ such that $u^\star = \text{BACKWARDSCHEDULER}(i, u^\star, r, d)$. The next lemma establishes the relationship between BACKWARDSCHEDULER and BACKWARDSCHEDULE.

Lemma 3. *For any task $i \in \mathcal{T}$, release dates and deadlines vectors r and d, the value $u^\star = \text{BACKWARDSCHEDULER}(i, u^\star, r, d)$ exists if and only if the value $t^\star = \text{BACKWARDSCHEDULE}(i, r, d)$ exists. Moreover, $u^\star = t^\star$.*

Proof. Assume first that $u^\star$ exists, then $\text{EXISTENCER}(i, u^\star, u^\star, r, d) =$ true and thus $\text{EXISTENCE}(i, u^\star, r, d) =$ true. $\text{BACKWARDSCHEDULE}(i, r, d)$ then returns an optimal value $t^\star \geq u^\star$. Conversely, let us suppose that $t^\star$ exists, then $\text{EXISTENCE}(i, t^\star, r, d) =$ true. The consequence is that $\text{EXISTENCER}(i, t^\star, t^\star, r, d) =$ true, thus $u^\star$ exists and $t^\star \leq u^\star$. $\qquad\square$

The following lemma shows an important property of the function BACKWARDSCHEDULER.

Lemma 4. *Let us consider $i \in \mathcal{T}$, release dates and deadlines vectors r and d. For $u \in \{r_i, \ldots, d_i - p_i\}$, the function $u \to \text{BACKWARDSCHEDULER}(i, u, r, d)$ is non decreasing (if it returns an integer).*

Proof. Assume that u and u' are two integers in $\{r_i, \ldots, d_i - p_i\}$ with $u' < u$. If $\text{BACKWARDSCHEDULER}(i, u', r, d) = v' \in \{r_i, \ldots, u'\}$, then we get $\text{EXISTENCER}(i, u', v', r, d) =$ true. Since $u' < u$, $\text{EXISTENCER}(i, u, v', r, d) =$ true and $v' \leq u' < u$. Thus, $\text{BACKWARDSCHEDULER}(i, u, r, d)$ exists and $v' \leq v$. $\qquad\square$

We now show how to compute the value u^*. This can be done by computing a sequence of upper bounds u_β, $\beta \geq 0$ of u^* that converges to u^*. Indeed, let us consider the sequence of integers u_β defined as:

1. $u_0 = d_i - p_i$;
2. For any $\beta > 0$, $u_\beta = \text{BACKWARDSCHEDULER}(i, u_{\beta-1}, r, d)$.

The next theorem shows the convergence of this sequence to t^*.

Theorem 3. *If t^* exists, the sequence u_β tends to t^* (ie. there exists $\beta^* \in \mathbb{N}$ such that $u_{\beta^*} = t^*$).*

Proof. We first prove that, for any value $\beta \in \mathbb{N}$, $u_\beta = t^*$ or $u_0 > u_1 > \ldots > u_\beta \geq t^*$. Indeed, $u_0 = d_i - p_i \geq t^*$. Now, let us suppose by recurrence that $u_0 > u_1 > \ldots > u_\beta \geq t^*$ for $\beta \geq 1$. By Lemma 4, the function BACKWARDSCHEDULER is non decreasing with respect to u, thus since $u_{\beta-1} > u_\beta$, we get $u_\beta \geq u_{\beta+1}$.

1. If $u_\beta = u_{\beta+1}$, then by definition of u^*, $u_\beta = u^* = u_{\beta+1}$ and thus by Lemma 3, $u_{\beta+1} = t^*$;
2. Let us suppose now that $u_\beta > u_{\beta+1}$. Then, since $u_\beta > t^*$, we get by Lemma 4 that $u_{\beta+1} \geq t^*$.

Lastly, since the sequence u_β is strictly decreasing until it reaches t^*, there exists a minimum integer β^* such that $u_{\beta^*} = t^*$, and the theorem is proved. $\square$

The implementation of BACKWARDSCHEDULE based on BACKWARDSCHEDULER simply consists of computing the sequence u_β until a fixed point is reached. Alas, we do not have any polynomial upper bound of the time complexity of this algorithm.

5   Experiments

This section is devoted to the description of our experiments' results. Subsection 5.1 briefly describes the parameters considered for the data generation. Subsection 5.2 compares the running times of our four algorithms, while Subsect. 5.3 deals with their output analysis.

5.1   Data Generation

Random instances have been generated, using parameters that were fixed to keep the problem size manageable and to generate non trivial comparable instances with respect to the deadline reduction measures. The detailed description of the generation of instances can be found in [16].

The number of tasks n varied from 10 to 50, while $p_{\max} \in [1,5]$. High values of m make the problem trivial, thus we set $m \in [1,3]$. We generated 10 instances for each combination of input parameters.

The algorithms were implemented using Python 3.7.6 coupled with the packages numpy 1.18.1 and networkX 2.4. All our experiments were performed on an Acer Swift SF314-41 composed of an AMD Ryzen 5 3500U running at 2.1 Ghz with 4 cores, 8 Logical Processors and 8 MB RAM.

5.2 Complexity Analysis

The choice of the maximum flow algorithm is discussed to find the most efficient implementation for the functions EXISTENCE and EXISTENCER. As expected, these functions took up the majority (98%) of eLPP runtime, so these choices had significant impact on time complexity. We considered for our experiments the shortest augmenting path flow algorithm [10] with an initial flow built using Jackson's preemptive algorithm [20] to solve these two problems. This choice was experimentally motivated in [16] against preflow push [12] and Edmonds Karp [10] algorithms.

The runtime of each of the four algorithms is compared, as well as how they change with each of the problem parameters.

As shown in Fig. 2, up to $n = 30$, eGJ performed similarly to eLPP, but eLPP was faster for higher values of n. The weak version of eGJ made no appreciable difference to the speed, but for eLPP the weak version was faster by about 33% on average.

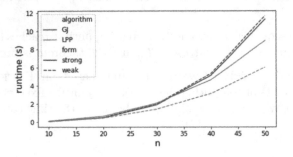

Fig. 2. Comparison of algorithm runtimes following the number of tasks

The runtime for each problem instance was regressed against n, m, p_{max}, $\frac{pw}{n}$ and the choice of algorithm. The pathwidth pw of an instance $\mathcal{I}$ is the maximum number of tasks that can be executed simultaneously considering only release dates and deadlines [26]. The runtime and number of tasks were both log transformed to help meet the model assumptions and identify the order of complexity. Up to 95% of the variation in runtime was able to be explained by these variables. The results were as follows:

- Though neither strong form is proven to be polynomial, in practice both eGJ forms were about $\mathcal{O}(n^{3.4})$, while the eLPP forms were about $\mathcal{O}(n^{2.8})$.
- Increasing m by 1 resulted in a 38% decrease in runtime for eGJ, and a 30% decrease for eLPP.
- Increasing p_{max} by 1 increased runtime by 26% for eGJ, and 18% for eLPP.
- Increasing pathwidth by $\frac{n}{10}$ increased runtime by 2% for eGJ, and 9% for eLPP.

As far as speed is concerned, eLPP is clearly the stronger option. It was also less sensitive to increases in most of the parameters, so the trend can be expected to continue beyond the range tested. The usefulness of the weak form of eLPP depends on the results of the following section.

5.3 Output Analysis

The aim of our experiments was to compare the strong and the weak versions of the eGJ and eLPP algorithms. For this purpose, we defined for each instance $\mathcal{I}$ and each algorithm A, the value $\delta_A^{\star}(\mathcal{I}) \in \mathbb{Z}$ which is the smallest value such that the algorithm A does not detect infeasibility if, for each task $i \in \mathcal{T}$, $d_i + \delta_A^{\star}(\mathcal{I})$ is used as the deadline. For each couple $(A, \mathcal{I})$, this value can be found by binary search. We measured it for each algorithm (eLPP and eGJ, weak and strong forms).

If we compare the results of the strong forms s-eGJ vs. s-eLPP, in all but 0.2% of problem instances, the final $\delta_A^{\star}$ values matched; in 96%, each individual date matched as well. Most of the 4% mismatches were due to one or two slightly higher s-eLPP deadlines. Even when the dates did not match, the differences were few and small, and overall, the s-eGJ dates dominated a majority of the time (but not all the time). When $n = 50$, neither algorithm has a significant advantage over the other in terms of date reduction.

The weak forms w-eGJ and w-eLPP had a $\delta_A^{\star}$ value that was 1 lower in about 1% of cases for each compared with the respective strong forms s-eGJ and s-eLPP. This decreased with n. At the vector level only 55% of the final deadlines vectors matched entirely their strong counterparts. The difference between the dates produced was often in several tasks and with values larger than 1.

Four metrics, defined below, were calculated for each instance and averaged across all instances.

Percentage of modified instances a 0/1 flag indicates whether any date was modified in an instance;

Percentage of modified tasks the proportion of individual deadline values modified;

Interval shrinkage the reduction as a proportion of the available intervals, so $1 - \frac{\sum_{i=1}^{n}(\tilde{d}_i - \tilde{r}_i)}{\sum_{i=1}^{n}(d_i - r_i)}$ where r, d are the initial dates and $\tilde{d}, \tilde{r}$ the final dates;

Pathwidth reduction the reduction in pathwidth pw as a percentage of the original.

Table 1 summarises the metrics discussed above by algorithm, across all test instances and instances for which $n = 50$. As the previous results suggested, the strong forms were nearly identical. The amount of reduction made by the weak forms was slightly less, though the number of modified dates was similar.

Dates were also reduced more with smaller m, and with larger pathwidth pw (as a proportion of n). In particular, the interval shrinkage was approximately

Table 1. Date modifications across all tests and for instances with $n = 50$

	All tests				$n = 50$			
	s-eGJ	s-eLPP	w-eGJ	w-eLPP	s-eGJ	s-eLPP	w-eGJ	w-eLPP
% instances modified	81%	81%	79%	79%	98%	98%	97%	97%
% tasks modified	34.2%	34.1%	32.4%	32.4%	52.8%	52.9%	52.1%	52.2%
Interval shrinkage	12.0%	12.0%	9.4%	9.5%	18.7%	18.8%	16.9%	16.9%
Pathwidth reduction	8.9%	8.9%	7.2%	7.2%	13.6%	13.6%	12.9%	12.9%

halved with each addition of a machine. The gap between the weak and strong forms narrowed slightly as n increased.

With such little difference between the outputs of the strong forms, eLPP maintains its advantage from the runtime results. The weak form offers a trade-off; while it is considerably faster, less reduction is performed. It may be useful in time-sensitive contexts where incremental reductions are relatively less valuable, particularly if n is large.

6   Conclusions

We developed in this paper several extensions of the GJ and the LPP algorithms to handle tasks with different processing times and precedence relations. The aim here was to evaluate whether considering at the same time precedence and resource constraints in deadline reduction algorithm was an interesting approach.

Two versions of each algorithm was developed: the weak ones (of polynomial time complexity), and the strong ones (non polynomially time bounded complexity). The strong version of the two extensions improves slightly the results, with experimentally the same complexity as their weak counterpart. As the LPP extensions outperforms the GJ ones in terms of theoretical as well as experimental complexity, it should be preferred.

However, their time complexity remains still large, and further improvement should be investigated, inspired by the recent improvements of the computational complexity of energetic reasoning for problems without precedence constraints [3, 27].

Our approach that embeds precedence and resources should be experimentally compared to a process of usual precedence relaxation to compute reduced deadlines followed by precedence propagation, repeated iteratively.

An interesting further study would compare the results of several interval reduction techniques, in particular the one proposed by Haouari et al. [18, 19].

Most of the algorithms that have been proposed for problems with parallel processors extend quite naturally to cumulative resources. The extension of eGJ and eLPP to such problems might be easier for eGJ.

Finally, the aim of the algorithms presented in this paper is to improve the efficiency of either branch and bound or constraint programming algorithms. This should be experimentally investigated in subsequent research.

References

1. Baptiste, P., Le Pape, C., Nuijten, W.: Satisfiability tests and time-bound adjustments for cumulative scheduling problems. Ann. Oper. Res. **92**, 305–333 (1999)
2. Bellenguez-Morineau, O.: Methods to solve multi-skill project scheduling problem. 4OR **6**(1), 85–88 (2008)
3. Bonifas, N.: A $\mathcal{O}(n^2 \log(n))$ propagation for the energy reasoning. In: Congrès ROADEF, February 2016
4. Carlier, A., Hanen, C., Kordon, A.M.: The equivalence of two classical list scheduling algorithms for dependent typed tasks with release dates, due dates and precedence delays. J. Sched. **20**(3), 303–311 (2017). https://doi.org/10.1007/s10951-016-0507-8
5. Carlier, J., Latapie, B.: Une méthode arborescente pour résoudre les problèmes cumulatifs. RAIRO - Oper. Res. Rech. Opérationnelle **25**(3), 311–340 (1991)
6. Carlier, J., Pinson, E., Sahli, A., Jouglet, A.: An $\mathcal{O}(n^2)$ algorithm for time-bound adjustments for the cumulative scheduling problem. Eur. J. Oper. Res. **286**(2), 468–476 (2020)
7. Carlier, J., Pinson, E., Sahli, A., Jouglet, A.: Comparison of three classical lower bounds for the cumulative scheduling problem. (submitted) (2021)
8. Cormen, T.H., Leiserson, C.E., Rivest, R.L., Stein, C.: Introduction to Algorithms, Third Edition. The MIT Press, 3rd edn., Cambridge (2009)
9. Derrien, A., Petit, T.: A new characterization of relevant intervals for energetic reasoning. In: O'Sullivan, B. (ed.) CP 2014. LNCS, vol. 8656, pp. 289–297. Springer, Cham (2014). https://doi.org/10.1007/978-3-319-10428-7_22
10. Edmonds, J., Karp, R.M.: Theoretical improvements in algorithmic efficiency for network flow problems. J. ACM (JACM) **19**(2), 248–264 (1972)
11. Garey, M.R., Johnson, D.S.: Two-processor scheduling with start-time and deadlines. SIAM J. Comput. **6**, 416–426 (1977)
12. Goldberg, A.V., Tarjan, R.E.: A new approach to the maximum-flow problem. J. ACM (JACM) **35**(4), 921–940 (1988)
13. Goldberg, A.V., Tarjan, R.E.: Efficient maximum flow algorithms. Commun. ACM **57**(8), 82–89 (2014)
14. Graham, R., Lawler, E., Lenstra, J., Kan, A.: Optimization and approximation in deterministic sequencing and scheduling: a survey. In: Hammer, P., Johnson, E., Korte, B. (eds.) Discrete Optimization II, Annals of Discrete Mathematics, vol. 5, pp. 287–326. Elsevier (1979)
15. Hanen, C., Munier Kordon, A.: Two deadline reduction algorithms for scheduling dependent typed-tasks systems. In: ROADEF conference (2020)
16. Hanen, C., Munier Kordon, A., Pedersen, T.: Two deadline reduction algorithm for scheduling dependent tasks on parallel processors (extended version) (2021). https://hal.archives-ouvertes.fr/hal-03200297
17. Hanen, C., Zinder, Y.: The worst-case analysis of the Garey-Johnson algorithm. J. Sched. **12**(4), 389–400 (2009)
18. Haouari, M., Kooli, A., Néron, E.: Enhanced energetic reasoning-based lower bounds for the resource constrained project scheduling problem. Comput. Oper. Res. **39**(5), 1187–1194 (2012)
19. Haouari, M., Kooli, A., Néron, E., Carlier, J.: A preemptive bound for the resource constrained project scheduling problem. J. Sched. **17**(3), 237–248 (2014)
20. Jackson, J.R.: Scheduling a production line to minimize maximum tardiness. management science research project (1955)

21. Laborie, P.: Algorithms for propagating resource constraints in AI planning and scheduling: Existing approaches and new results. Artif. Intell. **143**(2), 151–188 (2003)

22. Laborie, P., Nuijten, W.: Constraint Programming Formulations and propagation Algorithms, chap. 4, pp. 63–72. Wiley, Hoboken (2008)

23. Leung, A., Palem, K.V., Pnueli, A.: Scheduling time-constrained instructions on pipelined processors. ACM Trans. Program. Lang. Syst. **23**, 73–103 (2001)

24. Lombardi, M., Milano, M.: Optimal methods for resource allocation and scheduling: a cross-disciplinary survey. Constr. An Int. J. **17**(1), 51–85 (2012)

25. Martel, C.: Preemptive scheduling with release times, deadlines, and due times. J. Assoc. Comput. Mach. **29**(3), 812–829 (1982)

26. Munier Kordon, A.: A fixed-parameter algorithm for scheduling unit dependent tasks on parallel machines with time windows. Discret. Appl. Math. **290**, 1–6 (2021)

27. Ouellet, Y., Quimper, C.-G.: A $O(n \log^2 n)$ checker and $O(n^2 \log n)$ filtering algorithm for the energetic reasoning. In: van Hoeve, W.-J. (ed.) CPAIOR 2018. LNCS, vol. 10848, pp. 477–494. Springer, Cham (2018). https://doi.org/10.1007/978-3-319-93031-2_34

28. Tesch, A.: Improving energetic propagations for cumulative scheduling. In: Hooker, J. (ed.) CP 2018. LNCS, vol. 11008, pp. 629–645. Springer, Cham (2018). https://doi.org/10.1007/978-3-319-98334-9_41

29. Ullman, J.: NP-complete scheduling problems. J. Comput. Syst. Sci. **10**, 384–393 (1975)

Improving the Filtering
of Branch-and-Bound MDD Solver

Xavier Gillard[1]([envelope])[iD], Vianney Coppé[1][iD], Pierre Schaus[1][iD],
and André Augusto Cire[2][iD]

[1] Université Catholique de Louvain, Ottignies-Louvain-la-Neuve, Belgium
{xavier.gillard,vianney.coppe,pierre.schaus}@uclouvain.be
[2] Rotman School of Management, University of Toronto Scarborough,
Toronto, Canada
andre.cire@rotman.utoronto.ca

Abstract. This paper presents and evaluates two pruning techniques
to reinforce the efficiency of constraint optimization solvers based on
multi-valued decision-diagrams (MDD). It adopts the branch-and-bound
framework proposed by Bergman et al. in 2016 to solve dynamic programs to optimality. In particular, our paper presents and evaluates the
effectiveness of the local-bound (LocB) and rough upper-bound pruning
(RUB). LocB is a new and effective rule that leverages the approximate MDD structure to avoid the exploration of non-interesting nodes.
RUB is a rule to reduce the search space during the development of
bounded-width-MDDs. The experimental study we conducted on the
Maximum Independent Set Problem (MISP), Maximum Cut Problem
(MCP), Maximum 2 Satisfiability (MAX2SAT) and the Traveling Salesman Problem with Time Windows (TSPTW) shows evidence indicating
that rough-upper-bound and local-bound pruning have a high impact
on optimization solvers based on branch-and-bound with MDDs. In particular, it shows that RUB delivers excellent results but requires some
effort when defining the model. Also, it shows that LocB provides a
significant improvement automatically; without necessitating any user-
supplied information. Finally, it also shows that rough-upper-bound and
local-bound pruning are not mutually exclusive, and their combined benefit supersedes the individual benefit of using each technique.

1 Introduction

Multi-valued Decision Diagrams (MDD) are a generalization of *Binary Decision Diagrams* (BDD) which have long been used in the verification, e.g., for
model checking purposes [10]. Recently, these graphical models have drawn the
attention of researchers from the CP and OR communities. One of the research
streams which emerged from this increased interest about MDDs is *decision-
diagram-based optimization* (DDO) [5]. Its purpose is to efficiently solve combinatorial optimization problems by exploiting problem structure through DDs.
This paper belongs to the DDO sub-field and intends to further improve the

© Springer Nature Switzerland AG 2021
P. J. Stuckey (Ed.): CPAIOR 2021, LNCS 12735, pp. 231–247, 2021.
https://doi.org/10.1007/978-3-030-78230-6_15

efficiency of DDO solvers through the introduction of two bounding techniques: local-bounds pruning (LocB) and rough-upper-bound pruning (RUB).

This paper starts by covering the necessary background on DDO. Then, it presents the local-bound and rough-upper-bound pruning techniques in Sects. 3.1 and 3.2. After that, it presents an experimental study which we conducted using 'ddo' [17][1], our open source fast and generic MDD-based optimization library. This experimental study investigates the relevance of RUB and LocB through four disinct NP-hard problems: the Weighted Maximum Independent Set Problem (MISP), Maximum Cut Problem (MCP), Maximum 2 Satisfiability Problem (MAX2SAT) and the Traveling Salesman Problem with Time Windows (TSPTW). Finally, Sect. 5 discusses previous related work before drawing conclusions.

2 Background

The coming paragraphs give an overview of discrete optimization with decision diagrams. Most of the formalism presented here originates from [8]. Still, we reproduce it here for the sake of self-containedness.

Discrete Optimization. A discrete optimization problem is a constraint *satisfaction* problem with an associated objective function to be maximized. The discrete optimization problem $\mathcal{P}$ is defined as $\max\{f(x) \mid x \in D \wedge C(x)\}$ where C is a set of constraints, $x = \langle x_0, \ldots, x_{n-1} \rangle$ is an assignment of values to variables, each of which has an associated finite domain D_i s.t. $D = D_0 \times \cdots \times D_{n-1}$ from where the values are drawn. In that setup, the function $f : D \to \mathbb{R}$ is the objective to be maximized.

Among the set of feasible solutions $Sol(\mathcal{P}) \subseteq D$ (i.e. satisfying all constraints in C), we denote the optimal solution by x^*. That is, $x^* \in Sol(\mathcal{P})$ and $\forall x \in Sol(\mathcal{P}) : f(x^*) \geq f(x)$.

Dynamic Programming. Dynamic programming (DP) was introduced in the mid 50's by Bellman [3]. This strategy is significantly popular and is at the heart of many classical algorithms (e.g., Dijkstra's algorithm [12, p.658] or Bellman-Ford's [12, p.651]).

Even though a dynamic program is often thought of in terms of recursion, it is also natural to consider it as a labeled transition system. In that case, the *DP model* of a given discrete optimization problem $\mathcal{P}$ consists of:

- a set of state-spaces $S_0, \ldots, S_n$ among which one distinguishes the *initial state* r, the *terminal state* t and the *infeasible state* $\perp$.
- a set of transition functions $t_i : S_i \times D_i \to S_{i+1}$ for $i = 0, \ldots, n-1$ taking the system from one state s^i to the next state s^{i+1} based on the value d assigned to variable x_i (or to $\perp$ if assigning $x_i = d$ is infeasible). These functions should never allow one to recover from infeasibility ($t_i(\perp, d) = \perp$ for any $d \in D_i$).

---

[1] https://github.com/xgillard/ddo.

- a set of transition cost functions $h_i : S_i \times D_i \to \mathbb{R}$ representing the immediate reward of assigning some value $d \in D_i$ to the variable x_i for $i = 0, \ldots, n-1$.
- an initial value v_r.

On that basis, the objective function $f(x)$ of $\mathcal{P}$ can be formulated as follows:

$$\text{maximize } f(x) = v_r + \sum_{i=0}^{n-1} h_i(s^i, x_i)$$

subject to

$$s^{i+1} = t_i(s^i, x_i) \text{ for } i = 0, \ldots, n-1; x_i \in D_i \land C(x_i)$$

$$s^i \in S_i \text{ for } i = 0, \ldots, n$$

where $C(x_i)$ is a predicate that evaluates to *true* when the partial assignment $\langle x_0, \ldots, x_i \rangle$ does not violate any constraint in C.

The appeal of such a formulation stems from its simplicity and its expressiveness which allows it to effectively capture the problem structure. Moreover, this formulation naturally lends itself to a DD representation; in which case it represents an exact DD encoding the complete set $Sol(\mathcal{P})$.

2.1 Decision Diagrams

Because DDO aims at solving constraint *optimization* problems and not just constraint *satisfaction* problems, it uses a particular DD flavor known as reduced weighted DD – DD as of now. As initially posed by Hooker [21], DDs can be perceived as a compact representation of the search trees. This is achieved, in this context, by superimposing isomorphic subtrees.

To define our DD more formally, we will slightly adapt the notation from [5]. A DD $\mathcal{B}$ is a layered directed acyclic graph $\mathcal{B} = \langle n, U, A, l, d, v, \sigma \rangle$ where n is the number of variables from the encoded problem, U is a set of nodes; each of which is associated to some state $\sigma(u)$. The mapping $l : U \to \{0 \ldots n\}$ partitions the nodes from U in disjoint layers $L_0 \ldots L_n$ s.t. $L_i = \{u \in U : l(u) = i\}$ and the states of all the nodes belonging to the same layer pertain to the same DP state-space ($\forall u \in L_i : \sigma(u) \in S_i$ for $i = 0, \ldots, n$). Also, it should be the case that no two distinct nodes of one same layer have the same state ($\forall u_1, u_2 \in L_i : u_1 \neq u_2 \implies \sigma(u_1) \neq \sigma(u_2)$, for $i = 0, \ldots, n$).

The set $A \subseteq U \times U$ from our formal model is a set of directed arcs connecting the nodes from U. Each such arc $a = (u_1, u_2)$ connects nodes from subsequent layers ($l(u_1) = l(u_2) - 1$) and should be regarded as the materialization of a branching decision about variable $x_{l(u_1)}$. This is why all arcs are annotated via the mappings $d : A \to D$ and $v : A \to \mathbb{R}$ which respectively associate a decision and value (weight) with the given arc.

Example 1. An arc a connecting nodes $u_1 \in L_3$ to $u_2 \in L_4$, annotated with $d(a) = 6$ and $v(a) = 42$ should be understood as the assignment $x_3 = 6$ performed from state $\sigma(u_1)$. It should also be understood that $t_3(\sigma(u_1), 6) = \sigma(u_2)$ and the benefit of that assignment is $v(a) = h_3(\sigma(u_1), 6) = 42$.

Because each r-t path describes an assignment that satisfies $\mathcal{P}$, we will use $Sol(\mathcal{B})$ to denote the set of all the solutions encoded in the r-t paths of DD $\mathcal{B}$. Also, because unsatisfiability is irrecoverable, r-$\perp$ paths are typically omitted from DDs. It follows that a nice property from using a DD representation $\mathcal{B}$ for the DP formulation of a problem $\mathcal{P}$, is that finding x^* is as simple as finding the longest r-t path in $\mathcal{B}$ (according to the relation v on arcs).

Exact-MDD. For a given problem $\mathcal{P}$, an exact MDD $\mathcal{B}$ is an MDD that exactly encodes the solution set $Sol(\mathcal{B}) = Sol(\mathcal{P})$ of the problem $\mathcal{P}$. In other words, not only do all r-t paths encode valid solutions of $\mathcal{P}$, but no feasible solution is present in $Sol(\mathcal{P})$ and not in $\mathcal{B}$. An exact MDD for $\mathcal{P}$ can be compiled in a top-down fashion[2]. This naturally follows from the above definition. To that end, one simply proceeds by a repeated unrolling of the transition relations until all variables are assigned.

2.2 Bounded-Size Approximations

In spite of the compactness of their encoding, the construction of DD suffers from a potentially exponential memory requirement in the worst case[3]. Thus, using DDs to exactly encode the solution space of a problem is often intractable. Therefore, one must resort to the use of *bounded-size* approximation of the exact MDD. These are compiled generically by inserting a call to a width-bounding procedure to ensure that the width (the number $|L_i|$ of distinct nodes belonging to the L_i) of the current layer L_i does not exceed a given bound W. Depending on the behavior of that procedure, one can either compile a restricted-MDD (= an under-approximation) or a relaxed-MDD (= an over-approximation).

Restricted-MDD: Under-Approximation. A restricted-MDD provides an under-approximation of some exact-MDD. As such, all paths of a restricted-MDD encode valid solutions, but some solutions might be missing from the MDD. This is formally expressed as follows: given the DP formulation of a problem $\mathcal{P}$, $\mathcal{B}$ is a restricted-MDD iff $Sol(\mathcal{B}) \subseteq Sol(\mathcal{P})$.

To compile a restricted-MDD, it is sufficient to simply delete certain nodes from the current layer until its width fits within the specified bound W. To that end, the width-bounding procedure simply selects a subset of the nodes from L_i which are heuristically assumed to have the less impact on the tightness of the bound. Various heuristics have been studied in the literature [7], and $minLP$ was shown to be the heuristic that works best in practice. This heuristic decides to select (hence remove) the nodes having the shortest longest path from the root.

---

[2] An incremental refinement *a.k.a. construction by separation* procedure is detailed in [11, pp. 51–52] but we will not cover it here for the sake of conciseness.

[3] Consequently, it also suffers from a potentially exponential time requirement in the worst case. Indeed, time is constant in the final number of nodes (unless the transition functions themselves are exponential in the input).

Relaxed-MDD: Over-Approximation. A relaxed-MDD $\mathcal{B}$ provides a bounded-width over-approximation of some exact-MDD. As such, it may hold paths that are no solution to $\mathcal{P}$, the problem being solved. We have thus formally that $Sol(\mathcal{B}) \supseteq Sol(\mathcal{P})$.

Compiling a relaxed-MDD requires one to be able to *merge* several nodes into an inexact one. To that end, we use two operators:

- $\oplus$ which yields a new node combining the states of a selection of nodes so as to over-approximate the states reachable in the selection.
- $\varGamma$ which is used to possibly relax the weight of arcs incident to the selected nodes.

These operators are used as follows. Similar to the restricted-MDDs case, the width-bounding procedure starts by heuristically selecting the least promising nodes and removing them from layer L_i. Then the states of these selected nodes are combined with one another so as to create a merged node $\mathcal{M} = \oplus(selection)$. After that, the inbound arcs incident to all selected nodes are $\varGamma$-relaxed and redirected towards $\mathcal{M}$. Finally, the result of the merger ($\mathcal{M}$) is added to the layer in place of the initial selection of nodes.

Summary. Figure 1 summarizes the information from Sects. 2.1 and 2.2. It displays the three MDDs corresponding to one same example problem having four variables. The exact MDD (a) encodes the complete solution set and, equivalently, the state space of the underlying DP encoding. One easily notices that the restricted DD (b) is an under approximation of (a) since it achieves its width boundedness by removing nodes d and e and their children (i, j). Among others, it follows that the solution $[x_0 = 0, x_1 = 0, x_2 = 0, x_3 = 0]$ is not represented in (b) even though it exists in (a). Conversely, the relaxed diagram (c) achieves a maximum layer with of 3 by merging nodes d, e and h into a new inexact node $\mathcal{M}$ and by relaxing all arcs entering one of the merged nodes. Because of this, (c) introduces solutions that do not exist in (a) as is for instance the case of the assignment $[x_0 = 0, x_1 = 0, x_2 = 3, x_3 = 1]$. Moreover, because the operators $\oplus$ and $\varGamma$ are correct[4], the length of the longest path in (c) is an upper bound on the optimal value of the objective function. Indeed, one can see that the length of the longest path in (a) (= the exact optimal solution) has a value of 25 while it amounts to 26 in (c).

2.3   The Dynamics of Branch-and-Bound with DDs

Being able to derive good lower and upper bounds for some optimization problem $\mathcal{P}$ is useful when the goal is to use these bounds to strengthen algorithms [13, 31,32]. But it is not the only way these approximations can be used. A complete and efficient branch-and-bound algorithm relying on those approximations was proposed in [8] which we hereby reproduce (Algorithm 1).

---

[4] The very definition of these operators is problem-specific. However, [22] formally defines the conditions that are necessary to correctness.

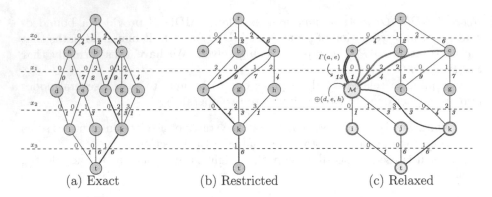

(a) Exact   (b) Restricted   (c) Relaxed

Fig. 1. The exact (a), restricted (b) and relaxed (c) versions of an MDD with four variables. The width of MDDs (b) and (c) have been bounded to a maximum layer width of three. The decision labels of the arcs are shown above the layers separation lines (dashed). The arc weights are shown below the layer separation lines. The longest path of each MDD is boldfaced. In (c), the node $\mathcal{M}$ is the result of merging nodes d, e and h with the $\oplus$ operator. Arcs that have been relaxed with the Γ operator are pictured with a double stroke. Note, because these arcs have been Γ-relaxed, their value might be greater than that of corresponding arcs in (a), (b). Similarly, all "inexact" nodes feature a double border.

This algorithm works as follows: at start, the node r is created for the initial state of the problem and placed onto the *fringe* – a global priority queue that tracks all nodes remaining to explore and orders them from the most to least promising. Then, a loop consumes the nodes from that fringe (line 1), one at a time and explores it until the complete state space has been exhausted. The *exploration* of a node u inside that loop proceeds as follows: first, one compiles a restricted DD $\underline{\mathcal{B}}$ for the sub-problem rooted in u (line 5). Because all paths in a restricted DD are feasible solutions, when the lower bound $v^*(\underline{\mathcal{B}})$ derived from the restricted DD $\underline{\mathcal{B}}$ improves over the current best known solution $\underline{v}$; then the longest path of $\underline{\mathcal{B}}$ (best sol. found in $\underline{\mathcal{B}}$) and its length $v^*(\underline{\mathcal{B}})$ are memorized (lines 7-9).

In the event where $\underline{\mathcal{B}}$ is exact (no restriction occurred during the compilation of $\underline{\mathcal{B}}$), it covers the complete state space of the sub-problem rooted in u. Which means the processing of u is complete and we may safely move to the next node. When this condition is not met, however, some additional effort is required. In that case, a *relaxed* DD $\overline{\mathcal{B}}$ is compiled from u (line 11). That relaxed DD serves two purposes: first, it is used to derive an upper bound $v^*(\overline{\mathcal{B}})$ which is compared to the current best known solution (line 12). This gives us a chance to prune the unexplored state space under u when $v^*(\overline{\mathcal{B}})$ guarantees it does not contain any better solution than the current best. The second use of $\overline{\mathcal{B}}$ happens when $v^*(\overline{\mathcal{B}})$ cannot provide such a guarantee. In that case, the exact cutset of $\overline{\mathcal{B}}$ is used to enumerate residual sub-problems which are enqueued onto the fringe (lines 13–14).

A cutset for some relaxed DD $\overline{B}$ is a subset C of the nodes from $\overline{B}$ such that any $r - t$ path of $\overline{B}$ goes through at least one node $\in C$. Also, a node u is said to be exact iff all its incoming paths lead to the same state $\sigma(u)$. From there, an exact cutset of $\overline{B}$ is simply a cutset whose nodes are all exact. Based on this definition, it is easy to convince oneself that an exact cutset constitutes a frontier up to which the relaxed DD $\overline{B}$ and its exact counterpart B have not diverged. And, because it is a cutset, the nodes composing that frontier cover all paths from both B and $\overline{B}$; which guarantees the completeness of Algorithm 1 [8].

Any relaxed-MDD admits at least one exact cutset – e.g. the trivial $\{r\}$ case. Often though, it is not unique and different options exist as to what cutset to use. It was experimentally shown by [8] that most of the time, the Last Exact Layer (LEL) is superior to all other exact cutsets in practice. LEL consists of the *deepest* layer of the relaxed-MDD having all its nodes exact.

Example 2. In Fig. 1 (c), the first inexact node $\mathcal{M}$ occurs in layer L_2. Hence, the LEL cutset comprises all nodes (a, b, c) from the layer L_1. Because $\mathcal{M}$ is inexact, and because it is a parent of nodes i, j and k, these three nodes are considered inexact too.

Algorithm 1. Branch-And-Bound with DD	**Algorithm 2.** Local bound pruning
1: Create node r and add it to *Fringe*	1: Create node r and add it to *Fringe*
2: $\underline{x} \leftarrow \bot$	2: $\underline{x} \leftarrow \bot$
3: $\underline{v} \leftarrow -\infty$	3: $\underline{v} \leftarrow -\infty$
4: **while** *Fringe* is not empty **do**	4: **while** *Fringe* is not empty **do**
5: $u \leftarrow Fringe.pop()$	5: $u \leftarrow Fringe.pop()$
6: $\underline{B} \leftarrow Restricted(u)$	6: **if** $v|_u^* \leq \underline{v}$ **then**
7: **if** $v^*(\underline{B}) > \underline{v}$ **then**	7: **continue**
8: $\underline{v} \leftarrow v^*(\underline{B})$	8: $\underline{B} \leftarrow Restricted(u)$
9: $\underline{x} \leftarrow x^*(\underline{B})$	9: **if** $v^*(\underline{B}) > \underline{v}$ **then**
10: **if** $\underline{B}$ is not exact **then**	10: $\underline{v} \leftarrow v^*(\underline{B})$
11: $\overline{B} \leftarrow Relaxed(u)$	11: $\underline{x} \leftarrow x^*(\underline{B})$
12: **if** $v^*(\overline{B}) > \underline{v}$ **then**	12: **if** $\underline{B}$ is not exact **then**
13: **for all** $u' \in \overline{B}.exact\_cutset()$ **do**	13: $\overline{B} \leftarrow Relaxed(u)$
14: $Fringe.add(u')$	14: **if** $v^*(\overline{B}) > \underline{v}$ **then**
15: **return** $(\underline{x}, \underline{v})$	15: **for all** $u' \in \overline{B}.exact\_cutset()$ **do**
	16: **if** $v|_{u'}^* > \underline{v}$ **then**
	17: $Fringe.add(u')$
	18: **return** $(\underline{x}, \underline{v})$

3   Improving the Filtering of Branch-and-Bound MDD

In the forthcoming paragraphs, we introduce the local bound and present the rough upper bound: two reasoning techniques to reinforce the pruning strength of Algorithm 1.

3.1   Local Bounds (LocB)

Conceptually, pruning with local bounds is rather simple: a relaxed MDD $\overline{B}$ provides us with *one* upper bound $v^*(\overline{B})$ on the optimal value of the objective function for some given sub-problem. However, in the event where $v^*(\overline{B})$

is greater than the best known lower bound $\underline{v}$ (best current solution) nothing guarantees that all nodes from the exact cutset of $\overline{B}$ admit a longest path to t with a length of $v^*(\overline{B})$. Actually, this is quite unlikely. This is why we propose to attach a *"local" upper bound* to each node of the cutset. This local upper bound – denoted $v|_u^*$ for some cutset node u – simply records the length of the longest r-t path passing through u in the relaxed MDD $\overline{B}$.

In other words, LocB allows us to refine the information provided by a relaxed DD $\overline{B}$. On one hand, $\overline{B}$ provides us with $v^*(\overline{B})$ which is the length of the longest r-t path in $\overline{B}$. As such, it provides an upper bound on the optimal value that can be reached from the root node of $\overline{B}$. With the addition of LocB, the relaxed DD provides us with an additional piece of information. For each individual node u in the exact cutset of $\overline{B}$, it defines the value $v|_u^*$ which is an upper bound on the value attainable from that node.

As shown in Algorithm 2, the value $v|_u^*$ can prove useful at two different moments. First, in the event where $v|_u^* \leq \underline{v}$, this value can serve as a justification to not enqueue the subproblem u (line 16) since exhausting this subproblem will yield no better solution than $\underline{v}$. More formally, by definition of a cutset and of LocB, it must be the case that the longest r-t path of $\overline{B}$ traverses one of the cutset nodes u and thus that $v^*(\overline{B}) = v|_u^*$ (where $v|_u^*$ is the local bound of u). Hence we have: $\exists u \in$ cutset of $\overline{B} : v^*(\overline{B}) = v|_u^*$. However, because $v^*(\overline{B})$ is the length of the *longest* r-t path of $\overline{B}$, there may exist cutset nodes that only belong to r-t paths shorter than $v^*(\overline{B})$. That is: $\forall u' \in$ cutset of $\overline{B} : v^*(\overline{B}) \geq v|_{u'}^*$. Which is why $v|_{u'}^*$ can be stricter than $v^*(\overline{B})$ and hence let LocB be stronger at pruning nodes from the frontier.

The second time when $v|_u^*$ might come in handy occurs when the node u is popped out of the fringe (line 6). Indeed, because the fringe is a global priority queue, any node that has been pushed on the fringe can remain there for a long period of time. Thus, chances are that the value $\underline{v}$ has increased between the moment when the node was pushed onto the fringe (line 17) and the moment when it is popped out of it. Hence, this gives us an additional chance to completely skip the exploration of the sub-problem rooted in u.

Let us illustrate that with the relaxed MDD shown on Fig. 2, for which the exact cutset comprises the highlighted nodes a and b. Please note that because this scenario may occur at any time during the problem resolution, we will assume that the fringe is not empty when it starts. Assuming that the current best solution $\underline{v}$ is 20 when one explores the pictured subproblem, we are certain that exploring the subproblem rooted in a is a waste of time, because the local bound $v|_a^*$ is only 16. Also, because the fringe was not empty, it might be the case that b was left on the fringe for a long period of time. And because of this, it might be the case that the best known value $\underline{v}$ was improved between the moment when b was pushed on the fringe and the moment when it was popped out of it. Assuming that $\underline{v}$ has improved to 110 when b is popped out of the fringe, it may safely be skipped because $v|_b^*$ guarantees that an exploration of b will not yield a better solution than 102.

Algorithm 3 describes the procedure to compute the local bound $v|_u^*$ of each node u belonging to the exact cutset of a relaxed MDD $\overline{\mathcal{B}}$. Intuitively, this is achieved by doing a bottom-up traversal of $\overline{\mathcal{B}}$, starting at t and stopping when the traversal crosses the last exact layer (line 5). During that bottom-up traversal, the algorithm marks the nodes that are reachable from t. This way, it can avoid the traversal of dead-end nodes. Also, Algorithm 3 maintains a value $v_{\uparrow t}^*(u)$ for each node u it encounters. This value represents the length of the longest u-t path. Afterwards (line 13), it is summed with the length of the longest r-u path v_{r-u}^* to derive the exact value of the local bound $v|_u^*$.

Algorithm 3. Computing the local bounds

1: *lel* ← Index of the last exact layer
2: $v_{\uparrow t}^*(u) \leftarrow -\infty$ **for each node** $u \in \overline{\mathcal{B}}$ // init. longest u-t path
3: $mark(t) \leftarrow$ true
4: $v_{\uparrow t}^*(t) \leftarrow 0$ // longest t-t path
5: **for all** $i = n$ to *lel* **do**
6:    **for all** node $u \in L_i$ **do**
7:      **if** $mark(u)$ **then**
8:        **for all** arc $a = (u', u)$ incident to u **do**
9:          $mark(u') \leftarrow$ true
10:          $v_{\uparrow t}^*(u') \leftarrow \max(v_{\uparrow t}^*(u'), v_{\uparrow t}^*(u) + v(a))$ // longest u'-t path
11: **for all** node $u \in \overline{\mathcal{B}}.exact\_cutset()$ **do**
12:    **if** $mark(u)$ **then**
13:      $v|_u^* \leftarrow v_{r-u}^* + v_{\uparrow t}^*(u)$ // longest r-u path + longest u-t path
14:    **else**
15:      $v|_u^* \leftarrow -\infty$

3.2 Rough Upper Bound (RUB)

Rough upper bound pruning departs from the following observation: assuming the knowledge of a lower bound $\underline{v}$ on the value of v^*, and assuming that one is able to swiftly compute a rough upper bound $\overline{v_s}$ on the optimal value v_s^* of the subproblem rooted in state s; any node u of a MDD having a rough upper bound $\overline{v_{\sigma(u)}} \leq \underline{v}$ may be discarded as it is guaranteed not to improve the best known solution. This is pretty much the same reasoning that underlies the whole branch-and-bound idea. But here, it is used to prune portions of the search space explored *while compiling* approximate MDDs.

To implement RUB, it suffices to adapt the MDD compilation procedure (top-down, iterative refinement, ...) and introduce a check that avoids creating a node u' with state *next* when $\overline{v_{next}} \leq \underline{v}$.

The key to RUB effectiveness is that RUB is used while compiling the restricted and relaxed DDs. As such, its computation does not directly appear in Algorithm 1, but rather is accounted within the compilations of $Restricted(u)$ and $Relaxed(u)$ from Algorithm 1. Thus, it really is not used as yet-an-other-bound competing with that of line 12, but instead to speed up the computation of restricted and relaxed DDs. More precisely, this speedup occurs because the compilation of the DDs discards some nodes that would otherwise be added to the next layer of the DD and then further expanded, which are ruled out by RUB. A second benefit of using RUBs is that it helps tightening the bound derived from a relaxed DD (Algorithm 1 line 12). Because the layers that are generated in a relaxed DD are narrower when applying RUB, there are fewer nodes exceeding the maximum layer width. The operator $\oplus$ hence needs to merge a smaller set of nodes in order to produce the relaxation.

The dynamics of RUB is graphically illustrated by Fig. 3 where the set of highlighted nodes can be safely elided since the (rough) upper bound computed in node s is lesser than the best lower bound.

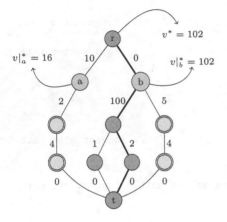
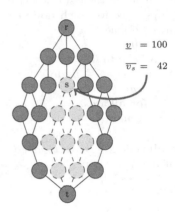

Fig. 2. An example relaxed-MDD having an exact cutset $\{a, b\}$ with local bounds $v|_a^*$ and $v|_b^*$. The nodes with a simple border represent exact nodes and those with a double border represent "inexact" nodes. The edges along the longest path are displayed in bold.

Fig. 3. Assuming a lower bound $\underline{v}$ of 100 and a rough upper bound $\overline{v_s}$ of 42 for the node s, all the highlighted nodes (in red, with a dashed border) may be pruned from the MDD. (Color figure online)

Important Note. It is important to understand that because the RUB is computed at each node of each restricted and relaxed MDD compiled during the instance resolution, it must be extremely inexpensive to compute. This is why RUB is best obtained from a fast and simple problem specific procedure.

4   Experimental Study

In order to evaluate the impact of the pruning techniques proposed above, we conducted a series of experiments on four problems. In particular, we conducted experiments on the Maximum Independent Set Problem (MISP), the

Maximum Cut Problem (MCP), the Maximum Weighted 2-Satisfiablility Problem (MAX2SAT) and the Traveling Salesman Problem with Time Windows (TSPTW). For the first three problems, we generated sets of random instances which we attempted to solve with different configurations of our own open source solver written in Rust [17][5]. For TSPTW, we reused openly available sets of benchmarks which are usually used to assess the efficiency of new solvers for TSPTW [27]. Thanks to the generic nature of our framework, the model and all heuristics used to solve the instances were the same for all experiments. This allowed us to isolate the impact of RUB and LocB on the solving performance and neutralize unrelated factors such as variable ordering. Indeed, the only variations between the different solver flavors relate to the presence (or absence) of RUB and LocB. All experiments were run on the same physical machine equipped with an AMD6176 processor and 48GB of RAM. A maximum time limit of 1800 1800seconds was allotted to each configuration to solve each instance.

The details of the DP models and RUBs we formulated for all four problems are given in the appendices to the extended version of this paper[6].

MISP. To assess the impact of RUB and LocB on MISP, we generated random graphs based on the Erdos-Renyi model $G(n, p)$ [15] with the number of vertices n = 250, 500, 750, 1000, 1250, 1500, 1750 and the probability of having an edge connecting any two vertices $p = 0.1, 0.2, ... , 0.9$. The weight of the edges in the generated graphs were drawn uniformly from the set $\{-5, -4, -3, -2, -1, 1, 2, 3, 4, 5\}$. We generated 10 instances for each combination of size and density (n, p).

MCP. In line with the strategy used for MISP, we generated random MCP instances as random graphs based on the Erdos-Renyi model $G(n, p)$. These graphs were generated with the number of vertices n = 30, 40, 50 and the probability p of connecting any two vertices = 0.1, 0.2, 0.3, .., 0.9. The weights of the edges in the generated graphs were drawn uniformly among $\{-1, 1\}$. Again, we generated 10 instances per combination n, p.

MAX2SAT. Similar to the above, we used random graphs based the Erdos-Renyi model $G(n, p)$ to derive MAX2SAT instances. To this end, we produced graphs with n = 60, 80, 100, 200, 400, 1000 (hence instances with 30, 40, 50, 100, 200 and 500 variables) and $p = 0.1, 0.2, 0.3, .. , 0.9$. For each combination of size (n) and density (p), we generated 10 instances. The weights of the clauses in the generated instances were drawn uniformly from the set $\{1, 2, 3, 5, 6, 7, 8, 9, 10\}$.

TSPTW. To evaluate the effectiveness of our rules on TSPTW, we used the 467 instances from the following suites of benchmarks, which are usually used to assess the efficiency of new TSPTW solvers. AFG [2], Dumas [14], Gendreau-Dumas [16], Langevin [26], Ohlmann-Thomas [28], Solomon-Pesant [29] and Solomon-Potvin-Bengio [30].

---

[5] https://github.com/xgillard/ddo.
[6] Available online at: http://hdl.handle.net/2078.1/245322.

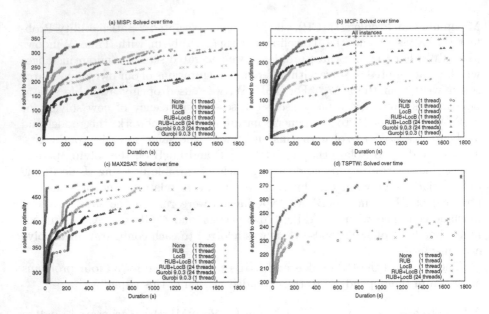

Fig. 4. Number of solved instances over time for each considered problem

Figure 4 gives an overview of the results from our experimental study. It respectively depicts the evolution over time of the number of instances solved by each technique for MISP (a), MCP (b) and MAX2SAT (c) and TSPTW (d).

As a first step, our observation of the graphs will focus on the differences that arise between the single threaded configurations of our *ddo* solvers. Then, in a second phase, we will incorporate an existing state-of-the-art ILP solver (Gurobi 9.0.3) in the comparison. Also, because both Gurobi and our *ddo* library come with built-in parallel computation capabilities, we will consider both the single threaded and parallel (24 threads) cases. This second phase, however, only bears on MISP, MCP and MAX2SAT by lack of a Gurobi TSPTW model.

DDO Configurations. The first observation to be made about the four graphs in Fig. 4, is that for all considered problems, both RUB and LocB outperformed the 'do-nothing' strategy; thereby showing the relevance of the rules we propose. It is not clear however which of the two rules brings the most improvement to the problem resolution. Indeed, RUB seems to be the driving improvement factor for MISP (a) and TSPTW (d) and the impact of LocB appears to be moderate or weak on these problems. However, it has a much higher impact for MCP (b) and MAX2SAT (c). In particular, LocB appears to be the driving improvement factor for MCP (b). This is quite remarkable given that LocB operates in a purely black box fashion, without any problem-specific knowledge. Finally, it should also be noted that the use of RUB and LocB are not mutually exclusive. Moreover, it turns out that for all considered problems, the combination RUB+LocB improved the situation over the use of any single rule.

Furthermore, Fig. 5 confirms the benefit of using both RUB *and* LocB together rather than using any single technique. For each problem, it measures the "performance" of using RUB+LocB vs the best single technique through the end gap. The end gap is defined as $\left(100 * \frac{|UB|-|LB|}{|UB|}\right)$. This metric allows us to account for all instances, including the ones that could not be solved to optimality. Basically, a small end gap means that the solver was able to confirm a tight confidence interval of the optimum. Hence, a smaller gap is better. On each subgraphs of Fig. 5, the distance along the x-axis represents the end gap for reach instance when using both RUB and LocB whereas the distance along y-axis represents the end gap when using the best single technique for the problem at hand. Any mark above the diagonal shows an instance for which using both RUB and LocB helped reduce the end gap and any mark below that line indicates an instance where it was detrimental.

From graphs 5-a, 5-c and 5-d it appears that the combination RUB+LocB supersedes the use of RUB only. Indeed the vast majority of the marks sit above the diagonal and the rest on it. This indicates a beneficial impact of using both techniques even for the hardest (unsolved) instances. The case of MCP (graph 5-b) is less clear as most of the marks sit on the diagonal. Still, we can only observe three marks below the diagonal and a bit more above it. Which means that even though the use of RUB in addition to LocB is of little help in the case of MCP, its use does not degrade the performance for that considered problem.

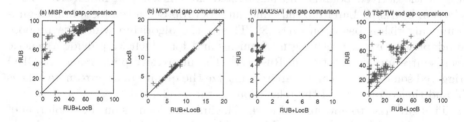

Fig. 5. End gap: the benefit of using both techniques vs the best single one

Comparison with Gurobi 9.0.3. The first observation to be made when comparing the performance of Gurobi vs the DDO configurations, is that when running on a single thread, ILP outperforms the basic DDO approach (without RUB and LocB). Furthermore, Gurobi turns out to be the best single threaded solver for MCP by a fair margin. However, in the MISP and MAX2SAT cases, Fig. 4 shows that the DDO solvers benefitting from RUB and LocB were able to solve more instances and to solve them faster than Gurobi. Which underlines the importance of RUB and LocB.

When lifting the one thread limit, one can see that the DD-based approach outperform ILP on each of the considered problems. In particular, in the case of MCP for which Gurobi is the best single threaded option; our DDO solver was able to find and prove the optimality of all tested instances in a little less than 800 s. The ILP solver, on the other end, was not able to prove the optimality of the 9 hardest instances within 30 min. Additionally, we also observe that in spite of the performance gains of MIP when running in parallel, Gurobi fails to solve as many MISP and MAX2SAT instances and to solve them as fast as the single threaded DDO solvers with RUB and LocB. This emphasizes once more the relevance of our techniques. It also shows that the observation from [9] still hold today: despite the many advances of MIP the DDO approach still scales better than MIP on the considered problems when invoked in parallel.

5   Previous Work

DDO emerged in the mid' 2000's when [24] proposed to use decision diagrams as a way to solve discrete optimization problems to optimality. More or less concomitantly, [1] devised relaxed-MDD even though the authors envisioned its use as a CP constraint store rather than a means to derive tight upper bounds for optimization problems. Then, the relationship between decision diagrams and dynamic programming was clarified by [21].

Recently, Bergman, Ciré and van Hoeve investigated the various ways to compile decision diagrams for optimization (top-down, construction by separation) [11]. They also investigated the heuristics used to parameterize these DD compilations. In particular, they analyzed the impact of variable ordering in [7,11] and node selection heuristics (for merge and deletion) in [7]. Doing so, they empirically demonstrated the crucial impact of variable ordering on the tightness of the derived bounds and highlighted the efficiency of minLP as a node selection heuristic. Later on, the same authors proposed a complete branch-and-bound algorithm based on DDs [8]. This is the algorithm which we propose to adapt with extra reasoning mechanisms and for which we provide a generic open-source implementation in Rust [17]. The impressive performance of DDO triggered some theoretical research to analyze the quality of approximate MDDs [5] and the correctness of the relaxation operators [22].

This gave rise to new lines of work. The first one focuses on the resolution of a larger class of optimization problems; chief of which multi-objective problems [4] and problems with a non-linear objective function. These are either solved by decomposition [4] or by using DDO to strengthen other IP techniques [13]. A second trend aims at hybridizing DDO with other IP techniques. For instance, by using Lagrangian relaxation [23] or by solving a MIP [6] to derive with very tight bounds. But the other direction is also under active investigation: for example, [31,32] use DD to derive tight bounds which are used to replace LP relaxation in a cutting planes solver. Very recently, a third hybridization approach has been proposed by Gonzàlez et al. [18]. It adopts the branch-and-bound MDD perspective, but whenever an upper bound is to be derived, it uses a trained classifier to decide whether the upper bound is to be computed with ILP or by developing a fixed-width relaxed MDD.

The techniques (ILP-cutoff pruning and ILP-cutoff heuristic) proposed by Gonzalez et al. [18] are related to RUB and LocB in the sense that all techniques aim at reducing the search space of the problem. However, they fundamentally differ as ILP-cutoff pruning acts as a replacement for the compilation of a relaxed

MDD whereas the goal of RUB is to speed up the development of that relaxed MDD by removing nodes *while* the MDD is being generated. The difference is even bigger in the case of ILP-cutoff heuristic vs LocB: the former is used as a primal heuristic while LocB is used to filter out sub-problems that can bear no better solution. In that sense, LocB belongs more to the line of work started by [1,19,20]: it enforces the constraint $lb \leq f(x) \leq ub$ and therefore provokes the deletion of nodes and arcs that cannot lead to the optimal solution.

More recently, Horn et al. explored an idea in [25] which closely relates to RUB. They use "fast-to-compute dual bounds" as an admissible heuristic to guide the compilation of MDDs in an A* fashion for the prize-collecting TSP. It prunes portions of the state space during the MDD construction, similarly to when RUB is applied. Our approach differs from that of [25] in that we attempt to incorporate problem specific knowledge in a framework that is otherwise fully generic. More precisely, it is perceived here as a problem-specific pruning that exploits the combinatorial structure implied by the state variables. It is independent of other MDD compilation techniques, e.g., our techniques are compatible with node merge ($\oplus$) operators and other methodologies defined in the DDO literature. We also emphasize that, as opposed to more complex LP-based heuristics that are now typical in A* search, we investigate quick methodologies that are also easy to incorporate in a MDD branch and bound.

6   Conclusion and Future Work

This paper presented and evaluated the impact of the local bound and rough upper bound techniques to strengthen the pruning of the branch-and-bound MDD algorithm. Our experimental study on MISP, MCP, MAX2SAT and TSPTW confirmed the relevance of these techniques. In particular, our experiments have shown that devising a fast and simple rough upper bound is worth the effort as it can significantly boost the efficiency of a solver. Similarly, our experiments showed that the use of local bound can significantly improve the efficiency of DDO solver despite its problem agnosticism. Furthermore, it revealed that a combination of RUB and LocB supersedes the benefit of any single reasoning technique. These results are very promising and we believe that the public availability of an open source DDO framework implementing RUB and LocB might serve as a basis for novel DP formulation for classic problems.

References

1. Andersen, H.R., Hadzic, T., Hooker, J.N., Tiedemann, P.: A constraint store based on multivalued decision diagrams. In: Bessière, C. (ed.) CP 2007. LNCS, vol. 4741, pp. 118–132. Springer, Heidelberg (2007). https://doi.org/10.1007/978-3-540-74970-7_11
2. Ascheuer, N.: Hamiltonian path problems in the on-line optimization of flexible manufacturing systems (1996)
3. Bellman, R.: The theory of dynamic programming. Bull. Am. Math. Soc. **60**(6), 503–515 (1954). https://projecteuclid.org:443/euclid.bams/1183519147

4. Bergman, D., Cire, A.A.: Multiobjective optimization by decision diagrams. In: Rueher, M. (ed.) CP 2016. LNCS, vol. 9892, pp. 86–95. Springer, Cham (2016). https://doi.org/10.1007/978-3-319-44953-1_6

5. Bergman, D., Cire, A.A.: Theoretical insights and algorithmic tools for decision diagram-based optimization. Constraints 21(4), 533–556 (2016). https://doi.org/10.1007/s10601-016-9239-9

6. Bergman, D., Cire, A.A.: On finding the optimal BDD relaxation. In: Salvagnin, D., Lombardi, M. (eds.) CPAIOR 2017. LNCS, vol. 10335, pp. 41–50. Springer, Cham (2017). https://doi.org/10.1007/978-3-319-59776-8_4

7. Bergman, D., Cire, A.A., van Hoeve, W.J., Hooker, J.N.: Optimization bounds from binary decision diagrams. INFORMS J. Comput. 26(2), 253–268 (2014). https://doi.org/10.1287/ijoc.2013.0561

8. Bergman, D., Cire, A.A., van Hoeve, W.J., Hooker, J.N.: Discrete optimization with decision diagrams. INFORMS J. Comput. 28(1), 47–66 (2016). https://doi.org/10.1287/ijoc.2015.0648

9. Bergman, D., Cire, A.A., Sabharwal, A., Samulowitz, H., Saraswat, V., van Hoeve, W.J.: Parallel combinatorial optimization with decision diagrams. In: International Conference on AI and OR Techniques in Constraint Programming for Combinatorial Optimization Problems, pp. 351–367 (2014)

10. Burch, J.R., Clarke, E.M., McMillan, K.L., Dill, D.L., Hwang, L.J.: Symbolic model checking: 10^{20} states and beyond. Inf. Comput. 98(2), 142–170 (1992). https://doi.org/10.1016/0890-5401(92)90017-A

11. Cire, A.A.: Decision diagrams for optimization. Ph.D. thesis, Carnegie Mellon University Tepper School of Business (2014)

12. Cormen, T.H., Leiserson, C.E., Rivest, R.L., Stein, C.: Introduction to Algorithms. MIT Press, Cambridge (2009)

13. Davarnia, D., van Hoeve, W.J.: Outer approximation for integer nonlinear programs via decision diagrams (2018)

14. Dumas, Y., Desrosiers, J., Gelinas, E., Solomon, M.M.: An optimal algorithm for the traveling salesman problem with time windows. Oper. Res. 43(2), 367–371 (1995)

15. Erdős, P., Rényi, A.: On random graphs i. Publicationes Mathematicae Debrecen 6, 290 (1959)

16. Gendreau, M., Hertz, A., Laporte, G., Stan, M.: A generalized insertion heuristic for the traveling salesman problem with time windows. Oper. Res. 46(3), 330–335 (1998)

17. Gillard, X., Schaus, P., Coppé, V.: Ddo, a generic and efficient framework for MDD-based optimization. Accepted at the International Joint Conference on Artificial Intelligence (IJCAI-20); DEMO track (2020)

18. Gonzalez, J.E., Cire, A.A., Lodi, A., Rousseau, L.M.: Integrated integer programming and decision diagram search tree with an application to the maximum independent set problem. Constraints 1–24 (2020)

19. Hadžić, T., Hooker, J., Tiedemann, P.: Propagating separable equalities in an MDD store. In: CPAIOR, pp. 318–322 (2008)

20. Hoda, S., van Hoeve, W.-J., Hooker, J.N.: A systematic approach to MDD-based constraint programming. In: Cohen, D. (ed.) CP 2010. LNCS, vol. 6308, pp. 266–280. Springer, Heidelberg (2010). https://doi.org/10.1007/978-3-642-15396-9_23

21. Hooker, J.N.: Decision diagrams and dynamic programming. In: Gomes, C., Sellmann, M. (eds.) CPAIOR 2013. LNCS, vol. 7874, pp. 94–110. Springer, Heidelberg (2013). https://doi.org/10.1007/978-3-642-38171-3_7

22. Hooker, J.N.: Job sequencing bounds from decision diagrams. In: Beck, J.C. (ed.) CP 2017. LNCS, vol. 10416, pp. 565–578. Springer, Cham (2017). https://doi.org/10.1007/978-3-319-66158-2_36

23. Hooker, J.N.: Improved job sequencing bounds from decision diagrams. In: Schiex, T., de Givry, S. (eds.) CP 2019. LNCS, vol. 11802, pp. 268–283. Springer, Cham (2019). https://doi.org/10.1007/978-3-030-30048-7_16

24. Hooker, J.: Discrete global optimization with binary decision diagrams. In: GICO-LAG 2006 (2006)

25. Horn, M., Maschler, J., Raidl, G.R., Rönnberg, E.: A*-based construction of decision diagrams for a prize-collecting scheduling problem. Comput. Oper. Res. **126**, 105125 (2021). https://doi.org/10.1016/j.cor.2020.105125, http://www.sciencedirect.com/science/article/pii/S0305054820302422

26. Langevin, A., Desrochers, M., Desrosiers, J., Gélinas, S., Soumis, F.: A two-commodity flow formulation for the traveling salesman and the makespan problems with time windows. Networks **23**(7), 631–640 (1993)

27. López-Ibáñez, M., Blum, C.: Benchmark instances for the travelling salesman problem with time windows. Online (2020). http://lopez-ibanez.eu/tsptw-instances

28. Ohlmann, J.W., Thomas, B.W.: A compressed-annealing heuristic for the traveling salesman problem with time windows. INFORMS J. Comput. **19**(1), 80–90 (2007)

29. Pesant, G., Gendreau, M., Potvin, J.Y., Rousseau, J.M.: An exact constraint logic programming algorithm for the traveling salesman problem with time windows. Transp. Sci. **32**(1), 12–29 (1998)

30. Potvin, J.Y., Bengio, S.: The vehicle routing problem with time windows part ii: genetic search. INFORMS J. Comput. **8**(2), 165–172 (1996)

31. Tjandraatmadja, C.: Decision diagram relaxations for integer programming. Ph.D. thesis, Carnegie Mellon University Tepper School of Business (2018)

32. Tjandraatmadja, C., van Hoeve, W.J.: Target cuts from relaxed decision diagrams. INFORMS J. Comput. **31**(2), 285–301 (2019). https://doi.org/10.1287/ijoc.2018.0830

On the Usefulness of Linear Modular Arithmetic in Constraint Programming

Gilles Pesant[1(✉)], Kuldeep S. Meel[2], and Mahshid Mohammadalitajrishi[1]

[1] Polytechnique Montréal, Montreal, Canada
gilles.pesant@polymtl.ca
[2] National University of Singapore, Singapore, Singapore
meel@comp.nus.edu.sg

Abstract. Linear modular constraints are a powerful class of constraints that arise naturally in cryptanalysis, checksums, hash functions, and the like. Given their importance, the past few years have witnessed the design of combinatorial solvers with native support for linear modular constraints, and the availability of such solvers has led to the emergence of new applications. While there exist global constraints in CP that consider congruence classes over domain values, linear modular arithmetic constraints have yet to appear in the global constraint catalogue despite their past investigation in the context of model counting for CSPs. In this work we seek to remedy the situation by advocating the integration of linear modular constraints in state-of-the-art CP solvers.

Contrary to previous belief, we conclude from an empirical investigation that Gauss-Jordan Elimination based techniques can provide an efficient and scalable way to handle linear modular constraints. On the theoretical side, we remark on the pairwise independence offered by hash functions based on linear modular constraints, and then discuss the design of hashing-based model counters for CP, supported by empirical results showing the accuracy and computational savings that can be achieved. We further demonstrate the usefulness of native support for linear modular constraints with applications to checksums and model counting.

1 Introduction

Given a set of variables $\mathcal{X} = \{x_1, x_2, \ldots, x_n\}$ with their associated domains of values $\mathcal{D} = \{D_1, D_2, \ldots, D_n\}$ and set of constraints $\mathcal{C}$ over $\mathcal{X}$, the Constraint Satisfaction Problem (CSP), denoted $\varphi = (\mathcal{X}, \mathcal{D}, \mathcal{C})$, seeks to assign to each variable $x_i \in \mathcal{X}$ a value from D_i such that every constraint in $\mathcal{C}$ is satisfied. It is often convenient and effective to use constraints that can succinctly express recurring relations of arbitrary arity. The global constraints catalogue [2] has grown over the years to encompass a wide variety of such constraints, including the case of values considered modulo a given parameter (e.g. ALLDIFFERENT_MODULO, AMONG_MODULO, MAXIMUM_MODULO). But despite the investigation of linear modular arithmetic constraints by Gomes et al. [11] in the context of model

© Springer Nature Switzerland AG 2021
P. J. Stuckey (Ed.): CPAIOR 2021, LNCS 12735, pp. 248–265, 2021.
https://doi.org/10.1007/978-3-030-78230-6_16

counting for CSPs, the latter constraints seem to have gone largely unnoticed in the CP community and indeed do not appear in that catalogue.[1]

The purpose of this paper is to advocate the inclusion of modular arithmetic constraints in CP solvers, motivated by important applications such as model counting, and to investigate the algorithmic opportunities currently available for efficient inference on such constraints as well as remaining challenges, through an empirical evaluation featuring linear modular arithmetic constraints.

In this paper we address the question of how to efficiently integrate linear modular constraints in a CP solver. As mentioned before, Gomes et al. [11] studied linear modular equalities in the context of model counting and remarked that a system based on Gauss-Jordan Elimination (GJE) would be inefficient. As a result, they proposed an adaptation of Trick's dynamic programming algorithm [18] to handle individual constraints. Their empirical evaluation was limited to short constraints (i.e. on about six variables). We take advantage of the compact table implementation for extensional constraints [9] to revisit GJE and thus reach the opposite conclusion: GJE applied to a system of linear modular constraints achieves significantly better performance than the alternative dynamic programming algorithm on individual constraints.

We demonstrate the scalability of our framework through an empirical evaluation on large linear modular constraints and show the opportunities offered by a solver with native support for linear modular constraints. Here we can draw a parallel with the availability of CryptoMiniSat, a SAT solver with native support for linear modular constraints in the Boolean domain (i.e., XOR constraints), which has opened up several applications. Linear modular arithmetic constraints naturally occur in several domains such as checksums, error correcting codes, cryptography, learning parity without noise, and model counting. In this paper we present applications to checksums and model counting.

The rest of the paper is organized as follows. We first present background and formal definition and representation of linear modular arithmetic constraints in Sect. 2. We present domain filtering algorithms for linear modular constraints in Sect. 3. We then present applications to checksums and model counting in Sects. 4 and 5. Finally, we conclude in Sect. 6.

2   Background

An integer *modulus* $p > 1$ defines a congruence equivalence relation on the set of all integers $\mathbb{Z}$: integers i and j are said to be congruent if there exists an integer k such that $i - j = kp$. Thus it partitions $\mathbb{Z}$ into p congruence classes, the ring of integers modulo p, on which addition and multiplication are defined in the obvious way.

We are interested in linear modular arithmetic constraints of the general form

$$\ell \leq ax \leq u \pmod{p}$$

---

[1] One exception is the work on bit-vector domains, involving some modular arithmetic, with applications to software verification and cryptography [1,8,13].

where x is a vector of n integer finite-domain variables, a a vector of integer coefficients, ℓ and u two integers, and p the modulus. We will also be interested in systems of m linear modular equalities in n integer finite-domain variables,

$$Ax = b \pmod{p}.$$

An integer i in such a ring has a multiplicative inverse if and only if i and p are coprime. When p is prime then clearly every $0 < i < p$ is coprime with p. In fact the ring of integers modulo p is a finite field $\mathbb{F}_p$—every non-zero element having a multiplicative inverse—if and only if p is prime.

Gauss-Jordan Elimination can solve systems of linear equations not only over the real numbers but also over any field, such as $\mathbb{F}_p$. We take advantage of this in Sect. 3.

The linear modular equations are closely related to the universal hash functions. Given two finite sets N and M, let $\mathcal{H}(N, M) \triangleq \{h : N \to M\}$ be a family of hash functions mapping N to M. We use $h \xleftarrow{R} \mathcal{H}(N, M)$ to denote the probability space obtained by choosing a function h uniformly at random from $\mathcal{H}(N, M)$.

Definition 1. *A family of hash functions $\mathcal{H}(N, M)$ is k-wise independent if $\forall \alpha_1, \alpha_2, \ldots \alpha_k \in M$ and for distinct $y_1, y_2, \ldots y_k \in N$, $h \xleftarrow{R} \mathcal{H}(N, M)$,*

$$\Pr\left[(h(y_1) = \alpha_1) \wedge (h(y_2) = \alpha_2) \ldots \wedge (h(y_k) = \alpha_k)\right] = \left(\frac{1}{M}\right)^k \tag{1}$$

Note that every k-wise independent hash family is also $k - 1$ wise independent. The phrase *strongly 2-universal* is also used to refer to 2-wise independent as noted by Vadhan in [19], although the concept of 2-universal hashing proposed by Carter and Wegman [3] only required that $\Pr[h(x) = h(y)] \leq \frac{1}{2^m}$.

3   Domain Filtering for Linear Modular Constraints

Gomes et al. [11] proposed filtering algorithms for linear finite-domain constraints over $\mathbb{F}_p$—in this section we describe our implementation,[2] including some important improvements. On equality constraints one can apply both GJE (provided p is prime) to simplify the system and optionally reach domain consistency, and also the dynamic programming representation for individual constraints to reach domain consistency. On inequality constraints only the latter applies. Note that domain values belonging to the same congruence class in $\mathbb{F}_p$ can be managed as a single one since their supports for these constraints will always be identical.

---

[2] Available at https://github.com/PesantGilles/MiniCPBP.

Algorithm 1: Filtering algorithm for system $Ax = b \pmod{p}$

$\tau^{ub} \leftarrow 1$
for $i \leftarrow 1$ to n_p do
  | if $x_{\mathbf{p}[i]}$ *is bound* then
  | | transfer index $\mathbf{p}[i]$ into $\mathbf{b}$
  | else
  | | $\tau^{ub} \leftarrow \tau^{ub} \times |D(x_{\mathbf{p}[i]})|$
if $\tau^{ub} \leq \tau^{\max}$ then
  | $\mathcal{T} \leftarrow \emptyset$
  | enumParamVars(1)
  | if $\mathcal{T}$ *is empty* then
  | | *fail*
  | else
  | | post TABLE($\langle x_i \rangle_{i \in \mathbf{p}} \langle x_i \rangle_{i \in \mathbf{d}}, \mathcal{T}$)
  | | set Algorithm 1 as inactive

3.1 Gauss-Jordan Elimination for Systems of Linear Modular Equality Constraints with a Prime Modulus

When p is prime every element of the finite field has a multiplicative inverse, which is required to apply GJE in order to simplify and solve systems of linear equations over $\mathbb{F}_p$. We precompute multiplicative inverses using the Extended Euclidean algorithm, which also allows us to confirm that p is prime. We do not reproduce these two algorithms here as they are well known.

Because our variables are not free but each have a finite domain restricting their value, deciding satisfiability for the system is not immediate given the reduced row echelon form. We may find that the system is inconsistent in which case we report it. Otherwise the resulting parametric form yields a more efficient domain consistency algorithm and smaller (i.e. with fewer variables) individual equality constraints to feed potentially to the dynamic programming filtering algorithm.

3.2 Domain Consistency for a System of Linear Modular Equality Constraints in Parametric Form

Recall that Gomes et al. [11] chose not to implement GJE. We present a straight-forward algorithm to achieve domain consistency on such systems and which is tractable when the number of parametric variables is small enough. Basically we enumerate the combinations of values for the parametric variables and check that each equation in the parametric form is satisfiable, i.e. that the required value belongs to the domain of the corresponding nonparametric variable. Any unsupported value in the domain of a parametric variable should be removed— any never-required value in the domain of a nonparametric variable should also be removed. Actually there already exists a constraint that can enforce this for us and even provide an efficient incremental algorithm: a TABLE constraint on

---

Algorithm 2: enumParamVars(r)

if $r \leq n_p$ then
    foreach $v \in D(x_{\mathbf{p}[r]})$ do
        $\tau[r] \leftarrow v$
        enumParamVars($r + 1$)
else
    for $i \leftarrow 1$ to n_d do
        $s \leftarrow b[\mathbf{d}[i]]$
        for $j \leftarrow 1$ to n_b do
            $s \leftarrow s - A[\mathbf{d}[i]][\mathbf{b}[j]] \times x_{\mathbf{b}[j]}$
        for $j \leftarrow 1$ to n_p do
            $s \leftarrow s - A[\mathbf{d}[i]][\mathbf{p}[j]] \times \tau[j]$
        $s \leftarrow s \pmod{p}$
        if $s \notin D(x_{\mathbf{d}[i]})$ then
            return
        $\tau[n_p + i] \leftarrow s$
    $\mathcal{T} \leftarrow \mathcal{T} \cup \{\tau\}$

---

the enumerated tuples using the compact table implementation. However as the number of tuples grows exponentially with the number of parametric variables we only enforce domain consistency once the number of tuples falls below a given threshold $\tau^{\max}$ as variables become bound and domains are reduced.

Let $\mathbf{p}$, $\mathbf{b}$, and $\mathbf{d}$ denote the array of indices of unbound parametric, bound parametric, and non-parametric (dependent) variables respectively, and n_p, n_b, n_d their size. We call Algorithm 1 whenever a parametric variable becomes bound. It first transfers newly-bound variables from $\mathbf{p}$ to $\mathbf{b}$ while at the same time computing the size of the Cartesian product of the domains of the remaining parametric variables, which is an upper bound on the number of valid tuples we would enumerate. If that upper bound does not exceed our threshold $\tau^{\max}$ we proceed to enumerate valid tuples (see Algorithm 2) and then post a TABLE constraint on the unbound variables. Once this happens, Algorithm 1 will no longer be called until we backtrack over that posted TABLE constraint.[3]

Theorem 1. *Algorithm 1 has a worst-case running time in* $\Theta(m(n - m)p^{n_p})$.

Proof. Its time complexity is dominated by that of Algorithm 2. We map domains to the set $\{0, 1, \ldots, p - 1\}$ and there are n_p parametric variables, so we have at most p^{n_p} tuples to enumerate. For each tuple there are at most m equations (the rank of the row-reduced matrix) on $n - m + 1$ variables to evaluate. We can check whether a value belongs to a domain in constant time (sparse set representation). □

In practice our choice of threshold $\tau^{\max}$ keeps the exponential factor p^{n_p} in check.

---

[3] In practice we actually implement the compact table filtering algorithm and apply it directly instead of repeatedly posting and retracting TABLE constraints.

3.3   Dynamic Programming for a Single Linear Modular Constraint

We next describe a simple adaptation of an existing filtering algorithm for individual linear constraints to be used when dealing with an inequality constraint or in conjunction with the previous algorithm for systems of linear equalities, as previously proposed [11].

First observe that the usual bounds consistency algorithm for linear constraints does not work correctly here. Consider for example

$$2x + y = 4 \qquad x, y \in \{1, 2, 3, 4\}.$$

Reasoning from the smallest value in the domain of x allows us to determine that the largest feasible value for y is 2, thereby declaring values 3 and 4 unsupported and filtering them out of the domain of y. But if we have instead

$$2x + y = 4 \pmod 5 \qquad x, y \in \{1, 2, 3, 4\}$$

then that same reasoning is incorrect since, for example, value 3 for y is supported by value 3 for x since $2 \cdot 3 + 3 = 9 \equiv 4 \pmod 5$. Note that the domain value yielding the smallest contribution of a variable with positive coefficient to the equation is not necessarily the smallest one: here value 3 gives the smallest contribution, 1, for x.

Consider the general linear modular constraint $\ell \le ax \le u \pmod p$ with an equality constraint corresponding to the special case $\ell = u$. The pseudo-polytime domain consistency algorithm based on dynamic programming that was originally proposed for knapsack constraints [18] can be easily adapted for modular arithmetic, leading to a worst-case time complexity in $\Theta(np \min(d, p))$ where d stands for the domain size. It potentially becomes less time- and space-consuming than its original version if the modulus is not too large, which is typically the case in many applications, and even truly polynomial if p is polynomially-related to the domain size or to the number of variables.[4] If there are several equality constraints of same prime modulus, GJE will have reduced the number of variables in each constraint, making the algorithm even faster.

We use that algorithm once the number of unbound variables falls below some chosen threshold $v^{\max}$: modular arithmetic makes the state space very densely connected which makes it hard to filter anything in the presence of several variables providing many degrees of freedom. That same observation led Gomes et al. [11] to apply it with at most six variables.

4   Application to Checksums

Checksums are commonly used to ensure data integrity of various identifiers such as social security and medicare numbers. This section is meant as an illustration of the usefulness of CP equipped with linear modular constraints, here for checksums.

---

[4] For example if p is chosen as the smallest prime number larger than the domain size, as one can always find a prime between d and $2d$ for any $d > 1$.

The International Standard Book Number (ISBN) is a unique identifier for books that uses a checksum in order to ensure its integrity. Originally ISBNs append a check digit (actually ranging from 0 to 10) to a nine-digit identifier. That check digit x_{10} is determined through a weighted sum with the other digits $x_1, \ldots, x_9$ in modular arithmetic:

$$\sum_{i=1}^{10} (11 - i)x_i \equiv 0 \pmod{11}$$

This added redundancy helps detect some common transcription errors: one can detect any single digit mistake as well as any pair of swapped digits. However double digit mistakes may go undetected. Arguably some digit mistakes are more likely than others, particularly from a handwritten version. For example digit "1" is easily confused with a "7" but not with an "8". So a natural question is: If we restrict double digit mistakes to such easily confused pairs, can they still go undetected?

We can write a CP model to help investigate this. Consider the very conservative set of confused ordered pairs $\mathcal{P} = \{(1, 7), (7, 1), (3, 5), (5, 3), (5, 8), (8, 5)\}$ and let $a_k = 11 - k$ $(1 \le k \le 10)$, the coefficients of the ISBN checksum. Sequences of variables $\langle x_1, x_2, \ldots, x_{10} \rangle$ and $\langle y_1, y_2, \ldots, y_{10} \rangle$ each model an ISBN. For every two digit positions $\langle i, j \rangle_{1 \le i < j \le 10}$ we ask whether, given a valid ISBN, replacing each digit at these positions by another from a confused pair can yield another valid ISBN:

$$Sum\_modulo(\langle a_k \rangle_{1 \le k \le 10}, \langle x_k \rangle_{1 \le k \le 10}, 0, 11)$$
$$Sum\_modulo(\langle a_k \rangle_{1 \le k \le 10}, \langle y_k \rangle_{1 \le k \le 10}, 0, 11)$$
$$Table(\langle x_i, y_i \rangle, \mathcal{P})$$
$$Table(\langle x_j, y_j \rangle, \mathcal{P})$$

$$y_k = x_k \qquad\qquad\qquad 1 \le k \le 10, \ k \ne i, k \ne j$$
$$x_i < y_i$$
$$x_k \in \{0, 1, \ldots, 9\} \qquad\qquad 1 \le k \le 9$$
$$x_{10} \in \{0, 1, \ldots, 10\}$$

The validity of each ISBN is enforced by a linear modular constraint SUM_MODULO. The close relationship of these ISBNs is enforced by using TABLE constraints for positions i and j constrained to exchange digits from a confused pair and by setting the other digits to be equal. We also add an inequality between the digits at position i in order to avoid symmetric solutions. Because many of the digits in the two ISBNs are identical and since we only seek to know whether or not the model is satisfiable, prior to search we arbitrarily set most of them to zero while leaving enough degrees of freedom, which greatly accelerates search.

Solving this model we find many solutions, indicating a real risk that such mistakes go undetected even when we consider few pairs of confused digits. Inspecting these solutions we find for example that if the leading digit is a "1" or a "7" being exchanged, the second exchanged digit yielding an undetected mistake (i.e. a valid ISBN) must occur at a position among the set $\{2, 3, 8, 9, 10\}$. We also notice that any confused pair can be used twice at positions $\langle 1, 10 \rangle$, $\langle 2, 9 \rangle$, $\langle 3, 8 \rangle$ and so forth. This is actually true for any arbitrary pair of digits (d_1, d_2) and can be derived analytically:

$$a_k d_\ell + a_{11-k} d_\ell \neq (11 - k) d_\ell + k d_\ell = 11 d_\ell \equiv 0 \pmod{11} \quad 1 \leq k \leq 10, \; 1 \leq \ell \leq 2$$

So even a single allowed pair of exchangeable digits, occurring at the right combination of positions, can lead to an undetected mistake.

Now what if we added a second checksum? For example we rotate left by one position the vector of coefficients $\langle a_k \rangle_{1 \leq k \leq 10}$ and add the corresponding SUM_MODULO constraint with a new check digit. Solving this augmented model reveals that all double digit mistakes are now detected. Can it even detect triple digit mistakes? No—even restricting to the set $\mathcal{P}$ of confused pairs, each triplet of digit positions admits exactly one combination of three pairs hiding the mistake. If we restrict further the confused pairs solely to "1" and "7" then any such mistake will be detected (i.e. we find no solution for any triplet). However if we had chosen instead a left rotation by three positions for the second checksum, we discover by solving the corresponding CP model that there is a single (though unlikely) undetected mistake at positions $\langle 5, 8, 9 \rangle$:

$$a_5 + a_8 + a_9 = 6 + 3 + 2 = 11 \equiv 0 \pmod{11}$$
$$a_8 + a_1 + a_2 = 3 + 10 + 9 = 22 \equiv 0 \pmod{11}$$

Again this serves only as an illustration of the kind of analysis made easier with CP.

5 Application to Model Counting

We now focus on the problem of model counting and demonstrate how the native support of linear modular constraints can lead to the development of scalable model counting techniques.

Given a CSP φ, let $\mathsf{sol}(\varphi)$ represent the set of solutions of φ. The problem of model counting is to estimate $|\mathsf{sol}(\varphi)|$. An approximate model counter takes in a CSP instance φ, tolerance parameter ε, and confidence parameter δ as input and returns an estimate c such that $\Pr[\frac{|\mathsf{sol}(\varphi)|}{1+\varepsilon} \leq c \leq (1 + \varepsilon)|\mathsf{sol}(\varphi)|] \geq 1 - \delta$.

The seminal work of Valiant [20] showed that this problem is #P-complete and the hardness manifests itself in the practical implementations of exact counting. Consequently, there has been a surge of interest in the design of approximate techniques. Hashing-based techniques have emerged as a dominant approach over the past few years with its promise of scalability and rigorous (ε, δ)-guarantees. The core idea is to employ pairwise independent hash functions to partition the solution space of φ into *roughly equal small* cells of solutions. To this end, the standard family of pairwise independent hash functions in the context of Boolean variables consists of linear polynomials over $\mathbb{F}_2$. The past few years have witnessed the development of scalable approximate model counters such as ApproxMC [5,6,10]. The availability of CryptoMiniSat [15,17], a solver with native support for XORs has been crucial for the scalability of these hashing-based techniques. The importance of CryptoMiniSat can be witnessed in Soos and Meel's recent work [15,16] that shows runtime improvements of two to three orders of magnitude solely in the handling of CNF-XORs drastically improved the performance of the underlying model counter, ApproxMC.

Gomes et al. [11] generalize the XOR counting framework for CSPs by using linear modular constraints. Their approach, which repeatedly tests satisfiability in cells defined by randomly-generated linear modular constraints, provides lower bounds on the solution count of a given problem φ. Another approach to counting for variables over finite domains is due to Chakraborty et al. [4] in the context of SMT constraints. They proposed the idea of using a conjunction of hash functions defined over a set of distinct primes $\{p_1, p_2, \ldots p_k\}$ to ensure that one can partition the solution space into the desired number of cells M by considering the prime factorization of M.

The primary focus of our work is to showcase the potential of linear modular arithmetic constraints, not (yet) to design a scalable approximate model counter for CSPs. Accordingly we focus on a simple procedure proposed by Chakraborty et al. [7]: Let d be the maximum size of the domain of a variable in φ and let n be the number of variables in φ. Then, let $N = d^n$. Chakraborty et al. [7] proposed the following simple algorithmic procedure that takes in a formula φ and c and returns $Y = 1$ if $|\text{sol}(\varphi)| \geq c$ and returns $Y = 0$ otherwise. The procedure is guaranteed to be correct with confidence at least $1 - \delta$.

The procedure is as follows: Repeat the following $\mathcal{O}(\log 1/\delta)$ times: at iteration i, choose a hash function $h \in \mathcal{H}(N, 2^{\lceil c \rceil})$ and check if $\varphi \wedge h^{-1}(0)$ is satisfiable, then set $Z_i = 1$ else $Z_i = 0$. Now we return $Y = 1$ if the median of $\{Z_1, Z_2, \ldots Z_i \ldots\}$ is 1, else we return $Y = 0$. We refer the reader to [7] for the proof.

For the purpose of this paper, we make a simple observation that the analysis of Chakraborty et al. can be extended with respect to any bound on the number of solutions of $\varphi \wedge h^{-1}(0)$, i.e., the current analysis checks whether the number of solutions of $\varphi \wedge h^{-1}(0)$ is greater than 1 but one could substitute any fixed threshold, as is also done in the context of (ε, δ) approximate counting algorithms. We refer the reader to [12] for a longer discussion. The implementation of this scheme is simple enough to illustrate the power of our framework yet retains the core aspect of the (ε, δ)-counter, thereby allowing one to extrapolate the importance of results in the context of approximate model counting for CSPs.

In the rest of this section we present experiments using linear modular *equality* constraints in order to evaluate both GJE on a system of constraints and the dynamic programming algorithm on individual constraints, in the context of approximately counting the solutions of CSPs. All experiments were run on a cluster of dual core AMD Opteron 275 @ 2.2 GHz processors running Java SE 11 on Linux CentOS 7.6 using the MiniCP 1.0 solver. For search we branch on the parametric variables identified during GJE (since the rest are dependent on them) using variable ordering heuristic *min-domain*. Individual entries in the tables of results are the average of thirty runs.

5.1   Synthetic Problem

We conducted a controlled experiment using a CSP on n variables and ten domain values (with $p = 11$). For half of the variables, one third of them must take value 0 (modeled using an EXACTLY constraint, which is decomposed into a SUM constraint over indicator variables); for the other half, all values must be different (modeled using an ALLDIFFERENT constraint enforcing domain consistency). The clean combinatorial nature of such a CSP allows us to derive analytically the exact number of solutions without the need to enumerate them, thus making it possible to measure the accuracy of an approximate count even when the full exploration of the search space is computationally prohibitive.

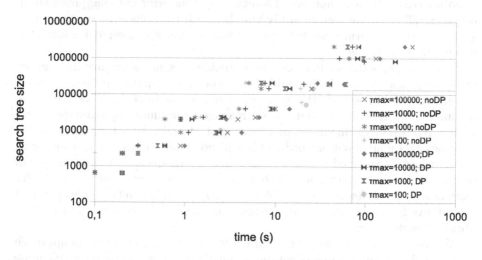

Fig. 1. Search tree size and computation time for different choices of $\tau^{\max}$ and using the dynamic programming algorithm for individual constraints ("DP" with $v^{\max} = 6$) or not. Each data point corresponds to an instance (n, m).

Figure 1 first evaluates the impact on efficiency of some choices of threshold $\tau^{\max}$ about the number of tuples that can be included in a TABLE constraint and of using or not the filtering algorithm in Sect. 3.3 for individual constraints. Though instances are not identified on the plot, search tree size for a given instance does not tend to vary a lot across configurations—hence its data points appear at about the same height. We make two observations from the horizontal spread of the points for a given instance: the dynamic programming algorithm's occasional small reduction in search effort does not make up for the frequent significant increase in computation time; a choice of $\tau^{\max} = 1000$ generally works best here. These pragmatic choices were also confirmed on several of the instances from Sect. 5.2. Accordingly all remaining experiments use these settings.

It is interesting to note that our more efficient implementation of GJE using a TABLE constraint leads to a conclusion that is the opposite of [11]: it is better to use GJE alone instead of the dynamic programming algorithm on each constraint and without GJE.

As mentioned before, the objective of this paper is not to build a full-blown approximate model counter but develop the underlying techniques to support such a model counter in the context of CP. Therefore, we demonstrate the effectiveness of our techniques via the approach due to Chakraborty et al. [7]. To simulate such an experiment, for a fixed number m of linear modular constraints, we seek to enumerate the solutions. It is perhaps worth recalling that the core idea of hashing-based counting is to enumerate solutions in a cell after adding a certain number of constraints, and then extrapolate the count of the original formula by scaling the count in a cell by the number of cells. In the context of hashing-based counters for SAT, one often needs to balance the tradeoff of handling cells with large number of solutions and the error in the approximation due to small cells. We seek to study whether such tradeoffs exist in CP as well, so as to allow the future developers of CP-based approximate model counters to make informed choices.

Table 1 reports the accuracy of our approximate count as we vary the number m of linear modular constraints added. As expected computation time decreases as m increases but so does the accuracy of our approximation. Nevertheless on instances with trillions of solutions we manage to produce approximate counts with a relative error under 1% in a matter of seconds. To achieve comparable accuracy we spend about one order of magnitude more time for an instance with three orders of magnitude more solutions in a search space that is five orders of magnitude larger. Consider as well that computing an exact count by exhaustive enumeration for as few as ten variables ($n = 10$; $m = 0$ line in the table) required almost five hours whereas an approximation with a relative error under 1% is obtained under a second.

We also report in Table 1 the accuracy of a much simpler yet naive approach to extrapolating the number of solutions enumerated in a subspace (cell): choose m variables uniformly at random and fix them to some value in their domain, also chosen uniformly at random (note that after each variable is fixed we perform constraint propagation to filter the remaining domains). We see clearly that the relative error is much larger and does not improve much as m decreases. It illustrates how the theoretical guarantees of linear modular constraints do make a difference.

These instances admit many solutions, with a ratio of the number of solutions to the size of the search space ranging from 1e−1 ($n = 10$) to 1e−5 ($n = 20$). We will see in the next section that the gain in performance may not always be as spectacular when that ratio is lower.

5.2 Benchmarks from [11]

In order to make some comparisons, we now consider the benchmark problems used in Gomes et al. [11]: the n-queens problem, DIMACS graph colouring

Table 1. Impact of the number of linear modular constraints m on our approximate count for synthetic instances. We also report the accuracy of a naive approach that simply fixes m variables at random.

n	Total #solns	m	Linear modular constraints			Naive
			Time (s)	#solns in cell	Error (%)	Error (%)
10	9.92×10^8	0	17829.8	992023200	–	–
		3	52.9	745356	0.04	20.49
		4	5.3	67806	0.26	36.14
		5	0.6	6171	0.92	45.38
		6	0.1	564	3.76	52.91
		7	0.0	50	11.42	57.32
15	2.25×10^{12}	6	373.2	1269977	0.07	76.73
		7	41.1	115414	0.15	84.51
		8	6.8	10497	0.75	86.80
		9	1.2	950	2.45	99.01
		10	0.2	85	9.28	118.33
20	2.08×10^{15}	9	2379.2	883113	0.09	79.04
		10	232.5	80301	0.24	101.13
		11	54.2	7322	1.12	79.31
		12	13.9	665	3.64	94.41
		13	3.3	59	9.72	101.79

instances, and the Spatially Balanced Latin Square problem. For each instance we set p to the smallest prime number greater or equal to the domain size.

Spatially Balanced Latin Squares. A Latin square of order n is an $n \times n$ matrix in which each cell is assigned one of n distinct symbols such that each row and column contains each symbol. A *spatially balanced Latin square* (SBLS) additionally requires that for each pair of symbols, the sum of their distance in each row be equal to a given constant. These find applications in experimental design. There are very few such combinatorial objects of any given order, i.e. their search space is very sparsely populated with solutions. As in [14] we consider particular *streamlined* SBLS, a subclass restricted to column order permutations of a cyclically-constructed Latin square. Our CP model uses n variables with identical domain $\{1, 2, \ldots, n\}$ to specify the order of the columns, an ALLD-IFFERENT constraint over them to enforce a permutation, and combinations of SUM and TABLE constraints to enforce spatial balance. We further fix the first column in order to break some amount of symbol symmetry. Its search space is much smaller yet solutions are still very sparsely distributed: the solution-to-search-space ratio ranges from 2.4e−9 to 1.2e−12 for the order −12, −14, and −15 instances we consider (there can be no solution whenever $n \equiv 1 \pmod 3$).

Table 2. Impact of the number of linear modular constraints m on our approximate count for several benchmark problems.

Instance	m	Time (s)	Number of solutions		
			In cell	Extrapolated	Error (%)
sbls12	0	1252	672	672	–
	1	994	53	685	1.95
	2	356	4	682	1.43
sbls14	0	170860	1968	1968	–
	1	145781	116	1968	0.00
	2	46170	7	2123	7.86
sbls15	0	2411411	13248	13248	–
	2	668312	45	13019	1.73
	3	140114	3	13101	1.11
queens13	0	98	73712	73712	–
	2	49	433	73115	0.81
	3	33	33	71767	2.64
queens15	0	3231	2279184	2279184	–
	3	918	464	2278649	0.02
	4	482	28	2360860	3.58
queens17	0	175140	95815104	95815104	–
	4	23415	1141	95319733	0.52
	5	11876	65	92432691	3.53
myciel4	0	224	142282920	1.423e+8	–
	3	79	1138181	1.423e+8	0.01
	5	36	45554	1.424e+8	0.05
	7	9	1821	1.422e+8	0.04
	9	3	73	1.417e+8	0.39
2_insertions_3	0	??	??	??	–
	11	29844	116705	5.698e+12	??
	13	8776	4662	5.691e+12	??
	15	1665	187	5.692e+12	??
	17	285	8	5.824e+12	??

Table 2 reports our results on these instances. Because there are so few solutions, after adding one or two linear modular constraints the cell does not contain many solutions. As a result the computational savings are modest. Still, for the order-15 instance we obtain approximations to within 1% in under two days whereas enumerating the solutions took 28 days. Recall that the focus of [11] was to compute lower bounds with high confidence using short (i.e. on at most six variables) linear modular constraints. For instances sbls14 and sbls15 they

Table 3. Searching the space of vertex colourings for graphs.

Instance	Vertices	Edges	Colours	Search space	Solutions	Ratio
myciel4	23	71	5	5^{21}	142282920	3e−7
myciel5	47	236	6	6^{45}	??	??
2_insertions_3	37	72	4	4^{35}	??	??

report lower bounds of 591 and 1748 respectively, computed in a few minutes. Though correct, these significantly underestimate the true counts, 27552 and 198720, obtained by multiplying our $m = 0$ counts by n to account for fixing the first column.

n-Queens Problem. One must place n queens on an $n \times n$ chessboard so that no two queens can attack each other. As usual we model this problem using n variables and three ALLDIFFERENT constraints. The solution-to-search-space ratio ranges from 2.4e−10 to 1.2e−13, slightly lower than that of the previous problem. One can get an approximate count with relative error under 1% at a computational cost reduced by a growing factor ranging here from 2 to 7.5 (Table 2). For queens15, [11] report 3.9e+5 as lower bound whereas the true count is close to 2.3e+6.

Graph Colouring. Given an undirected graph, assign a colour from a given set to each vertex so that vertices linked by an edge bear distinct colours. Our CP model has one variable per vertex whose domain is the set of colours and a binary disequality for each edge. We considered the four instances used by [11]. Because in our case we ultimately explore an entire cell, most of these instances were out of reach: we report on instance 2_insertions_3 and on the next smaller instance from myciel5, whose characteristics are given in Table 3. Here the search space is much more densely populated with solutions but the number of variables in the model is also significantly higher. Despite breaking some colour symmetry by arbitrarily colouring both endpoints of some edge, we only managed to enumerate the solutions of myciel4. On this instance we obtain an approximation with relative error under 1% at a computational cost reduced by close to two orders of magnitude (Table 2). While we cannot measure the error on instance 2_insertions_3 our converging results suggest that the true count is near 6.83e+13 ($4 \times 3 \times$ 5.69e+12, factoring in the pre-coloured edge) and that a close approximation can be computed in under 30 min ($m = 15$) by enumerating solutions in one of $5^{15} \approx$ 3e+10 cells—exploring the whole search space would take much much longer. The lower bound computed in [11] is 2.3e+12.

5.3 Towards a Practical Scalable Model Counter

The encouraging empirical evaluation in the preceding section leads one to ask: *what would be needed to design a practical efficient model counter?* To this end, we believe a general recipe would be the one followed by Chakraborty et al. in their design of SMTApproxMC but a direct translation of their approach would induce linear modulo constraints over different primes. In this context, one wonders whether there is an alternate approach that can ensure all the constraints are over the same modulus. We sketch out a promising direction below by observing the construction of hash functions based on inequalities. Instead of the usage of hash functions with different primes in order to partition the solution space into the desired number of cells, we seek to use inequalities. In particular we propose hash functions such that all the items x that map to a cell α are represented using: $Ax + b \leq \alpha \pmod{p}$ wherein p is a prime, and A, x, b, and α are matrices of sizes $m \times n$, $n \times 1$, $m \times 1$, and $m \times 1$ respectively with entries in $[0, p-1]$. Let $\alpha[i]$ represent the value of the i-th coordinate of α. We now state the desired properties of pairwise independence:

Lemma 1. *For $x, y \in [p]^n$, we have*

$$\Pr[Ax + b \leq \alpha] = \frac{\prod_m(\alpha[i] + 1)}{p^m} \tag{2}$$

$$\Pr[Ax + b \leq \alpha \mid Ay + b \leq \alpha] = \frac{\prod_m(\alpha[i] + 1)}{p^m} \tag{3}$$

Proof. Chakraborty et al. [4] showed

$$\Pr[Ax + b = \alpha] = \Pr[Ax + b = \alpha \mid Ay + b = \alpha] = \frac{1}{p^m} \tag{4}$$

For $u, v \in [p]^n$, we define $u \prec v$ if for all i, $u[i] \leq v[i]$.

$$\Pr[Ax + b \leq \alpha] = \Pr[\bigcup_{\beta \prec \alpha} Ax + b = \beta]$$

$$= \Pr[Ax + b = 0] \prod_m (\alpha[i] + 1) = \frac{\prod_m(\alpha[i] + 1)}{p^m}$$

Similarly, we have

$$\Pr[Ax + b \leq \alpha \mid Ay + b \leq \alpha] = \Pr[\bigcup_{\beta \prec \alpha} Ax + b = \beta \mid Ay + b \leq \alpha] \quad = \frac{\prod_m(\alpha[i] + 1)}{p^m}$$

□

The expected number of solutions is $\frac{|sol(\varphi)| \times \prod_m(\alpha[i]+1)}{p^m}$. Since $\alpha[i] \in [0, p-1]$, similar to the case of random XORs, there exists an appropriate assignment to $\alpha[i]$ such that the expected number of solutions is in the desired range.

6   Conclusion and Future Outlook

Motivated by the recent surge of interest in applications based on model counting in the SAT domain and the concurrent development of efficient hashing-based model counting, we examined the key enabling factors for such a development. We observed that the availability of solvers with native support for hashing constraints was a crucial contributing factor to the aforementioned development. In the context of CSPs, the hashing constraints with pairwise independence can be represented by linear modular arithmetic constraints. We provided an efficient implementation of such constraints in a CP solver, reversing previous choices of approach, and demonstrated their usefulness for model counting but also for other applications such as checksums. Our empirical evaluation highlighted the potential computational savings it can bring as well as the tradeoffs that should be taken into account when developing hashing-based techniques for approximate model counting on CSPs.

From our experiments in Sect. 5, despite our success in being able to reach close approximate counts at a fraction of the computational cost, we currently see two obstacles to the widespread use of hashing-based techniques for model counting in CP. The first is when the total number of solutions s is relatively small with respect to p (e.g. for SBLS): m must be smaller than $\log_p s$ to expect the resulting cell to contain solutions so if that quantity does not exceed two or three we cannot gain much speedup. The approach outlined in Sect. 5.3 may remove that first obstacle. The second is when the number of variables n is large (e.g. for graph coloring): $\log_p s$ may be large enough for us to add many linear modular constraints but having $n - m$ parametric variables may still be too many to fix before any GJE filtering can occur (recall threshold $\tau^{\max}$) and so the process remains time consuming even though in principle we are limiting our search to a single cell. This relates more generally to the lack of filtering opportunities for linear modular constraints on a large number of variables, as mentioned in Sect. 3: propagation will only appear late in the search tree, once enough variables have been instantiated.

Follow-up work in the short term includes building on this work to implement approximate model counting schemes, and improving the filtering capability of our GJE algorithm by replacing our simple $\tau^{\max}$ threshold by a more sophisticated mechanism and possibly by introducing smart tables [21] in order to attempt earlier propagation in the search tree.

The broader objective of this paper is to initiate discussion among the CP community on the development of solvers with native support for linear modular arithmetic constraints. Akin to the SAT community where the initial framework proposed by Soos et al. [17] in CryptoMiniSat received widespread attention and subsequent studies improved the framework considerably, we hope the same would hold true with respect to our work.

Acknowledgements. We thank the anonymous reviewers for their constructive criticism which helped us improve the original version of the paper. Financial support for this research was provided in part by NSERC Discovery Grant 218028/2017 and by National Research Foundation Singapore under its NRF Fellowship Programme [NRF-NRFFAI1-2019-0004].

References

1. Bardin, S., Herrmann, P., Perroud, F.: An alternative to SAT-based approaches for bit-vectors. In: Esparza, J., Majumdar, R. (eds.) TACAS 2010. LNCS, vol. 6015, pp. 84–98. Springer, Heidelberg (2010). https://doi.org/10.1007/978-3-642-12002-2_7

2. Beldiceanu, N., Carlsson, M., Demassey, S., Petit, T.: Global constraint catalogue: past, present and future. Constraints Int. J. **12**(1), 21–62 (2007). https://doi.org/10.1007/s10601-006-9010-8

3. Carter, J.L., Wegman, M.N.: Universal classes of hash functions. In: ACM Symposium on Theory of Computing, pp. 106–112. ACM (1977)

4. Chakraborty, S., Meel, K.S., Mistry, R., Vardi, M.Y.: Approximate probabilistic inference via word-level counting. In: Proceedings of AAAI (2016)

5. Chakraborty, S., Meel, K.S., Vardi, M.Y.: A scalable approximate model counter. In: Schulte, C. (ed.) CP 2013. LNCS, vol. 8124, pp. 200–216. Springer, Heidelberg (2013). https://doi.org/10.1007/978-3-642-40627-0_18

6. Chakraborty, S., Meel, K.S., Vardi, M.Y.: Algorithmic improvements in approximate counting for probabilistic inference: from linear to logarithmic SAT calls. In: Proceedings of IJCAI (2016)

7. Chakraborty, S., Meel, K.S., Vardi, M.Y.: On the hardness of probabilistic inference relaxations. In: Proceedings of AAAI (2019)

8. Chihani, Z., Marre, B., Bobot, F., Bardin, S.: Sharpening constraint programming approaches for bit-vector theory. In: Salvagnin, D., Lombardi, M. (eds.) CPAIOR 2017. LNCS, vol. 10335, pp. 3–20. Springer, Cham (2017). https://doi.org/10.1007/978-3-319-59776-8_1

9. Demeulenaere, J., et al.: Compact-table: efficiently filtering table constraints with reversible sparse bit-sets. In: Rueher, M. (ed.) CP 2016. LNCS, vol. 9892, pp. 207–223. Springer, Cham (2016). https://doi.org/10.1007/978-3-319-44953-1_14

10. Gomes, C.P., Sabharwal, A., Selman, B.: Model counting: a new strategy for obtaining good bounds. Proc. AAAI **21**, 54–61 (2006)

11. Gomes, C.P., van Hoeve, W.J., Sabharwal, A.Selman, B.: Counting CSP solutions using generalized XOR constraints. In: AAAI, pp. 204–209. AAAI Press (2007)

12. Meel, K.S., Akshay, S.: Sparse hashing for scalable approximate model counting: theory and practice. In: Proceedings of Logic in Computer science (LICS), July 2020

13. Michel, L.D., Van Hentenryck, P.: Constraint satisfaction over bit-vectors. In: Milano, M. (ed.) CP 2012. LNCS, pp. 527–543. Springer, Heidelberg (2012). https://doi.org/10.1007/978-3-642-33558-7_39

14. Smith, C., Gomes, C., Fernández, C.: Streamlining local search for spatially balanced Latin squares. In: Kaelbling, L.P., Saffiotti, A. (eds.) IJCAI-05, Proceedings of the Nineteenth International Joint Conference on Artificial Intelligence, Edinburgh, Scotland, UK, 30 July–5 August 2005, pp. 1539–1540. Professional Book Center (2005)

15. Soos, M., Gocht, S., Meel, K.S.: Tinted, detached, and lazy CNF-XOR solving and its applications to counting and sampling. In: Lahiri, S.K., Wang, C. (eds.) CAV 2020. LNCS, vol. 12224, pp. 463–484. Springer, Cham (2020). https://doi.org/10.1007/978-3-030-53288-8_22

16. Soos, M., Meel, K.S.: BIRD: engineering an efficient CNF-XOR SAT solver and its applications to approximate model counting. In: Proceedings of AAAI Conference on Artificial Intelligence (AAAI), January 2019

17. Soos, M., Nohl, K., Castelluccia, C.: Extending SAT solvers to cryptographic problems. In: Kullmann, O. (ed.) SAT 2009. LNCS, vol. 5584, pp. 244–257. Springer, Heidelberg (2009). https://doi.org/10.1007/978-3-642-02777-2_24
18. Trick, M.A.: A dynamic programming approach for consistency and propagation for knapsack constraints. Ann. OR **118**(1–4), 73–84 (2003). https://doi.org/10.1023/A:1021801522545
19. Vadhan, S.P., et al.: Pseudorandomness. Found. Trends® Theor. Comput. Sci. **7**(1–3), 1–336 (2012)
20. Valiant, L.G.: The complexity of enumeration and reliability problems. SIAM J. Comput. **8**(3), 410–421 (1979)
21. Verhaeghe, H., Lecoutre, C., Deville, Y., Schaus, P.: Extending compact-table to basic smart tables. In: Beck, J.C. (ed.) CP 2017. LNCS, vol. 10416, pp. 297–307. Springer, Cham (2017). https://doi.org/10.1007/978-3-319-66158-2_19

Injecting Domain Knowledge in Neural Networks: A Controlled Experiment on a Constrained Problem

Mattia Silvestri[✉], Michele Lombardi, and Michela Milano

University of Bologna, Bologna, Italy
{mattia.silvestri4,michele.lombardi2,michela.milano}@unibo.it

Abstract. Recent research has shown how Deep Neural Networks trained on historical solution pools can tackle CSPs to some degree, with potential applications in problems with implicit soft and hard constraints. In this paper, we consider a setup where one has offline access to symbolic, incomplete, problem knowledge, which cannot however be employed at search time. We show how such knowledge can be generally treated as a propagator, we devise an approach to distill it in the weights of a network, and we define a simple procedure to extensively exploit even small solution pools. Rather than tackling a real-world application directly, we perform experiments in a controlled setting, i.e. the classical Partial Latin Square completion problem, aimed at identifying patterns, potential advantages, and challenges. Our analysis shows that injecting knowledge at training time can be very beneficial with small solution pools, but may have less reliable effects with large solution pools. Scalability appears as the greatest challenge, as it affects the reliability of the incomplete knowledge and necessitates larger solution pools.

1 Introduction

Given enough data, Deep Neural Networks (DNNs) are capable of learning complex input-output relations with high accuracy. Recent work has shown how this applies also to the solution process of Constraint Satisfaction Problems, at least to some degree: examples include the approach from [26], relying on a pool of solutions, or Reinforcement Learning approaches inspired by [2], relying on solution checkers/evaluators. This class of approaches, while still not close to the state of the art in combinatorial decision making, may have advantages in terms of robustness and when implicit soft or hard constraints are present. For example, course timetables often need to take into account both explicit constraints (e.g. preferences, capacities) and informal agreements or manually enforced rules. A second, less explored, area of application concerns problems with well-defined sources of symbolic knowledge, which cannot however be easily exploited at search time. Examples include simulators, complex nonlinear equations, or particularly expensive (e.g. NP-hard) propagators. In this context, a Deep Learning approach may learn to satisfy such constraints without the need

© Springer Nature Switzerland AG 2021
P. J. Stuckey (Ed.): CPAIOR 2021, LNCS 12735, pp. 266–282, 2021.
https://doi.org/10.1007/978-3-030-78230-6_17

for a propagator at search time. In this paper, we focus on the latter use case and investigate methods for injecting offline information into DNNs designed to tackle combinatorial problems. Specifically, we will consider training a network for identifying variable-value assignments that are likely to be feasible. We will assume the availability of both implicit knowledge (from data), and explicit symbolic knowledge that can be accessed prior to the search process. Rather than tackling a real-world problem directly, *we perform experiments in a controlled setting*, with the aim to gauge the potential of the approach and identify the key challenges. The idea, in the spirit of [6], is to test the ground before starting the complex and time-consuming endeavor of applying such methods in a real-world use case.

In detail, we use as a benchmark the Partial Latin Square (PLS) completion problem, which requires to complete a partially filled $n \times n$ square with values in $\{1..n\}$, such that no value appears twice on any row or column. Despite its simplicity, the PLS is NP-hard, unless we start from an empty square, it has practical applications (e.g. in optical fiber routing), and serves as the basis for more complex problems (e.g. timetabling). We focus on the only PLS due to its clear structure, availability of multiple solutions that can be easily generated, and its single defining parameter (size). Using a classical constrained problem as a case study grants access to domain knowledge (the declarative formulation), and facilitates the generation of empirical data (problem solutions). This combination enables controlled experiments that are impossible to perform on real-world datasets.

As a baseline, we train on a pool of solutions *a problem-agnostic, data-driven, approach*. We devise *a simple method to extract multiple training examples* from a finite set of solutions, and we define a technique, building over Semantic Based Regularization [9] to *inject at training time domain knowledge coming from constraint propagators*. We then adjust the amount of initial data (empirical knowledge) and of injected constraints (domain knowledge) and assess the ability of the approach to identify feasible assignments. Our results show that even very small solution pools, provided they are coupled with offline knowledge injection, are enough for the DNN to identify feasible assignments with reliability comparable to a propagator at search time. When training solutions are plentiful, conversely, injecting offline knowledge has a less pronounced (or even deleterious) effect. Scalability appears as the greatest challenge, as it affects the reliability of the incomplete knowledge and necessitates larger solution pools.

The paper is organized as follows: Sect. 2 briefly surveys the related literature and motivates the choice of our baseline techniques; Sect. 3 discusses the details of the problem and methods we use; Sect. 4 presents the results of our analysis, while Sect. 5 provides concluding remarks.

2   Related Works and Baseline Choice

The analysis that we aim to perform requires 1) a data-driven technique that can solve a constrained problem, *with no access to its structure*; moreover, we need 2) methodologies for injecting domain knowledge in such a system. In this section, we briefly survey methods available in the literature for such tasks and we motivate our selection of techniques.

Neural Networks for Solving Constrained Problems. The integration of Machine Learning methods for the solution of constrained problems is an active research topic, recently surveyed in [3]. Many such approaches consider how ML can improve specific steps of the solution process: here, however, we are interested in methods that use learning to replace (entirely or in part) the modeling activity itself. These include Constraint Acquisition techniques (e.g. [4]), which attempt to learn a declarative problem description from feasible/infeasible variable assignments. These approaches may however have trouble dealing with implicit knowledge (e.g. preferences) that cannot be easily stated in a well-defined constraint language. Techniques for encoding Machine Learning models in constrained problems (e.g. [11,16,20,25]) are capable of integrating empirical and domain knowledge, but not at training time; additionally, they require to know a-priori which variables are involved in the constraints to be learned.

Some approaches (e.g. [1,5]) rely on carefully structured Hopfield Networks to solve constrained problems, but designing these networks (or their training algorithms) requires full problem knowledge. Recently, Reinforcement Learning and Pointer Networks [2] or Attention [14] have been used for solving specific classes of constrained problems, with some measure of success. These approaches also require a high degree of problem knowledge to generate the reward signal, and to some degree for the network design. The method from [26] applies Neural Networks to predict the feasibility of a binary CSP, with a very high degree of accuracy; the prediction is however based on a representation of the allowed variable-value pairs, and hence requires explicit information about the problem.

In the approach from [12], from some of the authors of this paper, a Neural Network is used to learn how to extend a partial variable assignment so as to retain feasibility. Despite its limited practical effectiveness, this method shares the best properties of Constraint Acquisition (no explicit problem information), without being restricted to constraints expressed in a classical declarative language. *This last approach was chosen as our baseline*, since it represents (to the best of our knowledge) the data driven method for constraint problems that requires the least amount of problem knowledge. In particular, it requires neither information about the problem constraints (like e.g. [26]), nor a fully known (or at least evaluable) problem model like all Reinforcement Learning approaches.

Domain Knowledge in Neural Networks. There are several approaches for incorporating external knowledge in Neural Networks, none of which has been applied so far on constrained decision problems. One method to take into account domain knowledge *at training time* is Semantic Based Regularization (SBR) [8], which is based on the idea of converting (logical) constraints into regularizing terms in the loss function used by a gradient-descent algorithm. Differentiability is achieved by means of fuzzy logic. In a similar way, [27] describes a semantic loss function that quantifies how much the network is satisfying constraints defined as sentences of propositional logic.

The approach can be pushed to an extreme by entirely replacing the loss function with a logical formula (again in fuzzy form), such as in Logic Tensor Networks (LTNs) [23]. LTNs are connected to Differentiable Reasoning [24], which uses relational background knowledge to benefit from unlabeled data.

Domain knowledge has also been introduced in differentiable Machine Learning (mainly Deep Networks) by adjusting their structure, rather than the loss function: examples include Deep Structured Models, e.g. [15] and [17], the latter integrating deep learning with Conditional Random Fields. The authors of [7] have developed a method to inject the domain knowledge encoded as First Order Logic formulas in Neural Networks generating an additional final layer that modifies the predictions according to the knowledge. Integration of external knowledge in Neural Networks *after training* is considered for example in DeepProbLog [18], where DNNs with probabilistic output (classifiers in particular) are treated as predicates. Markov Logic Networks achieve similar results via the use of Markov Fields defined over First Order Logic formulas [21], which may be defined via probabilistic ML models. [22] presents a Neural Theorem Prover using differentiable predicates and the Prolog backward chaining algorithm.

Some works attempt to both learn symbolic knowledge and enable reasoning with predicates represented by ML models. The method in [19] are similar in spirit to SBR or LTN, but they enable learning the weights of constraint terms (based on compatibility with the data), rather than having them fixed by an expert. This connects the approach to Differentiable Inductive Logic Programming, which attempts to learn (soft) logic problem from noisy data [10], by building over Inductive Logic Programming ideas.

We use a method loosely based on SBR for injecting knowledge at training time, as it offers a good compromise between flexibility and simplicity. In addition, since we regularize the propagator output rather than the constraint itself, our predicates are unary and hence we have no relational terms, making approaches like [7,23] and [27] extremely similar to SBR in our setup.

3   Basic Methods

We reimplemented the approach from [12] and extended it via a number of techniques, described in this section together with our evaluation procedure.

Neural Network for the Data Driven Approach. The baseline approach is based on training a Neural Network to extend a partial assignment (also called a *partial solution*) by making one additional assignment, so as to preserve feasibility. Formally, the network is a function:

$$f : \{0,1\}^m \to [0,1]^m \tag{1}$$

whose input and output are m dimensional vectors. Each element in the vectors is associated to a variable-value pair $\langle z_j, v_j \rangle$, where z_j is the associated variable

---

Algorithm 1. DECONSTRUCT(x)

$D = \emptyset$
while $\|x\|_1 > 0$ **do**
    Let $y = \mathbf{0}$ # zero vector
    Select a random index i s.t. $x_i = 1$
    Set $x_i = 0$, set $y_i = 1$
    Add the pair $\langle x, y \rangle$ to D
return D

---

and v_j is the associated value. We refer to the network's input as x, assuming that $x_j = 1$ iff $z_j = v_j$. Each component $f_j(x)$ of the output is proportional to the probability that pair $\langle z_j, v_j \rangle$ is chosen for the next assignment. This is achieved in practice by using an output layer with m neurons with a sigmoid activation function. The setup makes no assumptions on the constraint structure but requires a fixed problem size and variables with finite domains.

Dataset Generation Process. The input of each training example corresponds to a partial solution x, and the output to a single variable value assignment (represented as a vector y using a one-hot encoding). The training set is constructed by repeatedly calling the randomized deconstruction procedure of Algorithm 1 on an initial set of full solutions (referred to as *solution pool*). Each call generates a number of examples that are used to populate a dataset. At the end of the process, we discard multiple copies of identical examples. Two examples may have the same input, but different output, since a single partial assignment may have multiple viable completions.

Unlike [12], here we sometimes perform *multiple calls to Algorithm 1 for the same starting solution*. This simple approach enables to investigate independently the effect of the training set size and of the actual amount of empirical knowledge (the size of the solution pool).

Training and Knowledge Injection. The basic training for the NN is the same as for neural classifiers. Since the network output can be assimilated to a class, we process the network output through a softmax operator, and then we use as a loss function the categorical cross-entropy H. Additionally, we inject domain knowledge at training time via an approach that combines ideas of Semantic Based Regularization (SBR) and Constraint Programming.

Without loss of generality, we assimilate *domain knowledge to a constraint propagator*, in the sense that it can be used to flag specific variable-value pairs as either feasible or infeasible. In our experimentation, we indeed use a classical propagator (Forward Checking) as the source of domain knowledge.

Formally, given a constraint (or a collection of constraints) C, here we will treat its associated propagator as a multivariate function such that $C_j(x) = 1$ iff assignment $z_j = v_j$ has not been marked as infeasible by the propagator, while $C_j(x) = 0$ otherwise. Given that, we formulate three different approaches to augment the loss function with an SBR inspired term.

The first one relies on the usual assumption that pruned values are supposed to be provably infeasible. Given an example $\langle x, y \rangle$, we have:

$$L_{sbr}^{negative}(x) = \sum_{j=0}^{m-1} ((1 - C_j(x)) \cdot f_j(x)) \qquad (2)$$

i.e. increasing the output of a neuron corresponding to a pair flagged as infeasible incurs in a penalty that grows with $f_j(x)$.

For the other two methods, we just acknowledge that the domain knowledge may be incomplete, discouraging provably infeasible pairs, and encouraging the remaining ones. The only difference is in the cost function. In one instance the cost function is the binary cross-entropy, since for each partial solution there may exist many global viable completions, and the SBR inspired term is:

$$L_{sbr}^{bce}(x) = \sum_{j=0}^{m-1} (C_j(x) \cdot log(f_j(x)) + (1 - C_j(x)) \cdot log(1 - f_j(x)) \qquad (3)$$

In the other case instead we employ the mean squared error as cost function for the SBR inspired regularization:

$$L_{sbr}^{mse}(x) = \sum_{j=0}^{m-1} (C_j(x) - f_j(x))^2 \qquad (4)$$

Our full loss is hence given by:

$$L(x, y) = H\left(\frac{1}{Z}f(x), y\right) + \lambda L_{sbr}(x) \qquad (5)$$

where Z is the partition function and the scalar λ controls the balance between the cross-entropy term H and the SBR term, i.e. the amount of trust we put in the incomplete domain knowledge. Since we assume the domain knowledge/propagator to be incomplete, there is a risk of injecting incorrect information into the model. In practice, this is balanced by the presence of the categorical cross-entropy term in the loss: only the single pair that comes from the deconstruction of a full solution will be associated with a non-null component, and this pair is guaranteed to be *globally feasible*.

The method can be applied for all known propagators with discrete, finite domain, variables. By adapting the structure of the SBR term, it can be made to work for important classes of numerical propagators (e.g. those that enforce Bound Consistency).

Evaluation and Knowledge Injection. We evaluate the approach via a constraint solver, a classical PLS model, and a randomized search strategy. Formally, we assume access to a function SOLVE(x, C, h), where x is the starting partial assignment, C is the considered (sub)set of problem constraints, and h is a

---

Algorithm 2. FEASTEST(X, C, h)

$J^* = \arg\max\{h_j(x) \mid C_j(x) = 1\}$ # *Most likely assignments*

Pick j^* uniformly at random from J^*

Set $x_{j^*} = 1$

if SOLVE(x, C_{pls}, h_{rnd}) $\neq \bot$ **then**

   return 1   # *Globally feasible*

else

   return 0   # *Globally infeasible*

---

probability estimator for variable-value pairs (e.g. our trained NN). The function runs a Depth First Search using the Google or-tools constraint solver: the variable-value pair for the left branch is chosen at random with probabilities proportional to $h(x')$, where x' is the current state of assignments. The SOLVE function returns either a solution, or $\bot$ in case of infeasibility.

Our main evaluation method tests the ability of the NN to identify individual assignments that are globally feasible, i.e. that can be extended into full solutions. This is done via Algorithm 2, which 1) starts from a given partial solution; 2) relies on a constraint propagator C (if supplied) to discard some of the provably infeasible assignments; 3) uses the NN to make a (deterministic) single assignment; 4) attempts to complete it into a full solution (taking into account all problem constraints, i.e. C_{pls}). Replacing the NN with a uniform probability estimator provides an uninformed search strategy. We repeat the process on all partial solutions from a test set and collect statistics. This approach is identical to one of those in [12], with one major difference, i.e. the ability to use a constraint propagator for "correcting" the output of the probability estimator. This enables us to assess the impact of using the offline knowledge directly during the search, something that is allowed in our controlled setting, but that would be impossible (e.g.) with an actual simulator.

Unlike in typical Machine Learning evaluations, accuracy is not a meaningful metric in our case, as it is tied to the (practically irrelevant) ability to replicate the same sequence of assignments observed at training time. Incidentally, accuracy is very low when measured in the traditional way in all our experiments.

4   Empirical Analysis

In this section we discuss our experimental analysis, which is designed around three key questions:

Q1: Does injecting knowledge at training time improve the network's ability to identify feasible assignments?

Q2: What is the effect of adjusting the amount of available empirical knowledge?

Q3: Can knowledge injection improve the ability to satisfy constraints in a soft fashion, i.e. in terms of the number of violations?

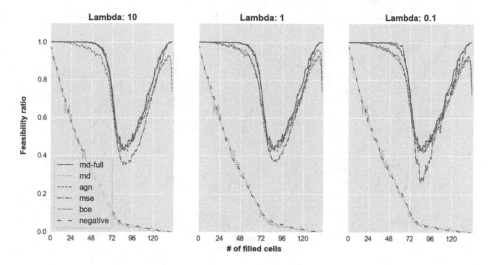

Fig. 1. Effect of the injection of the all constraints at training time comparing the regularization methods for different λ values, on the PLS-12. The dataset is generated from a 10,000 solutions pool.

While Q1 and Q2 focus on the feasibility of individual assignments, Q3 assumes that some degree of infeasibility can be tolerated. We present a series of experiments in our controlled use case that investigate such research directions. Details about the rationale and the setup of each experiment are reported in dedicated sections, but some common configurations can be immediately described.

We perform different experiments on 7×7, 10×10 and 12×12 PLS instances, resulting respectively in input and output vectors with 343, 1000 and 1728 elements. For all the experiments, we use a feed-forward, fully-connected Neural Network with three hidden layers, each with 512 units having ReLU activation function. This setup is considerably simpler than the one we used in [12], but manages to reach very similar results. We employ the Adam optimizer from Keras-TensorFlow 2.0, with default parameters. We use a batch size of 2048 for experiments on the PLS-7, whereas we adopt a batch size of 50,000 for the ones on PLS-10 and PLS-12.

4.1 Regularization Methods Comparison and λ-tuning

As a first step to evaluate the impact of knowledge injection at training time, we compare the regularization methods and evaluate how the λ value affects the performance of each of them. We focus on the PLS-12, which is the greatest dimension among the ones examined in this work so that advantages and limitations for each method can easily emerge. We refer as NEGATIVE, BCE and MSE to the methods which respectively employ the SBR inspired loss functions described in Eq. (2), eq. (3) and Eq. (4).

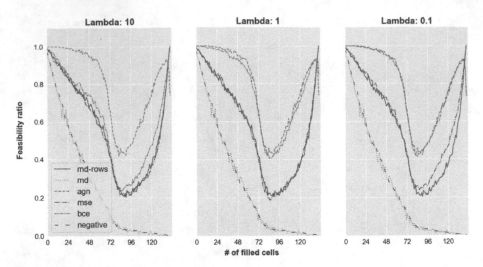

Fig. 2. Effect of the injection of the only rows constraints at training time comparing the regularization methods for different λ values, on the PLS-12. The dataset is generated from 10,000 solutions pool.

The evaluation concerns whether injecting domain knowledge *at training time* may help the NN in the identification of feasible assignments, *assuming the same knowledge is not available at search time*. We also assume in this instance that a large number of historical solutions is available.

This experimentation is motivated by practical situations in which: 1) a domain expert has only partial information about the problem structure, but a pool of historical solutions is available; 2) some constraints (e.g. from differential equations or discrete event simulation) cannot be enforced at search time. In detail, the training set is generated using the deconstruction approach from Sect. 3, starting from a set of 10,000 PLS solutions, 75% of which are used for training and the remaining ones for testing. Each solution is then deconstructed exactly once, yielding a training set of 1,000,000 examples. An additional validation set of 5,000 partial solutions is adopted to assess the improvements during training via the FEASTEST procedure, using the network as the heuristic h and an empty set of constraints as C (no propagation when choosing the assignment to be checked). Since this computation is really expensive, we perform the assessment every 10 epochs. If for 10 successive checks the best global feasibility ratio found so far is not improved then we stop the training.

For each regularization approach, we train two neural networks: one trained with knowledge about row constraints and another trained with knowledge about row and column constraints. For the first network, we use the SBR-inspired methods (and a Forward Checking propagator) to inject knowledge that both assigning a variable twice and assigning a value twice on the same row is forbidden. For the second one, we do the same, applying the Forward Checking propagator also to column constraints (i.e. no value can appear twice on the

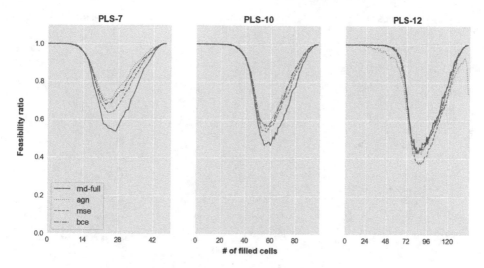

Fig. 3. Full constraints injection at training time on different problem dimensions.

same column). Due to the use of an incomplete propagator, both the networks make use of incomplete knowledge.

In addition, we train a model-agnostic neural network that lacks even the basic knowledge that a variable cannot be assigned twice, since this is not enforced by our input/output encoding, and must infer that from data.

We evaluate the resulting approaches via the FEASTEST procedure, using the separated test set as X, the trained networks as h, and an empty set of constraints (i.e. no propagation at test time). We compare them with methods that randomly choose an assignment with an uniform probability distribution *but that can rely on a set of constraints C during the evaluation*. We consider the two scenarios in which C is the set of the row constraints (RND-ROWS) and the one in which C is the set of column and row constraints (RND-FULL). These methods are representative of the behavior (at each search node) of a Constraint Programming solver having access to either only row constraints or the full problem definition. It allows us to gauge the ideal effect of the offline symbolic knowledge. Finally, we consider a very pessimistic baseline, referred to as RND, which again randomly chooses an assignment with an uniform probability distribution but does not rely on the propagation of any constraints (i.e. C is the empty set). We then produce "feasibility plots" that report on the x-axis the number of assigned variables (filled cells) in the considered partial solutions and on the y-axis the ratio of suggested assignments that are globally feasible. Since RND-ROWS and RND-FULL methods are the only ones that can rely on *online* constraints propagation, *we have highlighted them using solid lines*. In Fig. 1, we show results when all the constraints are employed by the Forward Checking constraints propagator, whereas in Fig. 2 we do not propagate the columns constraints. The balance between learning the constraints from empirical data and the Forward Checking propagator is tuned by λ: reducing its value means giving more emphasis on the

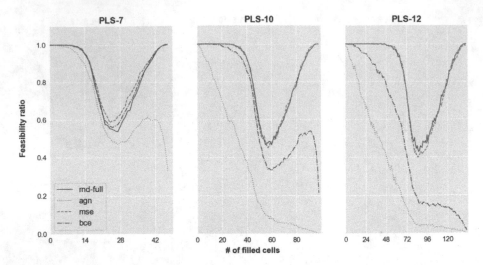

Fig. 4. Full constraints injection at training time when the dataset is reduced to the 10% of its initial size.

global feasible assignments obtained by deconstruction of the complete solutions rather than on the incomplete knowledge. We report results for λ equal to 10, 1 and 0.1

For all the λ values, the NEGATIVE approach's behavior is hardly distinguishable from RND. A reasonable explanation is that it encourages the network to keep the output the lowest as possible instead of discouraging the network to make provably infeasible assignments. Since this approach is not effective at all, we do not consider it for further analysis.

We choose the best λ parameters for the BCE and MSE regularization methods with the aim of distilling the constraints propagator in the neural network's weights, finding a tradeoff between learning from correct knowledge and the incomplete one. Considering the overall performance, the MSE regularization method provides better results with $\lambda = 1$, so this value is chosen for the successive analysis. The BCE approach provides the best performance with $\lambda = 10$. Despite in Fig. 2 lower values of λ provide better feasibility ratios, these results are not preferable since they make the regularization not effective, i.e. the methods collapse to AGN. The BCE method provides a little improvement over the MSE one but, as we will see when answering question 2, it is not robust when only a limited amount of empirical knowledge is available.

4.2 Domain Knowledge at Training Time for Different Problem Dimensions

Unlike the previous section, here we extend the analysis to the PLS of dimensions 7 and 10, considering the only MSE and BCE regularization methods together with their best λ values. The datasets are generated as described in the previous

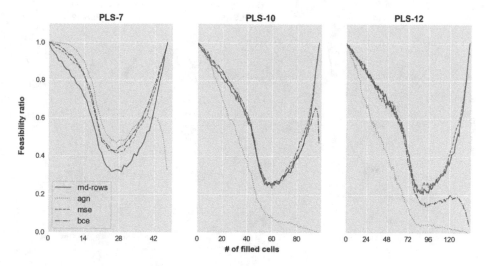

Fig. 5. Rows constraints injection at training time when the dataset is reduced to the 10% of its initial size.

section, yielding training sets of size $350,000$ and $700,000$ for respectively the PLS-7 and PLS-10.

In Fig. 3, we show results when all the constraints are employed by the forward checking constraints propagator. As long as the problem size is small enough, AGN performs considerably better than RND-FULL, even if no propagation is employed at evaluation time: this is symptomatic of the network actually managing to learn the problem constraints from the available data, which (unlike the propagator output) is guaranteed feasible. As the problem size grows, the gap decreases, until it almost disappears for PLS-12.

For PLS-7, injecting incomplete symbolic knowledge appears to have an *adverse* effect, as it biases the network toward trusting too much the incomplete propagator. With a large problem dimension (i.e. PLS-12) the benefits introduced by knowledge injection become more visible, especially when using the BCE regularization method. The decreasing performance of the data driven methods is likely a consequence of the training set size staying constant, in the face of a search space that becomes increasingly large. In all cases, the feasibility ratio is high for almost empty and almost full squares, with a noticeable drop when $\sim$60% of the square is filled. The trend may be connected to a known phase transition in the complexity of this problem [13].

4.3   Training Set Size and Empirical Information

Next, we proceed to tackle Question 2, by acting on the training set generation process. In classical Machine Learning approaches, the amount of available information is usually measured via the training set size: this is a reasonable approach since the number of training examples has a strong impact on the ability of an

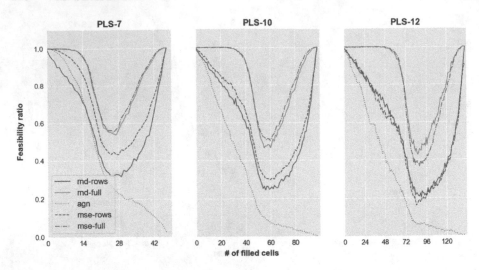

Fig. 6. Effect of reducing the solution pool size from 10,000 to 100.

ML method to learn and generalize. We performed experiments to probe the effect of the training set size on the performance of the data-driven approaches: the training sets are reduced to the 10% of the initial size, i.e. 35,000, 70,000 and 100,000 for respectively PLS of size 7, 10 and 12. In Fig. 4 and Fig. 5, we show results when respectively all the constraints and the only rows constraints are injected via the regularization methods. *In this case, knowledge injection at training time has a dramatic effect*: the AGN approach is very sensitive to the available number of examples and it has a great drop in performance. Despite being less pronounced, the BCE method has a major drop in performance too. Instead, the MSE approach provides much more robust results.

In our setup, we have also the possibility to apply the deconstruction process multiple times, so that the number of different examples that can be obtained from a single solution grows with the number of possible permutations of the variable indices (i.e. $O(n^2!)$ for the PLS). The approach opens up the possibility to *generate large training sets from very few starting solutions*. This is scientifically interesting since the "actual" empirical information depends on how many solutions are available; it is also very useful in practice since in many practical applications only a relatively small number of historical solutions exists.

The results of this evaluation are shown in Fig. 6 for a solution pool of 100 elements, rather than the original 10,000. Due to the bad results provided with the reduced datasets, we do not further investigate the BCE regularization approach but we examine the only MSE method. For this analysis, we collapse the feasibility results of the neural network trained with full knowledge injection (referred to as MSE-FULL) and of the network trained without the columns constraints knowledge injection (MSE-ROWS) in a single plot. The size of the generated training set is comparable to the original. Despite the dramatically reduced number of training solutions, the MSE-ROWS and MSE-FULL methods perform really close to respectively RND-ROWS and RND-FULL, i.e. they behave similarly to what the propagator would if employed at search time. Instead, the performance of the AGN drops dramatically, stressing again its sensitivity to the available empirical information.

From a practical point of view, it seems that injecting constraints during training can be a very effective strategy when only a small number of training solutions is available. Constraint injection tends to be redundant if the same type of propagation can be performed at search time, but can be very useful in cases when this is not possible.

4.4   Constraint Violation Assessment

In the last set of our experiments, we investigate the effectiveness of the trained NNs at guiding a search process toward solutions that are close to being feasible, but not necessarily so. This is equivalent to treating constraints as soft and may be of practical relevance on overconstrained problems (e.g. many real-world timetabling applications). This setup tends to be more challenging for the ML models, since chains of variable-value assignments may lead to partial solutions that are remarkably different from those observed at training time.

In detail, we used each trained neural network as a value selection heuristic in Depth First Search, once again for PLS of sizes 7, 10 and 12; we used for this experiment a fixed variable ordering. As a baseline for the comparison, we consider (uniformly) random value selection referred to as RND, while for the NNs we select a random value with probability proportional to the network output. We generate a fixed number of solutions (500) from an empty square, rather than starting from partially filled ones. When generating the solutions, we never propagate the entirety of the PLS constraints: this setup serves as a controlled experiment for use cases where some constraints are either unknown or cannot be enforced at search time. We measure the degree of feasibility of the generated solutions by quantifying the violations for the constraints that were not propagated at search time. For this purpose, we measure violations by counting how many times a value is not appearing exactly once in the same row or column, depending on which constraint is being considered.

We train two model-agnostic neural networks: one on the dataset obtained by random deconstruction of 10,000 solutions (referred to as AGN-10K) and the other one on the dataset obtained by multiple random deconstructions of 100 solutions (referred to as AGN-100). Similarly, we train two neural networks with knowledge injection at *training time* of all the constraints by means of the mean squared error version of the SBR-inspired method and the Forward Checking propagator (referred to as SBR-10K and SBR-100). Neither row nor column constraints are propagated during the search, and therefore we count the violations of both in the final solutions. Results are shown in Table 1: *the SBR-inspired approach allows to significantly reduce the number of violations, and it achieves very similar results even when only a small amount of empirical knowledge is available.* The AGN approach performs considerably better than RND, as long as

Table 1. Number of soft constraints violations per generated solution.

	Rnd		Agn-10k		Sbr-10k		Agn-100		Sbr-100	
	Rows	Cols	Rows	Cols	Rows	Cols	Rows	Cols	Rows	Cols
PLS-7	29	29	11	9	4	3	20	20	4	4
PLS-10	61	61	28	25	8	7	52	53	7	7
PLS-12	88	88	56	53	22	30	70	76	17	20

a large pool of solutions is available, but the gap narrows when trained on examples generated from 100 solutions. It is interesting to see how, when constraints are interpreted in a soft fashion, injecting full problem knowledge at training time has a much more robust effect compared to the analysis in Sect. 4.2.

5 Conclusion

We considered injecting domain knowledge in Deep Neural Networks to account for domain knowledge that cannot be easily enforced at search time. We chose the PLS as a case study and extended an existing NN approach to enable knowledge injection. We performed controlled experiments to investigate three main questions, drawing the following conclusions:

Q1: As long as enough empirical data is available w.r.t. the problem size, an agnostic data-driven approach can be better at identifying feasible assignments than random choice supported by propagation at search time. However, the performance gap narrows quickly as the problem size grows. Injecting incomplete domain problem knowledge at training time does not appear to provide reliable advantages.

Q2: A pure data-driven approach is very sensitive to the available empirical information. Injecting knowledge at training time significantly improves robustness: if both row and column constraints are considered, only a limited performance drop is observed with as few as 100 historical solutions.

Q3: If constraints are relaxed and treated as soft, injecting domain knowledge can be very effective.

As a side product of our analysis, we have formulated and tested different regularization approaches to develop an SBR-inspired method to constraint propagators into a source of training-time information, plus a technique to extract multiple training examples from a few historical solutions. An open question and future research direction is the experimentation with different problem types to make sure that our results hold in general.

References

1. Adorf, H.M., Johnston, M.D.: A discrete stochastic neural network algorithm for constraint satisfaction problems. In: Proceedings of IJCNN, vol. 3, pp. 917–924, June 1990. https://doi.org/10.1109/IJCNN.1990.137951

2. Bello, I., Pham, H., Le, Q.V., Norouzi, M., Bengio, S.: Neural combinatorial optimization with reinforcement learning. arXiv preprint arXiv:1611.09940 (2016)
3. Bengio, Y., Lodi, A., Prouvost, A.: Machine learning for combinatorial optimization: a methodological tour d'horizon. arXiv preprint arXiv:1811.06128 (2018)
4. Bessiere, C., Koriche, F., Lazaar, N., O'Sullivan, B.: Constraint acquisition. Artif. Intell. **244**, 315–342 (2017). https://doi.org/10.1016/j.artint.2015.08.001
5. Bouhouch, A., Chakir, L., Qadi, A.E.: Scheduling meeting solved by neural network and min-conflict heuristic. In: Proceedings of IEEE CIST, pp. 773–778, October 2016. https://doi.org/10.1109/CIST.2016.7804991
6. Van Cauwelaert, S., Lombardi, M., Schaus, P.: Understanding the potential of propagators. In: Michel, L. (ed.) CPAIOR 2015. LNCS, vol. 9075, pp. 427–436. Springer, Cham (2015). https://doi.org/10.1007/978-3-319-18008-3_29
7. Daniele, A., Serafini, L.: Neural networks enhancement through prior logical knowledge. arXiv preprint arXiv:2009.06087 (2020)
8. Diligenti, M., Gori, M., Sacca, C.: Semantic-based regularization for learning and inference. Artif. Intell. **244**, 143–165 (2017)
9. Diligenti, M., Gori, M., Saccà, C.: Semantic-based regularization for learning and inference. Artificial Intelligence 244, 143–165 (2017). https://doi.org/10.1016/j.artint.2015.08.011, http://www.sciencedirect.com/science/article/pii/S0004370215001344. Combining Constraint Solving with Mining and Learning
10. Evans, R., Grefenstette, E.: Learning explanatory rules from noisy data. J. Artif. Intell. Res. **61**, 1–64 (2018)
11. Fischetti, M., Jo, J.: Deep neural networks as 0–1 mixed integer linear programs: A feasibility study. In: Proceedings of CPAIOR (2018)
12. Galassi, A., Lombardi, M., Mello, P., Milano, M.: Model agnostic solution of CSPs via deep learning: a preliminary study. In: van Hoeve, W.-J. (ed.) CPAIOR 2018. LNCS, vol. 10848, pp. 254–262. Springer, Cham (2018). https://doi.org/10.1007/978-3-319-93031-2_18
13. Gomes, C.P., Selman, B., et al.: Problem structure in the presence of perturbations. AAAI/IAAI **97**, 221–226 (1997)
14. Kool, W., Hoof, H., Welling, M.: Attention solves your tsp, approximately. Statistics **1050**, 22 (2018)
15. Lin, G., Shen, C., Van Den Hengel, A., Reid, I.: Efficient piecewise training of deep structured models for semantic segmentation. In: Proceedings of the IEEE CVPR, pp. 3194–3203 (2016)
16. Lombardi, M., Milano, M., Bartolini, A.: Empirical decision model learning. Artif. Intell. **244**, 343–367 (2017). https://doi.org/10.1016/j.artint.2016.01.005
17. Ma, X., Hovy, E.: End-to-end sequence labeling via bi-directional lstm-cnns-crf. In: Proceedings of ACL, pp. 1064–1074. Association for Computational Linguistics (2016). https://doi.org/10.18653/v1/P16-1101, http://aclweb.org/anthology/P16-1101
18. Manhaeve, R., Dumančić, S., Kimmig, A., Demeester, T., De Raedt, L.: Deepproblog: neural probabilistic logic programming. arXiv preprint arXiv:1805.10872 (2018)
19. Marra, G., Giannini, F., Diligenti, M., Gori, M.: Integrating learning and reasoning with deep logic models. In: Brefeld, U., Fromont, E., Hotho, A., Knobbe, A., Maathuis, M., Robardet, C. (eds.) ECML PKDD 2019. LNCS (LNAI), vol. 11907, pp. 517–532. Springer, Cham (2020). https://doi.org/10.1007/978-3-030-46147-8_31

20. Mišić, V.V.: Optimization of tree ensembles. arXiv preprint arXiv:1705.10883 (2017)
21. Richardson, M., Domingos, P.: Markov logic networks. Mach. Learn. **62**(1–2), 107–136 (2006)
22. Rocktäschel, T., Riedel, S.: End-to-end differentiable proving. In: Advances in Neural Information Processing Systems, pp. 3788–3800 (2017)
23. Serafini, L., Garcez, A.D.: Logic tensor networks: deep learning and logical reasoning from data and knowledge. arXiv preprint arXiv:1606.04422 (2016)
24. Van Krieken, E., Acar, E., Van Harmelen, F.: Semi-supervised learning using differentiable reasoning. J. Appl. Logic (2019)
25. Verwer, S., Zhang, Y., Ye, Q.C.: Auction optimization using regression trees and linear models as integer programs. Artif. Intell. **244**(Suppl. C), 368–395 (2017). Combining Constraint Solving with Mining and Learning. https://doi.org/10.1016/j.artint.2015.05.004, http://www.sciencedirect.com/science/article/pii/S0004370215000788
26. Xu, H., Koenig, S., Kumar, T.K.S.: Towards effective deep learning for constraint satisfaction problems. In: Hooker, J. (ed.) CP 2018. LNCS, vol. 11008, pp. 588–597. Springer, Cham (2018). https://doi.org/10.1007/978-3-319-98334-9_38
27. Xu, J., Zhang, Z., Friedman, T., Liang, Y., Broeck, G.: A semantic loss function for deep learning with symbolic knowledge. In: International Conference on Machine Learning, pp. 5502–5511. PMLR (2018)

Learning Surrogate Functions
for the Short-Horizon Planning
in Same-Day Delivery Problems

Adrian Bracher[✉], Nikolaus Frohner, and Günther R. Raidl

Institute of Logic and Computation, TU Wien, Favoritenstraße 9–11/192-01,
1040 Vienna, Austria
{nfrohner,raidl}@ac.tuwien.ac.at

Abstract. Same-day delivery problems are challenging stochastic vehicle routing problems, where dynamically arriving orders have to be delivered to customers within a short time while minimizing costs. In this work, we consider the short-horizon planning of a problem variant where every order has to be delivered with the goal to minimize delivery tardiness, travel times, and labor costs of the drivers involved. Stochastic information as spatial and temporal order distributions is available upfront. Since timely routing decisions have to be made over the planning horizon of a day, the well-known sampling approach from the literature for considering expected future orders is not suitable due to its high runtimes. To mitigate this, we suggest to use a surrogate function for route durations that predicts the future delivery duration of the orders belonging to a route at its planned starting time. This surrogate function is directly used in the online optimization replacing the myopic current route duration. The function is trained offline by data obtained from running full day-simulations, sampling and solving a number of scenarios for each route at each decision point in time. We consider three different models for the surrogate function and compare with a sampling approach on challenging real-world inspired artificial instances. Results indicate that the new approach can outperform the sampling approach by orders of magnitude regarding runtime while significantly reducing travel costs in most cases.

Keywords: Same-day delivery · Dynamic and stochastic vehicle routing · Sampling · Surrogate function optimization · Supervised learning

1 Introduction

Short delivery times are essential when it comes to selling goods online, especially during the COVID-19 pandemic when many physical stores had to close temporarily. An increasing number of online retailers are offering same-day delivery

This project is partially funded by the Doctoral Program "Vienna Graduate School on Computational Optimization", Austrian Science Foundation (FWF) Project No. W1260-N35.

P. J. Stuckey (Ed.): CPAIOR 2021, LNCS 12735, pp. 283–298, 2021.
https://doi.org/10.1007/978-3-030-78230-6_18

to satisfy the demand for quickly available goods, further intensifying the need for cost and labor efficient dynamic vehicle routing. Same-day delivery problems [12] are stochastic and dynamic in nature and are a subcategory of vehicle routing problems. In this work a problem variant with additional constraints arising from practice is considered. Orders arrive dynamically over the day and are due only a short time after arrival. The orders have to be assigned to drivers and routes are generated with the goal to minimize delivery tardiness, travel times, and labor costs of the drivers involved. The fleet is homogeneous and the orders are served from a single depot. Each driver has a predefined shift, however the shift end times can be advanced or postponed to some extent to account for the uncertainty of the actual load.

This paper builds upon a double-horizon approach that was proposed in [3], which is further explained in Sect. 2.

However, we are unsatisfied with the existing short-horizon optimization, which we declare myopic, due to the following aspect: Routes that are optimal regarding all available orders sometimes have to start soon due to some orders, but also include one or more less urgent orders with a delivery deadline relatively far in the future. If these currently available orders with later deadlines introduce a significant travel overhead, it would frequently be wiser to postpone their delivery as they can likely be combined with future orders resulting in more efficient routes overall. Thus, it would be beneficial to split routes between urgent and less urgent orders. Routes can only be changed up to their departure, and possible future improvements for routes with a later starting time are not considered in the static, myopic optimization. The aim of this work is to present our adaptations to improve on this aspect.

The basic idea of our approach is to craft a function that discounts travel times based on the aforementioned observations, making separate routes with later starting times more attractive. We will refer to that function as surrogate, since it replaces the normal route duration in the objective function and also is used instead of a classical sampling approach, which is the de facto standard for stochastic considerations. This surrogate function is trained offline in a supervised learning fashion, reducing the computational effort in the online application in comparison to a sampling approach substantially. The necessary training data is generated by full-day simulations, in which we sample and solve 100 scenarios for each route at every decision point in time. Three different models for the surrogate are considered, a manually crafted exponential function, a linear regression, and a multi-layer perceptron.

In Sect. 2, an overview of related work is given and discussed. Section 3 defines and formalizes the problem at hand and aims to provide a better understanding with an illustrative example. Then, in Sect. 4, we explain our new approach in detail and describe the training data acquisition and training process in a step-by-step manner. Details on our test setup, and a comparison of our approaches with a sampling approach on real-world inspired artificial instances, can be found in Sect. 5. We observe that on our benchmark instances, the new approach reduces route travel costs by $\approx 6.1\%$ in the median compared to the myopic optimization with similar tardiness. The sampling approach, in comparison, achieves a similar route duration reduction but requires a computation time that is larger by a factor of ≈540. We finally conclude in Sect. 6.

2   Related Work

For a review on dynamic and stochastic vehicle routing problems, see Ritzinger et al. [9]. Our underlying problem variant is introduced in [3] and derived from an online store with promised delivery durations of one or two hours. The problem is fully dynamic with a planning horizon of one day, where stochastic information regarding the hourly number and spatial distribution of orders is available upfront. A distinguishing feature is a flexibility of the shift ending times of the drivers, which is considered in the objective function together with route durations and a penalty for delivery deadline violations.

So far, the pillars of solving this problem are an adaptive large neighborhood search (ALNS) [1,7,10] for the repeated point-in-time optimization runs to obtain routes for currently available orders and a dual horizon approach inspired by Mitrović-Minić et al. [6]. At every decision epoch, a simplified assignment problem is solved for the larger horizon (i.e., the whole day) using expected values for the orders and driver performance. This allows an estimation of the required labor time which is subsequently fed back into the objective function used in the point-in-time optimization runs considering only short horizons. Near real-time decisions regarding planned assignments of orders to multiple trips of drivers, when to send drivers home, and when to start routes are then derived from the result of this optimization.

Due to the short delivery deadlines within few hours after customers place their orders, the problem falls into the class of *same-day delivery problems* (SDDP). In a recent notable work, Voccia et al. [12] present a SDDP variant with hard time windows where orders can also be delegated to a third-party, apart from delivering it with the in-house fleet. The number of orders to be delivered in-house is to be maximized and formulated as reward of a Markov decision process (MDP). A multiple scenario sampling approach (MSA) [2] is tailored to the problem, where at every decision epoch a multi-trip team orienteering problem with time windows is solved heuristically and a consensus solution is derived. This method increases the number of filled requests for some instance classes substantially compared to a simple delay strategy, where decisions are postponed to gather more information until an impact on the number of filled requests occurs. Still, a relevant drawback is that at every decision a couple of minutes computation time is required, making it unsuitable for our near real-time setting.

Despite the fact that we do not model our problem as an MDP explicitly, we perform implicit state transitions where actions (for each driver in the depot either wait, start an unalterable delivery route, or end shift) are derived from the heuristic solution. The goal of this work is to further adapt the objective function so that the implied actions lead to states with a higher expected reward in the near future.

Using the approximate dynamic programming paradigm [8], Ulmer et al. [11] solve a single-vehicle SDDP with preemptive depot returns using an Approximate Value Iteration (AVI) scheme where the value function is learned in an offline training phase over full-day simulations. Furthermore, a dynamic state space aggregation is used to create a lookup table facilitating near real-time online decision making.

Joe and Lau [4] build upon this approach for a dynamic vehicle routing problem with stochastic customers and different degrees of dynamism where re-routing decisions have to be performed on routing plans over the day. They replace the lookup table with a deep Q network employing value function approximation via temporal difference learning with experience replay. A heuristic search in the decision space is performed via simulated annealing. This approach is compared with AVI [11] and MSA [2], and the authors report reductions of the costs in the range of 10% for higher degrees of dynamism.

In a similar spirit, we approximate the value of states by predicting the future costs of orders that are in a currently planned route by means of parametric functions to be learned in an offline training phase with training data derived from multiple realizations in the short horizon—we consider the routes separately and do not roll-out until the end of the day, hence we make use of a vehicle-based and temporal decomposition. This learned function is then incorporated as surrogate in the objective function to be solved heuristically using our ALNS, resulting in more anticipatory online decision making.

3   Problem Definition and Formalization

In this Section we first give a formal description of the considered dynamic and stochastic vehicle routing problem and then show an illustrative example where myopic optimization in the short horizon planning leads to inefficient routes.

We follow the notation of the preceding article [3] and distinguish between three different problem variants: the offline problem with full knowledge of the day in advance (OFF), the dynamic problem at a specific time $\tilde{t}$ (DYN-$\tilde{t}$), and the full dynamic problem for a whole day (DYN-DAY). In this paper we will mostly focus on the dynamic variants.

3.1   Instance Specification

A DYN-DAY instance consists of many DYN-$\tilde{t}$ instances for increasing times $\tilde{t}$ that are solved iteratively over the whole day. A DYN-DAY instance contains n orders collectively denoted by V, each of which has a release time t_v^{rel} at which the order is ready to be delivered by a driver and a due time t_v^{due} at which the order should be delivered the latest, $v \in V$. Moreover, for each order an availability time t_v^{avail} is provided which corresponds to the time the customer places the order and tells us when we are allowed to consider the order in our planning. We further assume to have for dynamic instances a function $\omega(t_1, t_2)$ available that yields the expected numbers of orders within the time interval

$[t_1, t_2]$ within any relevant business times. We also have an idea of the mean order duration, i.e., the mean active labor time by a driver to deliver an order, a good DYN-DAY solution in a particular application typically has and denote this by $\hat{\phi}$.

DYN-$\tilde{t}$ instances occur and are solved every time an order is released, i.e., at times $\{\tilde{t} \mid \exists v \in V : t = t_v^{\text{rel}}\}$.

Any problem instance also provides information about its relevant vehicles/drives, denoted by set U, with $m = |U|$, including each driver's planned shift time interval $[q_u^{\text{start}}, q_u^{\text{end}}]$ and earliest shift end $q_u^0 \in [q_u^{\text{start}}, q_u^{\text{end}}]$, $u \in U$. The drivers' shift ends are subject to flexibility and therefore also part of the decision process and objective function. Last but not least, order locations loc_v, $\forall v \in V$, expected travel times $\delta(v, v')$ from loc_v to $loc_{v'}$, where $v, v' \in V \cup \{0\}$, 0 denotes the warehouse, are given. The travel times include necessary delays like average stop time at the customers, average times for loading a vehicle at the warehouse and postprocessing times when returning to the warehouse. We also assume that the triangle inequality holds for δ.

3.2 Feasible Solutions

A candidate solution is a tuple $\langle R, \tau, q \rangle$ where

- $R = (R_u)_{u \in U}$ denotes the *ordered sequence of routes* $R_u = \{r_{u,1}, \ldots, r_{u,\ell_u}\}$ to be performed by each vehicle $u \in U$, and each *route* $r \in R_u$ is an ordered sequence $r = \{v_0^r = 0, v_1^r, \ldots, v_{l_r}^r, v_{l_r+1}^r = 0\}$ with $v_i^r \in V$, $i = 1, \ldots, l_r$, being the i-th order to be delivered and 0 representing the warehouse at which each tour starts and ends,
- $\tau = (\tau_r)_{r \in R_u, u \in U}$ are the (planned) *departure times* of the routes, and
- $q = (q_u)_{u \in U}$ are the *shift end times* of the vehicles.

The time at which the i-th order v_i^r of route r, $i = 1, \ldots, l_r$, is delivered is

$$a(r, i) = \tau_r + \sum_{j=0}^{i-1} \delta(v_j^r, v_{j+1}^r). \tag{1}$$

The total duration of a route $r \in R_u$ of a vehicle $u \in U$ is

$$d(r) = \sum_{i=0}^{l_r} \delta(v_i^r, v_{i+1}^r), \tag{2}$$

and the route therefore is supposed to end at time $\tau_r + d(r)$.

Let $\tau^{\min}(r) = \max_{i=1,\ldots,l_r} t_{v_i^r}^{\text{rel}}$ be the earliest feasible starting time of a route r, which corresponds to the maximum release time of the orders served in the route. In our planning all routes $r \in R$ can be changed up to their respective departure time τ_r, after which the route is fixed. Furthermore, let $\tau^{\max}(r)$ be the latest starting time without violating any due time, i.e.,

$$\tau^{\max}(r) = \min_{i=1,\ldots,l_r} \left(t_{v_i^r}^{\text{due}} - \sum_{j=0}^{i-1} \delta(v_j^r, v_{j+1}^r) \right). \tag{3}$$

A solution is feasible when

- each order $v \in V$ appears exactly once in all the routes in $\bigcup_{u \in U} R_u$,
- each route $r \in R_u$, $u \in U$, is started in the planned shift time of the assigned vehicle, i.e., $\tau_r \in [q_u^{\text{start}}, q_u^{\text{end}}]$, and not started before all orders are released, i.e., $\tau_r \geq \tau^{\min}(r)$,
- the routes in each R_u, $u \in U$, start at increasing times and do not overlap, i.e., $\tau_{r_{u,i}} + d(r_{u,i}) \leq \tau_{r_{u,i+1}}$, $i = 1, \ldots, |R_u| - 1$,
- and the actual shift end time is not smaller than the finishing time of the last route (if there is one) and the minimum shift time, i.e., $q_u \geq \max(q_u^0, \sup_{r \in R_u}(\tau_r + d(r)))$, $u \in U$.

3.3   Objective Function

The utmost goal is to minimize and balance any tardiness of deliveries. As secondary objectives, the route durations and the excess labor times are to be minimized. We model this via an objective function to be minimized consisting of a linear combination of a quadratic tardiness penalty term and linear cost terms for the secondary objectives:

$$f(\langle R, \tau, q \rangle) = \alpha \cdot \sum_{r \in R_u, u \in U} \sum_{i=1}^{l_r} \max(0, a(r,i) - t_{v_i^r}^{\text{due}})^2 + \gamma \cdot \sum_{u \in U}(q_u - q_u^0) + \sum_{r \in R_u, u \in U} d(r), \tag{4}$$

A tardiness penalty factor of $\alpha = 1000$ is chosen to approximate a lexicographic approach. The excess labor times cost factor γ is set to 4.

The objective values are not easily interpretable by humans. To give us another solution quality indicator we use the *mean order duration* $\phi(r) = \frac{d(r)}{l_r}$ for a route $r \in R$, measured in minutes, which can be understood as mean active labor time by a driver to deliver an order. Calculating the mean over all the routes in a solution yields $\bar{\phi}$, which is an important figure of merit of a whole solution.

3.4   Illustrative Example

To make the issue we address in this work clear, we present a simple example of a DYN-DAY instance, in which an optimal solution to a first DYN-$\tilde{t}$ instance leads to a situation so that an overall suboptimal solution for DYN-DAY is achieved.

Let us assume orders 1–6 become available at $\tilde{t} = 0$ and we are thus considering DYN-0. Orders 1–5 are supposed to have the same due time 60 min later. The remaining order 6 is located far away from orders 1 to 5 and has a substantially

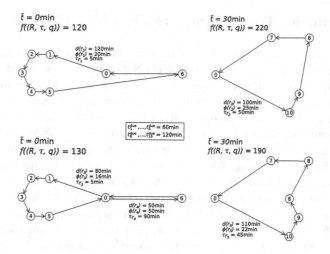

Fig. 1. Myopic solution (top) vs. optimal solution (bottom). Node 0 represents the warehouse, all other nodes orders.

later due time of 120 min. Considering only these orders an optimal solution to DYN-0 would be to pack all orders into one route, since only then the total route duration is minimal. This single route r_1 has to start at time $\tau_{r_1} = 5$ to avoid any tardiness. This solution is depicted in the top half of Fig. 1. The problem with this solution arises when half an hour later new orders 7–10 are placed with some delivery locations close to order 6, which itself, however, has been included in the already started first route. An optimal solution for DYN-30min will then be a route r_2 with the remaining orders 7–10 as also pictured in Fig. 1. The resulting total objective value, which in this case is equal to the sum of all route durations, is 220 min.

A better solution to this example can be seen in the second half of Fig. 1. The important difference is that the first route from the previous solution is split into two, resulting in one route r_3 comprising orders 1–5, starting at time τ_{r_3} = 5 and having a mean order duration better than the former single route, and one route r_4 containing only order 6. This latter route has a bad mean order duration of 50 min per order, but also a much later starting time of $\tau_{r_4} = 90$. Even though this results in a worse short-term objective value for DYN-0, this second route has now a lot of slack left and considering expected future orders can likely be improved later. In our example this happens when the new orders 7–10 become available in DYN-30, and order 6 can be delivered in one route r_5 together with the new orders. Overall the objective value and sum of all route durations for this solution is 190, which is an improvement of approximately 14% over the myopic solution.

In conclusion, when orders are expected in the near future, it makes sense to postpone to a certain degree the delivery of orders with due times farther in the future when they cannot be well integrated in soon-to-start routes.

4   Discounting Travel Times to Consider Expected Orders

As pointed out in Sect. 2, to at least partially avoid traps like the one sketched above arising from the myopic view of the DYN-$\tilde{t}$ instances, the standard method from literature is to sample scenarios into the near future by creating artificial orders from expected spatial and temporal distributions, to solve these scenarios, and then to derive a consensus solution [2, 12]. We propose the simpler approach of discounting durations of routes in the objective function of the DYN-$\tilde{t}$ instances in dependence of their starting times, the number of expected future orders, and further features. We make use of supervised learning to come up with a surrogate function for the route durations to move the computational effort into a one-time offline training phase. This function is then directly used in the optimization of a given DYN-$\tilde{t}$. We now describe our approach in detail.

In the definition of $f(\langle R, \tau, q \rangle)$ in (4) on page 6 we replace the route duration $d(r)$ of each route $r \in R_u$, $u \in U$ with a *discounted duration* $d'(r)$ acting as a surrogate for the future delivery time of the orders belonging to r. We define the following aims to guide us to a sensible discounting function.

- Routes that are already efficient, i.e., have a low mean order duration $\phi(r)$, should not be modified.
- The discounting of the duration should in general be stronger when more orders are expected in the near future. On the contrary, we should not reduce $d(r)$ if there are no further orders expected until route r should start.
- Routes that are inefficient and combine orders that are due soon with orders that have significantly more time left should be avoided in particular.
- In conclusion, the discounted route duration $d'(r)$ should approximate the *expected total time it will take to perform the deliveries of that route in the future*, taking into account expected new orders and assuming optimal routing decisions also in the future.

A current route that will be started soon cannot be expected to be improved much as not many new orders are expected. This includes routes with small slack $\max(0, \tau^{\max}(r) - \tilde{t})$ but also any other case in which the route is started soon, e.g., due to an earliest starting time strategy. In contrast, larger improvements are likely for any route that is planned to be started much later and which is not yet efficient, particularly if many orders are expected in the near future, more precisely in the time interval from the current time $\tilde{t}$ to the route's planned starting time τ_r. Thus, this duration is an important parameter of the discounting.

Another important parameter is the expected number of arriving new orders until the start of the route, i.e., $\omega(\tilde{t}, \tau_r)$. Moreover, the estimated mean order duration of a good DYN-DAY solution $\hat{\phi}$ is also important for the following consideration. A route r to be started at some distant time τ_r and whose mean order duration $\phi(r)$ is worse than $\hat{\phi}$ can be expected to be adapted and combined with future orders so that the average times for delivering the orders in r approaches $\hat{\phi}$.

Formally, we model this by the *discounted route duration function*

$$d'(r) = \begin{cases} g_\Theta(d(r), l_r, \omega(\tilde{t}, \tau_r), \hat{\phi}, \ldots) & \text{if } \tau_r > \tilde{t} \wedge \phi(r) > \hat{\phi} \\ d(r) & \text{else.} \end{cases} \quad (5)$$

where function g_Θ represents a machine learning model with trainable parameters Θ and input features that include at least $d(r)$, $l_r, \omega(\tilde{t}, \tau_r)$ and $\hat{\phi}$. This model is supposed to yield reduced durations within $[\hat{\phi} \cdot l_r, d(r)]$ for routes that are not started immediately ($\tau_r > \tilde{t}$) and where the current mean order duration $\phi(r)$ is worse than $\hat{\phi}$. In Sect. 4.2 we will consider three different approaches for realizing g_Θ, which are an exponential function, a linear regression, and a multilayer perceptron.

An aspect of this approximation that deserves mentioning is that multiple routes of the current solution may be scheduled at overlapping times in the future and may compete for new orders. This may slow down improvement of inefficient routes but may also create new possibilities for more efficient combinations. As we do not see any meaningful and efficient way to consider this aspect and also conjecture that the benefits and disadvantages of concurrent routes in conjunction with the route improvement potential may outweigh each other at least to a certain degree, we do not explore this further here. Moreover, the actual impact may be partially mitigated by suitably tuning Θ.

4.1 Obtaining Training Data

To obtain training date for our route duration discounting models, we apply the following sampling-based approach on a set of representative historic or artificial DYN-DAY training instances.

1. We consider a DYN-DAY instance and iteratively solve the implied DYN-$\tilde{t}$ instances in the classical way without any route distance discounting. For each obtained DYN-$\tilde{t}$ solution, we apply a decomposition approach, in which we consider each route independently by the following steps.
2. Each route r to be started not immediately, i.e., at some time $\tau_r > \tilde{t}$, and for which $\phi(r) > \hat{\phi}$, we create n_{sample} scenarios, with n_{sample} being a strategy parameter. Each scenario consists of the original orders of route r and $n_{\text{orders}} \sim \mathcal{P}(\omega(\tilde{t}, \tau_r))$ additional artificial orders, where n_{orders} is a random number always sampled anew from the Poisson distribution $\mathcal{P}(\omega(\tilde{t}, \tau_r))$ with mean $\omega(\tilde{t}, \tau_r)$. The motivation here is that the arrival of orders can be seen as a Poisson process. Each artificial order is assigned a randomly sampled geographical location from a set of sufficient size representing the delivery area, a random availability time in $(\tilde{t}, \tau_r]$, and a due time that corresponds to the availability time plus the maximum delivery duration promised to the customers. Each scenario created this way is then solved as an independent OFF instance.
3. In each obtained scenario solution we consider each original (i.e., not sampled) order and take its route's mean order duration. The sum of these times over

all original orders is then said to be the scenario's total delivery duration for the original orders of route r. Ultimately we average these total delivery durations over all scenarios to obtain the target duration $\hat{d}(r)$ which we want to approximate by our discounted route duration $d'(r)$.

4. We store the original route r together with $d(r)$, $\tilde{t}$, τ_r, $\omega(\tilde{t}, \tau_r)$, $\hat{\phi}$ and the obtained $\hat{d}(r)$ for training and continue by processing all further routes in the same way.

4.2 Models for the Discounting

As introduced in Eq. (5), function $g_\Theta(d(r), l_r, \omega(\tilde{t}, \tau_r), \hat{\phi}, \ldots)$ is a trainable model that yields the discounted route duration when $d(r) > \hat{\phi} \cdot l_r$. For training this model we apply the mean squared error (MSE) in respect to the training targets, i.e., $\hat{d}(r)$, as loss function. We investigate here three alternative models presented in the following.

Exponential Function

$$g_\rho^{\exp}(d(r), l_r, \omega(\tilde{t}, \tau_r), \hat{\phi}) = d(r) - (d(r) - \hat{\phi} \cdot l_r) \cdot (1 - e^{-\rho \cdot \omega(\tilde{t}, \tau_r)}) \qquad (6)$$

This function was manually crafted based on the previously explained considerations that the mean order delivery time of orders in a current route with a distant starting time can be expected to improve up to a certain amount. The expected maximum improvement is assumed to be equal to $d(r) - \hat{\phi} \cdot l_r$. However, actual improvement can only occur with additional orders in the interval $(\tilde{t}, \tau_r]$. This expressed by the last term in the function, where parameter ρ controls the speed of approaching $\hat{\phi} \cdot l_r$ in dependence of the number of expected upcoming new orders $\omega(\tilde{t}, \tau_r)$ until the route's starting time τ_r in an exponential manner. The parameter that needs to be learned here is just $\Theta = \rho$, and we apply grid search to find a value minimizing the MSE.

Linear Regression.

Our second approach is a linear combination of a larger set of manually selected features, i.e., linear regression, with the trainable parameters vector Θ being the respective regression coefficients. We initially consider the following features in addition to a constant bias.

1. The basic features $d(r)$, l_r, $\omega(\tilde{t}, \tau_r)$ and $\hat{\phi}$ as in the exponential function.
2. The relative starting time of the route $\tau_r - \tilde{t}$.
3. The difference $\phi(r) - \hat{\phi}$, i.e., how far off the route's mean delivery duration is from the assumed target value $\hat{\phi}$.
4. The *variance of the geographic locations of the orders for each route*, denoted by var(r); the farther apart the delivery locations are, the more likely it seems that a new order fits nicely in between two existing orders.
5. The square and the logarithm of each of the above features to also accommodate nonlinear dependencies in a simple form.

To avoid the inclusion of features that do not significantly improve the prediction and reduce the danger of overfitting, we started off with just the basic features and iteratively added a feature from the remaining pool that reduced the MSE the most. This process of selecting features was continued until the MSE did not change by more than one percent. 5-fold cross validation was used in this feature selection process to reduce the risk of overfitting. Ultimately, we came up with the feature vector $(d(r), l_r, \omega(\tilde{t}, \tau_r), \log(\omega(\tilde{t}, \tau_r)), \hat{\phi}, \phi(r) - \hat{\phi},$ $(\phi(r) - \hat{\phi})^2, (\tau_r - \tilde{t})^2, \log(\tau_r - \tilde{t}), \mathrm{var}(r), \mathrm{var}(r)^2)$ used in all further experiments.

Multilayer Perceptron. Our third model for discounting travel durations is a multilayer perceptron (MLP). It is fully connected with two hidden layers and ReLU activation functions in all layers, and Adam [5] is used as optimizer. The considered pool of features was the same as in the linear regression, and the same selection process was performed leading to the feature vector $(d(r), \log(d(r)),$ $l_r, \hat{\phi}, \tau_r - \tilde{t}, \phi(r) - \hat{\phi})$ used in all following experiments. Note in particular that here the variance of the orders' geographic locations did not show a significant contribution and therefore was not included. Further details on the network configuration and training will be provided below in the experimental results.

5   Computational Study

All algorithms were implemented in Python 3.8. Training and evaluation of the regressors was performed with `scikit-learn` version 0.23.1. All tests were conducted on Intel Xeon E5-2640 2.40 GHz processors in single-threaded mode and a memory limit of 4 GB.

In all tests a driver is sent home as early as possible, i.e., after the driver's last so far planned route or at the earliest shift end, to minimize labor cost. Planned routes always start at the latest possible departure time that does not increase the costs for labor time and tardiness to utilize the full slack for possible improvement. The three different discounting models are compared with results using the myopic optimization as done in [3] and the sampling approach with consensus function. The ALNS, which is the fundamental optimization method for all mentioned approaches, stops after 100 non-improving iterations, and we refer to [3] for all further details concerning its operators and configuration.

5.1   Instances

We consider artificial DYN-DAY instances that are inspired by real-world instances of an online retailer. We consider steady, linearly rising, and falling load patterns over 11 h, where the average load over the day is either 10, 20, 30, or 40 arriving orders per hour. Orders are due in one hour with 60% probability and with 40% in two hours. The order locations are uniformly distributed in the unit square. Travel times between orders are determined by the Euclidean distance multiplied by 50 min, additional constant six minutes stop times at the customers, and small loading and postprocessing times from and to the warehouse. The warehouse location is randomly chosen from $\{0.25, 0.75\}^2$ inspired by the slight off-center location of the real-world situation. Since we focus on the route duration costs, sufficiently many drivers are available all the time to ensure zero or very little tardiness. We generated 240 instances in total, 20 for each of the 12 instance classes and perform a 50/50 training and test split. $\hat{\phi}$ is provided for each class. All instances were made available on GitHub[1].

5.2   Training of the Discounted Route Duration Models

Following the training and test data generation as described in Sect. 4.1, we end up with a 60% batch of 33790 training samples and a 40% batch of 22527 test samples to train and evaluate an estimator for $\hat{d}(r)$.

We train the learnable parameter ρ in the exponential model (6) by means of a grid search. The result can be seen in Fig. 2, which displays how the MSE changes depending on ρ. Moreover, the instance's $\hat{\phi}$ is reduced by 20%, which was empirically determined to produce better results in previous experiments. The single global optimum for ρ is 0.091, at which the test MSE is 154909 and 154265 for the training batch.

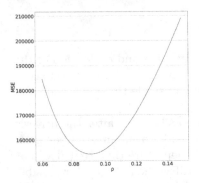

Fig. 2. Exponential model: MSE of predicted values $g_\Theta(r)$ with respect to labels $\hat{d}(r)$, i.e., the loss over ρ.

In case of the linear regression with the finally selected features as laid out in Sect. 4.2, MSEs of 144785 and 143329 were achieved on the training and test portions of the data, respectively.

Concerning the MLP, preliminary tests suggested that two hidden layers with 50 nodes each seem to be a reasonable choice, which we used further on. The learning rate that is used for training is a constant 0.001. To avoid overfitting we utilize early stopping, for which 300 iterations without improvement of a 10% validation set is the stopping criterion. The resulting training and test MSEs are 80032 and 79219 respectively, slightly less than half of the error of the exponential model.

Concerning the MSEs, we can conclude that the linear regression performs slightly better than the exponential model, but the MLP is clearly superior. As we considered separate training and test sets and the respective MSEs lie close together for all three models, we conclude that overfitting seems to be no issue for all three models.

---

[1] https://github.com/nfrohner/pdsvrpddsf.

5.3   Full-Day Simulation Results

The myopic short-horizon optimization serves as a baseline to quantify the improvement that is achieved. Furthermore, the three route duration discounting approaches are compared to a sampling approach with consensus function as the de facto standard for considering stochastic aspects. This approach creates for each DYN-$\tilde{t}$ instance 100 scenarios by augmenting the original instance with randomly sampled orders. These sampled orders are generated in the same manner as already explained in Sect. 4.1, except that the time interval $[\tilde{t}, \tilde{t} + 1h]$ is used instead of the slack of the route, i.e., samples are generated for up to one hour into the future. These scenarios are then solved with the myopic short-term optimization utilizing ALNS. Then, all sampled orders are removed from each scenario solution. Finally, a consensus solution is derived from the scenario solutions in a way that was inspired by [12]. The selection is done by counting identical scenario solutions and choosing the most frequent solution as consensus solution. We define identical in this context as two solutions that assign identical routes to the same drivers in the same sequence. Analogous to that identical routes are defined as routes that contain the same orders in the same sequence.

We use the original objective function $f(\langle R, \tau, q \rangle)$ as defined in (4) as the primary measure of success for comparing results, but also aim to gain a more in-depth understanding of the different approaches by observing the total duration of all routes in a solution, the total excess labor time of a solution, the mean order duration over the whole solution $\bar{\phi}$ and the running time on the specified test setup. Tardiness is not presented in this Section, because it is negligibly small for all instance classes and approaches alike, which was one of our aims when generating the test instances as explained in Sect. 5.1.

In Table 1 the median of the mentioned measures of success are compared for all instance classes and the median of the relative changes to results of the myopic approach is displayed for the most important measures as well. As the sampling approach did not terminate within a time limit of 700 h per full-day instance for average loads of 30 and higher, we only obtained results up to an average load of 20 for it. In Fig. 3 boxplots of $f(\langle R, \tau, q \rangle)$ are drawn over instance classes grouped by the average load as well as the load pattern.

As expected, all approaches that consider possible future orders outperform the myopic optimization, up to 8% in the median. We observe that the exponential approach outperformed the other approaches for average loads of 30 and 40. Furthermore, a positive correlation between the average load and the relative improvement over the myopic short-term optimization can be seen. Falling load solutions have higher $f(\langle R, \tau, q \rangle)$ in general, but the differences in relative improvement over the myopic optimization among load patterns is rather small, with steady load having a slight edge over falling and rising load.

Considering that the MLP has the smallest training MSE, it is unexpected to observe that some solutions are worse than the ones that utilize the exponential model. We suspect that the cause for this is attributed to the way in which the training data is generated. More specifically, we intentionally decided to restrict the training data generation to routes in final DYN-$\tilde{t}$ solutions obtained from

Table 1. The three discounting approaches, myopic optimization, and sampling applied to ten benchmark instances for each combination of average load and a falling, rising, or steady load as the day progresses.

Load Avg	Pattern	Approach	$f(\langle R,\tau,q\rangle)$ Median	Change	Trav. time [h] Median	Change	Labor Median	ϕ Median	Change	Runtime [min] Median
10	Falling	Myopic	4057.258	0.00%	67.197	0.00%	136.5	36.125	0.00%	4
		Exponential	3921.867	−2.92%	64.297	−3.14%	424.5	34.920	−3.14%	5
		Linear Regression	**3851.000**	−4.28%	**64.069**	**−4.32%**	141.0	35.170	**−4.33%**	8
		MLP	3912.190	**−4.96%**	64.349	−3.92%	147.9	**34.404**	−3.92%	11
		Classical Sampling	3928.320	−4.20%	64.634	−3.80%	201.4	35.157	−3.79%	2950
	Rising	Myopic	3816.300	0.00%	63.413	0.00%	172.5	35.485	0.00%	4
		Exponential	3695.408	−3.49%	61.564	−3.32%	**37.0**	33.920	−3.31%	5
		Linear Regression	**3662.175**	−3.57%	**61.036**	−3.59%	117.5	34.505	−3.58%	8
		MLP	3701.557	**−3.89%**	61.693	**−3.79%**	61.6	**33.976**	**−3.81%**	12
		Classical Sampling	3749.525	−1.14%	62.464	−0.85%	45.5	34.845	−0.85%	3347
	Steady	Myopic	3984.472	0.00%	66.040	0.00%	0.0	35.475	0.00%	5
		Exponential	3891.581	−4.31%	64.456	−5.01%	0.0	34.000	−5.00%	5
		Linear Regression	3938.683	−4.97%	65.258	−5.63%	135.0	33.795	−5.63%	7
		MLP	3845.711	−5.54%	63.349	−5.48%	0.0	33.064	−5.48%	8
		Classical Sampling	**3769.011**	**−8.56%**	**62.771**	**−8.41%**	3.5	**32.618**	**−8.41%**	2543
20	Falling	Myopic	7142.592	0.00%	118.993	0.00%	77.0	32.015	0.00%	24
		Exponential	6802.900	−5.90%	112.105	−5.87%	19.0	**30.025**	−5.87%	32
		Linear Regression	6823.300	−5.40%	112.755	−5.58%	25.0	30.265	−5.57%	49
		MLP	6884.071	−5.10%	112.935	−5.10%	**1.0**	30.360	−5.10%	51
		Classical Sampling	**6693.054**	**−6.32%**	**111.639**	**−6.18%**	4.0	30.127	**−6.18%**	30372
	Rising	Myopic	7027.803	0.00%	116.848	0.00%	380.0	32.365	0.00%	23
		Exponential	6482.008	−6.51%	107.817	**−6.53%**	177.0	**30.965**	−6.53%	32
		Linear Regression	**6419.892**	**−6.62%**	106.955	−6.52%	136.0	31.055	**−6.54%**	40
		MLP	6432.858	−6.54%	107.044	−6.22%	124.2	31.065	−6.20%	62
		Classical Sampling	6641.478	−3.79%	110.440	−3.72%	**102.0**	32.138	−3.72%	33728
	Steady	Myopic	7101.992	0.00%	117.963	0.00%	252.5	32.690	0.00%	26
		Exponential	6765.575	**−6.77%**	112.431	**−6.60%**	137.5	31.465	**−6.59%**	28
		Linear Regression	6830.842	−4.02%	113.435	−3.86%	127.5	31.585	−3.85%	37
		MLP	**6721.294**	−5.68%	**111.578**	−5.53%	95.1	31.519	−5.54%	52
		Classical Sampling	6735.598	−5.93%	112.078	−5.60%	**94.0**	**31.060**	−5.60%	25646
30	Falling	Myopic	10432.136	0.00%	172.496	0.00%	**487.5**	31.150	0.00%	77
		Exponential	**9657.147**	**−7.07%**	**159.930**	**−7.01%**	681.0	29.030	**−7.00%**	95
		Linear Regression	9721.894	−5.92%	160.721	−5.55%	713.0	**29.025**	−5.55%	135
		MLP	9689.704	−5.95%	160.690	−5.44%	714.7	29.072	−5.45%	176
	Rising	Myopic	10313.425	0.00%	170.299	0.00%	1308.5	31.055	0.00%	76
		Exponential	9687.325	−6.66%	160.255	**−6.66%**	802.5	**28.690**	**−6.68%**	99
		Linear Regression	9766.358	−6.26%	162.100	−6.14%	**385.5**	29.150	−6.14%	134
		MLP	**9599.087**	**−6.68%**	**159.226**	−6.06%	569.4	29.196	−6.04%	146
	Steady	Myopic	10378.903	0.00%	171.450	0.00%	867.5	31.460	0.00%	82
		Exponential	**9633.436**	**−6.94%**	**159.578**	−6.64%	629.0	**29.225**	**−6.62%**	76
		Linear Regression	9772.233	−5.61%	161.868	−5.32%	**305.5**	29.895	−5.34%	112
		MLP	9802.089	−4.79%	162.408	−4.82%	675.5	29.394	−4.83%	156
40	Falling	Myopic	12632.717	0.00%	209.611	0.00%	508.5	29.530	0.00%	149
		Exponential	**11713.483**	**−7.83%**	**194.497**	**−7.90%**	247.0	**27.420**	**−7.88%**	193
		Linear Regression	11970.428	−6.76%	198.876	−6.56%	241.5	27.465	−6.57%	295
		MLP	12031.577	−6.41%	199.944	−6.38%	414.8	27.629	−6.36%	425
	Rising	Myopic	12837.467	0.00%	212.832	0.00%	969.5	30.170	0.00%	195
		Exponential	**12005.597**	**−6.65%**	**199.567**	**−6.58%**	494.5	**27.675**	**−6.57%**	234
		Linear Regression	12238.508	−6.36%	203.612	−6.21%	**408.5**	28.025	−6.20%	311
		MLP	12042.505	−6.57%	199.835	−6.22%	615.2	27.862	−6.21%	332
	Steady	Myopic	12635.717	0.00%	209.439	0.00%	883.0	29.335	0.00%	178
		Exponential	**11715.214**	**−8.04%**	**194.479**	**−8.16%**	540.5	**27.000**	**−8.16%**	170
		Linear Regression	11798.203	−6.49%	196.503	−6.46%	**407.0**	27.560	−6.48%	259
		MLP	11836.576	−7.42%	196.776	−7.45%	535.3	27.341	−7.46%	309

the ALNS. The reasoning behind that decision is that we want to avoid an overwhelmingly large number of routes that are very bad, to derive finer tuned models for better routes, which usually end up in the solution. This is especially bad for the linear regression and the MLP that are more closely fitted to the training data, whereas the exponential function benefits in this regard from its simplicity and robustness.

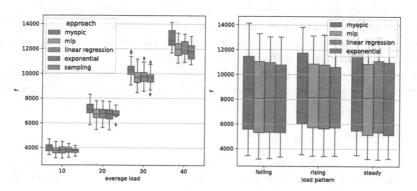

Fig. 3. Solution quality $f(\langle R, \tau, q \rangle)$ over average load and load pattern. The sampling approach is not included in the load pattern graphic due to missing data for average loads greater than 20.

6   Conclusions and Future Work

We considered a same-day delivery problem in which dynamically arriving orders have to be delivered within a short time span while minimizing travel times, labor costs, and tardiness. We focused on incorporating stochastic knowledge into the objective function of the point-in-time optimization runs, realized by an ALNS, by discounting route durations in dependence of diverse features. The most important features are the number of orders that can be expected up to the latest time the route would need to be started and the route's mean delivery duration, but several other factors were also considered and partly showed significant benefits.

Overall, our experiments clearly indicated that this approach is able to alleviate to a substantial degree the weaknesses of a myopic optimization, in particular in higher load situations. Of the three route duration discounting models the exponential function performs the best, reducing the travel time as well as the total objective by ≈6.1% on average over all instance classes. The more flexible neural net, in contrast, performed significantly weaker. We conjectured that the reason for this at the first glance surprising observation is the bias we have in the training data. The simpler exponential function seems to be more robust concerning candidate routes with properties that do not appear so frequently in the routes determined by the ALNS when generating training data. Moreover, the independent consideration of the routes is another source of potential errors. In our experiments, the exponential discounting even outperformed the sampling approach regarding solution quality in most cases and cuts down on runtime by several orders of magnitude.

Further work should consider alternative ways of generating training data to possibly reduce the bias. For example, intermediate solutions of the ALNS may occasionally also be used for data generation. Bootstrapping $\hat{\phi}$ from previous non-myopic runs could improve the accuracy of the parameter and lead to further improvement. Moreover, the variability of this mean order duration over the day due to varying load and traffic should be considered. Also, further tests with real-world inspired spatial order distributions (e.g., clustered instances) and load patterns could be helpful to evaluate practical aspects of the discounting models.

References

1. Azi, N., Gendreau, M., Potvin, J.Y.: An adaptive large neighborhood search for a vehicle routing problem with multiple routes. Comput. Oper. Res. **41**(1), 167–173 (2014)
2. Bent, R.W., Van Hentenryck, P.: Scenario-based planning for partially dynamic vehicle routing with stochastic customers. Oper. Res. **52**(6), 977–987 (2004)
3. Frohner, N., Raidl, G.R.: A double-horizon approach to a purely dynamic and stochastic vehicle routing problem with delivery deadlines and shift flexibility. In: Causmaecker, P.D., et al. (eds.) Proceedings of the 13th International Conference on the Practice and Theory of Automated Timetabling - PATAT 2021, Vol. I. Bruges, Belgium (2020)
4. Joe, W., Lau, H.C.: Deep reinforcement learning approach to solve dynamicvehicle routing problem with stochastic customers. In: Proceedings of theInternational Conference on Automated Planning and Scheduling, vol. 30, pp. 394–402 (2020)
5. Kingma, D.P., Ba, J.: Adam: A method for stochastic optimization. arXiv preprint arXiv:1412.6980 (2014)
6. Mitrović-Minić, S., Krishnamurti, R., Laporte, G.: Double-horizon based heuristics for the dynamic pickup and delivery problem with time windows. Trans. Res. Part B: Methodol. **38**(8), 669–685 (2004)
7. Pisinger, D., Ropke, S.: A general heuristic for node routing problems. Comput. Oper. Res. **34**, 2403–2435 (2007). https://doi.org/10.1007/978-3-642-46629-8_9
8. Powell, W.B.: Approximate Dynamic Programming: Solving the Curses of Dimensionality. Wiley, Hoboken (2007)
9. Ritzinger, U., Puchinger, J., Hartl, R.F.: A survey on dynamic and stochastic vehicle routing problems. Int. J. Prod. Res. **54**(1), 215–231 (2016)
10. Ropke, S., Pisinger, D.: An adaptive large neighborhood search heuristic for the pickup and delivery problem with time windows. Trans. Sci. **40**(4), 455–472 (2006)
11. Ulmer, Marlin W., Thomas, Barrett W., Mattfeld, Dirk C.: Preemptive depot returns for dynamic same-day delivery. EURO J. Trans. Logist. **8**(4), 327–361 (2018). https://doi.org/10.1007/s13676-018-0124-0
12. Voccia, S.A., Campbell, A.M., Thomas, B.W.: The same-day delivery problem for online purchases. Trans. Sci. **53**(1), 167–184 (2019)

Between Steps: Intermediate Relaxations Between Big-M and Convex Hull Formulations

Jan Kronqvist$^{(\boxtimes)}$ (iD), Ruth Misener (iD), and Calvin Tsay (iD)

Department of Computing, Imperial College London, London, UK
{j.kronqvist,r.misener,c.tsay}@imperial.ac.uk

Abstract. This work develops a class of relaxations in between the big-M and convex hull formulations of disjunctions, drawing advantages from both. The proposed "P-split" formulations split convex additively separable constraints into P partitions and form the convex hull of the partitioned disjuncts. Parameter P represents the trade-off of model size vs. relaxation strength. We examine the novel formulations and prove that, under certain assumptions, the relaxations form a hierarchy starting from a big-M equivalent and converging to the convex hull. We computationally compare the proposed formulations to big-M and convex hull formulations on a test set including: K-means clustering, P_ball problems, and ReLU neural networks. The computational results show that the intermediate P-split formulations can form strong outer approximations of the convex hull with fewer variables and constraints than the extended convex hull formulations, giving significant computational advantages over both the big-M and convex hull.

Keywords: Disjunctive programming · Relaxation comparison · Formulations · Mixed-integer programming · Convex MINLP

1 Introduction

There are well-known trade-offs between the big-M and convex hull formulations of disjunctions in terms of problem size and relaxation tightness. Convex hull formulations [4, 6, 9, 16, 20, 36] provide a *sharp formulation* for a single disjunction, *i.e.,* the continuous relaxation provides the best possible lower bound. The convex hull is often represented by so-called extended (a.k.a. perspective/multiple-choice) formulations [5, 7, 11, 14, 15, 17, 39], which introduce multiple copies of each variable in the disjunction(s). On the other hand, the big-M formulation only introduces one binary variable for each disjunct and results in a smaller problem in terms of both number of variables and constraints; however, in general it provides a weaker relaxation than the convex hull and may require a solver to explore significantly more nodes in a branch-and-bound tree [10, 39]. Even though the big-M formulation is weaker, in some cases it computationally outperforms extended convex hull formulations, as the simpler subproblems can

P. J. Stuckey (Ed.): CPAIOR 2021, LNCS 12735, pp. 299–314, 2021.
https://doi.org/10.1007/978-3-030-78230-6_19

offset the larger number of explored nodes. Anderson et al. [1] describe a folklore observation in mixed-integer programming (MIP) that extended convex hull formulations tend to perform worse than expected. The observation is supported by the numerical results in Anderson et al. [1] and in this paper.

This paper presents a framework for generating formulations for disjunctions between the big-M and convex hull with the intention of combining the best of both worlds: a tight, yet computationally efficient, formulation. The main idea behind the novel formulations is partitioning the constraints of each disjunct and moving most of the variables out of the disjunction. Forming the convex hull of the resulting disjunctions results in a smaller problem, while retaining some features of the convex hull. We call the new formulation the P-split, as the constraints are split into P parts. While many efforts have been devoted to computationally efficient convex hull formulations [3,11,19,33,37,40–42] and techniques for deriving the convex hull of MIP problems [2,22,25,31,35], our primary goal is not to generate the convex hull. Rather, we provide a straightforward framework for generating a family of relaxations that approximate the convex hull for a general class of disjunctions using a smaller problem formulation. Our experiments show that the P-split formulations can give a significant computational advantage over both the big-M and convex hull formulations.

This paper is organized as follows: the P-split formulation is presented in Sect. 2, together with properties of the P-split relaxations and how they compare to the big-M and convex hull relaxations. We also present a non-extended realization of the P-split formulation for the special case of a two-term disjunction. Finally, a numerical comparison of the formulations is presented in Sect. 3 using instances with both linear and nonlinear disjunctions.

1.1 Background

We consider optimization problems containing disjunctions of the form

$$\bigvee_{l \in \mathcal{D}} \left[g_k(\boldsymbol{x}) \le b_k \quad \forall k \in \mathcal{C}_l \right]$$
$$\boldsymbol{x} \in \mathcal{X} \subset \mathbb{R}^n, \tag{1}$$

where $\mathcal{D}$ contains the indices of the disjuncts, $\mathcal{C}_l$ the indices of the constraints in disjunct l, and $\mathcal{X}$ is a convex compact set. This paper assumes the following:

Assumption 1. *The functions $g_k : \mathbb{R}^n \to \mathbb{R}$ are convex additively separable functions, i.e., $g_k(\boldsymbol{x}) = \sum_{i=1}^n h_{ik}(x_i)$ where $h_{ik} : \mathbb{R} \to \mathbb{R}$ are convex functions, and each disjunct is non-empty on $\mathcal{X}$.*

Assumption 2. *All functions g_k are bounded over $\mathcal{X}$.*

Assumption 3. *Each disjunct contains significantly fewer constraints than the number of variables in the disjunction, i.e., $|\mathcal{C}_l| << n$.*

The first two assumptions are needed for the P-split formulation to be valid and result in a convex MIP. While the first assumption simplifies our analysis of P-split formulations, it could easily be relaxed to partially additively separable functions. Furthermore, the computational experiments only consider problems with linear or quadratic constraints, which ensures that the convex hull of the disjunction is representable by a polyhedron or (rotated) second-order cone constraints [6]. Assumption 3 characterizes problem structures favorable for the presented formulations. Problems with such a structure include, *e.g.*, clustering [28,32], mixed-integer classification [24,30], optimization over trained neural networks [1,8,12,13,34], and coverage optimization [18].

2   Relaxations Between Convex Hull and Big-M

The formulations in this section apply to disjunctions with multiple constraints per disjunct. However, to simplify the derivation, we only consider disjunctions with one constraint per disjunct, *i.e.*, $|\mathcal{C}_l| = 1 \ \forall l \in \mathcal{D}$. The extension to multiple constraints per disjunct simply applies the splitting procedure to each constraint.

To derive the new formulations, we partition the variables into P sets and form the corresponding index sets $\mathcal{I}_1, \ldots, \mathcal{I}_P$. The constraint for each disjunct is then split into P constraints, by introducing auxiliary variables $\alpha^j \in \mathbb{R}^P$

$$
\bigvee_{l \in \mathcal{D}} \left[g_l(\boldsymbol{x}) \leq b_l \right] \quad \longrightarrow \quad \bigvee_{l \subset \mathcal{D}} \left[\begin{array}{l} \sum_{i \in \mathcal{I}_1} h_{i,l}(x_i) \leq \alpha_1^l \\ \vdots \\ \sum_{i \in \mathcal{I}_P} h_{i,l}(x_i) \leq \alpha_P^l \\ \sum_{s=1}^{P} \alpha_s^l \leq b_l \\ \underline{\alpha}_s^l \leq \alpha_s^l \leq \bar{\alpha}_s^l \ \forall s \in \{1, \ldots, P\} \end{array} \right] \tag{2}
$$

$$
\boldsymbol{x} \in \mathcal{X} \qquad\qquad\qquad \boldsymbol{x} \in \mathcal{X}, \boldsymbol{\alpha}^l \in \mathbb{R}^P \ \forall \, l \in \mathcal{D}.
$$

Note that if the same function $h_{i,l}$ (or group of functions) appears in multiple disjuncts, then it can be represented by a single auxiliary variable in the disjuncts. By Assumption 2, function $h_{i,l}$ is bounded on $\mathcal{X}$, and bounds on the auxiliary variables are given by

$$
\underline{\alpha}_s^l := \min_{\boldsymbol{x} \in \mathcal{X}} \sum_{i \in \mathcal{I}_s} h_{i,l}(x_i), \qquad \bar{\alpha}_s^l := \max_{\boldsymbol{x} \in \mathcal{X}} \sum_{i \in \mathcal{I}_s} h_{i,l}(x_i). \tag{3}
$$

The P-split formulation does not require tight bounds, but weak bounds result in an overall weaker relaxation.

The splitting creates a lifted formulation by introducing $P \times |\mathcal{D}|$ auxiliary variables. Both formulations in (2) have the same feasible set in the $\boldsymbol{x}$ variables. We can then treat the splitted constraints as global constraints

$$
\bigvee_{l \in \mathcal{D}} \left[\begin{array}{l} \sum_{s=1}^{P} \alpha_s^l \leq b_l \\ \underline{\alpha}_s^l \leq \alpha_s^l \leq \bar{\alpha}_s^l \ \forall s \in \{1, \ldots, P\} \\ \sum_{i \in \mathcal{I}_s} h_{i,l}(x_i) \leq \alpha_s^l \end{array} \right] \qquad \forall s \in \{1, \ldots, P\}, \ \forall l \in \mathcal{D} \tag{4}
$$

$$
\boldsymbol{x} \in \mathcal{X}, \boldsymbol{\alpha}^l \in \mathbb{R}^P \qquad\qquad\qquad \forall \, l \in \mathcal{D}.
$$

Definition 1. *Formulation* (4) *is a P-split representation of the original disjunction in* (2).

Lemma 1 relates the P-split representation to the original disjunction. The property is rather simple, but for completeness we have stated it as a lemma.

Lemma 1. *The feasible set of P-split representation projected onto the x-space is equal to the feasible set of the original disjunctions in* (2).

Proof. An $\bar{x}$ that is feasible for (4) and violates (2) gives a contradiction. Similarly, an $\bar{x}$ that is feasible for (2) is also clearly feasible for (4). □

Using the extended formulation [4] to represent the convex hull of the disjunction in (4) results in the *P-split formulation*

$$
\begin{aligned}
&\alpha_s^l = \sum_{d \in \mathcal{D}} \nu_d^{\alpha_s^l} && \forall\, s \in \{1, \ldots, P\},\ \forall\, l \in \mathcal{D} \\
&\sum_{s=1}^{P} \nu_l^{\alpha_s^l} \leq b_l \lambda_l && \forall\, l \in \mathcal{D} \\
&\underline{\alpha}_s^l \lambda_d \leq \nu_d^{\alpha_s^l} \leq \bar{\alpha}_s^l \lambda_d && \forall\, s \in \{1, \ldots, P\}, \forall\, l, d \in \mathcal{D} && (P\text{-split}) \\
&\sum_{i \in \mathcal{I}_s} h_{i,l}(x_i) \leq \alpha_s^l && \forall\, s \in \{1, \ldots, P\},\ \forall\, l \in \mathcal{D} \\
&\sum_{l \in \mathcal{D}} \lambda_l = 1,\ \ \boldsymbol{\lambda} \in \{0, 1\}^{|\mathcal{D}|} \\
&\boldsymbol{x} \in \mathcal{X}, \boldsymbol{\alpha}^l \in \mathbb{R}^P,\ \boldsymbol{\nu}^{\alpha_s^l} \in \mathbb{R}^P\ \forall\, s \in \{1, \ldots, P\},\ \forall\, l \in \mathcal{D}\,,
\end{aligned}
$$

which forms a convex MIP problem. To clarify our terminology: a 2-split formulation is a formulation (P-split) where the constraints of the original disjunction are split up into two parts, *i.e.,* $P = 2$. We assume that the disjunction is part of a larger optimization problem that may contain multiple disjunctions. Therefore, we need to enforce integrality on the λ variables even if we recover the convex hull of the disjunction. Proposition 1 shows the correctness of the the (P-split) formulation of the original disjunction.

Proposition 1. *The set of feasible x variables in formulation (P-split) is equal to the feasible set of x variables in disjunction* (2).

Proof. By Lemma 1, (2) and (4) have equivalent x feasible sets. For $\lambda \in \{0, 1\}^{|\mathcal{D}|}$, the extended formulation (P-split) exactly represents the disjunction (4). □

Proposition 1 states that the P-split formulation is correct for integer feasible solutions, but it does not give any insight on the quality of the continuous relaxation. The following subsections further analyze the properties of the (P-split) formulation and its relation to the big-M and convex hull formulations.

Remark 1. A (P-split) formulation introduces $P \cdot (|\mathcal{D}|^2 + 1)$ continuous and $|\mathcal{D}|$ binary variables. Unlike the extended convex hull formulation (which introduces $|\mathcal{D}| \cdot n$ continuous and $|\mathcal{D}|$ binary variables), the number of "extra" variables is independent of n, *i.e.,* the number of variables in the original disjunction. As we later show, there are applications where $|\mathcal{D}| << n$ for which (P-split) formulations can be smaller and computationally more tractable than the extended convex hull formulation.

2.1 Properties of the P-Split Formulation

This section focuses on the strength of the continuous relaxation of the P-split formulation, and how it compares to convex hull and big-M formulations. To simplify the analyses, we only consider disjunctions with a single constraint per disjunct. However, the results again directly extend to the case of multiple constraints per disjunct by applying the same procedure to each constraint.

We first analyze the 1-split, as summarized in the following theorem.

Theorem 1. *The 1-split formulation is equivalent to the big-M formulation.*

Proof. We eliminate the disaggregated variables $\nu_d^{\alpha^l}$ from the 1-split formulation using Fourier-Motzkin elimination. Furthermore, we eliminate trivially redundant constraints, e.g., $\underline{\alpha}^l \lambda_d \leq \bar{\alpha}^l \lambda_d$, resulting in

$$
\begin{aligned}
& \alpha^l \leq b_l \lambda_l + \sum_{d \in \mathcal{D} \backslash l} \bar{\alpha}^l \lambda_d && \forall l \in \mathcal{D} \\
& \sum_{i=1}^{n} h_{i,l}(x_i) \leq \alpha^l && \forall\, l \in \mathcal{D} \\
& \sum_{l \in \mathcal{D}} \lambda_l = 1, \quad \boldsymbol{\lambda} \in \{0,1\}^{|\mathcal{D}|}, \boldsymbol{x} \in \mathcal{X}, \alpha^l \in \mathbb{R}\, \forall\, l \in \mathcal{D}.
\end{aligned} \tag{5}
$$

The auxiliary variables α^l are removed by combining the first and second constraints in (5). The smallest valid big-M coefficients are $M^l = \bar{\alpha}^l - b_l$, which enables us to write (5) as

$$
\begin{aligned}
& \sum_{i=1}^{n} h_{i,l}(x_i) \leq b_l + M^l (1 - \lambda_l) && \forall l \in \mathcal{D}_k \\
& \sum_{l \in \mathcal{D}} \lambda_l = 1, \quad \boldsymbol{\lambda} \in \{0,1\}^{|\mathcal{D}|}, \boldsymbol{x} \in \mathcal{X}.
\end{aligned} \tag{6}
$$

$\square$

Since the 1-split formulation introduces $|\mathcal{D}|^2 + 1$ auxiliary variables, but has the same continuous relaxation as the big-M formulation, there are no clear advantages of the 1-split formulation vs the big-M formulation.

We now examine the other extreme, where constraints are fully disaggregated, *i.e.*, the n-split. Its relation to the convex hull is given in the following theorem.

Theorem 2. *If all $h_{i,l}$ are affine functions, then the n-split formulation (where constraints are split for each variable) provides the convex hull of the disjunction.*

Proof. In the linear case, the original disjunction is given by

$$
\begin{aligned}
& \underset{l \in \mathcal{D}}{\vee} \left[(\mathbf{a}^l)^T \boldsymbol{x} \leq b_l \right] \\
& \boldsymbol{x} \in \mathcal{X},
\end{aligned} \tag{7}
$$

and the n-split formulation can be written compactly as

$$
\begin{aligned}
& \underset{l \in \mathcal{D}}{\vee} \left[\mathbf{B}^l \tilde{\alpha} \leq \tilde{b}_l \right] \\
& \tilde{\alpha} = \boldsymbol{\Gamma} \boldsymbol{x}, \quad \boldsymbol{x} \in \mathcal{X}, \tilde{\alpha} \in \mathbb{R}^{n \times |\mathcal{D}|}.
\end{aligned} \tag{8}
$$

The n-split formulation is given by the convex hull of (8) through the extended formulation. Here, $\mathbf{\Gamma}$ defines a bijective mapping between the $\boldsymbol{x}$ and $\tilde{\boldsymbol{\alpha}}$ variable spaces (only true for an n-split). A reverse mapping is given by $\boldsymbol{x} = \boldsymbol{\Psi}\tilde{\boldsymbol{\alpha}}$. The linear transformations preserve an exact representation of the feasible sets, *i.e.*,

$$\mathbf{B}^l\tilde{\boldsymbol{\alpha}} \leq \tilde{\boldsymbol{b}}_l \iff (\mathbf{a}^l)^T\boldsymbol{\Psi}\tilde{\boldsymbol{\alpha}} \leq b, \quad (\mathbf{a}^l)^T\boldsymbol{x} \leq b_l \iff \mathbf{B}^l\boldsymbol{\Gamma}\boldsymbol{x} \leq \tilde{\boldsymbol{b}}_l. \tag{9}$$

For any point $\boldsymbol{z}$ in the the convex hull of (8) $\exists\ \tilde{\boldsymbol{\alpha}}^1, \tilde{\boldsymbol{\alpha}}^2, \ldots \tilde{\boldsymbol{\alpha}}^{|\mathcal{D}|}$ and $\boldsymbol{\lambda} \in \mathbb{R}_+^{|\mathcal{D}|}$

$$\begin{aligned}
\boldsymbol{z} &= \sum_{l=1}^{|\mathcal{D}|} \lambda_l \tilde{\boldsymbol{\alpha}}^l \\
\sum_{l=1}^{|\mathcal{D}|} \lambda_l &= 1, \quad \mathbf{B}^l\tilde{\boldsymbol{\alpha}}^l \leq \tilde{\boldsymbol{b}}_l \quad \forall\, l \in \mathcal{D}.
\end{aligned} \tag{10}$$

Applying the reverse mapping to (10) gives

$$\boldsymbol{\Psi}\boldsymbol{z} = \sum_{l=1}^{|\mathcal{D}|} \lambda_l \boldsymbol{\Psi}\tilde{\boldsymbol{\alpha}}^l. \tag{11}$$

By construction, $(\mathbf{a}^l)^T \boldsymbol{\Psi}\tilde{\boldsymbol{\alpha}}^l \leq b_l \quad \forall l \in \mathcal{D}$. The point $\boldsymbol{\Psi}\boldsymbol{z}$ is given by a convex combination of points that all satisfy the constraints of one of the disjuncts in (7) and, therefore, belongs to the convex hull of (7). The same technique easily shows that any point in the convex hull of disjunction (7) also belongs to the convex hull of disjunction (8). $\qquad\square$

As mentioned before, non-tight bounds on the auxiliary variables $\boldsymbol{\alpha}^l$ can result in a weaker relaxation. Theorem 2 does not hold with nonlinear functions, since the mapping may be neither bijective nor a homomorphism. In general, the n-split formulation will not obtain the convex hull of nonlinear disjunctions, as Sect. 2.2 shows by example, but it can provide a strong outer approximation.

Two-Term Disjunctions. We further analyze the special case of a two-term disjunction for which we also present a non-lifted P-split formulation in the following theorem.

Theorem 3. *For a two-term disjunction, the P-split formulation has the following non-extended realization*

$$\begin{aligned}
\sum_{j \in \mathcal{S}_p}\left(\sum_{i \in \mathcal{I}_j} h_{i,1}(x_i)\right) &\leq \left(b_1 - \sum_{s \in \mathcal{S}\backslash\mathcal{S}_p} \underline{\alpha}_s^1\right)\lambda_1 + \sum_{s \in \mathcal{S}_p} \tilde{\alpha}_s^1\lambda_2 \quad \forall \mathcal{S}_p \subset \mathcal{S} \\
\sum_{j \in \mathcal{S}_p}\left(\sum_{i \in \mathcal{I}_j} h_{i,2}(x_i)\right) &\leq \left(b_2 - \sum_{s \in \mathcal{S}\backslash\mathcal{S}_p} \underline{\alpha}_s^2\right)\lambda_2 + \sum_{s \in \mathcal{S}_p} \tilde{\alpha}_s^2\lambda_1 \quad \forall \mathcal{S}_p \subset \mathcal{S} \\
\lambda_1 + \lambda_2 &= 1, \quad \boldsymbol{\lambda} \in \{0,1\}^2, \quad \boldsymbol{x} \in \mathcal{X},
\end{aligned} \tag{12}$$

where $\mathcal{S} = \{1, 2, \ldots P\}$.

Proof. The equality constraints for the disaggregated variables ($\alpha_s^l = \nu_1^{\alpha_s^l} + \nu_2^{\alpha_s^l}$) enable us to easily eliminate the variables $\nu_1^{\alpha_s^l}$ from (P-split), resulting in

$$\sum_{s=1}^{P} \left(\alpha_s^1 - \nu_2^{\alpha_s^1} \right) \leq b_1 \lambda_1 \tag{13}$$

$$\sum_{s=1}^{P} \nu_2^{\alpha_s^2} \leq b_2 \lambda_2 \tag{14}$$

$$\underline{\alpha}_s^l \lambda_1 \leq \alpha_s^l - \nu_2^{\alpha_s^l} \leq \bar{\alpha}_s^l \lambda_1 \qquad \forall s \in \{1,2,\ldots,P\}, \forall\, l \in \{1,2\} \tag{15}$$

$$\underline{\alpha}_s^l \lambda_2 \leq \nu_2^{\alpha_s^l} \leq \bar{\alpha}_s^l \lambda_2 \qquad \forall s \in \{1,2,\ldots,P\}, \forall\, l \in \{1,2\} \tag{16}$$

$$\sum_{i \in \mathcal{I}_s} h_{i,l}(x_i) \leq \alpha_s^l \qquad \forall\, s \in \{1,2,\ldots,P\},\ \forall\, l \in \{1,2\} \tag{17}$$

$$\lambda_1 + \lambda_2 = 1, \quad \boldsymbol{\lambda} \in \{0,1\}^2 \tag{18}$$

$$\boldsymbol{x} \in \mathcal{X}, \boldsymbol{\alpha}^l \in \mathbb{R}^P, \boldsymbol{\nu}^{\alpha_s^l} \in \mathbb{R}^P \qquad \forall\, l \in \{1,2\}, \forall\, s \in \{1,2,\ldots,P\}. \tag{19}$$

Next, we use Fourier-Motzkin elimination to project out the $\nu_2^{\alpha_s^1}$ variables. Combining the constraints in (15) and (16) only results in trivially redundant constraints, *e.g.*, $\alpha_s^l \leq \bar{\alpha}_s^l(\lambda_1 + \lambda_2)$. Eliminating the first variable $\nu_2^{\alpha_1^1}$ creates two new constraints by combining (13) with (15)–(16). The first constraint is obtained by removing $\nu_2^{\alpha_1^1}$ and α_1^1 from (13) and adding $\underline{\alpha}_1^1 \lambda_2$ to the left-hand side. The second constraint is obtained by removing $\nu_2^{\alpha_1^1}$ from (13) and subtracting $\bar{\alpha}_1^1 \lambda_2$ from the left-hand side. Eliminating the next variable is done by repeating the procedure of combining the two new constraints with the corresponding inequalities in (15)–(16). Each elimination step doubles the number of constraints originating from inequality (13). Eliminating all the variables $\nu_2^{\alpha_s^1}$ and α_s^1 results in the first set of constraints

$$\sum_{s \in \mathcal{S}_p} \alpha_s^1 \leq \left(b_1 - \sum_{s \in \mathcal{S} \setminus \mathcal{S}_p} \underline{\alpha}_s^1 \right) \lambda_1 + \sum_{s \in \mathcal{S}_p} \bar{\alpha}_s^1 \lambda_2 \quad \forall \mathcal{S}_p \subset \mathcal{S}. \tag{20}$$

The variables $\nu_2^{\alpha_s^2}$ and α_s^2 are eliminated by same steps, resulting in the second set of constraints in (12). $\qquad\square$

To further analyze the tightness of different P-split relaxations we require that the bounds on the auxiliary variables be *independent*, as defined below:

Definition 2. *We say that the upper and lower bounds for the constraint $\sum_{i=1}^n h_i(x_i) \leq 0$ are independent on $\mathcal{X}$ if*

$$\min_{x \in \mathcal{X}} (h_i(x_i) + h_j(x_j)) = \min_{x \in \mathcal{X}} h_i(x_i) + \min_{x \in \mathcal{X}} h_j(x_j)$$
$$\max_{x \in \mathcal{X}} (h_i(x_i) + h_j(x_j)) = \max_{x \in \mathcal{X}} h_i(x_i) + \max_{x \in \mathcal{X}} h_j(x_j), \tag{21}$$

hold for all $i, j \in \{1, 2, \ldots n\}$.

Independent bounds are not restricted to linear constraints, but the most general case of independent bounds are linear disjunctions with $\mathcal{X}$ defined as a box. Independent bounds enable us to establish a strict relation on the tightness of different P-split formulations, which is presented in the next corollary.

Corollary 1. *For a two-term disjunction with independent bounds, a $(P+1)$-split formulation, obtained by splitting one variable group in the P-split, is always as tight or tighter than the corresponding P-split formulation.*

Proof. The non-extended formulation (12) for the $(P+1)$-split comprises the constraints in the P-split formulation and some additional constraints.   □

From Corollary 1 it follows that the P-split formulations represent a hierarchy of relaxations, and we formally state this property in the following corollary.

Corollary 2. *For a linear two-term disjunction the P-split formulations form a hierarchy of relaxations, starting from the big-M relaxation $(P = 1)$ and converging to the convex hull relaxation $(P = n)$.*

Proof. Theorems 1 and 2 give equivalence to big-M and convex hull. By Corollary 1, the $(P+1)$-split is as tight or tighter than the P-split relaxation.   □

2.2   Illustrative Example

To see the differences between P-split formulations, consider the disjunction

$$\left[\textstyle\sum_{i=1}^4 x_i^2 \leq 1\right] \ \vee \ \left[\textstyle\sum_{i=1}^4 (3 - x_i)^2 \leq 1\right] \tag{ex-1}$$
$$x \in \mathbb{R}^4.$$

The tightest valid bounds on all the auxiliary variables are given by

$$\underline{\alpha}_s^l = 0, \quad \bar{\alpha}_s^l := \left(\sqrt{|\mathcal{I}_s| \cdot 3^2} + 1\right)^2 \quad \forall s \in \{1, 2, 3, 4\}, \ \forall l \in \{1, 2\}. \tag{22}$$

These bounds are derived from the fact that one of the two constraints in the disjunction must hold, and are symmetric for the two sets of α-variables. The continuously relaxed feasible sets of the P-split formulations of disjunction (ex-1) are shown in Fig. 1, which shows that the relaxations overall tighten with increasing number of splits P. The 4-split formulation does not give the convex hull, but provides a good approximation. For this example, the independent bound property does not hold and the relaxations do not form a proper hierarchy. To show why the independent bound property is needed, we compare the non-extended representations of the 1-split and 2-split formulations:

$$\sum_{i=1}^{4} x_i^2 \leq \lambda_1 + \left(\sqrt{36} + 1\right)^2 \lambda_2, \quad \sum_{i=1}^{4} (3 - x_i)^2 \leq \lambda_2 + \left(\sqrt{36} + 1\right)^2 \lambda_1 \tag{1-s}$$

$$\sum_{i=1}^{2} x_i^2 \leq \lambda_1 + \left(\sqrt{18} + 1\right)^2 \lambda_2, \quad \sum_{i=3}^{4} x_i^2 \leq \lambda_1 + \left(\sqrt{18} + 1\right)^2 \lambda_2 \tag{2-s1}$$

$$\sum_{i=1}^{2} (3 - x_i)^2 \leq \lambda_2 + \left(\sqrt{18} + 1\right)^2 \lambda_1, \sum_{i=3}^{4} (3 - x_i)^2 \leq \lambda_2 + \left(\sqrt{18} + 1\right)^2 \lambda_1 \tag{2-s2}$$

$$\sum_{i=1}^{4} x_i^2 \leq \lambda_1 + 2\left(\sqrt{18} + 1\right)^2 \lambda_2, \quad \sum_{i=1}^{4} (3 - x_i)^2 \leq \lambda_2 + 2\left(\sqrt{18} + 1\right)^2 \lambda_1. \tag{2-s3}$$

The 1-split formulation is given by (1-s), and the 2-split by (2-s1)–(2-s3). The 2-split contains additional constraints (2-s1) and (2-s2), but (2-s3) is a weaker version of (1-s). If the independent bound property were true, then (2-s3) and (1-s) would be identical and the relaxations would form a proper hierarchy.

3   Numerical Comparison

To compare how the formulations perform computationally, we apply the P-split, big-M, and convex hull formulations to several test problems. We consider three types of optimization problems that have a suitable structure for the P-split formulation (assumptions 1–3) and that are known to be challenging.

K-Means Clustering. Using the formulation by Papageorgiou and Trespalacios [28], the K-means clustering problem [26] is given by

$$\begin{aligned} \min_{r \in \mathbb{R}^L, x^j \in \mathbb{R}^n, \forall j \in \mathcal{K}} \quad & \sum_{i=1}^{L} r_i \\ \text{s.t.} \quad & \bigvee_{j \in \mathcal{K}} \left[\|x^j - d^i\|_2^2 \leq r_i \right] \quad \forall i \in \{1, 2, \ldots, L\}, \end{aligned} \tag{23}$$

where x^j are the cluster centers, $\{d^i\}_{i=1}^{L}$ are n-dimensional data points, and $\mathcal{K} = \{1, 2, \ldots k\}$. The tightest upper bound for the auxiliary variables in the P-split formulations are given by the largest squared Euclidean distance between any two data points in the subspace corresponding to the auxiliary variable. By introducing auxiliary variables for the differences $(x - d)$, we can express the convex hull of the disjunctions by rotated second order cone constraints [6] in a

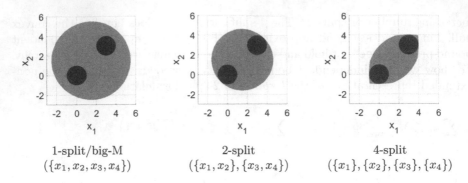

1-split/big-M   2-split   4-split
$(\{x_1, x_2, x_3, x_4\})$ $(\{x_1, x_2\}, \{x_3, x_4\})$ $(\{x_1\}, \{x_2\}, \{x_3\}, \{x_4\})$

Fig. 1. The dark circles show the feasible set of (ex-1) in the x_1, x_2 space. The light grey areas show the continuously relaxed feasible set of the P-split formulations. The sets in the parentheses show the partitioning of variables.

form suitable for Gurobi. We use the G2 data set [27] to generate low-dimensional test instances, and the MNIST data set [23] to generate high-dimensional test instances. For the MNIST-based problems, we select the first images of each class ranging from 0 to the number of clusters. Details about the problems are presented in Table 1.

P_ball Problems. The task is to assign p-points to n-dimensional unit balls such that the total ℓ_1 distance between all points is minimized and only one point is assigned to each unit ball [21]. Upper bounds on the auxiliary variables in the P-split formulation are given by the same technique as for the M-coefficients in [21], but in the subspace corresponding to the auxiliary variable. By introducing auxiliary variables for the differences between the points and the centers, we are able to express the convex hull by second order cone constraints [6] in a form suitable for Gurobi. We have generated a few larger instances to obtain more challenging problems and details of the problems are given in Table 1.

ReLU Neural Networks. Optimization over a ReLU neural network (NN) is used to quantify extreme outputs [1,8]. Each ReLU activation function $(y = \max\{0, \boldsymbol{w}^T \boldsymbol{x} + b\})$ can be expressed as a two-part disjunction using the P-split formulation, by separating $\boldsymbol{w}^T \boldsymbol{x} = \sum_{i \in \mathcal{S}_1 \cup \ldots \cup \mathcal{S}_P} w_i x_i$. Upper bounds on node outputs and auxiliary variables can be computed using simple interval arithmetic. We created several instances (Table 1) that minimize the prediction of single-output NNs trained on the d-dimensional Ackley/Rastrigin functions. All NNs were implemented in PyTorch [29] and trained for 1000 epochs, using a Latin hypercube of 10^6 samples. Note that more samples may be required to accurately represent the target functions, but here we are solely concerned with the performance of various optimization formulations. In later work, we explore techniques to tailor the P-split formulations proposed in this paper specifically to ReLU NNs [38].

Table 1. Details of the clustering, P_ball and neural network problems.

Name	Data points	Data dimension	Number of clusters
Cluster_g1	20	32	2
Cluster_g2	25	32	2
Cluster_g3	20	16	3
Cluster_m1	5	784	3
Cluster_m2	8	784	2
Cluster_m3	10	784	2
	Number of balls	Number of points	Ball dimension
P_ball_1	10	5	8
P_ball_2	10	5	16
P_ball_3	8	5	32
	Input dimension (d)	Hidden layers	Function
NN_1	2	[50, 50, 50]	Ackley
NN_2	10	[50, 50, 50]	Ackley
NN_3	3	[100, 100]	Rastrigin

Computational Setup. Optimization performance is dependent on both the tightness and the computational complexity of the continuous relaxation. The default (automatic) parameter selection in Gurobi causes large variations in the results that are due to different solution strategies rather than differences between formulations. Therefore, we used the parameter settings MIPFocus = 3, Cuts = 1, and MIQCPMethod = 1 for all problems. We found that using PreMIQCPForm = 2 drastically improves the performance of the extended convex hull formulations for the clustering and P_ball problems. However, it resulted in worse performance for the other formulations and, therefore, we only used it with the convex hull. Since the NN problems only contain linear constraints, only the MIPFocus and Cuts parameters apply to these problems The default values were used for all other settings. All problems were solved using Gurobi 9.0.3 on a desktop computer with an i7 8700k processor and 16 GB RAM.

Different variable partitionings can lead to differences in the P-split formulations. For all the problems, the variables are simply partitioned based on their ordered indices. For the K-means clustering and P_ball problems, we have used the smallest valid M-coefficients and tight bounds for the α-variables. The K-means clustering and P_ball problems both have analytical expressions for all the bounds. For the NN problems tight bounds are not easily obtained, and the bounds are obtained using interval arithmetic.

Table 2. CPU times [s] and numbers of nodes explored for test problems. In bold is the *winner* for each test instance with respect to both time and number of nodes. The grey shading shows the P-split times that strictly outperform both the big-M and convex hull formulations. The time limit was 1800 CPU seconds. Cells marked NA correspond to instances with fewer than P terms per disjunction. Note that, despite differences in the numbers of splits, the test problems have similar numbers of variables in the splitted variable sets. For example, the 8-split formulations for the Cluster_g instances and the 196-split formulations for the Cluster_m instances all have 4 variables per split.

Instance		Big-M	2-split	4-split	8-split	16-split	32-split	Convex hull
Cluster_g1	time	>1800	81.0	13.9	2.9	**1.7**	3.5	42.0
	nodes	>8998	2946	1096	256	98	91	**73**
Cluster_g2	time	>1800	106.3	7.7	4.3	**2.1**	4.5	40.6
	nodes	>10431	1736	481	217	104	86	**77**
Cluster_g3	time	>1800	>1800	870.6	**407.2**	597.5	NA	>1800
	nodes	>28906	>40820	19307	**14923**	16806		>7797
P_ball_1	time	403.0	235.4	285.1	**18.5**	NA	NA	42.2
	nodes	29493	7919	5518	2202			**1437**
P_ball_2	time	>1800	483.6	326.6	41.6	30.6	NA	**28.2**
	nodes	>19622	13602	5871	3921	1261		**531**
P_ball_3	time	>1800	>1800	>1800	149.3	91.1	**78.7**	114.0
	nodes	>7537	>6035	>6708	7042	3572	631	**554**
		big-M	14-split	28-split	56-split	196-split	392-split	convex hull
Cluster_m1	time	>1800	>1800	129.5	76.8	**32.0**	33.2	313.3
	nodes	>10680	>9651	2926	1462	524	**195**	228
Cluster_m2	time	>1800	1116.5	156.1	**27.1**	97.0	54.2	1260.1
	nodes	>4867	6220	1915	805	2752	1155	**131**
Cluster_m3	time	>1800	>1800	429.5	60.0	23.2	**19.8**	>1800
	nodes	>4419	>4197	3095	1502	741	**397**	>93
		1-split/ big-M	2-split	4-split	8-split	16-split	32-split	50-split/ convex hull*
NN_1	time	36.1	**29.4**	41.8	57.0	85.7	145.1	198.5
	nodes	24177	12377	11229	**7415**	11117	9793	11734
NN_2	time	**21.6**	35.5	50.7	131.4	287.3	776.1	>1800
	nodes	19746	20157	14003	11174	**6687**	12685	>4016
NN_3	time	**141.8**	210.6	206.5	275.5	305.8	429.1	556.6
	nodes	116996	101113	86582	84455	69022	56873	**48153**

*50-split is not the convex hull of each node for NN_3, which has layers of 100 nodes.

3.1 Numerical Results

Table 2 shows the elapsed CPU time and number of nodes explored to solve each problem. The results show that P-split formulations can drastically reduce the number of explored nodes compared to the big-M formulation, even with only a few splits. The differences are clearest for the nonlinear problems, where both the

CPU times and numbers of nodes are reduced by several orders of magnitude. As expected, the convex hull formulation results in the fewest explored nodes. However, the P-split formulations have a simpler[1] problem formulation, reducing the CPU times for all but one instance compared to the convex hull. The results clearly show the advantage of the intermediate P-split formulations, resulting in a tighter formulation than big-M and a computationally cheaper formulation than the extended convex hull.

Note that the P-split formulations are in general robust towards the choice of P. For the clustering and P_ball problems, all P-split formulations outperformed the big-M formulation both in terms of solution times and numbers of explored nodes. For the cases where the smallest P-split formulations timed out, Gurobi terminated with a much smaller gap compared to that of the big-M formulation. The P-split formulations also outperform the convex hull formulations in terms of solution time for a wide range of P in all but one of the test problems.

For the NN problems, which have linear disjunctions, the situation is somewhat different. Here, while increasing P still decreased the number of explored nodes, the improvements are less significant, and the trend is not completely monotonic. Note that bounds on the inputs to layers 2–3 are computed using interval arithmetic, resulting in overall weaker relaxations for all formulations. The weaker bounds in layers 2–3 reduce the benefits of both the P-split and convex hull formulations, and may favor the simpler big-M formulation. As the reduction in explored nodes is less drastic, smaller formulations perform the best in terms of CPU time, supporting claims that extended formulations may perform worse than expected [1,40]. This may also be a consequence of Gurobi efficiently handling linear problems when it detects big-M-type constraints. Ignoring the big-M (1-split), the 2- and 4-splits have the lowest CPU time for all NNs, and all the split formulations solve the problems significantly faster than the convex hull formulation.

4   Conclusions

We have presented a general framework for generating intermediate relaxations in between the big-M and convex hull. The numerical results show a great potential of the intermediate relaxations, by providing a good approximation of the convex hull through a computationally simpler problem. For several of the test problems, the intermediate relaxations result in a similar number of explored nodes as the convex hull formulation while reducing the total solution time by an order of magnitude.

Acknowledgements. The research was funded by a Newton International Fellowship by the Royal Society (NIF\R1\182194) to JK, a grant by the Swedish Cultural Foundation in Finland to JK, and by Engineering & Physical Sciences Research

---

[1] The extended convex hull formulations for the nonlinear problems require auxiliary variables and (rotated) second order cone constraints. All P-split formulations have fewer variables and constraints and only contain linear/convex-quadratic constraints.

Council (EPSRC) Fellowships to RM and CT (grant numbers EP/P016871/1 and EP/T001577/1). CT also acknowledges support from an Imperial College Research Fellowship.

References

1. Anderson, R., Huchette, J., Ma, W., Tjandraatmadja, C., Vielma, J.P.: Strong mixed-integer programming formulations for trained neural networks. Math. Program. **183**(1), 3–39 (2020). https://doi.org/10.1007/s10107-020-01474-5
2. Balas, E.: Disjunctive programming and a hierarchy of relaxations for discrete optimization problems. SIAM J. Algebraic Discrete Methods **6**(3), 466–486 (1985)
3. Balas, E.: On the convex hull of the union of certain polyhedra. Oper. Res. Lett. **7**(6), 279–283 (1988)
4. Balas, E.: Disjunctive programming: properties of the convex hull of feasible points. Discrete Appl. Math. **89**(1–3), 3–44 (1998)
5. Balas, E.: Disjunctive Programming. Springer International Publishing (2018). https://doi.org/10.1007/978-3-030-00148-3
6. Ben-Tal, A., Nemirovski, A.: Lectures on Modern Convex Optimization: Analysis, Algorithms, and Engineering Applications, vol. 2. Siam (2001)
7. Bonami, P., Lodi, A., Tramontani, A., Wiese, S.: On mathematical programming with indicator constraints. Math. Program. **151**(1), 191–223 (2015)
8. Botoeva, E., Kouvaros, P., Kronqvist, J., Lomuscio, A., Misener, R.: Efficient verification of ReLU-based neural networks via dependency analysis. In: AAAI-20 Proceedings, pp. 3291–3299 (2020)
9. Ceria, S., Soares, J.: Convex programming for disjunctive convex optimization. Math. Program. **86**(3), 595–614 (1999). https://doi.org/10.1007/s101070050106
10. Conforti, M., Cornuéjols, G., Zambelli, G.: Integer programming, volume 271 of graduate texts in mathematics (2014)
11. Conforti, M., Wolsey, L.A.: Compact formulations as a union of polyhedra. Math. Program. **114**(2), 277–289 (2008). https://doi.org/10.1007/s10107-007-0101-0
12. Fischetti, M., Jo, J.: Deep neural networks and mixed integer linear optimization. Constraints **23**(3), 296–309 (2018). https://doi.org/10.1007/s10601-018-9285-6
13. Grimstad, B., Andersson, H.: ReLU networks as surrogate models in mixed-integer linear programs. Comput. Chem. Eng. **131**, 106580 (2019)
14. Grossmann, I.E., Lee, S.: Generalized convex disjunctive programming: nonlinear convex hull relaxation. Comput. Optim. Appl. **26**(1), 83–100 (2003). https://doi.org/10.1023/A:1025154322278
15. Günlük, O., Linderoth, J.: Perspective reformulations of mixed integer nonlinear programs with indicator variables. Math. Program. **124**(1–2), 183–205 (2010)
16. Helton, J.W., Nie, J.: Sufficient and necessary conditions for semidefinite representability of convex hulls and sets. SIAM J. Optim **20**(2), 759–791 (2009)
17. Hijazi, H., Bonami, P., Cornuéjols, G., Ouorou, A.: Mixed-integer nonlinear programs featuring "on/off" constraints. Comput. Optim. Appl. **52**(2), 537–558 (2012). https://doi.org/10.1007/s10589-011-9424-0
18. Huang, C.F., Tseng, Y.C.: The coverage problem in a wireless sensor network. Mob. Netw. Appl. **10**(4), 519–528 (2005)
19. Jeroslow, R.G.: A simplification for some disjunctive formulations. Eur. J. Oper. Res. **36**(1), 116–121 (1988)

20. Jeroslow, R.G., Lowe, J.K.: Modelling with integer variables. In: Korte, B., Ritter, K. (eds.) Mathematical Programming at Oberwolfach II, pp. 167–184. Springer, Berlin (1984). https://doi.org/10.1007/BFb0121015
21. Kronqvist, J., Misener, R.: A disjunctive cut strengthening technique for convex MINLP. Optim. Eng. 1–31 (2020). https://doi.org/10.1007/s11081-020-09551-6
22. Lasserre, J.B.: An explicit exact SDP relaxation for nonlinear 0-1 programs. In: Aardal, K., Gerards, B. (eds.) IPCO 2001. LNCS, vol. 2081, pp. 293–303. Springer, Heidelberg (2001). https://doi.org/10.1007/3-540-45535-3_23
23. LeCun, Y., Cortes, C., Burges, C.: Mnist handwritten digit database. ATT Labs (2010). http://yann.lecun.com/exdb/mnist2
24. Liittschwager, J., Wang, C.: Integer programming solution of a classification problem. Manage. Sci. 24(14), 1515–1525 (1978)
25. Lovász, L., Schrijver, A.: Cones of matrices and set-functions and 0–1 optimization. SIAM J. Optim. 1(2), 166–190 (1991)
26. MacQueen, J., et al.: Some methods for classification and analysis of multivariate observations. In: Proceedings of the fifth Berkeley symposium on mathematical statistics and probability. vol. 1, pp. 281–297. Oakland, CA, USA (1967)
27. Mariescu-Istodor, P.F.R., Zhong, C.: XNN graph LNCS 10029, 207–217 (2016)
28. Papageorgiou, D.J., Trespalacios, F.: Pseudo basic steps: bound improvement guarantees from Lagrangian decomposition in convex disjunctive programming. EURO J. Comput. Optim. 6(1), 55–83 (2018)
29. Paszke, A., et al.: Pytorch: An imperative style, high-performance deep learning library. In: Advances in Neural Information Processing Systems, pp. 8026–8037 (2019)
30. Rubin, P.A.: Solving mixed integer classification problems by decomposition. Ann. Oper. Res. 74, 51–64 (1997). https://doi.org/10.1023/A:1018990909155
31. Ruiz, J.P., Grossmann, I.E.: A hierarchy of relaxations for nonlinear convex generalized disjunctive programming. Eur. J. Oper. Res. 218(1), 38–47 (2012)
32. Sağlam, B., Salman, F.S., Sayın, S., Türkay, M.: A mixed-integer programming approach to the clustering problem with an application in customer segmentation. Eur. J. Oper. Res. 173(3), 866–879 (2006)
33. Sawaya, N.W., Grossmann, I.E.: Computational implementation of non-linear convex hull reformulation. Comput. Chem. Eng. 31(7), 856–866 (2007)
34. Serra, T., Kumar, A., Ramalingam, S.: Lossless compression of deep neural networks. In: Hebrard, E., Musliu, N. (eds.) CPAIOR 2020. LNCS, vol. 12296, pp. 417–430. Springer, Cham (2020). https://doi.org/10.1007/978-3-030-58942-4_27
35. Sherali, H.D., Adams, W.P.: A hierarchy of relaxations between the continuous and convex hull representations for zero-one programming problems. SIAM J. Discrete Math. 3(3), 411–430 (1990)
36. Stubbs, R.A., Mehrotra, S.: A branch-and-cut method for 0–1 mixed convex programming. Math. Program. 86(3), 515–532 (1999)
37. Trespalacios, F., Grossmann, I.E.: Algorithmic approach for improved mixed-integer reformulations of convex generalized disjunctive programs. INFORMS J. Comput. 27(1), 59–74 (2015)
38. Tsay, C., Kronqvist, J., Thebelt, A., Misener, R.: Partition-based formulations for mixed-integer optimization of trained relu neural networks. arXiv preprint arXiv:2102.04373 (2021)
39. Vielma, J.P.: Mixed integer linear programming formulation techniques. SIAM Rev. 57(1), 3–57 (2015)
40. Vielma, J.P.: Small and strong formulations for unions of convex sets from the cayley embedding. Math. Program. 177(1–2), 21–53 (2019)

41. Vielma, J.P., Ahmed, S., Nemhauser, G.: Mixed-integer models for nonseparable piecewise-linear optimization: unifying framework and extensions. Oper. Res. **58**(2), 303–315 (2010)
42. Vielma, J.P., Nemhauser, G.L.: Modeling disjunctive constraints with a logarithmic number of binary variables and constraints. Math. Program. **128**(1–2), 49–72 (2011)

Logic-Based Benders Decomposition for an Inter-modal Transportation Problem

Ioannis Avgerinos[✉], Ioannis Mourtos, and Georgios Zois

ELTRUN Research Lab, Department of Management Science and Technology,
Athens University of Economics and Business, Athens 104 34, Greece
{iavgerinos,mourtos,georzois}@aueb.gr

Abstract. This paper studies a real-life inter-modal freight transportation problem, comprised by three consecutive stages: *disposition* where orders are picked up by trucks, transferred and unloaded to a set of warehouses in Central and Eastern Europe, *inter-region transport* where the orders are packed into trailers, which are shipped to warehouses in Turkey using different inter-region transport modes, and *last-mile delivery* where the orders unloaded at the warehouses in Turkey are picked up by vans and delivered to their final destination. The objective is to minimise total transport and tardiness cost. After restricting the routes in the disposition stage, we formulate a mixed-integer linear program that captures the entire delivery process. Then, we propose a Benders' decomposition and prove the validity of a set of optimality cuts and of a subproblem relaxation. We show the impact of this approach on large-scale real instances under user-imposed time limits. We further strengthening the master problem with a set of valid inequalities and speed up the solution of the subproblem using Constraint Programming.

Keywords: Inter-modal transportation · Pickup and delivery planning · Benders decomposition · Integer programming · Constraint programming

1 Introduction

It is known that Benders Decomposition [4] is quite effective for complex transportation problems [8,10], often occurring in inter-modal transportation (e.g., [3,7]). Although heuristic approaches are popular in such problems (e.g., [2,9]), we are interested in exact methods, because cost reduction appears to be of utmost importance for logistics providers. In addition, we are also motivated by the practical considerations of a real use case, arising from EKOL that is a

This research has been supported by the EU through the COG-LO Horizon 2020 project, grant number 769141 and the GSRI through the i@transport project, grant number T2EΔK00345.

P. J. Stuckey (Ed.): CPAIOR 2021, LNCS 12735, pp. 315–331, 2021.
https://doi.org/10.1007/978-3-030-78230-6_20

major such provider in Europe. EKOL implements weekly planning using shipment requests arriving until the very last moment, therefore it is critical that execution time remains below 1–2 h.

Let us detail the three stages of the problem in hand, while also discussing some indicative background. The first stage, called *disposition*, includes the collection of orders from their pickup points and their unloading to a set of depots in Poland, Hungary, Czech Republic and Slovakia, using single-trailer trucks (i.e., trucks of identical capacity). Before the orders are transferred to the warehouses, a customs' clearance process has to be carried out by specific companies, while the time-span of loading, customs clearance and unloading is restricted by the operating hours (i.e., time-windows) of all companies involved. This setting already resembles the Vehicle Routing Problem with Time Windows (VRPTW). The existence of multiple depots in our case introduces the multi-depot variant in which each route begins and ends at the same depot [12]. Given the difficulty of solving optimally that variant in reasonable time, we opt for the alternative approach of pre-defining the candidate routes and obtaining a set-covering formulation [11]. This results in a reasonable number of feasible routes because of the strict operating hours of the companies involved, the small average number of orders fitting the capacity of a trailer and the fact that each pick-up point must be served by its nearest depot. Thus, solving a TSP with time-windows returns the shortest time per route in a few seconds.

The second stage includes the *inter-regional transportation* of the orders from Central and Eastern Europe to Turkey. Here orders, each with a known weight, are assigned to trailers, each of known capacity. Afterwards, trailers are assigned to given transport modes, each with a known origin and destination, departure and arrival time, capacity (in number of trailers), a fixed and a variable (per trailer) cost. Transport modes link all depots to the warehouses of Istanbul or Denizli in Turkey through shipment by trucks (each carrying multiple trailers) from any depot or by railway from Hungary only to Istanbul or by combining railway from the Czech Republic to the port of Trieste and seaway from Trieste to Istanbul or Mersin (the cost and travel times from the railway stations and ports to the two warehouses are included in the arrival times and the fixed costs of the concerned modes). A relevant definition of these transport modes appears in [9] and an analogous mathematical model appears in [13]. Overall, this second stage must assign orders to trailers and trailers to transport modes at minimum cost, where of course a mode is eligible for a trailer only if its departure time exceeds the packing time of that trailer, which in turn must start after the arrival of all its orders at the depot. Indicative work on the relevant single-depot and multi-depot cross-docking problems appears in [2,14].

The third stage is the last-mile delivery of the orders, after their arrival to Turkey and their unloading at the respective warehouse. Here, a large number of small vans serve each warehouse and often carry a single order. Therefore, we can reasonably apply the simplifying assumption that each van delivers a single order, thus obtaining a resource-constrained parallel-scheduling formulation. Specifically, the delivery of each order is considered as a 'task' to be scheduled to any of the available vans, i.e., 'machines'. The problem aims to minimize the tardiness cost, as implied by the number of days of delayed delivery, compared to a predefined deadline, multiplied by an order-unique weight, referred to as *importance* by the EKOL dispatchers.

Our Contribution. Given the elaborate structure of the entire inter-modal setting we examine, our first offering is the construction of a unified mathematical model. We proceed with a Benders Decomposition, in which the master problem sustains the first two stages and the subproblem gets the third one. Apart from exploiting the Benders' approach, this decomposition allows us to add the redundant capacity constraints of [1] to the master problem, as that exhibits a facility location structure. We also use Constraint Programming for the solution of the subproblem. The proofs regarding the validity of both the optimality cuts and the subproblem relaxation added to the master problem follow the style of [5]. We experiment with several weekly instances of EKOL and establish the computational efficiency of our approach compared to a standard solver under time limits of $1/2$ or 1 h.

2   Modeling

Notation. Table 1 presents the notation of the MILP formulation. For each order $n \in N$, $pickup_n$, $customs_n$ and $delivery_n$ denote the corresponding locations; $demand_n$ is its weight and t_n^{due} its delivery deadline. The penalty imposed if t_n^{due} is exceeded equals the squared value of days of delay multiplied by the weight i_n.

Each trailer is linked with a single depot in Europe. The set of transport modes M includes pre-computed routes for disposition represented by the subset of pickup modes M^- and delivery modes M^+. If the origin of a delivery mode is a depot in Europe and its destination a warehouse in Turkey, the mode is considered as complete and is placed in M^+_{full}; otherwise, it is considered as partial and belongs to M^+_{part}. The distinction between *partial* and *complete* modes concerns the inter-regional modes only, therefore the subset M^- is not divided. seq_m denotes the sequence of nodes or partial modes that are linked with m. The departure time from the origin $origin_m$ of mode $m \in M$ and the arrival time to its destination $dest_m$, are $departure_m$ and $arrival_m$, respectively.

The travel cost of each trailer that is shipped by mode m is c_m^{travel} and the fixed cost of the mode, irrespective of the number of trailers, is c_m^{fixed}. The maximum number of trailers that can be shipped simultaneously by a single mode m is $Q_m^{trailers}$ and the availability of vehicles is denoted by b_m.

J is the set of depots in Europe and L is the set of warehouses in Turkey. T represents hourly time slots and D daily ones. The parameter α_{nm} is equal to 1 if order n is picked up by route $m \in M^-$ and 0 otherwise. The capacity of all trailers is q_g and the transportation cost from the warehouse $l \in L$ to the delivery location $delivery_n$ is $c_{nl}^{delivery}$. The number of available vans in warehouse l is

K_l and the travel duration from l to $delivery_n$ is t_{nl}. The time window per day $d \in D$ is $[opening_d, closing_d]$.

Variables h_{nm} determine the assignment of orders to modes, y_{ng} determine the assignment of orders to trailers and x_{gm} the transport mode of trailers. Variables u_{nlt} are equal to 1 if the delivery of order n from warehouse $l \in L$ begins at time $t \in T$. Integer variables w_m denote the number of times that mode m is used. The tardiness cost is defined by the integer variable T_n, day_{nd} indicates whether order n is handled in day d and o_g is a 'dummy' continuous variable, equal to the total payweight units in trailer g.

Preprocessing. We compute the candidate feasible pickup routes of disposition as a pre-processing step, by constructing sequences that do not exceed the capacity of a trailer and do not violate the time-windows imposed. The result of this process is the subset M^- (Table 1). If the constructed route collects an order, the respective parameter α_{nm} is fixed to 1. Each trailer is dedicated to a single depot, thus the rest of the depots are irrelevant as it cannot be shipped from them. Therefore, the variables x_{gm} that include trailers g and modes m that are featured with an irrelevant depot are fixed to 0. The MILP for the entire inter-modal transportation problem is as follows.

$\mathcal{P}$:
$$\min \sum_{g \in G} \sum_{m \in M} c_m^{travel} \cdot x_{gm} + \sum_{m \in M} c_m^{fixed} \cdot w_m + \sum_{n \in N} \sum_{l \in L} \sum_{t \in T} c_{nl} \cdot u_{nlt} + \sum_{n \in N} i_n \cdot T_n$$

subject to:

$$\sum_{m \in M^-} h_{nm} = 1 \qquad\qquad \forall n \in N \qquad\qquad (1)$$

$$\sum_{g \in G} \sum_{m \in M^-} \alpha_{nm} \cdot x_{gm} = 1 \qquad\qquad \forall n \in N \qquad\qquad (2)$$

$$\sum_{m \in M^-} x_{gm} \leq 1 \qquad\qquad \forall g \in G \qquad\qquad (3)$$

$$\sum_{m \in M_{full}^+} q_g \cdot x_{gm} \geq \sum_{n \in N} demand_n y_{ng} \qquad \forall g \in G \qquad\qquad (4)$$

$$\sum_{m \in M_{full}^+} x_{gm} \leq 1 \qquad\qquad \forall g \in G \qquad\qquad (5)$$

$$x_{gm} = x_{gm'} \qquad\qquad \forall g \in G, m \in M_{full}^+, m' \in seq_m \qquad (6)$$

$$Q_m^{trailers} \cdot w_m \geq \sum_{g \in G} x_{gm} \qquad\qquad \forall m \in M \qquad\qquad (7)$$

Table 1. MILP notation

Sets		
N		Orders
G		Trailers
M		Modes
$M^- \subset M$		Pickup Modes
$M^+ \subset M$		Delivery Modes
$M^+_{part} \subset M^+$		Partial Modes
$M^+_{full} \subset M^+$		Complete Modes
J		Depots in Central & Eastern Europe
L		Warehouses in Turkey
T		Hours
D		Days
Orders information		
$pickup_n$	$n \in N$	Pickup company of order n
$customs_n$	$n \in N$	Customs company of order n
$delivery_n$	$n \in N$	Delivery company of order n
$demand_n$	$n \in N$	Demand of order n in payweight units
t_n^{due}	$n \in N$	Delivery deadline of order n in hours
i_n	$n \in N$	Importance of order n
Modes information		
seq_m	$m \in M$	The ordered sequence of nodes that are included in mode m
$origin_m$	$m \in M$	The origin of mode m
$dest_m$	$m \in M$	The destination of mode m
$t_m^{departure}$	$m \in M$	The departure time of m in hours
$t_m^{arrival}$	$m \in M$	The arrival time of m in hours
c_m^{travel}	$m \in M$	Variable cost of each trailer shipped to m
c_m^{fixed}	$m \in M$	Fixed cost of each mode m
$Q_m^{trailers}$	$m \in M$	Maximum number of trailers of mode m
b_m	$m \in M$	Maximum number of times that m can be used
Other parameters		
α_{nm}	$n \in N, m \in M^-$	1 if order n is collected by m, 0 otherwise
q_g	$g \in G$	Capacity of trailer g in payweight units
$c_{nl}^{delivery}$	$n \in N, l \in L$	Delivery cost of order n from warehouse l
K_l	$l \in L$	Number of vans in warehouse l
t_{nl}	$n \in N, l \in L$	Travel duration from warehouse l to the final destination of n
$opening_d$	$d \in D$	Opening hour in day d
$closing_d$	$d \in D$	Closing hour in day d
Variables		
h_{nm}	$n \in N, m \in M$	1 if order n is shipped to mode m, 0 otherwise
x_{gm}	$g \in G, m \in M$	1 if trailer g is shipped to mode m, 0 otherwise
u_{nlt}	$n \in N, l \in L, t \in T$	1 if the delivery of order n starts at t from l, 0 otherwise
y_{ng}	$n \in N, g \in G$	1 if order n is packed to trailer g, 0 otherwise
w_m	$m \in M$	Integer number of times that mode m is used
r_n	$n \in N$	Continuous variable, release time of n
T_n	$n \in N$	Integer squared value of the number of days of tardiness of n
day_{nd}	$n \in N, d \in D$	1 if order n is handled in day d, 0 otherwise
o_g	$g \in G$	Continuous variable denoting the weight of trailer g

$$\sum_{m \in M^-} t_m^{arrival} \cdot h_{nm} \leq \sum_{m \in M^+} t_m^{departure} \cdot h_{nm} \qquad \forall n \in N \qquad (8)$$

$$h_{nm} + 1 \geq x_{gm} + y_{ng} \qquad \forall g \in G, n \in N, m \in M_{full}^+ \qquad (9)$$

$$\sum_{t \in T} u_{nlt} \geq h_{nm} \qquad \forall n \in N, m \in M_{full}^+, l = dest_m \qquad (10)$$

$$\sum_{m \in M_{full}^+} h_{nm} = 1 \qquad \forall n \in N \qquad (11)$$

$$\sum_{g \in G} y_{ng} = 1 \qquad \forall n \in N \qquad (12)$$

$$r_n \geq arrival_m \cdot h_{nm} \qquad \forall n \in N, m \in M \qquad (13)$$

$$\sum_{l \in L} \sum_{t \in T} u_{nlt} = 1 \qquad \forall n \in N \qquad (14)$$

$$\sum_{n \in N} \sum_{t' \in \{t' | t - t_{nl} < t' \leq t\}} u_{nlt} \leq K_l \qquad \forall l \in L, t \in T \qquad (15)$$

$$r_n - t \leq (1 - u_{nlt}) \cdot M \qquad \forall n \in N, l \in L, t \in T \qquad (16)$$

$$T_n \geq \sum_{l \in L} \sum_{t \in T} \lfloor 1 + \frac{t + t_{nl} - t_n^{due}}{24} \rfloor \cdot \lfloor |1 + \frac{t + t_{nl} - t_n^{due}}{24}| \rfloor \cdot u_{nlt} \qquad \forall n \in N$$
$$(17)$$

$$\sum_{d \in D} day_{nd} = 1 \qquad \forall n \in N \qquad (18)$$

$$\sum_{l \in L} \sum_{t \in T} t \cdot u_{nlt} \geq \sum_{d \in D} opening_d \cdot day_{nd} \qquad \forall n \in N \qquad (19)$$

$$\sum_{l \in L} \sum_{t \in T} t \cdot u_{nlt} \leq \sum_{d \in D} closing_d \cdot day_{nd} \qquad \forall n \in N \qquad (20)$$

$$x_{gm}, y_{ng}, u_{nlt}, h_{nm}, day_{nd} \in \{0,1\} \; \forall n \in N, g \in G, m \in M, l \in L, t \in T, d \in D$$
$$w_m \in \{0, ..., b_m\} \qquad\qquad\qquad \forall m \in M$$
$$r_n \in \mathcal{R}^+, T_n \in \mathcal{N} \qquad\qquad\qquad \forall n \in N$$

The objective function is the sum of travel and fixed costs per trailer, the costs of all orders during the last-mile delivery stage and the tardiness cost.

Constraints (1)–(3) refer to the *disposition* stage. Constraints (4)–(12) are imposed on the inter-modal transport stage, in which (4) is the Generalized Assignment Problem constraint of orders to trailers and (5)–(12) is the inter-modal routing problem. Constraints (13)–(19) are the last-mile delivery stage.

Constraints (1) ensure that each order is collected by exactly one pickup route and Constraints (2) ensure that the assigned trailer is shipped to the same route. Constraints (3) ensure that each trailer is assigned to one route at most. Constraints (4) restrict the total weight of each trailer to its capacity, while Constraints (5) are equivalent with (3), concerning the delivery transport modes. Constraints (6) enforce the assignment of each trailer to all corresponding partial modes. Constraints (7) define the number of times that each mode is used and Constraints (8) is a precedence constraint, ensuring that each order cannot be shipped to Turkey if the unloading to a European depot is not completed. The assignment variables are connected in (9), enforcing each order to be assigned to the same mode that its trailer is shipped to. (10) fixes the assignment of each order to the destination of the assigned mode, while (11) is equivalent with (1) for the delivery modes. Constraints (12) ensure that each order is loaded to one trailer. Constraints (13) define the release time of each order. The start of the delivery of each order occurs at a single discrete time instance in (14) and (15) is the parallel-scheduling constraint, restricting the maximum number of simultaneous deliveries to the number of available vehicles. (16) are big-M constraints that prevent each order to be delivered before it is released to Turkey and the squared value of tardiness in days. Constraints (18) ensure that the delivery of each order is completed in a single day and (19)–(20) restrict the delivery of all orders during the operating hours of the warehouses. Constraints (14)–(18) are a reformulation of the original minimum tardiness MILP of [5], adapted to our setting.

Constraints (17) ensure that T_n will receive the proper value of squared tardiness. While a straightforward computation of tardiness would imply a simpler constraint:

$$T_n \geq \sum_{l \in L} \sum_{t \in T} \frac{(t + t_{nl} - t_n^{due})}{24} \cdot u_{nlt} \qquad\qquad \forall n \in N$$

in which t_{nl} is the travel time from warehouse l to the delivery point of n, EKOL wishes to calculate the squared value of the number of days of tardy delivery. The hours of tardiness are divided by 24 and rounded down to obtain the actual days of tardy delivery:

$$T_n \geq \sum_{l \in L} \sum_{t \in T} \lfloor 1 + \frac{(t + t_{nl} - t_n^{due})}{24} \rfloor \cdot u_{nlt} \qquad \forall n \in N$$

As the squared value of that expression would incorrectly neglect the negative values, we multiply it with its absolute value to obtain a negative sign in the LHS if no tardiness occurs (imposing $T_n \geq 0$):

$$T_n \geq \sum_{l \in L} \sum_{t \in T} \lfloor 1 + \frac{(t + t_{nl} - t_n^{due})}{24} \rfloor \cdot \lfloor 1 + \frac{|t + t_{nl} - t_n^{due}|}{24} \rfloor \cdot u_{nlt} \qquad \forall n \in N$$

3 A Logic-Based Benders Decomposition Approach

Let $\mathcal{P} = \min\{f(x) + g(y) | x \in D_x, y \in D_y\}$ be an MILP formulation for an optimization problem $\mathcal{P}$, where x, y are groups of variables, $f(x)$, $g(y)$ are linear cost functions and D_x, D_y are the domains of x, y. Loosely speaking, a Benders Decomposition approach considers two smaller problems, the *master* problem $\mathcal{M} = \min\{z | z \geq f(x), x \in D_x\}$ and the *Subproblem* $\mathcal{S} = \min\{f(\hat{x}) + g(y) | y \in D_y\}$, where $\mathcal{M}$ iteratively provides its solution to $\mathcal{S}$. At each iteration, if the solution of $\mathcal{S}$ is greater than the solution of $\mathcal{M}$, a *Benders optimality cut* ensures that the same solution value will be computed only if a better one is not found. If $\mathcal{S}$ is infeasible, for an optimal solution $\hat{x}$ of $\mathcal{M}$, a *Benders feasibility cut* ensures that the solution will not be repeated. Therefore, the validity of the method depends on the construction of Benders cuts. An extended description of the Benders Decomposition approach can be found in [6].

Algorithm 1 provides an adaptation of the Benders Decomposition approach in our setting. It starts by solving the $\mathcal{M}$ to obtain an initial solution (x). The objective value of $\mathcal{M}$ is the incumbent lower bound $\hat{z}$ of the problem $\mathcal{P}$. In each iteration, after fixing all variables in x to the values of $\hat{x}$, Algorithm 1 solves $\mathcal{S}$. A *bounding function* denoted by $B_m^k(\hat{h_m})$ generates Benders cuts for each mode $m \in M_{full}^+$, which are added to $\mathcal{M}$ and considered for the current iteration. The optimal vector $\hat{x}_k$ of $\mathcal{M}$ in the current iteration is used to update the lower bound $\hat{z}$. If the sum of the objective value of $\mathcal{S}$ and the costs related to the x variables equal to $\hat{z}$, the algorithm terminates.

3.1 Constructing the Initial Master Problem

$\mathcal{M}$:

min z

subject to:

$$z \geq \sum_{g \in G} \sum_{m \in M} c_m^{travel} \cdot x_{gm} + \sum_{m \in M} c_m^{fixed} \cdot w_m + \sum_{n \in N} \sum_{l \in L} c_{nl}^{delivery} \cdot u_{nl} \qquad (21)$$

Algorithm 1. A Benders Decomposition Algorithm for an Inter-modal Transportation problem

1: Let $k = 0$ be an iteration number;
2: Solve $\mathcal{M}$ and let $\hat{x} = (\hat{h}_{nm}, \hat{x}_{gm}, \hat{u}_{nl}, \hat{y}_{ng}, \hat{w}_m, \hat{o}_g)$ be its optimal solution and $f(\hat{x})$ its objective value;
3: Let $\hat{z} = f(\hat{x})$ be the incumbent lower bound on the optimal solution of $\mathcal{P}$;
4: Let $\beta^0(\hat{x}) = 1$;
5: **while** $f(\hat{x}) + \beta^k(\hat{x}) > \hat{z}$ **do**
6:      Set $k = k + 1$;
7:      Solve $\mathcal{S}$ and let $\beta^k(\hat{x})$ be the optimal objective value;
8:      Construct a bounding function $B_m^k(\hat{h}_m)$ for each mode $m \in M_{full}^+$, where $\hat{h}_m$ are the values $\hat{h}_{nm}$ for all orders $n \in N$ assigned to mode m in $\hat{x}$;
9:      Add Optimality Cuts (37), (38) to $\mathcal{M}$;
10:     Solve $\mathcal{M}$ and let $\hat{x}_k$ be the optimal solution;
11:     Let $\hat{z}$ be equal to the objective value of the solution of $\mathcal{M}$;
12:     Set $\hat{x} = \hat{x}_k$;
13:     Let $f(\hat{x})$ be the total cost value in x;
14: Return $\hat{z}$;

$$(1)-(9) \text{ and } (11), (12) \text{ of } \mathcal{P}$$
$$u_{nl} \geq h_{nm} \qquad\qquad \forall n \in N, m \in M_{full}^+, l = dest_m \quad (22)$$

$$x_{gm}, y_{ng}, u_{nl}, h_{nm} \in \{0,1\} \qquad\qquad \forall n \in N, g \in G, m \in M, l \in L$$
$$w_m \in \{0, ..., b_m\} \qquad\qquad \forall m \in M$$

We formulate the master problem $\mathcal{M}$ by integrating the Disposition and Inter-region transportation stages and we define its objective by a continuous variable z that corresponds to a valid upper bound on the sum of travel and fixed costs over all trailers during these two stages (see Constraints (21)). We also replace variables u_{nlt} by u_{nl} that denotes just the assignment of orders to warehouses in Turkey. Constraints (1) to (9) and (11) to (12) are duplicated and Constraints (10) is replaced with Constraints (22).

Then, we add to $\mathcal{M}$ a set of redundant capacity constraints. Specifically, we adapt the valid inequalities proposed by [1] for two variants of the Capacitated Facility Location problem and induce a 'dummy' non-negative variable o_g per trailer $g \in G$, which is set equal to the total demand served by g. These inequalities are (23)–(25) and their validity is easy to show.

$$o_g \leq \sum_{n \in N} q_g \cdot y_{ng} \qquad\qquad \forall g \in G \qquad (23)$$

$$o_g = \sum_{n \in N} demand_n \cdot y_{ng} \qquad\qquad \forall g \in G \qquad (24)$$

$$\sum_{g \in G} o_g = \sum_{n \in N} demand_n \qquad\qquad (25)$$

$$o_g \in \mathbb{R}^+ \qquad\qquad \forall g \in G$$

Moreover, since the release time is not a part of the master problem, Constraints (13) are not included in the formulation of $\mathcal{M}$ and, based on the solution of $\mathcal{M}$, variables r_n are fixed to $\hat{r}_n$ as follows:

$$\hat{r}_n = \sum_{m \in M_{full}^+} \hat{h}_{nm} \cdot arrival_m \qquad\qquad \forall n \in N$$

3.2   Benders Subproblem

An MILP Formulation. We fix the solution values of the master problem, i.e., $\hat{u}_{nl}$ and $\hat{r}_n$, and we formulate the following subproblem $\mathcal{S}$:

$\mathcal{S}$:

$$\min \sum_{n \in N} i_n \cdot T_n$$

subject to:

$$\sum_{t \in T} u_{nlt} = \hat{u}_{nl} \qquad\qquad \forall n \in N, l \in L \qquad\qquad (26)$$

$$u_{nlt} = 0 \qquad\qquad \forall n \in N, l \in L, t < \hat{r}_n \qquad (27)$$

(14), (15) and (17)–(20) of $\mathcal{P}$

$$T_n \in \mathbb{N} \qquad\qquad \forall n \in N$$

$$u_{nlt}, day_{nd} \in \{0,1\} \qquad\qquad \forall n \in N, l \in L, t \in T, d \in D$$

Since the release time of each order is fixed, we apply Constraints (26) to enforce that the delivery of order n to its final destination starts at a single time slot $t \in T$ that has been already assigned by $\mathcal{M}$ (given by $\hat{u}_{nl}$). Note also that Constraints (16) can be replaced by the simple precedence Constraints (27), which ensure that no order will start its delivery before its release time. Constraints (14) to (15) and (17) to (20) of $\mathcal{P}$ are duplicated to $\mathcal{S}$.

A Constraint Programming Formulation. Due to the increased effectiveness of CP methods on scheduling problems, we propose an equivalent formulation, $\mathcal{S} - CP$, where we consider the sets N, L, D (of orders, locations and days, respectively) and the variables s_n, τ_n, day_n, denoting that start of delivery time, the tardiness (in days) and the day of delivery, for each order $n \in N$, respectively. We also consider the fixed values $\hat{u}_n \in L$ and $\hat{r}_n$, computed by the solution of $\mathcal{M}$.

Constraints (28) are global constraints (of the form $Cumulative(s, p, c, C)$) ensuring that if each order n uses a single vehicle and the maximum number of available vehicles in each warehouse l is K_l, then the start times s of orders in l, which are processed in t_{nl} time units, do not violate the number of available vehicles. Constraints (29) ensure that the delivery of each order starts after its release time. Constraints (30) ensure that, when an order n is delivered to the final destination by a warehouse l, the tardiness (in days) is defined by the difference of delivery time $(s_n + t_{nl})$ and deadline of n, while Constraints (31) ensure that squared tardiness will contribute to the objective value. Finally, Constraints (32), (33) ensure that the start time of an order delivery begins strictly during the operating hours of the warehouses in Turkey.

$\mathcal{S} - CP$:

$$\min \sum_{n \in N} i_n \cdot T_n$$

subject to:

$$Cumulative((s_n | \hat{u}_n = l), (t_{nl} | \hat{u}_n = l), 1, K_l) \qquad \forall l \in L \qquad (28)$$

$$s_n \geq \hat{r}_n \qquad \forall n \in N \qquad (29)$$

$$u_n = l \rightarrow \tau_n \geq \frac{s_n + t_{nl} - t_n^{due}}{24} \qquad \forall n \in N \qquad (30)$$

$$T_n \geq \tau_n^2 \qquad \forall n \in N \qquad (31)$$

$$day_n = d \rightarrow s_n \geq opening_d \qquad \forall n \in N \qquad (32)$$

$$day_n = d \rightarrow s_n \leq closing_d \qquad \forall n \in N \qquad (33)$$

$$T_n, s_n, \tau_n \in \mathbb{N} \qquad \forall n \in N$$

$$day_n \in D \qquad \forall n \in N$$

3.3   Adding Valid Benders Optimality Cuts

We consider two groups of variables X_k : $x_{gm}^k, y_{ng}^k, h_{nm}^k, u_{nl}^k, w_m^k, o_g^k$ and Y_k : $T_n^k, u_{nlt}^k, day_{nd}^k$ and let $f(X_k) = \sum_{g \in G} \sum_{m \in M} c_m^{travel} \cdot x_{gm}^k + \sum_{m \in M} (c_m^{fixed} \cdot w_m^k) + \sum_{n \in N} \sum_{l \in L} c_{nl}^{delivery} \cdot u_{nl}^k$, for each iteration k of Algorithm 1 and $g(Y_k) = \sum_{n \in N} i_n \cdot T_n^k$.

Let $\hat{X}_{k-1} = (\hat{h}_{nm}^{k-1}, \hat{x}_{gm}^{k-1}, \hat{u}_{nl}^{k-1}, \hat{y}_{ng}^{k-1}, \hat{w}_m^{k-1}, \hat{o}_g^{k-1})$ be an optimal solution of $\mathcal{M}$ in iteration $k-1$. Let also $N_m^{k-1} = \{n \in N | \hat{h}_{nm}^{k-1} = 1\}$ be the set of orders assigned to mode $m \in M_{full}^+$, for each $m \in M_{full}^+$. In the next iteration k, as described in Algorithm 1, the objective value of an optimal solution $\hat{Y}_k = (\hat{u}_{nlt}^k, \hat{T}_n^k, \hat{r}_n^k, \hat{day}_{nd}^k)$ of Subproblem $\mathcal{S}$ (which uses as fixed the values of $\hat{X}_{k-1}$) is denoted by $\beta^k(\hat{X}_{k-1})$. Let now $\hat{T}_m^k = \sum_{n \in N_m^{k-1}} i_n \cdot \hat{T}_n^k$ be the total tardiness cost of mode $m \in M_{full}^+$ in $\mathcal{S}$ in iteration k. Then, the objective value of $\mathcal{S}$, can be written as:

$$\beta^k(\hat{X}_{k-1}) = \sum_{m \in M_{full}^+} \hat{T}_m^k \qquad (34)$$

We now construct the bounding function, $B_m^k(\hat{h}_m^k)$ for each mode $m \in M_{full}^+$, where $\hat{h}_m^k$ are the values $\hat{h}_{nm}^k$ for all orders $n \in N$ assigned to m in $\hat{X}_k$. This function is equal to $\hat{T}_m^k$ if all orders in N_m^{k-1} will be assigned to the same mode also in iteration k, and 0 otherwise, i.e.,:

$$B_m^k(\hat{h}_m^k) = \begin{cases} \hat{T}_m^k & \text{if } \{n \in N_m^{k-1} | \hat{h}_{nm}^k = 0\} = \emptyset \\ 0 & \text{otherwise} \end{cases} \tag{35}$$

The following Lemma is crucial in order to guarantee the validity of our approach.

Lemma 1. *Let $B^k(X_k) = \sum_{m \in M_{full}^+} B_m^k(h_m^k) + f(X_k)$ and $\hat{X}_{k-1}$ be an optimal solution of $\mathcal{M}$ in iteration $k-1$. Then, the following two properties hold:*

P_1: *If $X_k = \hat{X}_{k-1}$, then $B^k(X_k) = f(\hat{X}_{k-1}) + \beta^k(\hat{X}_{k-1})$.*
P_2: *$f(X'_{k-1}) + g(Y'_k) \geq B^k(X'_k)$, for any feasible solutions X'_{k-1}, X'_k and Y'_k in iterations $k-1$, k respectively.*

Proof. For Property P_1, by replacing the variables in X_k with the corresponding solution values in $\hat{X}_{k-1}$ the set $\{n \in N_m^{k-1} | \hat{h}_{nm}^k = 0\}$ will be equal to the empty set. Thus, by (35), $\sum_{m \in M_{full}^+} B_m^k(\hat{h}_m) = \sum_{m \in M_{full}^+} \hat{T}_m^k$, which, by (34), is equal to $\beta^k(\hat{x})$. Hence, $B^k(X_k) = \beta^k(\hat{X}_{k-1}) + f(\hat{X}_{k-1}) = f(\hat{X}_{k-1})$.

For Property P_2, assume to the contrary that there is a feasible solution with values X'_k, Y'_k, for $\mathcal{M}, \mathcal{S}$ respectively, in iteration k, such that:

$$B^k(X'_k) > f(X'_{k-1}) + g(Y'_k) \tag{36}$$

If $X'_k = \hat{X}_{k-1}$, by Property P_1, $B^k(X'_k) = f(\hat{X}_{k-1}) + \beta^k(\hat{X}_{k-1})$. Since the objective value of $\mathcal{S}$ is $g(Y'_k) = \beta^k(\hat{X}_{k-1})$, we yield a contradiction. Therefore, it must hold that the solution values X'_k, Y'_k are different than the optimal solution values, $\hat{X}_{k-1}, \hat{Y}_k$, of $\mathcal{M}, \mathcal{S}$ in iteration $k-1$ and k respectively. If $X'_k \neq \hat{X}_{k-1}$, then the values h'^k_m are different from $\hat{h}_m^{k-1}$. Thus, there is at least an order $n \in N$ that is not assigned to the same mode $m \in M_{full}^+$ in iteration k, and thus the set $\{n \in N_m^{k-1} | h'^k_{nm} = 0\}$ in X'_k is not empty, $\{n \in N_m^{k-1} | h'^k_{nm} = 0\} \neq \emptyset$. Hence, by definition of our bounding function, $B_m^k(h'^k_m) = 0$, which is definitely less than or equal to the tardiness cost of mode m in Y'_k. Now, let order n be assigned to a mode $m' \neq m$ in X'_k. If the rest of the orders $n' \in N_{m'}^{k-1}$ are not displaced in X'_k, then the tardiness cost of m' must be greater or equal than $\hat{T}_{m'}^k$ in Y'_k, since the number of orders that have to be scheduled is increased, while $B_{m'}^k(h'^k_{m'}) = \hat{T}_{m'}^k$ by definition. If any order $n' \in N_{m'}^{k-1}$ is displaced in X'_k, then $\{n' \in N_{m'}^{k-1} | h'^k_{n'm'} = 0\} \neq \emptyset$, hence the value of $B_{m'}^k(h'^k_m)$ is less or equal to the tardiness cost of m' in Y'_k. Therefore, we can safely assume that $B_m^k(h'^k_m)$ is less than or equal than the new tardiness cost for all modes m and their sum is less or equal than the new total tardiness cost, i.e., $\sum_{m \in M_{full}^+} B_m^k(h'^k_{m'}) \leq g(Y'_k)$, which is equivalent to $f(X'_{k-1}) + \sum_{m \in M_{full}^+} B_m^k(h'_m) \leq f(X'_{k-1}) + g(Y'_m)$, and thus $B^k(X'_k) \leq f(X'_{k-1}) + g(Y'_k)$, contradicting (36). $\qquad\square$

Since by Lemma 1 the bounding function $B^k(\mathbf{x}_k)$ satisfies Properties $P1$ and $P2$ in each iteration k of Algorithm 1, and the domain of variables Y is finite, the following can be shown as [5, Theorem 1].

Theorem 1. *Algorithm 1, converges to the optimal value of $\mathcal{P}$ after finitely many steps.*

Now, based on the validity of our Bounding function, we integrate the following set of valid linear inequalities to the master problem $\mathcal{M}$ at iteration k.

$$z_m^r \geq \hat{T}_m^r - \hat{T}_m^r \cdot (|N_m^{r-1}| - \sum_{n \in N_m^{r-1}} h_{nm}) \qquad\qquad \forall m \in M_{full}^+, r = 1, ..., k$$

$$(37)$$

$$z \geq \sum_{m \in M_{full}^+} z_m^r + f(\mathbf{x}_r) \qquad\qquad \forall r = 1, ..., k \qquad (38)$$

where z_m^r are positive real variables for all $m \in M_{full}^+$. Constraints (37) ensure that if at least one order $n \in N_m^{r-1}$ is not assigned to m in iteration r, z_m^r will be equal to 0. On the contrary, if all orders are placed to the same warehouse again, z_m^r will be equal to $\hat{T}_m^r$. Constraints (38) ensure that the objective value of $\mathcal{M}$ will take into account the lower bound of the subproblem computed by the bounding function.

The iterative addition of (37) and (38) to $\mathcal{M}$ will converge to the optimal solution after a finite number of iterations, as it is proved by Theorem 1.

3.4   Subproblem Relaxation

Adding valid cuts to $\mathcal{M}$ is proved to be adequate for achieving optimality after a number of iterations, however converging to optimality is usually a long process. To fasten our algorithm a simple but quite effective valid subproblem relaxation $\mathcal{R}$ is added to $\mathcal{M}$, assuming that the number of single-driver vans in the warehouses of Turkey $l \in L$ is infinite, and thus each order n is delivered immediately after being unloaded.

$\mathcal{R}$:

$$z_n \geq \sum_{m \in M_{full}^+} \lfloor 1 + \tfrac{arrival_m + t_{nl} - t_n^{due}}{24} \rfloor \cdot \lfloor |1 + \tfrac{arrival_m + t_{nl} - t_n^{due}}{24} |\rfloor \cdot h_{nm} \; \forall n \in N$$

$$(39)$$

$$z \geq \sum_{n \in N} i_n \cdot z_n + f(\mathbf{x}) \qquad\qquad (40)$$

Note that $\mathcal{R}$ is a valid relaxation of $\mathcal{S}$, because the feasible solutions of T_n are a subset of the feasible solutions of z_n, and thus the corresponding squared tardiness is a lower bound on T_n. Constraints (39) ensure that the z_n is equal to the squared tardiness (in days), if we assume that each order is handled immediately after being released and Constraints (40) updates the objective function, adding also the tardiness costs.

4 Computational Work

The experiments are run on the Linux server (4 processors, 3.3 GHz CPU, 12 GB RAM) using CPLEX 12.10 (Python API for MILP and CP Optimizer in OPL for CP). We present the results for 20 consecutive weeks, from June 2016 to November 2016 and we impose a time limit of 1800, 3600 and 7200 s on the MILP and the master problem $\mathcal{M}$.

Each dataset concerns orders received during an entire week (i.e., 168 h). The number of orders ranges from 41 to 153, the number of pickup and customs companies from 61 to 201 and the number of pickup routes ($m \in M^-$) from 96 and 810. The number of delivery modes $m \in M^+$ equals 131 in all data sets. The fixed cost c_m^{fixed} lays in the range of €891-2179 and the travel cost of each trailer c_m^{travel} in €34-337. All trailers have a fixed capacity q_g of 27000 payweight units and the capacity of modes in number of trailers is equal to either 10 or 20 for roadway modes, 30 for railway modes and 60 or 100 for seaway shipments. Roadway transports can be carried out by either single-driver' vehicles (imposing a break of 9 h after each 9-h continuous travel) or by double-driver' vehicles (each driver is resting during the 9 h of the other one's shift). The importance weight i_n of orders is between 1 and 50, multiplied by the cost of 100 €. Last, the number of single-driver vans K_l is 15 per warehouse in Turkey.

Table 2. Weekly datasets results - 0.5 h limit

#	GAP (%)				TIME (s)			
	MILP	BENDERS (w/o REDIN)	BENDERS (w REDIN)	BENDERS (CP)	MILP	BENDERS (w/o REDIN)	BENDERS (w REDIN)	BENDERS (CP)
1	4.5	0.0	0.0	0.0	1813	141	30	28
2	-	22.1	3.2	3.2	-	1867	1867	1839
3	0.1	0.0	0.0	0.0	1815	85	31	26
4	-	-	2.1	2.1	-	-	1905	1866
5	-	8.9	6.6	6.6	-	1863	1863	1786
6	-	8.7	4.6	4.6	-	1853	1854	1847
7	-	10.8	1.8	1.8	-	1864	1865	1857
8	0.0	0.0	0.0	0.0	181	91	73	60
9	4.7	1.1	0.0	0.0	1845	1846	274	88
10	-	10.2	2.6	2.6	-	1882	1884	1863
11	6.3	3.6	0.0	0.0	1855	1857	1547	1403
12	-	8.0	3.0	3.0	-	2008	2008	1868
13	-	5.0	0.3	0.3	-	1920	1877	702
14	-	6.6	3.8	3.8	-	1955	1957	1865
15	9.4	4.5	3.3	3.3	1904	1909	1910	1903
16	9.1	3.8	3.0	3.0	1902	1934	1928	1852
17	-	7.1	3.2	3.2	-	1911	2005	1899
18	-	5.0	0.7	0.7	-	1963	1956	1833
19	-	4.4	1.3	1.3	-	1890	1938	1855
20	5.2	3.3	1.2	1.2	1977	2037	2041	1980

The number of variables ranges from 10^5 to $7.2 \cdot 10^5$ and the number of constraints between $4.0 \cdot 10^5$ and $6.0 \cdot 10^6$, i.e., instances are large compared to the ones that reported in the literature.

The results displayed in Tables 2 and 3 concern the solution of the original MILP (MILP) against the proposed decomposition approach with an MILP-formulated subproblem (BENDERS (w/o REDIN)), the same approach after adding the cuts (21)–(23) (BENDERS (w REDIN)) and, finally, the same with the cuts but with a CP formulation for the subproblem (BENDERS (CP)).

If the GAP value is more than 0.0, the time limit has been reached without achieving optimality. Symbol '-' denotes that no feasible solution was found. For decomposition algorithms, the time limit is imposed on the solution of the master problem $\mathcal{M}$. Since the solution of $\mathcal{M}$ is a valid lower bound of the optimal solution and the solution of the subproblem is an upper bound, we can safely assume that the computed GAP, equal to $\frac{S\ Solution - \mathcal{M}\ BestBound}{S\ Solution}$, is the worst possible optimality gap. As for the problem instances that could not be solved to optimality within the allotted time, the difference of TIME from the time limit is the duration of the preprocessing (and the subproblem, for the decomposition cases). Note that due to the complexity of the master problem, if the solution of the decomposition terminates strictly in 1800 s, the algorithm would never proceed to the subproblem. A GAP (%) that is exclusively computed by the master should not be compared with the GAP (%) of the MILP, since the incumbent solution of the master is not a valid upper bound on the solution of the overall problem. On the other hand, if we allow full MIP to run for equal time (i.e., an extra 180 s, for Instance 20) with Benders decomposition, as we can note, the results will be the same. Regarding the CP-styled variant, the master problem is identical for the Benders (w RedIn) and Benders (CP) experiments. Since the time limit is imposed on the master and the subproblem is solved optimally in both experiments, the GAP (%) is expected to remain identical as well. The supremacy of CP variant comes from shortening the (already quick) subproblem optimization. We believe that CP could be even more valuable in larger instances or in instances with tighter deadlines.

Table 3. Weekly datasets results - 1 h limit

#	GAP (%)				TIME (s)			
	MILP	BENDERS (W/O REDIN)	BENDERS (W REDIN)	BENDERS (CP)	MILP	BENDERS (W/O REDIN)	BENDERS (W REDIN)	BENDERS (CP)
1	4.3	0.0	0.0	0.0	3613	142	29	31
2	-	11.0	3.2	3.2	-	3674	3736	3633
3	0.0	0.0	0.0	0.0	1884	89	39	35
4	-	10.0	2.1	2.1	-	3798	3776	3676
5	-	8.8	6.6	6.6	-	3670	3686	3606
6	8.9	8.7	4.6	4.6	3655	3657	3664	3620
7	7.4	8.9	1.8	1.8	3666	3670	3865	3604
8	0.0	0.0	0.0	0.0	213	93	100	97
9	4.7	1.1	0.0	0.0	3649	3648	292	294
10	-	6.8	2.4	2.4	-	3690	3762	3645
11	6.3	3.6	0.0	0.0	3657	3663	2971	2853
12	-	10.1	3.0	3.0	-	3833	3869	3771
13	-	4.7	0.2	0.2	-	3772	3726	3686
14	-	6.0	3.8	3.8	-	3771	3802	3749
15	11.9	4.5	3.3	3.3	3712	3721	3731	3692
16	7.7	3.6	2.9	2.9	3754	3755	3761	3680
17	-	6.7	3.0	3.0	-	3780	3869	3793
18	9.9	4.7	0.6	0.6	3766	3768	3775	3694
19	9.7	4.0	1.3	1.3	3732	3724	3802	3701
20	3.9	2.8	0.3	0.3	3829	3747	3766	3688

These results show that the initial MILP formulation fails to compute a feasible solution for several datasets, although this is less frequent as the time limit is increased. Interestingly, despite the high complexity of the mathematical formulation, the structure of the problem and the reasonable assumption that each order is served by the nearest depot (thus variables y_{ng} and h_{nm} become 0, for all trailers g not dedicated to the nearest depot of order n and all modes that originate from a different depot respectively) helps MILP to feasible solve with quite good GAPS (%) some of our datasets (Table 3). The Benders Decomposition method computes solutions for all datasets, while the optimality gap is not remarkably affected by the extension of the time limit. The addition of the redundant capacity constraints to the master problem achieves lower optimality gaps, again irrespective of the time limit. Finally, the CP formulation of the subproblem solves the scheduling problem in a few seconds, i.e., quite faster than the MILP formulation. Let us conclude by noting that increasing the time limit to 2 h had no significant impact thus the corresponding results are omitted.

5   Conclusions

This study is motivated by a large inter-modal transportation problem that includes a pickup stage, a packing and shipment process and a last-mile delivery to the final recipients. As the MILP formulation of the entire problem is not computationally effective, we employ a Logic-Based Benders Decomposition, consider redundant constraints that tighten the master problem and a CP-styled formulation of the subproblem. The successful application of this method on real datasets of large size justify our approach.

A more general case would be to consider that each pick-up point could be served by any depot; then, the disposition routes might increase drastically but the total cost might be reduced. Moreover, lifting the assumption that each last-mile delivery vehicle handles each order individually augments the problem with a vehicle routing formulation. Both directions of future work could benefit by looking at alternative decompositions and considering the use of additional families of cutting planes.

References

1. Aardal, K.: Reformulation of capacitated facility location problems: how redundant information can help. Ann. Oper. Res. **82**, 289–308 (1998)
2. Ahkamiraad, A., Wang, Y.: Capacitated and multiple cross-docked vehicle routing problem with pickup, delivery, and time windows. Comput. Ind. Eng. **119**, 76–84 (2018)
3. Azizi, V., Hu, G.: Multi-product pickup and delivery supply chain design with location-routing and direct shipment. Int. J. Prod. Res. **226**, 107648 (2020)
4. Benders, J.F.: Partitioning procedures for solving mixed-variables programming problems. Numer. Math. **4**, 238–252 (1962)
5. Hooker, J.N.: Planning and scheduling by logic-based benders decomposition. Oper. Res. **55**(3), 588–602 (2007)
6. Hooker, J.N., Ottosson, G.: Logic-based Benders decomposition. Math. Program. A **96**, 33–60 (2003)
7. Li, J., Li, Y., Pardalos, P.M.: Multi-depot vehicle routing problem with time windows under shared depot resources. J. Comb. Optim. **31**(2), 515–532 (2014). https://doi.org/10.1007/s10878-014-9767-4

8. Mahéo, A., Kilby, P., Hentenryck, P.V.: Benders decomposition for the design of a hub and shuttle public transit system. Transp. Sci. **53**(1), 77–88 (2019)
9. Moccia, L., Cordeau, J.-F., Ropke, S., Valentini, M.P.: Modeling and solving a multimodal transportation problem with flexible-time and scheduled services. Networks **57**(1), 53–68 (2011)
10. Rahmaniani, R., Crainic, T.G., Gendreau, M., Rei, W.: The Benders decomposition algorithm: a literature review. Eur. J. Oper. Res. **259**(3), 801–817 (2017)
11. Rousseau, L.-M., Gendreau, M., Pesant, G., Focacci, F.: Solving VRPTWs with constraint programming based column generation. Ann. Oper. Res. **130**, 199–216 (2004)
12. Wu, T.-H., Low, C., Bai, J.-W.: Heuristic solutions to multi-depot location-routing problems. Comput. Oper. Res. **29**, 1393–1415 (2002)
13. Xiong, G., Wang, Y.: Best routes selection in multimodal networks using multi-objective genetic algorithm. J. Comb. Optim. **28**(3), 655–673 (2012). https://doi.org/10.1007/s10878-012-9574-8
14. Yu, V.F., Jewpanya, P., Redi, A.A.N.P.: Open vehicle routing problem with cross-docking. Comput. Ind. Eng. **94**, 6–17 (2016)

Checking Constraint Satisfaction

Victor Jung and Jean-Charles Régin[✉][iD]

Université Côte d'Azur, CNRS, I3S, Nice, France
{victor.jung,jean-charles.regin}@univ-cotedazur.fr

Abstract. We address the problem of verifying a constraint by a set of solutions S. This problem is present in almost all systems aiming at learning or acquiring constraints or constraint parameters. We propose an original approach based on MDDs. Indeed, the set of solutions can be represented by the MDD denoted by MDD_S. Checking whether S satisfies a given constraint C can be done using $MDD(C)$, the MDD that contains the set of solutions of C, and by searching if the intersection between $MDD(S)$ and $MDD(C)$ is equal to $MDD(S)$. This step is equivalent to searching whether $MDD(S)$ is included in $MDD(C)$. Thus, we give an inclusion algorithm to speed up these calculations. Then, we generalize this approach for the computation of global constraint parameters satisfying C. Next, we introduce the notion of properties on the MDD nodes and define a new algorithm allowing to compute in only one step the set of parameters we are looking for. Finally, we present experimental results showing the interest of our approach.

Keywords: Multi-valued decision diagram · Inclusion · Constraint learning

1 Introduction

Many works in Constraint Programming try to improve the quality of a model by adding new implicit constraints [11], redundant constraints [4] or global constraints [8]. All these works face a common problem: the verification of the satisfaction of constraints by a given set of solutions. Some choose a brute force approach [8,11], others prefer a more specific but ad-hoc approach [2]. In all cases, these methods go through the solutions to test if they satisfy constraints. Constraints are not necessarily tested individually, but the solutions can be considered one after the other.

In this paper, we propose a more global and efficient method to test whether a set of solutions verifies one or a set of constraints. Multi-valued decision diagrams (MDDs) are a very efficient data structure to represent a set of solutions in a compressed way and for which many operators are available to combine MDDs without decompressing them. We therefore propose to use $MDD(S)$ the MDD which corresponds to the set of solutions S. We show that we can simply test if S satisfies a constraint C, by using $MDD(C)$, the MDD that represents

© Springer Nature Switzerland AG 2021
P. J. Stuckey (Ed.): CPAIOR 2021, LNCS 12735, pp. 332–347, 2021.
https://doi.org/10.1007/978-3-030-78230-6_21

the solutions of C, and then by performing the intersection between $MDD(S)$ and $MDD(C)$. This method is simple to implement since it only requires constructing the two MDDs and performing their intersection, for which efficient algorithms are available, and then testing whether the resulting MDD is similar to $MDD(S)$. However it has an important flaw: it will create and calculate a MDD even if the intersection will not be equal to $MDD(S)$. It therefore risks doing many operations unnecessarily. To avoid this we introduce an inclusion operator between MDDs since this is what we want to test: is $MDD(S)$ included in $MDD(C)$?

This operator is efficient when it is a question of verifying a precise and unique constraint, but not very efficient when it is a question of searching for the parameters of a constraint such that the resulting constraint is satisfied by a set of solutions. Finding the parameters of a global constraint so that it is satisfied by a set of solutions is a recurrent problem at present [2,4,8, 11,12]. To solve this problem we propose to work with $MDD(S)$ which we enrich by introducing the notion of node properties. Then, a process called "the parent-child propagation of the parameters" is performed through the MDD. More precisely, the global constraint is expressed by properties including the parameters and these properties are propagated in $MDD(S)$ in order to compute for each sub-tree of the MDD those which are compatible with the constraint. Thus, we determine the most restrictive parameters of the constraint that are satisfied by the MDD.

This article is organized as follows. First, we give some basic definitions. Then, we present a general scheme to check the satisfaction of a constraint and improve it by defining a new operation between MDDs. Next, we address the problem of finding parameters of global constraints by introducing the notion of node properties. Finally, we provide benchmarks and results testing the different approaches described in this article, and we conclude.

2   Preliminaries

2.1   Constraint Programming

A finite constraint network $\mathcal{N}$. is defined as a set of n variables $X = \{x_1, \ldots, x_n\}$, a set of current domains $\mathcal{D} = \{D(x_1), \ldots, D(x_n)\}$ where $D(x_i)$ is the finite set of possible values for variable x_i, and a set C of constraints between variables. We introduce the particular notation $\mathcal{D}_0 = \{D_0(x_1), \ldots, D_0(x_n)\}$ to represent the set of initial domains of $\mathcal{N}$ on which constraint definitions were stated. A constraint C on the ordered set of variables $X(C) = (x_{i_1}, \ldots, x_{i_r})$ is a subset $T(C)$ of the Cartesian product $D_0(x_{i_1}) \times \cdots \times D_0(x_{i_r})$ that specifies the allowed combinations of values for the variables $x_{i_1}, \ldots, x_{i_r}$. An element of $D_0(x_{i_1}) \times \cdots \times D_0(x_{i_r})$ is called a tuple on $X(C)$. A value a for a variable x is often denoted by (x, a). Let C be a constraint. A tuple τ on $X(C)$ is valid if $\forall (x, a) \in \tau, a \in D(x)$. C is consistent iff there exists a tuple τ of $T(C)$ which is valid. A value $a \in D(x)$ is consistent with C iff $x \notin X(C)$ or there exists a valid tuple τ of $T(C)$ with

$(x, a) \in \tau$. We denote by $\#(a, \tau)$ the number of occurences of the value a in a tuple τ.

We present some constraints that we will use in the rest of this paper.

A global cardinality constraint (GCC) constrains the number of times every value can be taken by a set of variables. This is certainly one of the most useful constraints in practice. Note that the ALLDIFF constraint corresponds to a GCC in which every value can be taken at most once.

Definition 1. *A* **global cardinality constraint** *is a constraint C in which each value $a_i \in D(X(C))$ is associated with two positive integers l_i and u_i with $l_i \leq u_i$ defined by*
$\mathrm{GCC}(X, l, u) = \{\tau | \tau$ *is a tuple on $X(C)$ and $\forall a_i \in D(X(C)) : l_i \leq \#(a_i, \tau) \leq u_i\}$.*

Definition 2. *Given X a set of variables, l and u two integers with $l \leq u$ and V a set of values. The* **among** *constraint ensures that at least l variables of X and at most u will take a value in V, that is*
$\mathrm{AMONG}(X, V, l, u) = \{\tau \mid \tau$ *is a tuple on $X(C)$ and $l \leq \sum_{a \in V} \#(a, \tau) \leq u\}$*

This constraint has been introduced in CHIP [1].

The SEQUENCE constraint [1] is a conjunction of sliding AMONG constraints.

Definition 3. *Given X a set of variables, q, l and u three integers with $l \leq u$ and V a set of values. The* **sequence** *constraint holds if and only if for $1 \leq i \leq n_q + 1$ $\mathrm{AMONG}(\{x_i, ..., x_{i+q-1}\}, V, l, u)$ holds. More precisely*
$\mathrm{SEQUENCE}(X, V, q, l, u) = \{\tau \mid \tau$ *is a tuple on $X(C)$ and for each sequence S*
of q consecutive variables: $l \leq \sum_{v \in V} \#(v, \tau, S) \leq u\}$.

2.2 Multi-valued Decision Diagram

The decision diagrams considered in this paper are reduced, ordered multi-valued decision diagrams (MDD) [3,7,13], which are a generalization of binary decision diagrams [5]. They use a fixed variable ordering for canonical representation and shared sub-graphs for compression obtained by means of a reduction operation. An MDD is a rooted directed acyclic graph (DAG) used to represent some multi-valued functions $f : \{0...d - 1\}^n \rightarrow true, false$. Given the n input variables, the DAG contains $n + 1$ layers of nodes, such that each variable is represented at a specific layer of the graph. Each node on a given layer has at most d outgoing arcs to nodes in the next layer of the graph. Each arc is labeled by its corresponding integer. The arc (u, a, v) is from node u to node v and labeled by a. Sometimes it is convenient to say that v is a child of u. All outgoing arcs of the layer n reach tt, the true terminal node (the false terminal node is typically omitted). There is an equivalence between $f(a_1, ..., a_n) = true$ and the existence of a path from the root node to the tt whose arcs are labeled $a_1, ..., a_n$.

The reduction of an MDD is an important operation that may reduce the MDD size by an exponential factor. It consists in removing nodes that have no successor and merging equivalent nodes, i.e. nodes having the same set of neighbors associated with the same labels. This means that only nodes of the same layer can be merged.

MDD of a Constraint. Let C be a constraint defined on $X(C)$. The MDD associated with C, denoted by $MDD(C)$, is an MDD modeling the set of tuples of C. More precisely, $MDD(C)$ is defined on $X(C)$, such that the arc labels of the layer of the variable x correspond to values of x, and a path of $MDD(C)$ (that is a path from the root node to the tt node) where a_i is the label of layer i corresponds to a tuple $(a_1, ..., a_n)$ on $X(C)$.

Operators. We reproduce here the description of the generic Function APPLY [9,10] because we will see that the inclusion can be easily modelled thanks to it[1]. From the MDDs mdd_1 and mdd_2 it computes $mdd_r = mdd_1 \oplus mdd_2$, where $\oplus$ is union, intersection, difference, symmetric difference, complementary of union and complementary of intersection. Function APPLY is mainly based on the possible combinations of labeled arcs. It proceeds by associating nodes of the two MDDs operands. Each node x of the resulting MDD is associated with a node x_1 of the first MDD and a node x_2 of the second MDD represented by a pair (x_1, x_2). First, the root is created from the two roots. Then, the layers are successively built. From the nodes of layer $i - 1$ the nodes of layer i are built as follows. For each node $x = (x_1, x_2)$ of layer $i - 1$, the arcs outgoing from nodes x_1 and x_2 and labeled by the same value v are considered. We recall that there is only one arc leaving a node x with a given label. Thus, there are four possibilities depending on whether there are y_1 and y_2 such that (x_1, v, y_1) and (x_2, v, y_2) exist or not. The action that is performed for each of these possibilities will define the operation performed for the given layer. For instance, a union is defined by creating a node $y = (y_1, y_2)$ and an arc (x, v, y) each time one of the arcs (x_1, v, y_1) or (x_2, v, y_2) exists. An intersection is defined by creating a node $y = (y_1, y_2)$ and an arc (x, v, y) when both arcs (x_1, v, y_1) and (x_2, v, y_2) exist. Thus, these operations can be simply defined by expressing the condition for creating a node and an arc.

Function APPLY, given in Algorithm 1 takes as parameters the two MDDs, two arrays op, V having as many elements as layers, and $typeOp$ the operation type (i.e. intersection, union...). For each layer i, $op[i]$ contains 4 entries, each one representing the fact that we create an arc or not for a combination of arc existence in the two MDDs and $V[i]$ represents the set of values needed by the complementary set. If it is equal to nil then $V[i]$ will be equal to the union of the values of the neighbors of the considered nodes. At the end the resulting MDD is reduced by calling PREDUCE algorithm [9].

The values of $op[i]$ defining the binary operations are defined as follows for the different combinations:

---

[1] Unlike Perez and Régin [9], the complementary of an MDD M is computed by making the difference between the universal MDD and M. This avoids the need of a dedicated algorithm.

	op[0] $\neg a1 \wedge \neg a2$		op[1] $\neg a1 \wedge a2$		op[2] $a1 \wedge \neg a2$		op[3] $a1 \wedge a2$	
Layer	[1..r-1]	r	[1..r-1]	r	[1..r-1]	r	[1..r-1]	r
$A \cap B$	F	F	F	F	F	F	T	T
$A \cup B$	F	F	T	T	T	T	T	T
$A - B$	F	F	F	F	T	T	T	F

Algorithm 1: Generic Apply Function.

APPLY($mdd_1, mdd_2, op, V, typeOp$): MDD
  // $L[i]$ is the set of nodes in layer i.
  $root \leftarrow$ CREATENODE($root(mdd_1), root(mdd_2)$)
  $L[0] \leftarrow \{root\}$
  for each $i \in 1..r$ **do**
    $L[i] \leftarrow \emptyset$
    for each *node* $x \in L[i-1]$ **do**
      get x_1 and x_2 from $x = (x_1, x_2)$
      if $V[i] = nil$ **then** $V[i] \leftarrow$ VALUES($\omega^+(x_1) \cup \omega^+(x_2)$)
      for each $v \in V[i]$ **do**
        if $\nexists(x_1, v, y_1) \in \omega^+(x_1)$ **then**
          if $\nexists(x_2, v, y_2) \in \omega^+(x_2) \wedge op[0]$ **then** CREATEARC($L, i, x, v, w[i]$)
          if $\exists(x_2, v, y_2) \in \omega^+(x_2) \wedge op[1]$ **then** ADDARCANDNODE(L, i, x, v, nil, y_2)
        else
          if $\nexists(x_2, v, y_2) \in \omega^+(x_2) \wedge op[2]$ **then** ADDARCANDNODE(L, i, x, v, y_1, nil)
          if $\exists(x_2, v, y_2) \in \omega^+(x_2) \wedge op[3]$ **then** ADDARCANDNODE(L, i, x, v, y_1, y_2)
        if $typeOp = Inclusion$ **then**
          if $\exists(x_1, v, y_1) \in \omega^+(x_1) \wedge \nexists(x_2, v, y_2) \in \omega^+(x_2)$ **then** return *false*

  if $typeOp = Inclusion$ **then** return *true*
  merge all nodes of $L[r]$ into t
  PREDUCE(L)
  return *root*

ADDARCANDNODE(L, i, x, y_1, v, y_2)
  if $\nexists y \in L[i]$ s.t. $y = (y_1, y_2)$ **then**
    $y \leftarrow$ CREATENODE(y_1, y_2)
    add y to $L[i]$
  CREATEARC(L, i, x, v, y)

---

3   Checking Constraint Satisfaction

A first solution to test whether a set of solutions S satisfies a constraint C is to represent S by an MDD, denoted $MDD(S)$, then use $MDD(C)$ the MDD of the constraint C and calculate the intersection between $MDD(S)$ and $MDD(C)$. If this intersection is equal to $MDD(S)$ then this means that all solutions of S satisfy the constraint C. The proof of the soundness of this approach is quite immediate: an MDD is a set of solutions, so if the intersection does not modify $MDD(S)$ then it means that any solution of $MDD(S)$ is also a solution of $MDD(C)$ and therefore this solution satisfies the constraint.

This approach is not particularly efficient, because it systematically requires the intermediate calculation of an intersection between MDDs. However, we are not interested in this intersection[2]. What matters is to know whether this intersection is similar to the initial MDD. We can reduce what we are trying to do in a single step: we check a relation of inclusion. Indeed, answering the question $MDD(S) \cap MDD(C) = MDD(S)$? is equivalent to answering the following question: $MDD(S) \subseteq MDD(C)$? As this operator does not exist in the literature, we propose to create it.

3.1   Operator of Inclusion

The inclusion operator between MDDs is easily done using the generic function APPLY. Let's consider that we want to know if MDD_1 is included in MDD_2. We use the same rules as the intersection operator with a notable exception: if an edge a is in MDD_1 but not in MDD_2, then we end the algorithm by returning false. In this case there is at least one solution in MDD_1 which is not in MDD_2 so MDD_1 cannot be included in MDD_2. For the three other cases we can easily find those of the intersection. Using the terminology of the preliminaries, we have clearly: $\neg a_1$ implies that no arc is created and $a_1 \wedge a_2$ implies an arc creation, since the solutions are common.

We notice that it is not necessary to keep in memory the MDD that is built. Indeed, we just need to know if we can create each new level. To do this, only the last level that has just been built is useful and must be kept in memory, the others being no longer useful can be destroyed. The reduction of the built MDD is no longer useful either since we are only interested in the ability to build an MDD from the root to tt. Function APPLY must therefore return true instead of performing the reduction at the end. This allows us to save time compared to the previous method.

4   Inferring Parameters of Global Constraints

The inclusion operator allows to answer in an efficient way to the question of the satisfaction of a constraint by a set of solutions S. However, in practice, the question that is often asked is more general: given a global constraint C involving a set of parameters, for which parameters S satisfies this constraint?

Let us consider P a set of parameters and $C(P)$ a constraint defined using these parameters. Formally, we can present the problem in the following way: What are the sets P such that $\forall s \in S$, s satisfies $C(P)$?

A first way to proceed is to check for each set of parameters P if we have $MDD_S \subseteq MDD_{C(P)}$.

We propose to add additional information, called properties, to each node of an MDD. This idea has similarities with the scheme introduced by J. Hooker et al. [6]. This information is used to memorize the valid parameters from the root to the node in relation to the constraint under consideration. When we

---

[2] We could also perform the intersection between $MDD(S)$ and the negation of $MDD(C)$ and check whether it is empty or not. However the computation of the negation is required so it does not improve the classical intersection.

will reach tt, we will know the parameters that are checked by all the solutions. In addition, retaining all the parameter sets is superfluous and we can be satisfied with retaining the more restrictive parameters. Other acceptable parameters may be derived from these restrictive parameters. For example, if the constraint SEQUENCE$(X, V, q = 3, l = 1, u = 2)$ is satisfied then the constraints SEQUENCE$(X, V, q = 3, l = 0, u = 2)$, SEQUENCE$(X, V, q = 3, l = 1, u = 3,)$ and SEQUENCE$(X, V, q = 3, l = 0, u = 3)$ are also satisfied. So, the more restrictive parameters are $(q = 3, l = 1, u = 2)$.

We present on an example the ideas of our algorithm. We will use the binary representation of a sequence constraint. Indeed, for a sequence constraint, we can abstract any X and V into a binary problem with $V = \{1\}$. If $x_i = 1$ then it means that x_i takes its value in V $(x_i \in V)$ otherwise we have $x_i=0$. So we are in the presence of only binary variable and we are looking for the parameter values (q, u, l) which are satisfied by S.

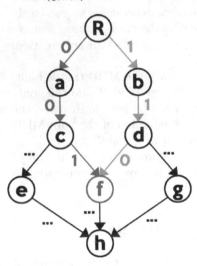

Fig. 1. Sequences for $MDD(S)$ (Color figure online)

Consider the red node f in $MDD(S)$ (Fig. 1). There are two paths to access this node from the root: take 0, 0 then 1 (which corresponds to the blue path -left- on the diagram) or take 1, 1 then 0 (which corresponds to the path in green -right-). In total, there are 5 sub-MDDs (i.e. smaller MDDs contained in the main MDD) having this red node f as the terminal node: 1 having as starting point the general root ($\{0\text{-}0\text{-}1, 1\text{-}1\text{-}0\}$), then 2 having as root the two nodes of the first layer ($\{0\text{-}1\}, \{1\text{-}0\}$) and finally 2 having as root the two nodes of the second layer ($\{1\}, \{0\}$). Strictly speaking, there are actually 6 sub-MDDs, since the MDD consisting only of the red node f exists.

Properties (i.e. satisfied sequences) are added to nodes:

- **Node R.** This node contains only the basic information, i.e. $(q = 0, l = 0, u = 0)$, since the only way to reach this node is to start from it and take no edge. This is the basic case, viable for all nodes.
- **Node a (blue).** In addition to the basic case, it is possible to reach this node starting from the root and taking the value 0. As we take 0 times the value 1, the satisfied SEQUENCE is $(q = 1, l = 0, u = 0)$.
- **Node b (green).** Same as for the previous node, except that we take once the value 1. The satisfied SEQUENCE is therefore $(q = 1, l = 1, u = 1)$.
- **Node c (blue).** We retrieve the information from the parent node (there is only one here). So we have $(q = 0, l = 0, u = 0)$ and $(q = 1, l = 0, u = 0)$.

There is only one edge that can be traversed, with a value of 0. If we add 0 to the preceding satisfied sequences, we obtain $(q = 1, l = 0, u = 0)$ and $(q = 2, l = 0, u = 0)$. We retain these sequences.

- **Node d (green).** By the same reasoning, we obtain $(q = 1, l = 1, u = 1)$ and $(q = 2, l = 2, u = 2)$.
- **Node f (red).** We start by looking at parent Node c (blue). By adding the fact that we can reach the red node f by taking the value 1, we have $(q = 1, l = 0, u = 1)$, $(q = 2, l = 0, u = 1)$ and $(q = 3, l = 1, u = 1)$. In the same way, looking at the side of parent Node d (green), we obtain $(q = 1, l = 0, u = 1)$, $(q = 2, l = 1, u = 2)$ and $(q = 3, l = 2, u = 2)$. We notice that SEQUENCE constraints of size 2 and size 3 are not compatible. In this case, the union of the two SEQUENCE constraints is performed (since both are satisfied). We thus obtain $(q = 1, l = 0, u = 1)$, $(q = 2, l = 0, u = 2)$ and $(q = 3, l = 1, u = 2)$. We can then check that for each path leading to the red node f, we take between 0 and 2 times the value 1 for a path of size 2, and between 1 and 2 times the value 1 for a path of size 3.

Figure 2 shows a slightly more complete example.

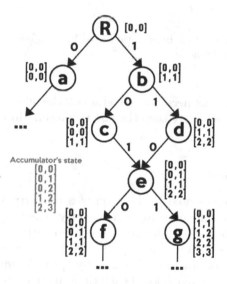

Fig. 2. Satisfied sequence constraints for each node. Value q corresponds to the index in the array associated with a node.

However, one thing is noticeable: it is possible to lose information (Fig. 3). For example, the node b contains the information $(q = 0, l = 0, u = 0)$, $(q = 1, l = 0, u = 0)$ and $(q = 2, l = 1, u = 1)$, but we can see that for paths of size 1 belonging to the MDD it is possible to take between 0 and 1 times the value 1. To solve this problem, we make the union of all the nodes of a layer (for each layer) that we store in an accumulator. This operation can be performed at the same time as the information is constructed.

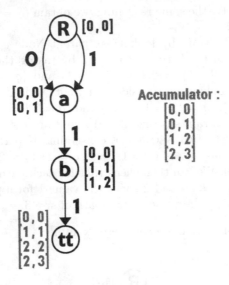

Fig. 3. Simple example to show information loss. Value q correspond to the index in the array associated with a node.

We remind that it is not necessary to retain *all* the information as represented on the diagram: we can simply keep the last two layers since the construction is done in a sliding way.

4.1 Implementation

The information is represented in the form of a property associated with each node. Each constraint will define its own property. The propagation of information from the MDD nodes is performed using a breadth-first approach, because we need all parent nodes to be correctly defined before propagating to the children. A depth-first approach would be strictly speaking impossible because all children would have to be re-explored each time the parents are updated (which is highly inefficient). During propagation, one looks to see if the child is already carrying a property or not. Two cases are possible.

- The child's property is already defined. In this case, the information already present must be merged with the new information provided by the parent. Function MERGEWITHPROPERTY of a property is in charge of this.
- The child property is not defined. In this case we simply create new information based only on the parents' information. Function CREATEPROPERTY of the property performs this operation.

Algorithm 2 is a possible implementation of this mechanism. The important Functions are MERGEWITHPROPERTY and CREATEPROPERTY. They depend on the type of constraint that the property represents, so it is difficult to define a general way to represent the information. Technically, Function CREATEPROPERTY is quite simple. We believe that the real difficulty lies in the definition of Function MERGEWITHPROPERTY, because the information must be complete and valid, i.e. it must represent the state of the constraint in a correct way for the node that contains it at any moment of the propagation.

Algorithm 2: PROPAGATION

PROPAGATEPROPERTYMDD, property
    MDD.*root.addProperty*(property)
    for each *layer L in MDD* **do**
        for each *node in L* **do**
            TRANSFERPROPERTY(*node*)
    return MDD.*tt.getProperty*().*getResult*()

TRANSFERPROPERTY(*node*) **for each** (*label, child*) *in node.children* **do**
    if *child.hasProperty*() **then**
        child.getProperty().MERGEWITHPROPERTY(node.*getProperty*(), *label*)
    else *child.addProperty*(node.*getProperty*().CREATEPROPERTY(*label*))

Time Complexity: in $O((|A| - |N|) \times$ O(MERGEWITHPROPERTY) + $|N| \times$ O(CREATEPROPERTY)), where $|A|$ is the number of edges and $|N|$ the number of nodes. Function CREATEPROPERTY is called only once for each node, that is $|N|$ times globally. Function MERGEWITHPROPERTY is globally called $|A| - |N|$ times. O(MERGEWITHPROPERTY) and O(CREATEPROPERTY) are the time complexity of the functions MERGEWITHPROPERTY and CREATEPROPERTY. As these functions depend on the constraint and the implementation, we can't give any further details.

Space Complexity. We retain (i+1) information for each node of layer i. The space complexity thus depends both on the layer where we are located (the last two layers in reality) and on the number of nodes present in the layer. As it is not possible to predict the number of nodes in a layer for any MDD, the complexity remains rather vague.

Let L_i be the number of nodes in the layer i and L the number of layers. The space complexity is in: $O(\text{MAX}(i \times |L_i| + (i + 1) \times |L_{i+1}|))$, $\forall i$ s.t. $0 \leq i < L$.

4.2 Properties Definitions

We provide the definition of properties for sum ($\sum_{x \in X} x = [min, max]$), sequence, and cardinality constraints as examples. These constraints have been chosen in relation to the problems we are working on, and not in relation to the difficulty of implementation. These are also quite common constraints.

Sum Constraint. The sum property is quite simple to model (Algorithm 3). As the result is an interval, we just need to use a pair (min, max) to represent the data. Switching from a parent to a child is simply the addition of an interval with an integer (the value of the arc), and the merge operation is a simple union between two intervals. Each node contains the minimum and maximum value that can be obtained by a path from the root to that node. The end node tt therefore contains the minimum and maximum value that can be obtained by taking any path through the MDD.

Algorithm 3: Sum Property

CREATEPROPERTY($label$)

   | $(min, max) \leftarrow (\text{this}_{min} + \text{label}, \text{this}_{max} + \text{label})$
   | return (min, max)

MERGEWITHPROPERTY($property, label$)

   | $\text{this}_{min} \leftarrow \text{MIN}(\text{this}_{min}, \text{property}_{min} + \text{label})$
   | $\text{this}_{max} \leftarrow \text{MAX}(\text{this}_{max}, \text{property}_{max} + \text{label})$

GETRESULT()

   L return this

Cardinality Constraint. We represent the property of a global cardinality constraint as a *values* matrix of size $|V| \times 2$. Each value considered in the constraint is associated with a pair (min, max) representing the minimum and maximum number of times the value is taken. Moving from a parent to a child through an arc labeled by a, amounts to incrementing by 1 the number of times the value a is taken, which is a simple addition operation on the intervals. On the other hand, performing the merge amounts, as for the sum, to performing a union operation. Algorithm 4 is a possible implementation.

Sequence Constraint. The sequence property is the most complex of the three (Algorithm 5). In itself, Function CREATEPROPERTY and MERGEWITHPROPERTY correspond, as for the cardinality constraint, to perform respectively an addition on intervals and a union operation. The difference comes from the fact that, for the sequence, it is necessary to have separate information, represented here by *accumulator*. This peculiarity comes from the sliding and global aspect of the sequence constraint: each node contains information about a local sequence, i.e. sequences that contain it as the final value. However, it is possible in this case to lose information - as shown in Fig. 3. This accumulator can be implemented in different ways: either by building it on the fly (memory gain but time loss), in which case it is not necessary to retain the information on more than two layers (the last two in a sliding manner), or by building it once the propagation of the property is completed, but all the information of all the nodes must be in memory (time gain but memory loss). Here, the accumulator is built on the fly.

Algorithm 4: Cardinality Property

CREATEPROPERTY(*label*)

    $values \leftarrow \emptyset$

    for each *value v in the GCC* **do**

        $values[v]_{min} \leftarrow \text{this.}values[v]_{min}$

        $values[v]_{max} \leftarrow \text{this.}values[v]_{max}$

    if $label \in values$ **then**

        $values[label]_{min} \leftarrow values[label]_{min} + 1$

        $values[label]_{max} \leftarrow values[label]_{max} + 1$

    $property.values \leftarrow values$

    return *property*

MERGEWITHPROPERTY(*property*, *label*)

    for each *value v in the GCC* **do**

        $add \leftarrow 0$

        if $v = label$ **then**

          $add \leftarrow 1$

        $values[v]_{min} \leftarrow \text{MIN}(property.values[v]_{min} + add, values[v]_{min})$

        $values[v]_{max} \leftarrow \text{MIN}(property.values[v]_{max} + add, values[v]_{max})$

GETRESULT

    return *this*

Algorithm 5: SEQUENCE Property

CREATEPROPERTY(*label*)

    $values \leftarrow \emptyset$

    $values[0] \leftarrow (0, 0)$

    $add \leftarrow$ (*label is in the sequence values*)? $1 : 0$

    for each *i* **from** 1 *to depth* + 1 **do**

        $values[i]_{min} \leftarrow \text{this.}values[i-1]_{min} + add$

        $values[i]_{max} \leftarrow \text{this.}values[i-1]_{max} + add$

    $property.values \leftarrow values$

    $property.depth \leftarrow depth + 1$

    ACCUMULATE(*property*)

    return *property*

MERGEWITHPROPERTY(*property*, *label*)

    $add \leftarrow$ (*label is in the sequence values*)? $1 : 0$

    for each *i* **from** 1 *to depth* **do**

        $values[i]_{min} \leftarrow \text{MIN}(property.values[i-1]_{min} + add, values[i]_{min})$

        $values[i]_{max} \leftarrow \text{MAX}(property.values[i-1]_{max} + add, values[i]_{max})$

    ACCUMULATE(this)

ACCUMULATE(*property*)

    for each *i* **from** 1 *to property.depth* **do**

        $accumulator[i]_{min} \leftarrow \text{MIN}(property.values[i]_{min}, accumulator[i]_{min})$

        $accumulator[i]_{max} \leftarrow \text{MAX}(property.values[i]_{max}, accumulator[i]_{max})$

GETRESULT()

    return *accumulator*

5   Experiments

5.1   Testing Environment

The algorithms have been implemented in Java 12. The experiments were performed on a Windows 10 machine using a Ryzen 2600 AMD CPU and 32 GB of RAM for the car sequencing problem and on a machine having four E7- 4870 Intel processors, each having 10 cores with 256 GB of memory and running under Scientific Linux for the nurse rostering problem.

The different tests comparing the methods presented in this paper were performed using solutions from instances of Car Sequencing and Nurse Rostering (represented as a MDD). These instances can be obtained on request from the authors. However, we give some important information:

Car Sequencing. The Car Sequencing MDD contains 25942 nodes, 53985 arcs and represents 2.6×10^{14} solutions. There are two options with capacity 1/2 and 2/3, 4 car classes (configurations) and a total of 100 cars.

Nurse Rostering. The Nurse Rostering MDD contains 128325 nodes, 220600 arcs and represents 1.2×10^{28} solutions. There are 6 nurses, 28 days and 3 shifts. The scheduling has some predefined data : 1 means that the nurse is working that day, - means that the nurse is not working that day and 0 means that it is yet to be determined. Each day, each shift has a minimum number required of nurse working that shift (we can have more, but never less). We have various constraints, such that a nurse cannot work more than 7 days straight and must have at least 2 free days in a row in a 2-weeks window.

Testing each raw solution individually requires testing 2.6×10^{14} and 1.2×10^{28} solutions (respectively for Car Sequencing and Nurse Rostering problems). Doing such benchmark in a reasonable amount of time is out of the question.

5.2   Comparison Between Inclusion and Intersection Based Inclusion

We will compare the two methods to compute the inclusion presented in this paper. We do not take into account the time and space needed to create and store the MDDs representing the constraints - we only focus on time and space needed to compute the inclusion between two MDDs.

We can see from results in Table 1 that the inclusion method is better in every way than the intersection based inclusion. We observe improvements by a factor between 2 and 3 in time and between 1.5 and 3 in memory for GCC and Sum constraints, and almost 5 in time for the sequence constraint.

This result was clearly expected, as the inclusion operation is at worst an intersection, without the reduce and compare part of the intersection based inclusion.

In the Car Sequencing problem, the GCC constraint expresses the number of time a car has to be produced and the Sequence constraint represents the maximum capacity of each option (that is, for any subsequence of q consecutive

Table 1. Intersection vs Inclusion (Average time in ms, memory in MB).

Constraints	Car sequencing				Nurse rostering			
	Intersection		Inclusion		Intersection		Inclusion	
	Time	Memory	Time	Memory	Time	Memory	Time	Memory
GCC	95	48	50	29	7 052	1 716	2 820	1 186
Sum	85	47	46	30	–	–	–	–
Sequence	187	98	38	27	15 320	3 244	5 039	1 402

cars, the maximum number of cars that can have this option). In the Nurse Rostering problem, the GCC constraint expresses the demands for a shift (that is the minimum number of nurses required for a given shift) and the Sequence constraint expresses the fact that a nurse must have at least a certain number of breaks in a sliding time window.

The Sum constraint does not have any powerful meaning in these problems, but we decided to test it on at least one problem. We did not test the sum constraint for the Nurse Rostering problem, hence the dash symbol.

5.3   Learning Parameters of a Global Constraint

We compare the different methods (inclusion, intersection based inclusion and properties) in order to determine the parameters of a global constraint. The tested constraints are the sequence, sum and GCC constraints. To determine the parameters with the inclusion and intersection based methods, a dichotomous search on the parameters is performed. For example, if the constraint SUM(a, b) is satisfied, we test if the constraint SUM($a, b/2$) is satisfied: if it is, we test SUM($a, b/4$), otherwise we test SUM($a, b * 3/4$). We modify the parameters one by one until they are fixed. However, for both methods, we do not compute the sequence parameters for all possible values of q because it would take too much time (we stop at $q = 11$). The time and memory used when building the MDDs are integrated into the results.

As we can see from these results, using properties to compute the parameters of a global constraint is better than performing successive inclusion checks by at least a factor 65 in time (1 391 ms vs 22 ms for the sum constraint for the Car Sequencing, in Table 2) and at most a factor 145 in time (283 994 ms vs 1 952 ms for the GCC constraint for the Nurse Rostering, in Table 3). For the sequence constraint, we reach a factor of 75 with the properties even if we do not compute all sequences with the other methods. Furthermore, a very big part of the process is to build all the MDDs of the constraints, resulting in an increase in both time and memory consumption, as expected. Once again we find the factor 2 that we had in our previous comparison between the two methods of inclusion.

Table 2. Car sequencing problem (time in ms, memory in MB).

Methods	GCC		Sum		Sequence	
	Time	Memory	Time	Memory	Time	Memory
Intersection	6 080	3 240	2 137	1 328	11 306	6 493
Inclusion	3 371	2 204	1 391	1 021	6 114	3 867
Properties	48	10	22	4	82	37

Table 3. Nurse rostering problem (average time in ms, memory in MB).

Methods	GCC		Sequence	
	Time	Memory	Time	Memory
Intersection	663 647	170 281	704 405	137 625
Inclusion	283 994	137 625	235 996	65 768
Properties	1 952	348	5 677	1 436

5.4 Conclusion

This article sheds light on a new aspect of the interest of using MDDs in the context of constraint programming. We have introduced a new inclusion operator that allows to answer the question of the satisfaction of a constraint more efficiently than by using the classical sequence of operations on MDDs (intersection, reduction, comparison). In addition, we have shown that adding properties to the nodes of a MDD allowing to represent locally the state of a constraint is a very efficient way to obtain the parameters of a global constraint (we presented GCC, sum and sequence), provided that this constraint can be formalized as a node property. This method is much more advantageous than a succession of inclusion operations, both from a temporal and spatial point of view, because it does not require any constraint construction in the form of an MDD. The use of inclusion is nevertheless of interest when the constraint is particularly complex, unique, or very difficult to formalize in the form of a property. The question of the use of properties for other constraints (other than global) seems to be the next step in order to answer in more detail the problem of extracting the parameters of a constraint from a set of solutions.

Acknowledgments. This work has been supported by the French government, through the 3IA Côte d'Azur Investments in the Future project managed by the National Research Agency (ANR) with the reference number ANR-19-P3IA-0002.

References

1. Beldiceanu, N., Contejean, E.: Introducing global constraints in CHIP. J. Math. Comput. Modell. **20**(12), 97–123 (1994)
2. Beldiceanu, N., Simonis, H.: A constraint seeker: finding and ranking global constraints from examples. In: Lee, J. (ed.) CP 2011. LNCS, vol. 6876, pp. 12–26. Springer, Heidelberg (2011). https://doi.org/10.1007/978-3-642-23786-7_4
3. Bergman, D., Ciré, A.A., van Hoeve, W., Hooker, J.N.: Decision Diagrams for Optimization. Artificial Intelligence: Foundations, Theory, and Algorithms. Springer, Heidelberg (2016). https://doi.org/10.1007/978-3-319-42849-9
4. Bessière, C., Coletta, R., Petit, T.: Learning implied global constraints. In: IJCAI 2007, Proceedings of the 20th International Joint Conference on Artificial Intelligence, Hyderabad, India, 6–12 January 2007, pp. 44–49 (2007)
5. Bryant, R.E.: Graph-based algorithms for boolean function manipulation **35**(8), 677–691 (1986)
6. Hoda, S., van Hoeve, W.-J., Hooker, J.N.: A systematic approach to MDD-based constraint programming. In: Cohen, D. (ed.) CP 2010. LNCS, vol. 6308, pp. 266–280. Springer, Heidelberg (2010). https://doi.org/10.1007/978-3-642-15396-9_23
7. Kam, T., Brayton, R.K.: Multi-valued decision diagrams. Technical report UCB/ERL M90/125, EECS Department, University of California, Berkeley, December 1990. http://www2.eecs.berkeley.edu/Pubs/TechRpts/1990/1671.html
8. Leo, K., Mears, C., Tack, G., Garcia de la Banda, M.: Globalizing constraint models. In: Schulte, C. (ed.) CP 2013. LNCS, vol. 8124, pp. 432–447. Springer, Heidelberg (2013). https://doi.org/10.1007/978-3-642-40627-0_34
9. Perez, G., Régin, J.C.: Efficient operations on MDDs for building constraint programming models. In: International Joint Conference on Artificial Intelligence, IJCAI-15, Argentina, pp. 374–380 (2015)
10. Perez, G.: Decision diagrams: constraints and algorithms. Ph.D. thesis, Université Nice Sophia Antipolis (2017)
11. Picard-Cantin, É., Bouchard, M., Quimper, C.-G., Sweeney, J.: Learning parameters for the sequence constraint from solutions. In: Rueher, M. (ed.) CP 2016. LNCS, vol. 9892, pp. 405–420. Springer, Cham (2016). https://doi.org/10.1007/978-3-319-44953-1_26
12. Picard-Cantin, É., Bouchard, M., Quimper, C.-G., Sweeney, J.: Learning the parameters of global constraints using branch-and-bound. In: Beck, J.C. (ed.) CP 2017. LNCS, vol. 10416, pp. 512–528. Springer, Cham (2017). https://doi.org/10.1007/978-3-319-66158-2_33
13. Srinivasan, A., Ham, T., Malik, S., Brayton, R.K.: Algorithms for discrete function manipulation. In: 1990 IEEE International Conference on Computer-Aided Design. Digest of Technical Papers, pp. 92–95 (1990). https://doi.org/10.1109/ICCAD.1990.129849

Finding Subgraphs with Side Constraints

Özgür Akgün[1], Jessica Enright[2], Christopher Jefferson[1],
Ciaran McCreesh[2(✉)], Patrick Prosser[2], and Steffen Zschaler[3]

[1] University of St Andrews, St Andrews, Scotland
[2] University of Glasgow, Glasgow, Scotland
ciaran.mccreesh@glasgow.ac.uk
[3] King's College London, London, UK

Abstract. The subgraph isomorphism problem is to find a small "pattern" graph inside a larger "target" graph. There are excellent dedicated solvers for this problem, but they require substantial programming effort to handle the complex side constraints that often occur in practical applications of the problem; however, general purpose constraint solvers struggle on more difficult graph instances. We show how to combine the state of the art Glasgow Subgraph Solver with the Minion constraint programming solver to get a "subgraphs modulo theories" solver that is both performant and flexible. We also show how such an approach can be driven by the Essence high level modelling language, giving ease of modelling and prototyping to non-expert users. We give practical examples involving temporal graphs, typed graphs from software engineering, and costed subgraph isomorphism problems.

1 Introduction

Finding small "pattern" graphs inside larger "target" graphs is a widely applicable hard problem, with applications including compilers [5], bioinformatics [6,16], chemistry [29], malware detection [8], pattern recognition [17], and the design of mechanical locks [35]. This has led to the development of numerous dedicated algorithms, with the Glasgow Subgraph Solver [24] being the current state of the art [33]. However, practitioners are often interested in versions of the problem with additional restrictions, or side constraints. Some of these, such as exact vertex labelling schemes, are trivial to include in a dedicated solver, but others currently require either extensive programming or inefficient postprocessing. This paper explores a different approach: by allowing the Glasgow Subgraph Solver to use the Minion constraint programming (CP) solver [19] for side constraints, we achieve both the performance only a dedicated solver can offer, with the flexibility of a full CP toolkit. This hybrid modelling system can be driven by the Essence high level modelling language [18] and the Conjure toolchain, making it accessible to non-specialists.

This research was supported by the Engineering and Physical Sciences Research Council [grant number EP/P026842/1].

P. J. Stuckey (Ed.): CPAIOR 2021, LNCS 12735, pp. 348–364, 2021.
https://doi.org/10.1007/978-3-030-78230-6_22

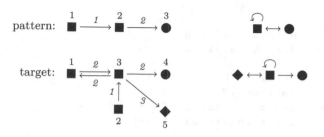

Fig. 1. A small pattern and a larger target graph used in examples throughout this paper. The plain text numbers are vertex names, and the shapes on vertices represent vertex labels. The graphs to the right are *type graphs*, which are used in Sect. 3.1. The italic labels on edges are used for *temporal graphs*, which are discussed in Sect. 3.2, and should otherwise be ignored.

1.1 Preliminaries

We begin with a look at the subgraph isomorphism problem, from a high level constraint modelling perspective. The basic non-induced subgraph isomorphism problem is to find an injective mapping from a pattern graph to a target graph, such that adjacent vertices in the pattern are mapped to adjacent vertices in the target. Variations on the problem are common, and are often combined. For example, in the induced version of the problem, non-edges must be mapped to non-edges; in the directed version, the input graphs have directed edges whose orientations must be preserved by the mapping; in the vertex labelled version, each vertex has a label, and the mapping must map vertices to like-labelled vertices; and in the edge-labelled version, edges have labels which must be preserved. It is also common to want to count or enumerate all solutions, rather than deciding whether at least one solution exists. Subsets of these variations are supported by many dedicated subgraph isomorphism algorithms, including the Glasgow Subgraph Solver.

We can express these problems in the Essence modelling language, as follows. We assume vertices take their labels from the set $L = \{1 \ldots \ell\}$ for some given ℓ, and edges from $E = \{1 \ldots e\}$ (and so ℓ and / or e may be 1, for applications that do not use labels on vertices and/or edges):

```
given l, e : int
letting L be domain int(1..l)
letting E be domain int(1..e)
```

We take as input a directed pattern graph which has p vertices (which we number from 1 to p, in the set P), and a directed target graph which has t vertices (numbered from 1 to t, the set T). Each graph is represented as total function from vertices to vertex labels, and a *partial* function from pairs of (not necessarily distinct) vertices to edge labels:

```
given p, t : int
letting P be domain int(1..p)
letting T be domain int(1..t)

given pat  : function (P, P) --> E
given tgt  : function (T, T) --> E
given plab : function (total) P --> L
given tlab : function (total) T --> L
```

Now we wish to find an injective mapping f:

```
find f : function (total, injective) P --> T
```

that preserves vertex labels,

```
such that forAll a : P .  plab(a) = tlab(f(a))
```

and directed edges, including their labels:

```
such that forAll ((a, b), lbl) in pat .
    ((f(a), f(b)), lbl) in toSet(tgt)
```

As a simple example, the following inputs show the problem instance represented in Fig. 1. We have three different vertex labels (circle, square, and diamond), and only a single edge type (which is directed; the numerical labels on edges are not used in this section):

```
letting l be 3
letting e be 1
```

We may now describe the pattern:

```
letting p be 3
letting pat be function ((1, 2) --> 1, (2, 3) --> 1)
letting plab be function (1 --> 1, 2 --> 1, 3 --> 2)
```

and the target:

```
letting t be 5
letting tgt be function ((1, 3) --> 1, (3, 1) --> 1,
    (2, 3) --> 1, (3, 4) --> 1, (3, 5) --> 1)
letting tlab be function (1 --> 1, 2 --> 1, 3 --> 1,
    4 --> 2, 5 --> 3)
```

Using the Conjure tool to compile Essence to a constraint programming model which is then solved by Minion, we find there are exactly two solutions to the problem, as we would expect:

```
(1 --> 1, 2 --> 3, 3 --> 4)
(1 --> 2, 2 --> 3, 3 --> 4)
```

But what if our application requires induced isomorphisms? Then we can easily add the constraint

```
such that forAll (a, b) : (P, P) .
    (f(a), f(b)) in defined(tgt) -> (a, b) in defined(pat)
```

And Conjure will now find us a single solution,

```
(1 --> 2, 2 --> 3, 3 --> 4)
```

As we will see in Sect. 3, supporting other problem variants and constraints is similarly straightforward, even if auxiliary variables are required. For example, if instead we want to allow relabelling on vertex labels (which is typically not supported by dedicated solvers), we could do the following:

```
find r : function (total, injective) L --> L
such that forAll a : P .   r(plab(a)) = tlab(f(a))
```

and we would find two additional solutions,

```
(1 --> 1, 2 --> 3, 3 --> 5)
(1 --> 2, 2 --> 3, 3 --> 5)
```

and if we removed the injective keyword for the relabelling, we would find a fifth mapping

```
(1 --> 2, 2 --> 3, 3 --> 1)
```

We return to relabelling in Sect. 3.1.

1.2   Initial Experiments and Motivation

Unfortunately, whilst elegant and flexible, the performance of this approach leaves a lot to be desired on basic subgraph isomorphism instances. The computational experiments in this paper are performed on a cluster of machines with dual Intel Xeon E5-2697A v4 processors and 512GBytes RAM running Ubuntu 18.04. The source code used for these experiments is released as part of the Glasgow Subgraph Solver[1], Minion[2], and Conjure[3] distributions, and we provide a separate archive for experimental scripts[4].

For graphs, we will be using the 14,621 unlabelled, undirected instances from Solnon's benchmark suite[5]. This benchmark suite was originally designed for algorithm portfolios work [21], and brings together several collections of application and randomly-generated instances with varying difficulties and solution counts (including many unsatisfiable instances). Some of the instances have up to 900 vertices and 14,420 edges in patterns and up to 6,671 vertices and 209,000 edges in targets. These lead to rather large models, by constraint programming standards: the largest generated table constraint has nearly half a million entries. However, these sizes are realistic from an applications perspective, and it would be desirable if solvers could handle even larger target graphs.

In Fig. 2 we plot the cumulative number of instances solved over time for the non-induced decision problem, comparing the high level approach to the

---

[1] https://github.com/ciaranm/glasgow-subgraph-solver/releases/tag/cpaior2021-finding-subgraphs-with-side-constraints.

[2] https://github.com/minion/minion/releases/tag/1.9.

[3] https://github.com/conjure-cp/conjure.

[4] https://github.com/ciaranm/cpaior2021-finding-subgraphs-with-side-constraints.

[5] https://perso.liris.cnrs.fr/christine.solnon/SIP.html.

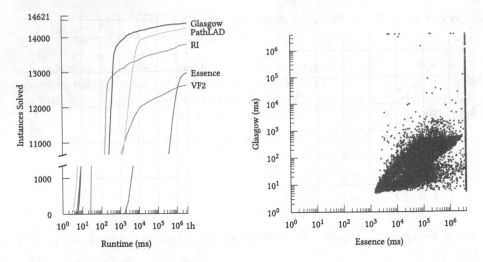

Fig. 2. Left, the cumulative number of instances solved over time, for the non-induced decision problem with no side constraints. Right, comparing the high level approach with the Glasgow Subgraph Solver on an instance by instance basis; points on the outer axes represent timeouts.

Glasgow Subgraph Solver [24] and PathLAD [21] (the two strongest CP-inspired approaches), and to VF2 [10] and RI [6] (simpler algorithms which perform well on easy instances). The high level approach has very slow startup times (which is to be expected as it involves launching a Java virtual machine and reading in a very large table constraint), but much more worryingly, only catches up with the worst other solver in number of instances solved as the timeout approaches. Worse, as the scatter plot in Fig. 2 shows, there are almost no instances where the high level approach does better than the Glasgow Subgraph Solver. (For induced problems, the results are even less favourable).

These first results motivate the remainder of this paper. We want to retain the convenience of the high level modelling approach, and to be able to add arbitrary side constraints to suit different applications, but we do not want to have to abandon the performance that dedicated solvers can get on hard instances. In Sect. 2 we evaluate several ways of using a CP solver in conjunction with the Glasgow Subgraph Solver, with a focus on low level implementation details. In Sect. 3 we then return to high level modelling, and look at the convenience it provides for retyping typed graphs, for temporal problems, and for optimisation problems.

2   Hybrid Solving with High-Level Modelling

Designing an effective hybrid solving system involved three major decisions: how the high-level modelling language would identify suitable problems to hybridise, how the solvers would communicate, and how often this communication would occur. The first two decisions were relatively straightforward to make, but the third required using computational experiments to evaluate different options. This section discusses all three of these decisions.

2.1 High-Level Modelling

We chose to use the Essence modelling language [18] because of its support for convenient high-level types like functions and relations, which can easily describe graphs and related abstractions. In a conventional modelling pipeline, problems are specified in Essence and then are converted to concrete models that can be solved by a CP solver (in our experiments Minion) via the Conjure and SavileRow tools. We augmented Conjure with a command line option that instructs it to generate an extra file which describes how the variables representing the graph are represented in the SavileRow input (known as Essence'). This is used by the graph solver so it can map between its internal state and the variables in Essence'. SavileRow converts the graph model to Minion input, and this conversion contains information which allows Minion to map its internal state back to the Essence' given to SavileRow. The graph solver and Minion then communicate mappings between the nodes and edges in the graph using the identifiers in the Essence' representation of the problem.

This design is based upon the notion that a CP solver and a subgraph solver can have enough of a shared understanding of a problem to solve it co-operatively. Indeed, the Glasgow Subgraph Solver [24] employs a CP approach to solve subgraph-finding problems, but using special data structures and algorithms— for example, rather than representing the adjacency constraint using an explicit table, it uses bitset adjacency matrices [22]. The solver also exploits various graph invariants involving degrees [36] and paths [2] to further reduce the search space, and employs special search order heuristics [1]. From this paper's perspective, the most important design aspect is that internally, the solver has a CP style variable for each vertex in the pattern graph, whose domains range over the vertices of the target graph. The solver performs a backtracking search with restarts and nogood recording, attempting to assign each variable a value from its domain, whilst respecting adjacency and injectivity constraints. At each recursive call of search, the solver performs *propagation* to eliminate infeasible values from domains. If any domain becomes empty, the solver backtracks; otherwise, it selects a variable, and tries assigning it each value from its domain in turn.

The high-level approach, then, gives us a way of setting up the subgraph solver and a CP solver such that they both have an equivalent set of variables and values for the graph part of the model, and tells us how to form a correspondence between their internal representations. Importantly, this allows the CP solver to have additional variables that the subgraph solver does not know about, and we do not specifically require the CP solver to be aware of all of the graph constraints. (Additionally, due to preprocessing, the CP solver may also sometimes have only a subset of values for some graph variables visible to it).

2.2 When to Communicate?

Having found a way to set up the two solvers, we must next ask when they should communicate. The simplest approach would be to use the CP solver as a *solution checker* for the subgraph solver. Whenever the subgraph solver finds a solution, it will pass it to the CP solver, which will treat the solution as a set of equality constraints. The CP solver will then attempt to find a satisfying assignment. If the CP solver does not have any additional variables, this is equivalent to simply checking that the remaining constraints hold, but in general this will require search. For a decision problem, the CP solver then communicates back to the subgraph solver either "yes, this is a valid solution", or "no, reject this solution and keep going". If we are solving a counting or enumeration problem, the CP solver must find *all* solutions and communicate this back to the subgraph solver.

For more power, but possibly also greater cost, we could additionally ask a CP solver at every stage of search to test whether the subgraph solver is in an obviously infeasible state. Whenever the subgraph solver has finished performing propagation, it can communicate the *trail* (that is, its current sequence of guessed assignments) to the CP solver, which again treats these as additional equality constraints. The CP solver then performs its own propagation (but not search), and communicates back either a "yes, keep going" or a "no, backtrack immediately". Finally, after this testing, we could also ask the CP solver to communicate any deletions it infers back to the subgraph solver. In other words, the subgraph solver would use the CP solver as an additional propagator.

Unfortunately, each of these approaches has drawbacks. The solution we will settle upon is based upon *rollbacks*; we will describe this below, after presenting experiments that demonstrate the difference between these approaches.

2.3 How to Communicate

To enable communication between the two solvers, we use FIFOs (named pipes), and a simple text-based protocol. Both solvers are run and initialised, and then the subgraph solver proceeds as normal, whilst the CP solver waits to be given commands. The subgraph solver then communicates its trail or a candidate solution as a set of assignment constraints to the CP solver. The CP solver then sets these assignments as its state and either performs a single propagation, or complete search, as directed. When finished the solver communicates its success or failure state, and any deletions (if requested), back to the subgraph solver and reverts any changes made by setting the assignments. This approach is designed to be solver-agnostic, and adding support for different CP solvers (or non-CP solvers) is simple, as long as they support performing a search or propagation from a given set of assignments.

2.4 Design Experiments

We now present the results of some computational experiments. The experiments in this section are designed to be hard, and to emphasise the difference

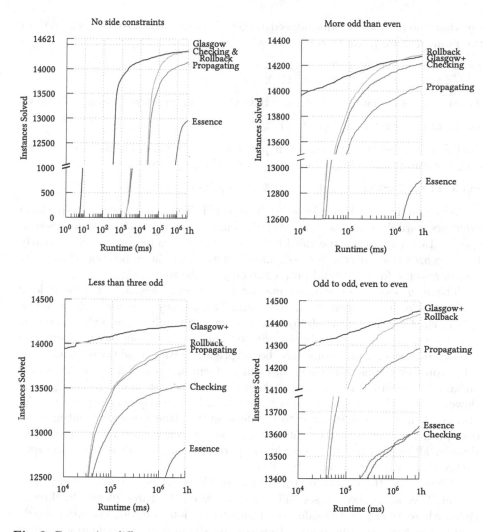

Fig. 3. Comparing different approaches to hybrid solving, showing the cumulative number of instances solved over time. On the top row, "no side constraints" then with the "more odd target vertices than even target vertices" side constraint; on the bottom row, the "mostly odd target vertices" side constraint on the left, and the "odd to odd, even to even" side constraint on the right. In the top left plot, the "Checking" and "Rollback" lines are indistinguishable. The Glasgow+ lines show the Glasgow Subgraph Solver with manually-implemented side constraints.

between the approaches, rather than to be realistic. We will continue to work with Solnon's non-induced subgraph isomorphism benchmark instances, but will consider four variations. Firstly, we will consider the problem with no side constraints. This is, in some sense, the worst case scenario, where we must pay the full price of hybrid solving, but cannot get any benefit from it. Secondly, let us say that the number of odd-indexed target vertices used must be greater than the number of even target vertices used:

```
such that (sum i : P . f(i) % 2) >
    (sum i : P . (f(i) + 1) % 2)
```

Thirdly, let us instead say that fewer than three odd target vertices may be used:

```
such that (sum i : P . f(i) % 2) < 3
```

And fourthly, let us say that even pattern vertices must be mapped to even target vertices, and odd pattern vertices to odd target vertices:

```
such that forAll a : P . (a % 2) = (f(a) % 2)
```

We chose these problems because the second is likely not to be able to perform inference until deep in a search tree (when most pattern vertices are mapped to specific target vertices), the third is likely to be able to perform inference early in the search tree (after a few assignments have been made, but not at the root node), and the fourth should propagate only at the root node.

The results of these experiments are presented in Fig. 3. Let us first look at the top left plot, where we do not actually have any side constraints. When calling the CP solver as a solution checker, we ultimately achieve the same performance as the Glasgow Subgraph Solver[6], although we can pay a substantial startup overhead. This should not be surprising: with no side constraints, the subgraph solver runs as normal, and will perform just one call to the CP solver on satisfiable instances. The propagating approach is over an order of magnitude slower: calling the CP solver at every search node is clearly very expensive. (We also tried calling the CP solver to test feasibility, without communicating deletions; this made no noticeable difference to performance, and so is not pictured.) Finally, all approaches substantially outperform using a CP solver on its own without help from the subgraph solver.

What about the remaining three plots in Fig. 3, where we do have side constraints? As we hoped, we see differences between the three plots. On the top right, where we expect the side constraints to fire late, solution checking clearly beats propagation during search. However, on the bottom left, where we expect side constraints to fire early, the propagating approach is much better than solution checking. Meanwhile, on the bottom right, where our constraints fire only at the root node, checking performs extremely poorly compared to propagating. In each case, any hybrid approach using the subgraph solver remains much better than a pure CP approach, except that in bottom right plot checking is slightly worse than just using the CP solver on its own.

As a point of comparison, we also implemented these three sets of side constraints natively inside the Glasgow solver: we label these as Glasgow+. For the "more odd than even" case, we did this through solution checking; for "less than three odd" we implemented checking during search; and for "odd to odd, even to even" we implemented initial domain filtering. Our implementation choices here

---

[6] Actually, because high level modelling can use different names for variables and values, we get slight differences due to changes to tiebreaking in search order heuristics.

are intended to reflect what a reasonable programmer would do if the high level approach were not available (and we intentionally selected side constraints that would not be too difficult to implement). In two of the three cases, the hand-crafted code does somewhat outperform the hybrid solving, but in the "more odd than even" case, hybrid solving actually beats the hand-crafted dedicated solver implementation when using the rollback approach, which we now describe.

2.5   A Rollback Approach to Communication

From what we have seen so far, it is obviously important to call the CP solver *some* of the time during search, but too expensive to call it *all* of the time. We will therefore introduce a new approach, which we call *rollback*. This approach is inspired by backjumping [27], as well as by the conflict analysis methods used in SAT and SMT solvers [32] and in lazy clause generating CP solvers [25,34]. The idea is as follows. Firstly, we call the CP solver with full propagation at the root node, in case we are dealing with a particularly rich labelling scheme. Secondly, we use the CP solver for solution checking, since this is required for correctness. Now, suppose the CP solver rejects a candidate solution: this will cause the subgraph solver to backtrack. At this point, we call the CP solver again, with full propagation. Either the CP solver indicates feasibility, in which case we proceed with search (potentially with a reduced set of domains), or the CP solver indicates failure, in which case we backtrack again, and do another attempt at full propagation, and so on until feasibility is reached.

The idea behind this approach is to avoid calling the CP solver when it is unlikely to do anything useful, but that once a failure has been encountered, we want to extract as much information as we can from the CP solver. If the failure encountered was due to a "local" property of the solution, such as in the "more odd than even" example, then we will quickly return to just using the subgraph solver for search. However, if the failure is due to only a few early assignments, as in the "fewer than three odd vertices" example, then we will jump back to nearly the root of the search tree.

The results in Fig. 3 demonstrate the success of this approach. When there are no side constraints, this approach has no overheads compared to solution checking. When constraints fire late, this approach is better than solution checking, and when constraints fire early, this approach is better than always propagating during search. In other words, rolling back from failures gives us all of the strengths and none of the weaknesses of the simpler approaches. We will therefore use this method for the remainder of the paper.

3   Subgraph Problems with Side Constraints

We now look at three classes of real-world subgraph-finding problems that, until now, have been solved using dedicated approaches. We show how easy it is to model these problems in Essence, demonstrating the usefulness of the high-level modelling approach for prototyping and development.

3.1 Retyping Problems

The basic notion of a graph conveys only adjacency information, and a subgraph isomorphism simply finds a certain structural pattern. In practice, this is often augmented with additional information—for example, we have seen how labels can be associated with vertices and edges, which can be used in chemical applications to represent different kinds of atom or bond. In this case, subgraph isomorphisms are also expected to preserve labels, so carbon atoms can only be mapped to carbon atoms, and double bonds must be mapped to double bonds. A richer labelling abstraction comes in the form of *typed graphs*, where the labels themselves also carry a graph structure [15]; we show an example in Fig. 1. In practice this labelling structure is specified by providing two graphs, together with a morphism from the main graph to its type graph.

For typed graphs, morphisms between the graphs are typically defined between graphs typed over the same type graph, but there are situations where we are interested in mapping between graphs typed over different type graphs. One such scenario from a software engineering context is described by Durán et al. [12,13], where graph transformation systems are composed by defining morphisms between the rules constituting the respective transformation systems. In this case, the source and target transformation systems will normally have different type graphs; a morphism must also be established between the two type graphs. Mappings between the various graphs making up the rules then need to preserve structure and typing subject to the morphism between type graphs. This approach to specification composition is implemented by the GTSMorpher tool.[7] A key objective is to minimise the amount of specification that needs to be written. For example, the tool allows morphisms between graph transformation systems to be only partially specified and then automatically completes the full morphism, if it can do so unambiguously—this requires solving a subisomorphism problem.

We may describe typed graph subisomorphism problems in Essence as follows. As before, we are given a pattern graph and a target graph, both of which carry labels; we will draw the vertex labels from different sets, to emphasise the relabelling.

```
given pl, tl, e : int
letting PL be domain int(1..pl)
letting TL be domain int(1..tl)
letting E be domain int(1..e)
```

We are also given labelled graphs,

```
given p, t : int
letting P be domain int(1..p)
letting T be domain int(1..t)

given pat : function (P, P) --> E
given tgt : function (T, T) --> E
```

---

[7] https://github.com/gts-morpher/gts_morpher.

```
given plab : function (total) P --> PL
given tlab : function (total) T --> TL
```

but now the labels also carry a graph structure,

```
given pattype : function (PL, PL) --> E
given tgttype : function (TL, TL) --> E
```

We are looking for an injective mapping from the pattern graph to the target graph,

```
find f : function
    (total, injective) P --> T
```

as well as an injective mapping between the label graphs,

```
find r : function
    (total, injective) PL --> TL
```

in such a way that graph structure and labels are preserved,

```
such that forAll ((a, b), lbl) in pat .
    ((f(a), f(b)), lbl) in toSet(tgt)
such that forAll a : P .
    r(plab(a)) = tlab(f(a))
```

and also requiring that the structure on the labels is preserved,

```
such that forAll (a,b) in defined(pattype) .
    pattype((a,b)) = tgttype((r(a),r(b)))
```

Consider again the example in Fig. 1, and now suppose they are equipped with the type structures shown to the right of each graph,

```
letting pattype be function (
    (1, 1) --> 1, (1, 2) --> 1, (2, 1) --> 1 )
letting tgttype be function (
    (1, 1) --> 1, (1, 2) --> 1,
    (1, 3) --> 1, (3, 1) --> 1 )
```

We now find two solutions:

```
(1 --> 1, 2 --> 3, 3 --> 5)
(1 --> 2, 2 --> 3, 3 --> 5)
```

because we can map pattern vertex 3 to target vertex 5 through retyping, but mapping pattern vertex 3 to target vertex 4 would not respect the type graph structure.

More generally, the field of model-driven software engineering includes numerous examples of using search and optimisation techniques to generate or transform graphs [7]. Existing approaches largely make use of ad-hoc [31] and metaheuristic methods [4,9,14], but we believe that with the help of suitably accessible high-level modelling tools, this could become a fruitful area for constraint programming research in the future.

3.2 Temporal Subgraph Problems

Another labelling scheme is used in *temporal graphs*, where edges are labelled with timestamps that denote times when edges are active—here we use integers as timestamps. Including information on the timing of edges substantially increases the modelling power of these graphs, allowing them to more accurately reflect the structure and dynamics of a wide variety of real-world systems (e.g. trade networks, changing contact networks, transport networks), and address optimisation questions in which the timing of edges is fundamental.

As algorithms and formalisms have become available for temporal graphs, examples of their application have become widespread [20], notably including applications within epidemiology [3] and computational social science [30]. Because the use of temporal graphs has spread beyond theoretical researchers, the ability to rapidly define and experiment with new problem definitions and constraints is valuable—practitioners are unlikely to define bespoke algorithms for novel problems as they arise.

There are at least three common kinds of temporal subgraph isomorphism. In an *exact* subisomorphism, times are simply labels that must match exactly. If we look at Fig. 1, now ignoring vertex labels but using the edge labels to carry the timestamps,

```
letting l be 1
letting e be 3

letting p be 3
letting pat be function ((1, 2) --> 1, (2, 3) --> 2)
letting plab be function (1 --> 1, 2 --> 1, 3 --> 1)

letting t be 5
letting tgt be function ((1, 3) --> 2, (3, 1) --> 2,
    (2, 3) --> 1, (3, 4) --> 2, (3, 5) --> 3)
letting tlab be function (1 --> 1, 2 --> 1, 3 --> 1,
    4 --> 1, 5 --> 1 )
```

then there are two solutions,

```
(1 --> 2, 2 --> 3, 3 --> 1)
(1 --> 2, 2 --> 3, 3 --> 4)
```

A less strict kind of subisomorphism is an *offset*, where edge labels must match exactly, but offset by an integer constant. In our example, this means "find a mapping where the event from 2 to 3 occurs one time unit after the event from 1 and 2". We can model this as follows:

```
find o : int(-e..e)
such that forAll (a,b) in defined(pat) .
    pat((a,b)) = o + tgt((f(a), f(b)))
```

and we find one additional solution,

```
(1 --> 1, 2 --> 3, 3 --> 5)
```

Finally, in an *order* embedding, the pattern edge labels simply define an order on events. We can model this as follows:

```
find o : function (total) E --> E
such that forAll x : int(1..e - 1) .  o(x) <= o(x + 1)
such that forAll (a,b) in defined(pat) .
    pat((a,b)) = o(tgt((f(a), f(b))))
```

which gets us yet another solution,

```
(1 --> 2, 2 --> 3, 3 --> 5)
```

Of course, when using a high level modelling approach, we are not restricted to these three problem variants, and could easily try out new models in an interactive setting. For example, it would take only a few minutes to write a model for a temporal problem where all edges must occur within a short but unspecified time period [28], whereas adapting a dedicated solver to check this constraint would be a substantial programming effort (and making the solver propagate rather than check this constraint would be even harder).

3.3 Subgraph Isomorphism with Costs

The system we created also support optimisation problems (and does not require that the subgraph isomorphism solver be aware that this is what is going on). If, for example, each target vertex has a cost associated with it,

```
given tcost : function (total) T --> int
```

then we can ask to find the cheapest solution,

```
minimising sum([ tcost(f(a)) | a : P])
```

We could also just as easily ask for the solution whose most expensive edge is cheapest, or that uses fewest vertices with a particular label. These kinds of problem occur widely in practice, including in skyline graph queries [26], labelled subgraph finding [11], and weighted clique problems [23].

4   Conclusion

The system we have presented shows that it is possible to combine the power of modern subgraph solvers with the flexibility of a general purpose constraint programming toolkit, although doing so efficiently requires careful consideration of how frequently the solvers communicate. We believe further research in this direction may be useful—for example, would it be possible to make use of some kind of conflict analysis rather than a backjumping approach?

When driven by a high level modelling approach, this system is particularly suitable for rapid prototyping and for dynamic queries where side constraints can be specified in response to user need. However, the high level modelling approach does come with a large startup cost, which makes it unsuitable for deployment in application contexts that involve solving many thousands of problem instances in real-time. Fortunately though, connecting the low level solvers manually is also an option once a design has been decided upon. We also expect that new approaches may be necessary to deal with the huge but sparse graphs that arise in some applications, since table constraints and conventional CP domain stores both struggle when moving beyond ten thousand of vertices in target graphs.

References

1. Archibald, B., Dunlop, F., Hoffmann, R., McCreesh, C., Prosser, P., Trimble, J.: Sequential and parallel solution-biased search for subgraph algorithms. In: Rousseau, L.-M., Stergiou, K. (eds.) CPAIOR 2019. LNCS, vol. 11494, pp. 20–38. Springer, Cham (2019). https://doi.org/10.1007/978-3-030-19212-9_2

2. Audemard, G., Lecoutre, C., Samy-Modeliar, M., Goncalves, G., Porumbel, D.: Scoring-based neighborhood dominance for the subgraph isomorphism problem. In: O'Sullivan, B. (ed.) CP 2014. LNCS, vol. 8656, pp. 125–141. Springer, Cham (2014). https://doi.org/10.1007/978-3-319-10428-7_12

3. Bansal, S., Read, J., Pourbohloul, B., Meyers, L.A.: The dynamic nature of contact networks in infectious disease epidemiology. J. Biol. Dyn. **4**(5), 478–489 (2010)

4. Bill, R., Fleck, M., Troya, J., Mayerhofer, T., Wimmer, M.: A local and global tour on MOMoT. Softw. Syst. Model. **18**(2), 1017–1046 (2017). https://doi.org/10.1007/s10270-017-0644-3

5. Hjort Blindell, G., Castañeda Lozano, R., Carlsson, M., Schulte, C.: Modeling universal instruction selection. In: Pesant, G. (ed.) CP 2015. LNCS, vol. 9255, pp. 609–626. Springer, Cham (2015). https://doi.org/10.1007/978-3-319-23219-5_42

6. Bonnici, V., Giugno, R., Pulvirenti, A., Shasha, D.E., Ferro, A.: A subgraph isomorphism algorithm and its application to biochemical data. BMC Bioinf. **14**(S-7), S13 (2013)

7. Boussaïd, I., Siarry, P., Ahmed-Nacer, M.: A survey on search-based model-driven engineering. Autom. Softw. Eng. **24**(2), 233–294 (2017). https://doi.org/10.1007/s10515-017-0215-4

8. Bruschi, D., Martignoni, L., Monga, M.: Detecting self-mutating malware using control-flow graph matching. In: Büschkes, R., Laskov, P. (eds.) DIMVA 2006. LNCS, vol. 4064, pp. 129–143. Springer, Heidelberg (2006). https://doi.org/10.1007/11790754_8

9. Burdusel, A., Zschaler, S., John, S.: Automatic generation of atomic consistency preserving search operators for search-based model engineering. In: Kessentini, M., Yue, T., Pretschner, A., Voss, S., Burgueño, L. (eds.) 22nd ACM/IEEE International Conference on Model Driven Engineering Languages and Systems, MODELS 2019, Munich, Germany, 15–20 September 2019, pp. 106–116. IEEE (2019). https://doi.org/10.1109/MODELS.2019.00-10

10. Cordella, L.P., Foggia, P., Sansone, C., Vento, M.: A (sub)graph isomorphism algorithm for matching large graphs. IEEE Trans. Pattern Anal. Mach. Intell. **26**(10), 1367–1372 (2004)

11. Dell'Olmo, P., Cerulli, R., Carrabs, F.: The maximum labeled clique problem. In: Adacher, L., Flamini, M., Leo, G., Nicosia, G., Pacifici, A., Piccialli, V. (eds.) Proceedings of the 10th Cologne-Twente Workshop on Graphs and Combinatorial Optimization. Extended Abstracts, Villa Mondragone, Frascati, Italy, 14–16 June 2011, pp. 146–149 (2011)

12. Durán, F., Moreno-Delgado, A., Orejas, F., Zschaler, S.: Amalgamation of domain specific languages with behaviour. J. Log. Algebraic Methods Program. **86**, 208–235 (2017). https://doi.org/10.1016/j.jlamp.2015.09.005

13. Durán, F., Zschaler, S., Troya, J.: On the reusable specification of non-functional properties in DSLs. In: Czarnecki, K., Hedin, G. (eds.) SLE 2012. LNCS, vol. 7745, pp. 332–351. Springer, Heidelberg (2013). https://doi.org/10.1007/978-3-642-36089-3_19

14. Efstathiou, D., Williams, J.R., Zschaler, S.: Crepe complete: multi-objective optimization for your models. In: Paige, R.F., Kessentini, M., Langer, P., Wimmer, M. (eds.) Proceedings of the First International Workshop on Combining Modelling with Search- and Example-Based Approaches co-located with 17th International Conference on Model Driven Engineering Languages and Systems (MODELS 2014), Valencia, Spain, 28 September 2014. CEUR Workshop Proceedings, vol. 1340, pp. 25–34. CEUR-WS.org (2014)

15. Ehrig, H., Ehrig, K., Prange, U., Taentzer, G.: Fundamentals of Algebraic Graph Transformation. Monographs in Theoretical Computer Science. An EATCSSeries. Springer, Heidelberg (2006). https://doi.org/10.1007/3-540-31188-2

16. Elhesha, R., Sarkar, A., Kahveci, T.: Motifs in biological networks. In: Yoon, B.-J., Qian, X. (eds.) Recent Advances in Biological Network Analysis, pp. 101–123. Springer, Cham (2021). https://doi.org/10.1007/978-3-030-57173-3_5

17. Foggia, P., Percannella, G., Vento, M.: Graph matching and learning in pattern recognition in the last 10 years. IJPRAI **28**(1), 1450001 (2014). https://doi.org/10.1142/S0218001414500013

18. Frisch, A.M., Harvey, W., Jefferson, C., Hernández, B.M., Miguel, I.: Essence: a constraint language for specifying combinatorial problems. Constraints Int. J. **13**(3), 268–306 (2008). https://doi.org/10.1007/s10601-008-9047-y

19. Gent, I.P., Jefferson, C., Miguel, I.: Minion: a fast scalable constraint solver. In: Brewka, G., Coradeschi, S., Perini, A., Traverso, P. (eds.) ECAI 2006, 17th European Conference on Artificial Intelligence, 29 August–1 September 2006, Riva del Garda, Italy, Including Prestigious Applications of Intelligent Systems (PAIS 2006), Proceedings. Frontiers in Artificial Intelligence and Applications, vol. 141, pp. 98–102. IOS Press (2006)

20. Holme, P., Saramäki, J.: Temporal networks. Phys. Rep. **519**(3), 97–125 (2012). https://doi.org/10.1016/j.physrep.2012.03.001

21. Kotthoff, L., McCreesh, C., Solnon, C.: Portfolios of subgraph isomorphism algorithms. In: Festa, P., Sellmann, M., Vanschoren, J. (eds.) LION 2016. LNCS, vol. 10079, pp. 107–122. Springer, Cham (2016). https://doi.org/10.1007/978-3-319-50349-3_8

22. McCreesh, C., Prosser, P.: A parallel, backjumping subgraph isomorphism algorithm using supplemental graphs. In: Pesant, G. (ed.) CP 2015. LNCS, vol. 9255, pp. 295–312. Springer, Cham (2015). https://doi.org/10.1007/978-3-319-23219-5_21

23. McCreesh, C., Prosser, P., Simpson, K., Trimble, J.: On maximum weight clique algorithms, and how they are evaluated. In: Beck, J.C. (ed.) CP 2017. LNCS, vol. 10416, pp. 206–225. Springer, Cham (2017). https://doi.org/10.1007/978-3-319-66158-2_14

24. McCreesh, C., Prosser, P., Trimble, J.: The glasgow subgraph solver: using constraint programming to tackle hard subgraph isomorphism problem variants. In: Gadducci, F., Kehrer, T. (eds.) ICGT 2020. LNCS, vol. 12150, pp. 316–324. Springer, Cham (2020). https://doi.org/10.1007/978-3-030-51372-6_19

25. Ohrimenko, O., Stuckey, P.J., Codish, M.: Propagation = lazy clause generation. In: Bessière, C. (ed.) CP 2007. LNCS, vol. 4741, pp. 544–558. Springer, Heidelberg (2007). https://doi.org/10.1007/978-3-540-74970-7_39

26. Pande, S., Ranu, S., Bhattacharya, A.: SkyGraph: retrieving regions of interest using skyline subgraph queries. Proc. VLDB Endow. **10**(11), 1382–1393 (2017). https://doi.org/10.14778/3137628.3137647

27. Prosser, P.: Hybrid algorithms for the constraint satisfaction problem. Comput. Intell. **9**, 268–299 (1993). https://doi.org/10.1111/j.1467-8640.1993.tb00310.x

28. Redmond, U., Cunningham, P.: Temporal subgraph isomorphism. In: Rokne, J.G., Faloutsos, C. (eds.) Advances in Social Networks Analysis and Mining 2013, ASONAM 2013, Niagara, ON, Canada, 25–29 August 2013, pp. 1451–1452. ACM (2013). https://doi.org/10.1145/2492517.2492586

29. Régin, J.: Développement d'outils algorithmiques pour l'Intelligence Artificielle. Application à la chimie organique. Ph.D. thesis, Université Montpellier 2 (1995)

30. Sekara, V., Stopczynski, A., Lehmann, S.: Fundamental structures of dynamic social networks. Proc. Natl. Acad. Sci. **113**(36), 9977–9982 (2016)

31. Semeráth, O., Nagy, A.S., Varró, D.: A graph solver for the automated generation of consistent domain-specific models. In: Chaudron, M., Crnkovic, I., Chechik, M., Harman, M. (eds.) Proceedings of the 40th International Conference on Software Engineering, ICSE 2018, Gothenburg, Sweden, 27 May–03 June 2018, pp. 969–980. ACM (2018). https://doi.org/10.1145/3180155.3180186

32. Silva, J.P.M., Sakallah, K.A.: GRASP - a new search algorithm for satisfiability. In: Rutenbar, R.A., Otten, R.H.J.M. (eds.) Proceedings of the 1996 IEEE/ACM International Conference on Computer-Aided Design, ICCAD 1996, San Jose, CA, USA, 10–14 November 1996, pp. 220–227. IEEE Computer Society/ACM (1996). https://doi.org/10.1109/ICCAD.1996.569607

33. Solnon, C.: Experimental evaluation of subgraph isomorphism solvers. In: Conte, D., Ramel, J.-Y., Foggia, P. (eds.) GbRPR 2019. LNCS, vol. 11510, pp. 1–13. Springer, Cham (2019). https://doi.org/10.1007/978-3-030-20081-7_1

34. Stuckey, P.J.: Lazy clause generation: combining the power of sat and CP (and MIP?) solving. In: Lodi, A., Milano, M., Toth, P. (eds.) CPAIOR 2010. LNCS, vol. 6140, pp. 5–9. Springer, Heidelberg (2010). https://doi.org/10.1007/978-3-642-13520-0_3

35. Vömel, C., de Lorenzi, F., Beer, S., Fuchs, E.: The secret life of keys: on the calculation of mechanical lock systems. SIAM Rev. **59**(2), 393–422 (2017). https://doi.org/10.1137/15M1030054

36. Zampelli, S., Deville, Y., Solnon, C.: Solving subgraph isomorphism problems with constraint programming. Constraints **15**(3), 327–353 (2010)

Short-Term Scheduling of Production Fleets in Underground Mines Using CP-Based LNS

Max Åstrand[1,2]([✉]), Mikael Johansson[1], and Hamid Reza Feyzmahdavian[2]

[1] Division of Decision and Control Systems, School of Electrical Engineering and Computer Science, KTH Royal Institute of Technology, Stockholm, Sweden
mikaelj@kth.se
[2] ABB Corporate Research, Västerås, Sweden
{max.astrand,hamid.feyzmahdavian}@se.abb.com

Abstract. Coordinating the mobile production fleet in underground mines becomes increasingly important as the machines are more and more automated. We present a scheduling approach that applies to several of the most important production methods used in underground mines. Our algorithm combines constraint programming with a large neighborhood search strategy that dynamically adjusts the neighborhood size. The resulting algorithm is complete and able to rapidly improve constructed schedules in practice. In addition, it has important benefits when it comes to the acceptance of the approach in real-life operations. Our approach is evaluated on public and private industrial problem instances representing different mines and production methods. We find significant improvements over the current industrial practice.

Keywords: Scheduling · Underground mining · Constraint programming · Large neighborhood search

1 Introduction

In today's modern underground mine, the excavation is increasingly performed by highly automated mobile machines. Several of these machines can now be operated simultaneously by a single human operator from a centralized control room. This transformation alleviates some traditionally limiting factors in mining operations, such as access to human resources and the safety of underground workers. Consequently, when the mining activities are increasingly automated, it is possible to coordinate the activities more efficiently. Instead of focusing on *how* individual activities should be performed, mine automation now turns to address *when* each activity should take place and by *which* machine. This coordination of activities is called short-term mine scheduling, and it is the process of allocating resources and determining feasible start and end times for the upcoming mining activities.

© Springer Nature Switzerland AG 2021
P. J. Stuckey (Ed.): CPAIOR 2021, LNCS 12735, pp. 365–382, 2021.
https://doi.org/10.1007/978-3-030-78230-6_23

Short-term mine scheduling is usually a manual process that depends critically on the skills of the manual scheduler [2, 30]. Even if heuristic constructive procedures have been adapted in some mines, the overall level of automation in mine scheduling is low. While constructive procedures can quickly provide feasible schedules for several hundreds of activities, they are typically cumbersome to customize and maintain over time as well as limited when it comes to optimization. Another approach effective at quickly constructing feasible schedules is Constraint Programming (CP). CP has been shown to perform well on scheduling benchmarks [16], in industrial manufacturing settings [29], and in the mining industry [3]. A benefit of CP compared to constructive procedures is due to the declarative paradigm: separating *what* a solution is from *how* to find it. In combination with the expressiveness of CP, these attributes make it easier to consider the plurality of process constraints that real-world industrial settings may exhibit. In complex environments, there is also often an opportunity for significant optimization over manually constructed schedules. The standard approach for optimization in CP (backtracking search using Branch and Bound) may however be unsatisfactory when it comes to the speed at which good solutions are found. In these cases, combining CP with Large Neighborhood Search (LNS) facilitates finding high-quality solutions faster [28].

In this paper, we present a CP model that applies to several of the most common production methods used in underground mines. The constraint model strengthens and generalizes previous approaches in mine scheduling. In addition, the general structure of the approach lends itself to production scheduling in other heavy industries, including forestry and agriculture, where the production fleet is increasingly automated. For these scheduling problems, we propose an algorithm that dynamically adjusts the size of the neighborhoods explored by LNS. In contrast to more common approaches, the algorithm is both complete and able to rapidly improve the constructed schedules. Its local optimality properties are also easy to explain to human operators, which is essential for real-life acceptance. Furthermore, we combine ideas from adaptive LNS and tabu-search to guide the neighborhood selection. Our joint approach is evaluated on both public and private industrial problem instances, representing different mines and production methods. We find that our algorithm compares favorably to both conventional restart-based backtracking search and to modern industrial practice.

2   Underground Mine Scheduling

Underground mining is the process of excavating ore from beneath the surface. The ore is located in deposits (ore-bodies) which may be located several kilometers underground. Depending on the shape and orientation of the ore-body, different mining (production) methods are used to extract the ore. A common factor for most mining methods is that they use drilling and blasting to separate the ore from the ore-body. The operation hence follows what is known as a *blast cycle*: a number of sequential process steps required to advance the exca-

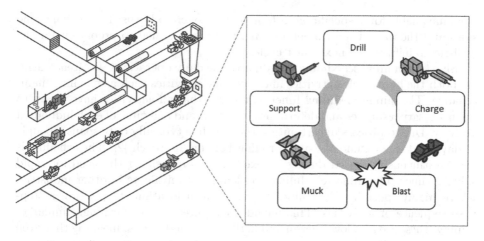

Fig. 1. An illustration of the main activities in a typical blast cycle taking place at a mining face.

vation process. Even in small operations, many blast cycles are simultaneously conducted at different *mining faces* (tunnel endings) throughout the mine.

The approach presented in this work applies to mining methods where the activities of each individual blast cycle take place at the same location. This includes the most common methods used in small- and medium-sized mines, as well as some large-scale mining methods. In addition, it covers the scheduling of mine development (tunneling) that occurs in all underground mines regardless of mining method. We will pay special attention to *room-and-pillar* mining and *cut-and-fill* mining [10]. Room-and-pillar is common for flat, near-horizontal, ore-bodies containing *e.g.* copper, coal, or potash. As the name suggests, it is based on excavating the ore such that pillars are left in the ore-body to support the surrounding rock mass. Cut-and-fill is more common for steep irregular ore-bodies containing high-value ore, such as gold or silver, and the ore is extracted in horizontal slices.

While the typical fleet and the number of mining faces vary between the considered mining methods, they follow roughly the same blast cycle (see Fig. 1). First, the mining face is drilled using a drill-rig. The drill-rig prepares multiple holes in the rock mass, which are then charged with explosives using a charging vehicle. The explosives are in turn detonated to separate the ore from the rock mass. Due to safety, blasting can typically only occur during pre-defined time slots called *blast windows*. After the toxic blast fumes have been ventilated, the ore is transported away from the face using tailored front-end loaders known as load-haul-dump machines. Later, material loosely attached to the insides of the mining tunnel is scaled using a hydraulic hammer mounted on a mobile machine. Next, the sides and the roof of the tunnel are reinforced by bolting the inside of the tunnel to the adjacent rock mass and spraying the inside with concrete to reduce the risk of sudden rock bursts. When concrete is being used, the face is left unavailable for a specific *cure time* as the concrete burns. Lastly, when the sides and the roof of the tunnel are safe, the face is scaled to prepare it for the drill-rig to initiate the next blast cycle.

Short-term mine scheduling coordinates the daily operations. It often considers both the blast cycle activities and auxiliary activities such as machine maintenance. The mine scheduling function ensures that all activities have access to the necessary resources and determines the start- and end-time of each individual activity. Due to process uncertainties, the schedule typically only spans a couple of days and it is extended using a rolling horizon approach [2]. The schedule is also continuously revised during its execution, emphasizing the need for rapid solving times. The mine scheduling problem has equivalents in other industries where highly specialized machines travel between locations to perform a stepwise sequence of activities. This includes *e.g.* forestry and farming. Similar to mining, these are also low-margin and high-volume industries, meaning that even minor efficiency improvements may have a significant impact on profitability.

In contrast to most works in the scheduling literature that use makespan as the objective function, we will use the sum of face makespans. The motivation is twofold. First, using the overall makespan in a rolling horizon approach puts too much emphasis on the face that currently has the most mining activities scheduled on it. In our experience, using the sum of face makespan makes the schedules less prone to rapid changes of production paces between faces when the horizon extends. Second, the sum of face makespan is easily converted into "blasting pace", which is a common key performance metric used in the mining industry.

Related Work. There are only a few related works on short-term mine scheduling. Schulze et al. [24] consider a room-and-pillar mine where they formulate a MILP model for small instances and present several constructive procedures suitable for larger settings. The constructive procedures are later [25] extended to include job selection and staffing with the goal of minimizing the deviation between the amount of mined potash and a targeted production rate. Seifi et al. [26] study the same mine but employ a two-stage approach: using MILP to solve a relaxed version of the scheduling problem and then incorporate all mining constraints by means of a heuristic algorithm. This approach is later [27] outperformed by a stronger MILP model. Our previous work [4], extending on [3], employs CP to schedule the blast cycle activities in a cut-and-fill mine. The model considers several mining-specific details such as machine travel times and the mix of interruptible and uninterruptible activities. In [4], a simple LNS strategy using fixed neighborhood sizes is also introduced and indicates that LNS appears to be a promising improvement strategy.

Our CP model draws inspiration from job-shop scheduling. Grimes et al. [13] compare two models for job-shop scheduling: one based on unary-constraints and tailored search heuristics, and one model based on simpler reified disjunctive constraints combined with learning search heuristics. Despite providing less propagation, the model based on reified disjunctive constraints compares favorably on some classical job shop instances due to, among other things, the combination of learning search heuristics and restart-based search. Grimes and Hebrard [12] note that the simpler model can easily be adapted to handle various side constraints and objective functions, including sum objectives. This model serves as an inspiration to the model we employ for mine scheduling. The standard job-shop does however not include task allocation. Booth et al. [7] remark that the literature on combining task allocation and scheduling using CP is limited. They compare a CP-based approach with a MILP-based approach to a multi-robot task allocation and scheduling problem. The authors conclude that CP works well for the general structure of these problems. The task allocation is implemented using optional interval variables, which are more recently also used to schedule truck-and-drone final mile delivery systems [14] and for simultaneous task allocation and motion scheduling of dual-arm robots [6].

Adaptive Large Neighborhood Search (ALNS) is a method to identify favorable neighborhood structures among a set of alternatives [23]. In the contemporary literature, ALNS is often combined with tabu-based techniques to avoid re-visiting solutions. A recent example is He et al. [15] that studies oversubscribed sequence-dependent scheduling problems. Unlike previous approaches, they use a tight integration of the two methods, which is demonstrated to work well on a real-world satellite scheduling problem. The authors attribute this to the diversifying capabilities of ALNS together with the intensifying proficiency of tabu-search.

3   Approach

We consider mine scheduling problems specified by the following data:

$\mathcal{L}$ A set of locations $\mathcal{L} = \{1, \cdots, |\mathcal{L}|\}$ representing the mining faces. Further, the travel time $\ell_{l,l'}$ is fully specified for all combinations of locations and it is the time needed to drive from location l to location l'.

$\mathcal{M}$ A machine set $\mathcal{M} = \{1, \cdots, |\mathcal{M}|\}$ of the machines available for production. Each machine $m \in \mathcal{M}$ belongs to a machine class c representing the machine-job eligibility. The subset $\mathcal{M}_c \subset \mathcal{M}$ contains all machines of class c.

$\mathcal{J}$ A set $\mathcal{J} = \{1, \cdots, |\mathcal{J}|\}$ of mining activities (jobs) to be scheduled. For each job $j \in \mathcal{J}$, the following 3 parameters are known. First, the machine class needed to perform the job (M_j). All machines $m \in \mathcal{M}_c$ can process job j as long as $M_j = c$. Second, the location $L_j \in \mathcal{L}$ where the job takes place. Third, the expected duration $D_j \in \mathbb{Z}^+$ that the machine is required to be present at the location.

$\mathcal{P}$ A set of tuples $\mathcal{P} = \{(j_1^p, j_1^s), \cdots, (j_{|\mathcal{P}|}^p, j_{|\mathcal{P}|}^s)\}$ denoting precedence constraints $j^p \prec j^s$ between preceding job j^p and succeeding job j^s.

For clarity, in this contribution we will assume that all machines travel with the same travel speed and that all machines of the same class exhibit an equal processing rate. While our approach can easily be adapted to relax these assumptions, our experience is that these assumptions are not limiting in the most common production scenarios.

3.1   Constraint Programming

Model.   Our constraint model uses the following decision variables:

s_j A nonnegative variable representing the start time of job $j \in \mathcal{J}$.

o_{jm} A boolean variable indicating whether machine $m \in \mathcal{M}$ is allocated to job $j \in \mathcal{J}$.

$\mathsf{b}_{jj'}$ A boolean sequencing variable for the machines. If $\mathsf{b}_{jj'} = 1$, then job j precedes job j' on the same machine. Of course, $\mathsf{b}_{jj'} = 0$ represents the opposite, *i.e.* $j' \prec j$.

$\mathsf{c}_{jj'}$ Similar to $\mathsf{b}_{jj'}$, but instead of denoting the order of jobs on machines, it describes the order of jobs on the locations.

O The objective function representing the sum of makespans over all locations.

Since exactly one eligible machine needs to be allocated to each job, we have the constraints

$$\mathsf{o}_{jm} = 0 \qquad \forall (m, j) : m \in \mathcal{M}_c \wedge c \neq M_j \tag{1}$$

$$\sum_{m \in \mathcal{M}} \mathsf{o}_{jm} = 1 \qquad \forall j \in \mathcal{J}. \tag{2}$$

Jobs that require no mobile machinery (*e.g.* blasting) are included by introducing a virtual machine class with infinite capacity.

For sequencing the jobs on the machines, we enforce the following:

$$\mathsf{b}_{jj'} = 1 \iff \mathsf{o}_{jm} = 1 \wedge \mathsf{o}_{j'm} = 1 \wedge$$
$$\mathsf{s}_j + D_j + \ell_{L_j L_{j'}} < \mathsf{s}_{j'} \quad \forall (m, j, j') : j < j'. \tag{3}$$

$$\mathsf{b}_{jj'} = 0 \iff \mathsf{o}_{jm} = 1 \wedge \mathsf{o}_{j'm} = 1 \wedge$$
$$\mathsf{s}_{j'} + D_{j'} + \ell_{L_{j'} L_j} < \mathsf{s}_j \quad \forall (m, j, j') : j < j'. \tag{4}$$

$$\mathsf{b}_{jj'} = \star \iff \neg(\mathsf{o}_{jm} = 1 \wedge \mathsf{o}_{j'm} = 1) \quad \forall (m, j, j'), \tag{5}$$

where $\mathsf{b}_{jj'} = \star$ is a dummy value introduced to represent that job j and j' are scheduled on different machines and hence require no sequencing.

Note that our earlier model [4] only considers production scenarios where the job order on each location is fully specified and can be implemented as precedence constraints. By introducing another sequencing variable $\mathsf{c}_{jj'}$, representing the job order on the locations, this model generalizes to more mining methods and the scheduling of vital auxiliary jobs such as maintenance and inspection. We constrain the location sequencing variables by

$$\mathsf{c}_{jj'} = 1 \iff \mathsf{s}_j + D_j < \mathsf{s}_{j'} \quad \forall (m, j, j') : j < j' \wedge L_j = L_{j'} \tag{6}$$

$$\mathsf{c}_{jj'} = 0 \iff \mathsf{s}_{j'} + D_{j'} < \mathsf{s}_j \quad \forall (m, j, j') : j < j' \wedge L_j = L_{j'} \tag{7}$$

$$\mathsf{c}_{jj'} = \star \qquad \forall (j, j') : L_j \neq L'_j. \tag{8}$$

Further, there are many precedence constraints in mining due to *e.g.* geomechanics, the confined environment, and the blast cycle. The start times of the preceding jobs j^p and the succeeding jobs j^s are restricted by

$$s_{j^p} + D_{j^p} \leq s_{j^s} \quad \forall (j^p, j^s) \in \mathcal{P}. \tag{9}$$

The objective is to minimize the sum of location makespans

$$0 = \sum_{l \in \mathcal{L}} \max_{j \in \mathcal{J}_l} (s_j + D_j), \tag{10}$$

where the set $\mathcal{J}_l = \{j \in \mathcal{J} \mid L_j = l\}$ holds all jobs on location l.

Extensions. Different mining methods exhibit different process constraints. If concrete cure-times are needed for some mining activities, they affect the subsequent availability of the location but not the machine. This is enforced by adding a time buffer D_{cure} to Eq. (6) such that $s_j + D_j + D_{cure} < s_{j'}$, and modifying Eqs. (7) and (9) similarly for the jobs that involve concrete.

Individual temporal constraints are supported by directly restricting the feasible domain of the start-times. Consider *e.g.* blast jobs $j \in \mathcal{J}^{(bl)} \subset \mathcal{J}$ where the start times need to be aligned with a blast window, then

$$s_j \in \{s_b \mid b = 1, \ldots, B\} \quad \forall j \in \mathcal{J}^{(bl)}. \tag{11}$$

Here, s_b denotes the start-time of blast window b. Further, using the compressed-time approach as described in [4] deals with the mix of interruptible and uninterruptible jobs efficiently.

While Eqs. (3) to (5) indeed provide disjunctive machine execution, we find it useful to overload them by an optional **unary**-constraint when searching highly constrained search spaces

$$\mathtt{unary}\Big(\{(s_j, D_j, o_{jm}) \mid j \in \mathcal{J}\}\Big) \quad \forall m \in \mathcal{M}. \tag{12}$$

This also holds for the sequencing of jobs on the locations, and hence

$$\mathtt{unary}\Big(\{(s_j, D_j) \mid j \in \mathcal{J} : L_j = l\}\Big) \quad \forall l \in \mathcal{L} \tag{13}$$

is also added as a redundant constraint.

Search. Our search strategy assigns machines to jobs, then branches on the sequencing variables, and lastly determines the start-times.

First, we assign the machine allocation variables $o_{jm} = 1$ such that machines of the same class are allocated an equal number of jobs (ties are broken index-wise). Moreover, the machine allocation variables are grouped by class, meaning that all jobs requiring the same machine class are branched on sequentially.

Second, we branch on the variables $c_{jj'}$ that sequence the jobs on the locations. The variables are selected in increasing order of $\min(s_j) + \min(s_{j'})$, where $\min(s_j)$ denotes the lowest value in the domain of s_j. The value selection is

$$c_{jj'} = \begin{cases} 1, & \text{if } \min(s_j) < \min(s_{j'}), \\ 0, & \text{otherwise.} \end{cases} \tag{14}$$

Note that this search strategy sequences jobs in order of how early they feasibly can be scheduled given the current state of branching, which resembles greedy constructive procedures that previously have been applied to the mine scheduling problem [20]. A fully determined sequence of the jobs on each location need not impose an order of the assigned jobs on each individual machine. Therefore, after the c's are determined, we branch on the b variables following the same strategy.

Finally, when a feasible job sequencing has been determined (for both locations and machines) a solution is instantiated by assigning the start times to their lowest feasible value $s_j = \min(s_j)$. This also determines the objective function O.

There may be several machines within each class that have the same processing rate and travel speed. In these cases, we eliminate symmetrical solutions by lightweight dynamic symmetry breaking [19] using that the columns of o_{jm} are partially exchangeable.

3.2 Large Neighborhood Search

Large Neighborhood Search (LNS) is a method for local search using CP [28]. Starting from a solution, LNS is based on relaxing a subset of the decision variables to their original domain and resolving within that relaxation in the hope of finding an improving solution. The relaxed variables imply the *neighborhood* that is being explored, and the choice of neighborhood is crucial for the performance of the method [9].

In the scheduling literature, neighborhoods are often machine-based, time-based, or completely random. However, restricting the neighborhoods based on these criteria performs poorly in our setting. We attribute this to the large number of precedence constraints between jobs on the same location. This results in effective neighborhoods that are too inflexible to contain improving solutions. Instead, we have found location-based relaxations to be effective. This means that we implicitly determine which decision variables to relax by selecting a subset of locations. For the selected locations, all decision variables ($s_j, o_{jm}, b_{jj'}$, and $c_{jj'}$) corresponding to jobs on those locations are relaxed.

Our model uses $b_{jj'}$ and $c_{jj'}$ to hold the structure of the schedule (inspired by [13]). This modeling structure resembles constructing a Partial Order Schedule, a concept that is used in the LNS by Laborie and Godard [17]. In contrast to their approach to relax all start times in the schedule, we have found it beneficial to only relax the start times of the jobs within the neighborhood. In our setting, we attribute this to a more advantageous trade-off between each iteration's computational complexity and the number of iterations performed.

Algorithm 1. LNS-IncLoc

Input: feasible schedule σ with objective O, initial neighborhood size $n = 2$
1: **while** not terminated **do**
2: $\quad N \leftarrow$ draw and remove a neighborhood from $\mathcal{N}_n$ according to strategy S
3: $\quad p_N \leftarrow$ relax $(\sigma, \{s_j, o_{jm}, b_{jj'}, c_{jj'} | \, \forall m, j, j' : \, m \in \mathcal{M}, L_j \in N, L_{j'} \in N\})$
4: $\quad \sigma' \leftarrow$ solve p_N requiring $O' < O$
5: $\quad$ **if** solution found **then**
6: $\quad\quad n \leftarrow 2$, restore $\mathcal{N}_n$
7: $\quad\quad \sigma \leftarrow \sigma'$
8: $\quad$ **else if** $N_n = \emptyset$ **then**
9: $\quad\quad n \leftarrow n + 1$, restore $\mathcal{N}_n$
10: $\quad$ **end if**
11: **end while**

Dynamic Neighborhood Size. Let $\mathcal{N}_n$ denote the set of all combinations of n locations. For $n = 2$, then

$$\mathcal{N}_2 = \Big\{ \{l, l'\} \mid l, l' \in \mathcal{L}, \, l < l' \Big\} \tag{15}$$

constitutes the set of all potential neighborhoods. For small n, the search space induced by relaxing n locations is relatively small and can be rapidly explored. While the computational burden increases with n, so does the potential of finding larger improvements. In our earlier work [4] we compared $n = 2$ with $n = 3$ and found no strategy to be strictly dominating.

Therefore, in this work, we introduce LNS-IncLoc: an algorithm that adjusts n dynamically throughout the solution process. The algorithm is based on increasing the number of locations when all current neighborhoods are proven not to contain improving solutions. If an improving solution is found, n is decreased, and the search continues in smaller neighborhoods. In Algorithm 1, solve p_N refers to using the CP model to solve the problem p_N constructed by relaxing the variables corresponding to the locations in neighborhood $N \in \mathcal{N}_n$. Note that the majority of the explored neighborhoods do not contain any improving solution. Therefore, after finding the first feasible solution, the search strategies for the sequencing variables are replaced by the weighted degree search heuristic [8] to show infeasibility faster.

Our algorithm resembles variable-depth neighborhood search [1] in the sense that the neighborhoods grow as the solution process progresses. However, instead of partially searching the neighborhoods using heuristics, as is most common in variable-depth neighborhood search, we search the neighborhoods exhaustively. Since eventually $n = |\mathcal{L}|$, our algorithm is in-fact complete; given enough time it will return the optimal solution. Unfortunately, there are $\binom{|\mathcal{L}|}{n}$ combinations for each n. Hence to prove optimality, $\sum \binom{|\mathcal{L}|}{n} = 2^{|\mathcal{L}|}$ neighborhoods need to be searched sequentially, making optimality only relevant for small instances. However, we observe empirically that our approach tends to find many improvements in small neighborhoods for the problem under study. These improvements strengthen the upper bound on the objective, which is crucial for tractable solution times when increasing n. Therefore, while completeness is only relevant for small instances, LNS-IncLoc is also efficient in rapidly improving large instances.

The dynamic adjustments of n in LNS-IncLoc also bring another benefit in real-life deployments. From our previous efforts working close with mine schedulers, we remark that it is important to be able to explain the rationale behind the automatically created schedules. This is a challenge we have faced both when using constructive heuristics [20] and CP [3]. A distinct feature of LNS-IncLoc is that if the algorithm terminates at *e.g.* $n = 5$ we can guarantee the manual scheduler that there is no better solution unless we rearrange activities on more than 4 locations. The fact that LNS-IncLoc removes these obvious improvements (that skilled manual schedulers may find by *e.g.* swapping two machines on two locations) increases the trust in the automation.

Neighborhood Selection Strategies. The neighborhood size n does not determine in what order the neighborhoods $N \in \mathcal{N}_n$ are explored. This choice is made by a separate selection strategy S at line 3 in Algorithm 1. The selection strategy is important for the combined performance of our approach. During preliminary experiments, we investigated several different strategies. For brevity, we will only present three of them: $S_\text{in-order}$, S_random, and S_tabu.

The first strategy $S_\text{in-order}$ uses a fixed permutation of $\mathcal{N}_n$ and selects neighborhoods index-wise using that permutation. As an example, for any given permutation π the strategy $S_\text{in-order}$ produces a static neighborhood sequence $N_{\pi_1}, \ldots, N_{\pi_{|\mathcal{N}_n|}}$. In general, this permutation can be calculated based on features of the incumbent. However, in our numerical experiments we use a fixed random permutation throughout all iterations. This strategy resembles a first-improvement local search strategy using a fixed neighborhood order [21].

The second strategy S_random selects a neighborhood from $\mathcal{N}_n$ at random using a uniform distribution over the remaining neighborhoods in $\mathcal{N}_n$. This strategy, in contrast to $S_\text{in-order}$, will produce different random sequences each time.

Inspired by previous work that tightly couple ALNS with tabu-based methods [15,31], we introduce the selection strategy S_tabu. When using a uniform distribution over the neighborhoods nothing prevents the strategy from accidentally selecting several similar neighborhoods in sequence. As an example, consider the three neighborhoods $\{l_1, l_2\}$, $\{l_1, l_3\}$ and $\{l_3, l_4\}$. If $\{l_1, l_2\}$ is proven not to contain any improving solutions, then in a tabu-fashion we should favor selecting the more disparate neighborhood $\{l_3, l_4\}$ over the more similar $\{l_1, l_3\}$. To this end, the strategy S_tabu implements a non-uniform distribution over the neighborhoods in each iteration. This is implemented by counting the number of times a location has been relaxed $c_l^\mathcal{L}, l = 1, \ldots, |\mathcal{L}|$ and calculating the sum of location counts for each individual neighborhood $N \in \mathcal{N}_n$

$$c_N^\mathcal{N} = \sum_{l \in N} c_l^\mathcal{L}, \quad N = 1, \ldots, |\mathcal{N}_n|. \tag{16}$$

Each neighborhood is then associated with a weight w_N which is inversely proportional to the number of times the constituent locations have been relaxed

$$w_N = \frac{1}{1 + c_N^{\mathcal{N}} - \min_{N'} c_{N'}^{\mathcal{N}}}. \tag{17}$$

By using a roulette wheel selection procedure [5] the probability p_N of selecting any neighborhood becomes

$$p_N = \frac{w_N^{\gamma}}{\sum_{N' \in \mathcal{N}_n} w_{N'}^{\gamma}}, \quad N = 1, \ldots, |\mathcal{N}_n| \tag{18}$$

where $\gamma \geq 0$ is a tuning parameter. Letting $\gamma = 0$ we recover S_{random}, while $\gamma > 0$ enforces tabu. In our problem setting, $\gamma = 2$ appears to work well.

4   Results

In this section, we first investigate the impact that the neighborhood selection strategy has on LNS-IncLoc. We then compare LNS-IncLoc with restart-based search and an ALNS-inspired method. In our evaluations, we will use instances from a room-and-pillar mine, a cut-and-fill mine, and synthetic instances exhibiting more general precedence constraints. More specifically, we use the 8 public instances studied separately in [27]. These instances come from a German potash mine that uses the room-and-pillar mining method. The instances vary between 91–120 jobs, 18–24 disjunctive locations, and 13–14 machines divided over 7 machine classes. Compared to the other datasets, these instances have fewer jobs per location and represent mines where the fleet is comparably small considering the number of available mining faces. We will also use data from [4] representing a Swedish cut-and-fill mine. These 12 instances vary between 100–105 jobs, 10–20 locations, and 21 to 23 machines divided non-uniformly over 7 machine classes. These instances have a fully specified order of execution for the jobs on each location. This is because the jobs on each location come from the same blast cycle, whereas in the instances from the room-and-pillar mine, each disjunctive location consists of several workplaces representing one blast cycle each.

We also generate synthetic instances of three different problem sizes: 100 jobs on 10 locations, 150 jobs on 15 locations, and 200 jobs on 20 locations (denoted *J100*, *J150*, and *J200*, respectively). By varying the number of available machines, we can evaluate instances where both locations and machines are the main bottlenecks. Therefore, for each problem size, we use three different machine setups: 8 machines uniformly divided on 4 machine classes, 10 machines on 5 classes, and 15 machines on 5 classes. For each combination of problem size and machine setup, we generate 10 randomized instances where the job durations are uniformly sampled between 10 and 50 time-units, the machine travel times are set to be on average 25% of the average job duration, and 25% of the jobs have a precedence relation. This results in $3 \times 3 \times 10 = 90$ generated instances.

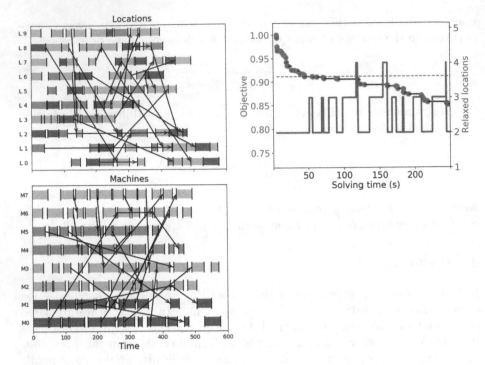

Fig. 2. A feasible schedule containing 100 jobs on 10 locations using 8 machines from 4 machine classes (left). The characteristic behavior of LNS-IncLoc where the blue dashed line correspond to the terminal solution using a fixed neighborhood size of $n = 2$ (above). (Color figure online)

The CP model has been implemented using Gecode 6.2 [11], where the solver runs in a single thread on an Intel Xeon 3.1 GHz processor. The LNS is implemented using Gecode's meta-search engines. Figure 2 shows a feasible solution to a *J100* instance with 8 available machines. The color of the job corresponds to the machine class needed to process it, while the arrows denote precedence constraints. The small time buffers between jobs in the machine-view represent the time needed for the corresponding machine to move between locations. Lastly, note that using only 8 machines produces a machine-constrained instance, which can be understood since fewer jobs are scheduled back-to-back in the location-view compared to the machine-view.

In Fig. 2 we can also see how the objective O and the neighborhood size n evolve using LNS-IncLoc. The red dots correspond to the objective when an improving solution is found, while the black dots correspond to neighborhoods that are proven not to contain any improving solution. We can observe that once an improving solution is found, *e.g.* when $n = 4$ at ~150 s, many subsequent solutions are found in smaller neighborhoods. The search space that needs to be explored in each iteration grows exponentially with n. Hence, restarting from $n = 2$ each time a solution is found ensures that the tightest upper bound on the

Table 1. Comparing different neighborhood selection strategies by the average improvement after 5 min (Δ_{5min}), the average and the range of the solutions at termination (Δ_{term} and R_{term}), the average number of iterations ($\#_{its}$) and the average number of improving solutions ($\#_{sols}$).

		8 machines			10 machines			15 machines		
		$S_{\text{in-order}}$	S_{random}	S_{tabu}	$S_{\text{in-order}}$	S_{random}	S_{tabu}	$S_{\text{in-order}}$	S_{random}	S_{tabu}
J100	Δ_{5min}	17.7	19.0	**19.4**	19.1	20.2	**20.7**	21.7	23.5	**24.1**
	Δ_{term}	22.6	22.4	**23.0**	23.9	**24.3**	23.7	26.0	26.1	**26.5**
	R_{term}	22.8	18.9	22.3	7.7	19.7	11.7	11.2	12.1	10.4
	$\#_{its}$	4758	4266	4366	4796	4307	4051	2857	2479	2667
	$\#_{sols}$	125	115	118	139	124	114	102	96	103
J150	Δ_{5min}	5.6	6.4	**7.3**	7.7	8.4	**9.1**	9.4	13.4	**13.7**
	Δ_{term}	10.5	**12.7**	12.2	13.7	13.1	**14.0**	17.0	18.4	**18.8**
	R_{term}	7.6	6.9	9.8	15.1	15.4	12.2	13.5	16.6	10.9
	$\#_{its}$	5409	5701	5499	5056	4906	5015	3080	3085	3033
	$\#_{sols}$	74	108	96	105	92	104	88	93	100
J200	Δ_{5min}	2.0	3.1	**3.3**	3.2	4.9	**5.0**	3.8	6.2	**7.0**
	Δ_{term}	5.2	**7.1**	7.0	7.2	**8.5**	8.3	9.2	11.2	**11.7**
	R_{term}	6.2	9.0	11.2	9.2	10.5	10.5	15.3	17.3	16.3
	$\#_{its}$	4604	4701	4603	4376	4337	4278	2664	2570	2631
	$\#_{sols}$	49	64	63	69	81	78	66	80	82

objective is used before increasing the neighborhood size. When using a static neighborhood size of $n = 2$, the value of the objective stagnates at a terminal solution indicated by the blue dashed line in Fig. 2.

To compare the neighborhood strategies introduced in Sect. 3.2 we solve all instances of *J100*, *J150*, and *J200* using 15, 30, and 45 min time-out, respectively. The result of the comparison is shown in Table 1, where we observe that S_{random} and S_{tabu} clearly outperform $S_{\text{in-order}}$. There is a notable difference in average performance on all problem sizes, but it is most prominent when we consider the improvement of the objective after 5 min of solve time (Δ_{5min}) on the largest problem instance (200 jobs and 15 machines). On these instances, the average improvement using S_{tabu} is almost twice as large as using $S_{\text{in-order}}$. The tabu-inspired strategy actually has the best average improvement after 5 min on all instances; however S_{random} seems to perform equally well concerning the solution quality at termination. Figure 3 contains a more detailed view of the average objective during the first 5 min of solving *J150*. Indeed, S_{tabu} is slightly better than S_{random} for all machine setups.

We now compare LNS-IncLoc using S_{tabu} with restart-based search on the synthetic data, the room-and-pillar data (RnP), and the cut-and-fill data (CnF). For the restart-based search, we use the same model as presented in Sect. 3.1. However, since learning heuristics and restarts have been proven to be a potent combination [8], we switch to a weighted degree heuristic for the sequencing

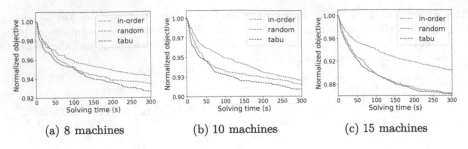

Fig. 3. Comparison of the average objective improvement during the first 5 min on all *J150* instances using different neighborhood strategies.

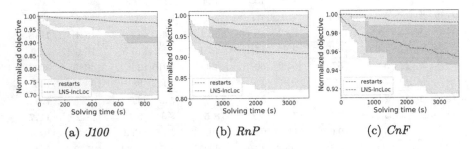

Fig. 4. Comparing LNS-IncLoc using S_{tabu} with restart-based search on a variety of instances. The dashed line depicts the average objective improvement and the shaded area corresponds to the range of all solving runs.

variables after finding the first feasible solution. We use a Luby cut-off sequence [18] with a scale factor of 256.

In Fig. 4 we compare the average, the maximum, and the minimum improvement of objective value on *J100*, *RnP*, and *CnF* using LNS-IncLoc and restart-based search. On all instances, we can see that LNS-IncLoc is better at quickly improving the solution quality. For the *RnP* data, it takes ∼700 s before *any* improving solution is found using restart-based search. In contrast, during the first 700 s of LNS-IncLoc it has on average already improved the schedule by ∼8%. On the *CnF* data, there are indeed situations where restart-based search may be competitive, considering that the best improvement using restart-based search is better than the average performance using LNS-IncLoc after ∼1000 s. However, it is also visible in Fig. 4 that LNS-IncLoc exhibits better average performance and more rapid convergence on all studied instances.

Finally, we evaluate an ALNS-inspired method to choose between the three different neighborhood selection strategies ($S_{\text{in-order}}$, S_{random}, and S_{tabu}). Similar to [17] we associate each strategy S with a weight ω_S proportional to the probability of being selected. The weights are updated as $\omega_S \leftarrow \lambda \omega_S + (1 - \lambda)r$ using the reward $r = \Delta0/\Delta t$, where $\Delta0$ and Δt denote the objective improvement and the computation time of that iteration. We heuristically set $\lambda = 0.99$ based on preliminary experiments. While ALNS proved better than using only $S_{\text{in-order}}$, we

Table 2. Comparison of different approaches using average improvement after 5 min (Δ_{5min}), the average and the range of the solutions at termination (Δ_{term} and R_{term}), and the average number of improving solutions ($\#_{sols}$).

		J100	J150	J200	RnP	CnF
restarts	Δ_{5min}	1.0	0.2	0.0	0.0	0.0
	Δ_{term}	3.0	0.6	0.0	3.1	0.9
	R_{term}	10.3	6.4	0.0	7.5	5.4
	$\#_{sols}$	10	3	1	14	4
ALNS	Δ_{5min}	19.9	7.4	3.3	4.6	1.2
	Δ_{term}	23.6	12.9	7.5	8.2	4.2
	R_{term}	22.3	15.4	14.4	9.9	5.8
	$\#_{sols}$	108	84	61	54	20
LNS-tabu	Δ_{5min}	**21.4**	**10.0**	**5.1**	**5.6**	1.2
	Δ_{term}	**24.4**	**15.0**	**9.0**	**9.3**	**4.6**
	R_{term}	22.3	17.2	18.3	13.4	7.4
	$\#_{sols}$	112	100	75	56	20

do not observe any consistent average performance improvement (see Table 2). An interesting challenge for ALNS in our setting is that the reward signal is very sparse: it is only non-zero for 1–2% of the iterations (cf. $\#_{its}$ and $\#_{sols}$ in Table 1). This means that the weights quickly decay after finding a solution and the resulting strategy converges to a uniform random choice over the available strategies (cf. [22]). An interesting remark is, however, that the ALNS-inspired method exhibits smaller average range on all studied instances.

Even though the vast majority of underground mines still rely on manual scheduling, there are some adopters of constructive procedures. As noted in Sect. 3.1, our search strategy finds a first feasible solution that resembles a schedule constructed by a greedy constructive procedure. In our earlier work [4] we showed that on *a particular* mining instance, a simple static LNS strategy with $n = 3$ improved by 7% over a common constructive algorithm. Referring to Table 2, we now observe that the *average* improvement of LNS-IncLoc on the synthetic, the room-and-pillar, and the cut-and-fill instances, is in the range 4–24%. Remember that our objective function roughly translates to production pace. Hence, for a high-volume industry such as mining, even a seemingly modest average objective improvement of 4% may result in large operational benefits.

5   Discussion and Conclusion

Optimizing the construction of short-term mine schedules holds great potential for future mining operations. In this work, we have presented a CP model for mine scheduling that strengthens and generalizes previous models to accommodate for characteristics found in several common underground mining methods. We also introduced `LNS-IncLoc`, a CP-based LNS strategy that dynamically adjusts the location-based neighborhood size. The approach is complete and able to rapidly improve constructed schedules, as demonstrated on real-life data from different mines that operate using different mining methods. Further, `LNS-IncLoc` has local optimality properties that are useful when interacting with manual schedulers. This facilitates the acceptance of the method in practice. We conclude that combining CP, LNS, and dynamic neighborhood sizes, appears to be a promising approach for short-term mine scheduling.

When evaluating the ALNS-inspired approach to adaptively choose the neighborhood selection strategy, we experienced that using the common reward $\Delta 0/\Delta t$ is challenging since improving solutions are only found in 1–2% of all iterations. Hence, ALNS with sparse reward is an interesting future research direction. Moreover, since the majority of the explored neighborhoods do not contain any improving solution, it would be interesting to see whether a cumulative relaxation could be leveraged to prove infeasibility faster.

Acknowledgements. This work was partially supported by the Wallenberg AI, Autonomous Systems and Software Program (WASP) funded by the Knut and Alice Wallenberg Foundation.

References

1. Ahuja, R.K., Ergun, Ö., Orlin, J.B., Punnen, A.P.: A survey of very large-scale neighborhood search techniques. Discret. Appl. Math. **123**(1–3), 75–102 (2002)
2. Åstrand, M., Johansson, M., Greberg, J.: Underground mine scheduling modeled as a flow shop - a review of relevant works and future challenges. J. Southern Afr. Inst. Min. Metall. **118**(12), 1265–1276 (2018)
3. Åstrand, M., Johansson, M., Zanarini, A.: Fleet scheduling in underground mines using constraint programming. In: van Hoeve, W.-J. (ed.) CPAIOR 2018. LNCS, vol. 10848, pp. 605–613. Springer, Cham (2018). https://doi.org/10.1007/978-3-319-93031-2_44
4. Åstrand, M., Johansson, M., Zanarini, A.: Underground mine scheduling of mobile machines using constraint programming and large neighborhood search. Comput. Oper. Res. **123**, 105036 (2020)
5. Back, T.: Evolutionary Algorithms in Theory and Practice: Evolution Strategies, Evolutionary Programming, Genetic Algorithms. Oxford University Press, Oxford (1996)
6. Behrens, J.K., Lange, R., Mansouri, M.: A constraint programming approach to simultaneous task allocation and motion scheduling for industrial dual-arm manipulation tasks. In: 2019 International Conference on Robotics and Automation (ICRA), pp. 8705–8711. IEEE (2019)
7. Booth, K.E.C., Nejat, G., Beck, J.C.: A constraint programming approach to multi-robot task allocation and scheduling in retirement homes. In: Rueher, M. (ed.) CP 2016. LNCS, vol. 9892, pp. 539–555. Springer, Cham (2016). https://doi.org/10.1007/978-3-319-44953-1_34
8. Boussemart, F., Hemery, F., Lecoutre, C., Sais, L.: Boosting systematic search by weighting constraints. In: ECAI, vol. 16, p. 146 (2004)

9. Carchrae, T., Beck, J.C.: Principles for the design of large neighborhood search. J. Math. Modell. Algorithms **8**, 245–270 (2009)
10. Darling, P.: SME Mining Engineering Handbook, vol. 1. SME (2011)
11. Gecode Team: Gecode: Generic constraint development environment (2019). https://www.gecode.org
12. Grimes, D., Hebrard, E.: Solving variants of the job shop scheduling problem through conflict-directed search. INFORMS J. Comput. **27**(2), 268–284 (2015)
13. Grimes, D., Hebrard, E., Malapert, A.: Closing the open shop: contradicting conventional wisdom. In: Gent, I.P. (ed.) CP 2009. LNCS, vol. 5732, pp. 400–408. Springer, Heidelberg (2009). https://doi.org/10.1007/978-3-642-04244-7_33
14. Ham, A.M.: Integrated scheduling of m-truck, m-drone, and m-depot constrained by time-window, drop-pickup, and m-visit using constraint programming. Transp. Res. Part C: Emerg. Technol. **91**, 1–14 (2018)
15. He, L., de Weerdt, M., Yorke-Smith, N.: Time/sequence-dependent scheduling: the design and evaluation of a general purpose tabu-based adaptive large neighbourhood search algorithm. J. Intell. Manuf. **31**(4), 1051–1078 (2019). https://doi.org/10.1007/s10845-019-01518-4
16. Laborie, P.: An update on the comparison of MIP, CP and hybrid approaches for mixed resource allocation and scheduling. In: van Hoeve, W.-J. (ed.) CPAIOR 2018. LNCS, vol. 10848, pp. 403–411. Springer, Cham (2018). https://doi.org/10.1007/978-3-319-93031-2_29
17. Laborie, P., Godard, D.: Self-adapting large neighborhood search: application to single-mode scheduling problems. Proceedings MISTA-07, Paris 8 (2007)
18. Luby, M., Sinclair, A., Zuckerman, D.: Optimal speedup of Las Vegas algorithms. Inf. Process. Lett. **47**(4), 173–180 (1993)
19. Mears, C., De La Banda, M.G., Demoen, B., Wallace, M.: Lightweight dynamic symmetry breaking. Constraints **19**(3), 195–242 (2014)
20. Mishchenko, K., Åstrand, M., Molander, M., Lindkvist, R., Viklund, T.: Developing a tool for automatic mine scheduling. In: Topal, E. (ed.) MPES 2019. SSGG, pp. 146–153. Springer, Cham (2020). https://doi.org/10.1007/978-3-030-33954-8_18
21. Papadimitriou, C., Steiglitz, K.: Combinatorial Optimization: Algorithms and Complexity. Courier Corporation (1998)
22. Pisinger, D., Ropke, S.: A general heuristic for vehicle routing problems. Comput. Oper. Res. **34**(8), 2403–2435 (2007)
23. Ropke, S., Pisinger, D.: An adaptive large neighborhood search heuristic for the pickup and delivery problem with time windows. Transp. Sci. **40**(4), 455–472 (2006)
24. Schulze, M., Rieck, J., Seifi, C., Zimmermann, J.: Machine scheduling in underground mining: an application in the potash industry. OR Spectrum **38**(2), 365–403 (2015). https://doi.org/10.1007/s00291-015-0414-y
25. Schulze, M., Zimmermann, J.: Staff and machine shift scheduling in a German potash mine. J. Scheduling 1–22 (2017)
26. Seifi, C., Schulze, M., Zimmermann, J.: A two-stage solution approach for a shift scheduling problem with a simultaneous assignment of machines and workers. In: The 39th International Symposium on Application of Computers and Operations Research in the Mineral Industry, Wroclaw, Poland (2019)
27. Seifi, C., Schulze, M., Zimmermann, J.: A new mathematical formulation for apotash-mine shift scheduling problem with a simultaneous assignment of machines and workers. Eur. J. Oper. Res. **292**, 27–42 (2020)
28. Shaw, P.: Using constraint programming and local search methods to solve vehicle routing problems. In: Maher, M., Puget, J.-F. (eds.) CP 1998. LNCS, vol. 1520, pp. 417–431. Springer, Heidelberg (1998). https://doi.org/10.1007/3-540-49481-2_30

29. Simonis, H.: Building industrial applications with constraint programming. In: Goos, G., Hartmanis, J., van Leeuwen, J., Comon, H., Marché, C., Treinen, R. (eds.) CCL 1999. LNCS, vol. 2002, pp. 271–309. Springer, Heidelberg (2001). https://doi.org/10.1007/3-540-45406-3_6
30. Song, Z., Schunnesson, H., Rinne, M., Sturgul, J.: Intelligent scheduling for underground mobile mining equipment. PloS One **10**(6), e0131003 (2015)
31. Žulj, I., Kramer, S., Schneider, M.: A hybrid of adaptive large neighborhood search and tabu search for the order-batching problem. Eur. J. Oper. Res. **264**(2), 653–664 (2018)

Learning to Reduce State-Expanded Networks for Multi-activity Shift Scheduling

Till Porrmann and Michael Römer[✉]

Decision Analytics Group, Bielefeld University, 33615 Bielefeld, Germany
{till.porrmann,michael.roemer}@uni-bielefeld.de

Abstract. For personnel scheduling problems, mixed-integer linear programming formulations based on state-expanded networks in which nodes correspond to rule-related states often have very strong LP relaxations. A challenge of these formulations is that they typically give rise to large model instances. If one is willing to trade in optimality for computation time, a way to reduce the size of the model instances is to heuristically remove unpromising nodes and arcs from the state-expanded networks.

In this paper, we propose to employ machine learning models for guiding the reduction of state-expanded networks for multi-activity shift scheduling problems. More specifically, we train a model that predicts the flow through a node from its state attributes, and based on this prediction, we decide whether to keep a node or not. In experiments with a well-known set of multi-activity shift scheduling instances, we show that our approach substantially reduces both the size of the model instances and their solution times while still obtaining optimal solutions for the vast majority of the instances. The results indicate that our approach is competitive with a state-of-the art Lagrangian-relaxation-based matheuristic for multi-activity shift scheduling problems.

1 Introduction

The shift scheduling problem consists in designing a set of employee work shifts that cover demands given in terms of number of persons needed per sub-period (e.g. 15 or 30 min) of a day. The composition of a shift needs to follow rules e.g. stemming from labor regulations, union agreements and work contracts. These rules govern aspects such as the total number of working hours per day, the number, placement and duration of breaks and the duration of consecutive blocks of work. In this paper, we focus on an important variant of the problem, the so-called multi-activity shift scheduling problem (MASSP). In this problem, there are multiple activities, that is, different types of demand to be covered in each period, and there are additional rules governing the minimum and maximum duration of consecutive periods an employee can be assigned a single activity and rules governing the sequencing of activities.

© Springer Nature Switzerland AG 2021
P. J. Stuckey (Ed.): CPAIOR 2021, LNCS 12735, pp. 383–391, 2021.
https://doi.org/10.1007/978-3-030-78230-6_24

Various publications propose mathematical programming approaches for solving the MASSP exactly, either directly solving monolithic MILP formulations [3,4,6] or using decomposition approaches such as Branch-and-Price [5,7]. Several of these approaches rely on formulating the set of feasible shifts in formal languages such as automata [3,7] or context-free grammars [4,5]. Heuristic approaches for the MASSP that have been proposed in the literature are a large neighbourhood search [10] and a recently published Lagrangian relaxation-based matheuristic [8] which can be considered the state-of-the-art in heuristic approaches for the MASSP.

In this paper, we formulate the MASSP as a MILP which handles the shift composition rules in a state-expanded network in which every source-to-sink path correspond to a feasible shift. This network, in which nodes are associated with rule-related states and arcs are associated with the assignment of activities and breaks, is embedded in in the MILP as a flow component. Formulations based on state-expanded networks were previously applied to other personnel scheduling problems such as airline crew scheduling [9] and nurse rostering [11].

The state-expanded network formulation for the MASSP presented here is very strong, but it leads to huge model instances and its solution time is not competitive with state-of-the-art exact approaches. Still, its structure allows us to employ heuristic network reduction techniques that make the models much smaller and faster to solve. Obviously, such a network reduction turns the formulation into a primal approximation, and an important (and difficult) question is which nodes and arcs can be removed without compromising the solution quality too much. Therefore, we propose to use machine learning to support this selection. More specifically, we train a regression model that predicts the flow through a node from its state attributes. Based on this prediction, we decide whether to keep a node or not. We observe that while there is a constantly growing amount of papers combining ML and OR, see e.g. the recent overview [1], we are not aware of any approach using ML techniques for reducing networks underlying large-scale MILP models. For an established set of MASSP instances, we show that compared to a simple network reduction heuristic, the ML-based reduction is clearly superior. Within a time limit of 10 min, the approach obtains the (known) optimal solution for most instances – showing that the new approach is competitive with the state-of-the art in heuristics for the MASSP.

The remainder of the paper is organized as follows: In the next section, we introduce the structure of state-expanded networks that encode all shift composition rules for a simple example and for the well-known Demassey instances first introduced in [7]. In Sect. 3, we present the MILP formulation based on the state-expanded network, and in Sect. 4, we describe how we use Machine Learning to guide the reduction of the state-expanded network underlying the formulation. Section 5 presents the results from our computational experiments.

2   Shifts as Paths in State-Expanded Networks

In the MASSP, we seek to design a minimum-cost set of shifts covering the demands for a set of activities A for each period p in a planning period P (as an example, P may consist of all 15 min-periods of a day). Typically, the number n of shifts to be composed is given a priori, and it is assumed that all employees to which these shifts will be assigned have the same skills. The objective is to minimize total costs composed of assignment costs incurred for each period and of penalties for under- or overcovering demand for an activity a in a period p. In this paper, we assume that all shift composition rules are hard, but it is straightforward to incorporate soft rules.

As mentioned in the introduction, key to our formulation is the idea that all shift composition rules are encoded in a (directed) state-expanded network $G = (N, E)$ in a way that each path from the source node to the sink v^{sink} corresponds to a feasible shift and the set of all source-sink paths in G corresponds to the set of all feasible shifts. Each node in $N^{\mathrm{state}} = N \backslash \{v^{\mathrm{source}}, v^{\mathrm{sink}}\}$ is associated with a rule related state s_v which typically is a tuple of state attributes. The arc $e^{\mathrm{circ}} = (v^{\mathrm{source}}, v^{\mathrm{sink}}) \in E$ is denoted as the flow circulation arc, and in case of a given number of employees n, its flow value is fixed to n. All arcs between the nodes in N^{state} represent state transitions induced by assigning a block of work or break with a length of at least one period.

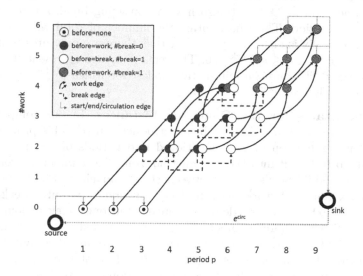

Fig. 1. State-expanded network for the example

Example. To illustrate the construction of a state-expanded network representing the set of feasible shifts, let us consider a small and simplified single-activity shift scheduling problem with a horizon of 9 periods. In this problem, a feasible shift needs to satisfy the following hard shift composition rules: A shift needs to contain either five or six periods of work spread across two blocks of work which have to be separated by a break lasting one period. A work block has a duration between two periods and four periods. To model this problem, we assign each node $v \in N^{\mathrm{state}}$ a state s_v consisting of four state attributes: s_v^p is the period index of the node, s_v^{before} is a categorical attribute denoting whether the last assignment was *none*, *break* or *work*. The attribute $s_v^{\#\text{break}}$ denotes the number of break periods taken so far and $s_v^{\#\text{work}}$ is a counter of the number of work periods assigned so far. The resulting network with all feasible nodes is displayed in Fig. 1. The state attributes of the nodes in N^{state} are visualized using the node position (s_v^{p} corresponds to the x-axis, $s_v^{\#\text{work}}$ to the y-axis); the combinations of the other attributes and the source and sink nodes are depicted with different colors and shapes, see the legend of the figure. The arcs between nodes in $v \in N^{\text{state}}$ represent state transitions induced by assigning activities and breaks. In particular, the arcs emanating from a node v for which $s_v^{\text{before}} =$ *none* or *break* represent assignments of two consecutive work periods, ensuring that the minimum work block duration rule is satisfied. A node v for which $s_v^{\text{before}} =$ *work* may have an outgoing work arc representing a single additional period of work assignment if the maximum duration of the work block of four will not be exceeded. Note in the special constellation of rules in the example, we do not need to store an extra attribute to measure the length of a work block, since we can infer feasibility of the maximum work block duration rule from a combination of the state attributes $s_v^{\#\text{work}}$ and $s_v^{\#\text{break}}$. The nodes for which $s_v^{\text{before}} =$ *work* and $s_v^{\#\text{break}} = 0$ also have an outgoing break arc representing a single break period. The rule limiting the number of total work periods to be either five or six is ensured by the fact that there are only arcs from nodes $v \in N^{\text{state}}$ to v^{sink} if $5 \leq s_v^{\#\text{work}} \leq 6$. The remaining arcs represent the connections between v^{source} and the initial state nodes (for which $v^{\text{before}} =$ *none* and the flow circulation arc e^{circ}.

Rules and States for the Demassey Instances. In our computational experiments, we use the MASS problem introduced in [7]. This problem deals with a planning horizon of 24 h partitioned into periods of 15 min, and each shift needs to fully fall into the planning horizon. The number of work hours performed per shift needs to be between three and eight. Short shifts with less than six hours of work need to have a single 15-minute break, longer shifts need to exhibit two single-period breaks and one break of four periods (one hour). The minimum number of consecutive periods for which an activity is performed is four, and a switch between two different activities is only allowed if there is a break in between. To model these shift composition rules, we need to slightly extend the state representation discussed in the example above: Instead of having a single break-related state attribute, we now have one attribute for the number of short breaks and one for the number of long breaks assigned so far. The full tuple of state attributes of a node v is as follows: $s_v = (s_v^{\text{p}}, s_v^{\text{before}}, s_v^{\#\text{work}}, s_v^{\#\text{shortbreak}}, s_v^{\#\text{longbreak}})$. Note that in the multi-activity case, the attribute s_v^{before} has one possible value for each activity (instead of only *work* in the single-activity case) in addition to the values *none* and *break*. Analogously, the two types of work arcs (long arcs representing a block with the minimum-duration activity assignment and short arcs representing single-period activity assignment) discussed above are also present in the network for each activity. Regarding the breaks, instead of only having single-break arcs, we also add arcs spanning four periods to represent long breaks.

3   MILP Formulation

The state-expanded network constitutes the core element of our MILP formulation for the MASSP. It enters the model in form of a network flow component. The flow on an arc $e \in E$ is represented by the integer decision variable X_e. The cost of a unit flow on arc e is denoted as c_e. As an example, depending on the assignment represented by arc e, this cost factor may include the cost of the working time represented by e, and be 0 otherwise. The other two sets of decision variables are $Y^u_{a,p}$ and $Y^o_{a,p}$ which model the under- and overcovering of the demand $d_{a,p}$ of activity a in period p; these variables are associated with penalties c^u and c^o for under- and overcovering. Note that in case of hard covering limits, the corresponding variables can be forced to be 0.

Using the described symbols, the MILP model can be written as follows:

$$\min \sum_{e \in E} c_e X_e + \sum_{a \in A} \sum_{p \in P} \left(c^o Y^o_{a,p} + c^u Y^u_{a,p} \right) \tag{1}$$

$$\sum_{e \in v^{\text{in}}} X_e = \sum_{e \in v^{\text{out}}} X_e \qquad \forall v \in N \tag{2}$$

$$X_{e^{\text{circ}}} = n \tag{3}$$

$$\sum_{e \in E^{\text{covers}}_{a,p}} X_e + Y^u_{a,p} - Y^o_{a,p} = d_{a,p} \qquad \forall a \in A, p \in P \tag{4}$$

$$X_e \in \mathbb{Z}^+_0 \qquad \forall e \in E \tag{5}$$

$$Y^o_{a,p} \geq 0, \quad Y^u_{a,p} \geq 0 \qquad \forall a \in A, p \in P \tag{6}$$

The objective function (1) contains the cost induced by the flow in the state-expanded network and the penalties for over- and undercovering demand. (2) are the flow balance constraints for each node ensuring that for each v, the flow on the incoming arcs v^{in} equals the flow on the outgoing arcs v^{out}, and constraint (3) fixes the flow on the circulation arc e^{circ} to the number n of employees. In case that the number of employees is a decision in the problem at hand, this constraint can be dropped, and the cost of an employee can be modeled in the cost coefficient c_e^{circ}. Constraint set (4) models the demand covering for each activity and period, the set $E^{\text{covers}}_{a,p} \subset E$ is the set of arcs representing an assignment that covers activity a in period p. In order to extract the set of n shifts from the solution of this model, we need to decompose the flow in G into n paths; observe that such a decomposition is not necessarily unique.

4   Learning to Reduce the Network

The formulation presented in the previous sections gives rise to huge model instances resulting in relatively long solution times which impairs the practical usefulness of the approach. If one is willing to sacrifice the guarantee of finding an optimal solution, one way to reduce both model size and computation time is to remove "unpromising" nodes and arcs from the state-expanded network. To decide which nodes to remove, a straightforward approach is to resort to simple heuristics such as removing arcs representing very long shifts. As we will see later, however, such simplistic rules may severely impact the solution quality.

To achieve better results, we propose to employ machine learning to guide the selection of the nodes to remove from the network. For this prediction, we exploit the fact that each node $v \in N^{\text{state}}$ is associated with a tuple of state attributes which can be used as predictors. The idea is to let a regression model predict the flow $f_v = \sum_{e \in v^{\text{in}}} X_e$ through a node in an optimal solution. The features used for the prediction $\hat{f}_v$ of the flow through a node v are the state attributes s_v of a node and the average workload in the problem instance, that is, the ratio between the total demand $d = \sum_{a \in A} \sum_{p \in P} d_{a,p}$ and the number of employees n. This means that for the Demassey instances used in the paper, we have six features (the state attributes described in Sect. 2 and the average workload) for each data point. One subtlety arises with respect to the attribute s_v^{before}: If the previous assignment was a work activity, its value corresponds to an activity id, and for each activity, there is a "copy" of v with all other attributes being equal. This leads to symmetry issues and to the problem that a regression model trained with a certain number of activities does not generalize to instances with different numbers of activities. To overcome these issues, we aggregate the node information of all nodes for which s^{before} is a work activity and all other attributes are identical; the flow label of this "aggregated node" is then the sum of the flows of all the flow labels of the original nodes.

To train the regression model, we use the solutions obtained by optimally solving the full model for a set of training instances. Note that this means that for each instance in the training set, the number of data points corresponds to the number of nodes $|N^{\text{state}}|$ of a single-activity instance corresponding to around 13,000 data points per instance. The reduced network is obtained as follows: For each node in the network we check if it should be kept or not by comparing the (real-valued) predicted flow value $\hat{f}_v$ to a threshold θ. If $\hat{f}_v < \theta$, we mark the node for removal. In a second step, we check if there are nodes without a removal mark for which all predecessors are marked. If there are such unreachable nodes, we make them reachable by (recursively) deleting removal marks from all predecessors of the unreachable unmarked nodes. After this procedure, we remove all marked nodes and their adjacent arcs.

5   Experimental Results

To evaluate our approach, we use the Demassey instances that were introduced and described in [7]. These instances are probably the most widely used instances for evaluating and comparing solution approaches for the MASSP. The instance set contains 100 instances with 1 to 10 activities and for each number of activities there are 10 instances. All instances have the same rules set and only vary with respect to the number of activities and with respect to work demand. To ensure that we only use out-of-sample predictions when evaluating our ML-based network reduction approach, we partition the instance set into three subsets (the first subset consists of the first four instances of each activity, the second of the next three and the third subset of the last three). Like in k−fold cross validation, we predict the flow values for the instances in each subset using a regression model that was trained on the instances in the union of the two other subsets. For the regression, we use the XGBoost-Regressor [2] with standard parameters. XGBoost uses an ensemble of regression trees, and in our tests outperformed other regression approaches. After training, the XGBoost-Regressor can be used to perform a feature importance analysis. For our problem, it turned out that by far the most important features were state attributes s^p and $s_v^{\#work}$ as well as the average workload per employee in the instance.

Table 1 provides information about the size of the model instances for the model using the full (unreduced) network. In addition, it shows the average number of nodes and the average reduction in number of nodes in percent that is obtained by a simple heuristic reduction rule and by the ML reduction. In the simple reduction heuristic, all nodes that represent states in which the number of work periods is higher than the average work load of an employee in the instance + four periods is removed. In the ML-based reduction, we use the strategy described in Sect. 4 using $\theta = 0.0005$ for the instance groups with up to 5 activities and $\theta = 0.0015$ for the others; these values were determined experimentally and could certainly be tuned in future work. The table shows that the ML-based strategy removes the nodes much more aggressively than the simple reduction rule: While the ML-based reduction on average removes about 62% of all nodes, the simple heuristic only removes about 20% on average.

Table 1. Instance sizes and effect of the network reduction

	Full network			Heur. reduction (avg)		ML reduction (avg)	
Act	Cols	Rows	Nodes	Nodes	% removed	Nodes	% removed
1	22204	13017	12920	11857.5	8.2	7082.2	45.2
2	44303	20178	19985	17339.4	13.2	9925.8	50.3
3	66402	27339	27050	23391.7	13.5	13975.9	48.3
4	88501	34500	34115	26635.7	21.9	15846	53.6
5	110600	41661	41180	33014.5	19.8	18117.7	56
6	132699	48822	48245	37038.8	23.2	13577.1	71.9
7	154798	55983	55310	37405.2	32.4	13192.6	76.1
8	176897	63144	62375	47136.8	24.4	15804.4	74.7
9	198996	70305	69440	52200.3	24.8	16860.8	75.7
10	221095	77466	76505	60930.1	20.4	21655.2	71.7
AVG:	121649.5	45241.5	44712.5	34695	20.2	14603.8	62.3

In Table 2, we present the average quality of the solutions and the average solution times of the approaches presented in Table 1. Our MILP models are solved using Gurobi 9.1.1 with the Barrier solver for solving the root relaxation. All computations are conducted on a notebook with an Intel CORE i7 - 10750H CPU (2.60 GHz) and 16 GB RAM. Table 1 compares these results to those from the state-of-the-art matheuristic approach presented in [8] which had been obtained on a server with an Intel Xeon E5-2687W 3.1 gigahertz processor and 64 GB RAM, using CPLEX 12.6. For solving the full model, we used a time limit of 60 min, and for the reduced models, we set a time limit of 10 min. The result shows that even the full model could be solved to optimality for all but nine instances within one hour. Regarding the reduced models, we can make the following observations: First, the quality of the solutions obtained with the simple reduction heuristic is much worse than the quality of the solutions obtained with the ML-based reduction approach; despite the fact that the ML-based model instances are only about half as big.

Table 2. Solution quality and solution times

	Full network		Heur. reduction			ML Reduction			Lagrangian matheuristic		
	NbS	Time	NbS		Time	NbS		Time	NbS		Time
Act	Opt	Avg s	Opt	1%	Avg s	Opt	1%	Avg s	0.01%	1%	Avg s
1	10	5.5	5	5	2.4	10	10	4	7	10	12
2	10	12.9	3	3	5.6	9	9	4.4	8	9	75.6
3	10	98	5	5	38.5	9	10	48.8	9	10	97.5
4	10	97.9	3	5	39.6	10	10	80.3	6	7	180.4
5	10	186.5	4	4	76.7	9	10	185.8	9	9	35.3
6	9	566.7	8	8	119.9	9	9	54	7	7	95.9
7	9	858.6	6	7	80	9	10	100.1	7	9	131.6
8	10	1170.9	6	8	97	9	10	63.6	4	7	113.8
9	7	1443.3	5	5	83.4	10	10	142	7	10	54.6
10	6	2148.4	1	3	319.8	6	6	324	6	8	172.1
AVG:	9.1	658.9	4.6	5.3	86.3	9	9.4	100.7	7	8.6	96.9

Second, the quality of the solutions obtained with the ML-based approach is not only good, but for the vast majority of the instances, it is able to find the optimal solution to the original problem. The quality of the solutions for the instance group 10 falls behind a bit here, but in that case, for all four instances for which no solution with a gap smaller than 1% was found, the 10-min time limit was hit. Third, the ML-based approach is clearly competitive with the matheuristic presented in [8]: For most instance groups, our approach finds more near-optimal solutions, but also, the average computation time of our approach is slightly worse. As mentioned above, however, their results were obtained on different hardware and with an older solver – the same is true for the results with state-of-the art exact approaches not reported here.

6 Conclusions

In this paper, we presented initial results from using ML to guide the reduction of state-expanded networks for solving MASSP problems. The results are very encouraging: Compared to a simple network reduction heuristic, the ML-based strategy is able to achieve much better results that are in the range of other state-of-the art heuristics for the MASSP.

References

1. Bengio, Y., Lodi, A., Prouvost, A.: Machine learning for combinatorial optimization: a methodological tour d'horizon. Eur. J. Oper. Res. **290**(2), 405–421 (2021)
2. Chen, T., Guestrin, C.: XGBoost: a scalable tree boosting system. In: Proceedings of the 22nd ACM SIGKDD International Conference on Knowledge Discovery and Data Mining, KDD 2016, pp. 785–794. ACM, New York (2016). http://doi.acm.org/10.1145/2939672.2939785
3. Côté, M.C., Gendron, B., Quimper, C.G., Rousseau, L.M.: Formal languages for integer programming modeling of shift scheduling problems. Constraints **16**(1), 54–76 (2011)
4. Côté, M.C., Gendron, B., Rousseau, L.M.: Grammar-based integer programming models for multiactivity shift scheduling. Manage. Sci. **57**(1), 151–163 (2010)
5. Côté, M.C., Gendron, B., Rousseau, L.M.: Grammar-based column generation for personalized multi-activity shift scheduling. INFORMS J. Comput. **25**(3), 461–474 (2013)
6. Dahmen, S., Rekik, M., Soumis, F.: An implicit model for multi-activity shift scheduling problems. J. Sched. **21**(3), 285–304 (2017). https://doi.org/10.1007/s10951-017-0544-y
7. Demassey, S., Pesant, G., Rousseau, L.-M.: Constraint programming based column generation for employee timetabling. In: Barták, R., Milano, M. (eds.) CPAIOR 2005. LNCS, vol. 3524, pp. 140–154. Springer, Heidelberg (2005). https://doi.org/10.1007/11493853_12
8. Hernández-Leandro, N.A., Boyer, V., Salazar-Aguilar, M.A., Rousseau, L.M.: A matheuristic based on Lagrangian relaxation for the multi-activity shift scheduling problem. Eur. J. Oper. Res. **272**(3), 859–867 (2019)
9. Mellouli, T.: A network flow approach to crew scheduling based on an analogy to a train/aircraft maintenance routing problem. In: Voss, S., Daduna, J. (eds.) Computer-Aided Scheduling of Public Transport. LNEMS, vol. 505, pp. 91–120. Springer, Berlin (2001). https://doi.org/10.1007/978-3-642-56423-9_6
10. Quimper, C.G., Rousseau, L.M.: A large neighbourhood search approach to the multi-activity shift scheduling problem. J. Heuristics **16**(3), 373–392 (2010)
11. Römer, M., Mellouli, T.: A direct MILP approach based on state-expanded network flows and anticipation for multi-stage nurse rostering under uncertainty. In: Burke, E.K., Di Gaspero, L., Özcan, E., McCollum, B., Schaerf, A. (eds.) PATAT 2016: Proceedings of the 11th International Conference of the Practice and Theory of Automated Timetabling, Udine, Italy, pp. 549–552 (2016)

SeaPearl: A Constraint Programming Solver Guided by Reinforcement Learning

Félix Chalumeau[1], Ilan Coulon[1], Quentin Cappart[2(✉)],
and Louis-Martin Rousseau[2]

[1] École Polytechnique, Institut Polytechnique de Paris, Palaiseau, France
{felix.chalumeau,ilan.coulon}@polytechnique.edu
[2] École Polytechnique de Montréal, Montreal, Canada
{quentin.cappart,louis-martin.rousseau}@polymtl.ca

Abstract. The design of efficient and generic algorithms for solving combinatorial optimization problems has been an active field of research for many years. Standard exact solving approaches are based on a clever and complete enumeration of the solution set. A critical and non-trivial design choice with such methods is the branching strategy, directing how the search is performed. The last decade has shown an increasing interest in the design of machine learning-based heuristics to solve combinatorial optimization problems. The goal is to leverage knowledge from historical data to solve similar new instances of a problem. Used alone, such heuristics are only able to provide approximate solutions efficiently, but cannot prove optimality nor bounds on their solution. Recent works have shown that reinforcement learning can be successfully used for driving the search phase of constraint programming (CP) solvers. However, it has also been shown that this hybridization is challenging to build, as standard CP frameworks do not natively include machine learning mechanisms, leading to some sources of inefficiencies. This paper presents the proof of concept for `SeaPearl`, a new CP solver implemented in *Julia*, that supports machine learning routines in order to learn branching decisions using reinforcement learning. Support for modeling the learning component is also provided. We illustrate the modeling and solution performance of this new solver on two problems. Although not yet competitive with industrial solvers, `SeaPearl` aims to provide a flexible and open-source framework in order to facilitate future research in the hybridization of constraint programming and machine learning.

Keywords: Reinforcement learning · Solver design · Constraint programming

1 Introduction

The goal of combinatorial optimization is to find an optimal solution among a finite set of possibilities. Such problems are frequently encountered in transportation, telecommunications, finance, healthcare, and many other fields [1, 5, 32, 42,

---

F. Chalumeau and I. Coulon—The two authors contributed equally to this paper.

© Springer Nature Switzerland AG 2021
P. J. Stuckey (Ed.): CPAIOR 2021, LNCS 12735, pp. 392–409, 2021.
https://doi.org/10.1007/978-3-030-78230-6_25

48, 49]. Finding efficient methods to solve them has motivated research efforts for decades. Many approaches have emerged in the recent years seeking to take advantage of learning methods to outperform standard solving approaches. Two types of approaches have been particularly successful while still showing drawbacks. First, machine learning approaches, such as deep reinforcement learning (DRL), have shown their promise for designing good heuristics dedicated to solve combinatorial optimization problems [6, 15, 16, 29]. The idea is to leverage knowledge from historical data related to a specific problem in order to solve rapidly future instances of the problem. Although a very fast computation time for solving the problem is guaranteed, such approaches only act as a heuristics and no mechanisms for improving a solution nor to obtain optimality proofs are proposed. A second alternative is to embed a learning-component inside a search procedure. This has been proposed for mixed-integer programming [22], local search [21, 56], SAT solvers [38, 45], and constraint programming [2, 11]. However, it has been shown that such a hybridization is challenging to build, as standard optimization frameworks do not natively include machine learning mechanisms, leading to some sources of inefficiencies. As an illustrative example, Cappart et al. [11] used a deep reinforcement learning approach to learn the value-selection heuristic for driving the search of a CP solver. To do so, they resorted to a *Python* binding in order to call deep learning routines in the solver Gecode [44], causing an important computational overhead.

Following this idea, we think that learning branching decisions in a constraint programming solver is an interesting research direction. That being said, we believe that a framework that can be used for prototyping and evaluating new ideas is currently missing in this research field. Based on this context, this paper presents SeaPearl (homonym of CPuRL, standing for *constraint programming using reinforcement learning*), a flexible, easy-to-use, research-oriented, and open-source constraint programming solver able to natively use deep reinforcement learning algorithms for learning value-selection heuristics. The philosophy behind this solver is to ease and speed-up the development process to any researcher desiring to design learning-based approaches to improve constraint programming solvers. Accompanying this paper, the code is available on Github[1], together with a tutorial showcasing the main functionalities of the solver and how specific design choices can be stated. Experiments on two toy-problems, namely the *graph coloring* and the *travelling salesman problem with time windows*, are proposed in order to highlight the learning aspect of the solver. Note also that compared to *impact-based search* [40], hybridization with ant colony optimization [46], or similar mechanisms where a learning component is used to improve the search of the solving process for a specific instance, the goal of our learning process is to leverage knowledge learned from other similar instances. This paper is built upon the proof of concept proposed by Cappart et al. [11]. Our specific and original contributions are as follows: (1) we propose an architecture able to solve CP models, whereas [11] was restricted to dynamic programming models, (2) the learning phase is fully integrated inside the CP

---

[1] https://github.com/corail-research/SeaPearl.jl.

solver, and (3) the reinforcement learning environment is different as it allows CP backtracking inside an episode. Finally, the solver is fully implemented in *Julia* language, avoiding the overhead of *Python* calls from a C++ solver.

The next section introduces reinforcement learning and graph neural network, a deep architecture used in the solver. The complete architecture of SeaPearl is then proposed, followed by an illustration of the modelling support for the learning component. Finally, experiments and discussions about research directions that can be carried out with this solver are proposed.

2   Technical Background

This section gives details about the two main concepts that make SeaPearl different from other CP solvers.

2.1   Reinforcement Learning

Reinforcement Learning (RL) [47] is a sub-field of machine learning dedicated to train agents to take actions in an environment in order to maximize an accumulated reward. The goal is to let the agent interacts with the environment and discovers which sequences of actions lead to the highest reward. Formally, let $\langle S, A, T, R \rangle$ be a tuple representing the environment, where S is the set of states that can be encountered in the environment, A is the set of actions that can be taken by the agent, $T : S \times A \mapsto S$ is the transition function leading the agent from a state to another one given the action taken and, $R : S \times A \mapsto \mathbb{R}$ is the reward function associated with a particular transition. The behaviour of an agent is driven by its policy $\pi : S \mapsto A$, deciding which action to take when facing a specific state S. The goal of an agent is to compute a policy maximizing the accumulated sum of rewards during its lifetime, referred to as an *episode*, and defined by a sequence of states $s_t \in S$ with $t \in [1, T]$ and s_T is the terminal state. Considering a discounting factor γ, the total return at step t is denoted by $G_t = \sum_{k=t}^{T} \gamma^{k-t} R(s_k, a_k)$.

In deterministic environments, the value of taking an action a from a state s under a policy π is defined by the action-value function $Q^{\pi}(s_t, a_t) = G_t$. Then, the problem consists in finding a policy that maximizes the final return: $\pi^* = \text{argmax}_{\pi} Q^{\pi}(s, a), \forall s, a \in S \times A$. However, the number of possibilities has an exponential increase with the number of states and actions, which makes solving this problem exactly intractable. Reinforcement learning approaches tackle this issue by letting the agent interact with the environment in order to learn information that can be leveraged to build a good policy. Many RL algorithms have been developed for this purpose, the most recent and successful ones are based on deep learning [23] and are referred to as *deep reinforcement learning* [3]. The idea is to approximate either the policy π, or the action-value function Q by a neural network in order to scale up to larger state-action spaces. For instance, *value-based methods*, such as DQN [36], have the following approximation: $\hat{Q}^{\pi}(\theta, s_t, a_t) \approx Q^{\pi}(s_t, a_t)$; whereas *policy-based methods* approximates the policy: $\hat{\pi}(\theta, s) \approx \pi(s)$, where θ are parameters of a trained neural network.

2.2   Graph Neural Network

Learning on graph structures is a recent and active field of research in the machine learning community. It has plenty of applications such as molecular biology [27], social sciences [37], and physics [41]. It has also been considered for solving combinatorial optimization problems [10]. Formally, let $G = (V, E)$ be a graph with V the set of vertices, E the set of edges, $\mathbf{f}_v \in \mathbb{R}^k$ a vector of k features attached to a vertex $v \in V$, and similarly, $\mathbf{h}_{v,u} \in \mathbb{R}^q$ a vector of q features attached to an edge $(v, u) \in V$. Intuitively, the goal of *graph neural networks* (GNN) is to learn a p-dimensional representation $\mu_v \in \mathbb{R}^p$ for each node $v \in V$ of G. Similar to convolutional neural networks that aggregate information from neighboring pixels of an image, GNNs aggregate information from neighboring nodes using edges as conveyors. The features f_v are aggregated iteratively with the neighboring nodes in the graph. After a predefined number of aggregation steps, the node embedding are produced and encompass both local and global characteristics of the graph.

Such aggregations can be performed in different ways. A simple one has been proposed by Dai et al. [14] and used by Khalil et al. [28]. It works as follows. Let T be the number of aggregation steps, μ_v^t be the node embedding of v obtained after t steps and $\mathcal{N}(v)$ the set of neighboring nodes of $v \in V$ in G. The recursive computation of μ_v^t is shown in Eq. (1), where vectors $\theta_1 \in \mathbb{R}^{p \times k}$, $\theta_2 \in \mathbb{R}^{p \times p}$, $\theta_3 \in \mathbb{R}^{p \times p}$, $\theta_4 \in \mathbb{R}^{p \times q}$ are vectors of parameters that are learned, and σ a non-linear activation function such as ReLU. The final embedding μ_v^{T+1} obtained gives a representation for each node v of the graph, that can consequently be used as input of regular neural networks for any prediction tasks.

$$\mu_v^{t+1} = \sigma \left(\theta_1 \mathbf{f}_v + \theta_2 \sum_{u \in \mathcal{N}(v)} \mu_u^t + \theta_3 \sum_{u \in \mathcal{N}(v)} \sigma \left(\theta_4 \mathbf{h}_{v,u} \right) \right) \quad \forall t \in \{1, \dots, T\} \quad (1)$$

Many variants and improvements have been proposed to this framework. A noteworthy example is the *graph attention network* [50], that uses an attention mechanism [4], commonly used in recurrent neural networks. Detailed information about GNNs are proposed in the following surveys [10,12,53,55] and an intensive comparisons on the computational results of the different architectures have been proposed by Dwivedi et al. [20].

3   Embedding Learning in Constraint Programming

This section describes the architecture and the design choices behind SeaPearl. A high-level overview is illustrated in Fig. 1. Mainly inspired by [11], the architecture has three parts: a constraint programming solver, a reinforcement learning model, and a common representation acting as a bridge between both modules.

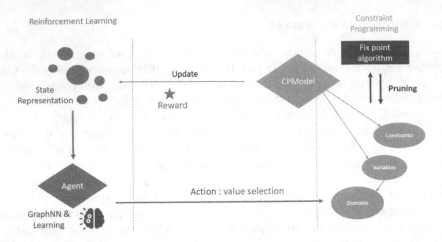

Fig. 1. Overview of `SeaPearl` architecture

Constraint Programming Solver. A CP model is a tuple $\langle X, D, C, O \rangle$ where X is the set of variables we are trying to assign a value to, $D(X)$ is the set of domains associated with each variable, C the set of constraints that the variables must respect and O an objective function. The goal of the solver is to assign a value for each variable $x \in X$ from $D(x)$ which satisfy all the constraints in C and that optimize the objective function O. The design of the solving process is heavily inspired by what has been done in modern *trailing-based* solvers such as OscaR [39], or Choco [26]. It also takes inspiration from MiniCP [30] in its philosophy. The focus is on the extensibility and flexibility of the solver, especially for the learning component. The goal is to make learning easy and fast to prototype inside the solver. That being said, the solver is *minimalist*. At the time of writing, only few constraints are implemented.

Reinforcement Learning Model. The goal is to improve the CP solving process using knowledge from previously solved problems. It is done by learning an appropriate value-selection heuristic and using it at each node of the tree search. Following Bengio et al. [7], this kind of learning belongs to the third class (*machine learning alongside optimization algorithms*) of ML approaches for solving combinatorial problems, and raises many challenges. To do so, a generic reinforcement learning environment genuinely representing the behaviour of the solving process for solving a CP model must be designed. Let $\mathcal{Q}^p$ be a specific instance of a combinatorial problem p we want to solve, $\mathcal{C}_i^p$ be the associated CP model at the i-th explored node of the tree search, and $\mathcal{S}_i^p$ be statistics of the solving process at the i-th node (number of bactracks, if the node has been already visited, etc.). The environment $\langle S, A, T, R \rangle$ we designed is as follows.

State. We define a state $s_i \in S$ as the triplet $(\mathcal{Q}^p, \mathcal{C}_i^p, \mathcal{S}_i^p)$. By doing so, each state contains (1) information about the instance that is solved, (2) the current state of the CP model and (3) the current state of the solving process. In practice, each state is embedded into a d-dimensional vector of features, that serves as input for a neural network. This can be done in different manners and two possible representations are proposed in the case studies.

Action. Each action corresponds to a value that can be assigned to a variable of the CP model. An action $a \in A$ at a state $s_i \in S$ is available if and only if it is in the domain of the variable x that has been selected for branching on at step i ($a \in D(X)$ for $\mathcal{C}_i^p$).

Transition Function. The transition updates the current state according to the action that has been selected. In our case, it updates the domains of the different variables. It is important to highlight that the transition encompasses everything that is done inside the CP solver at each branching step, such as the constraint propagation during the fix-point computation, or the trailing in case of backtrack. This is an important difference with [11] where the transition only consists in the assignation of a value to a variable and is disconnected from the internal mechanisms of the CP solving process.

Reward Function. The reward is a key component of any reinforcement learning environment [17]. In our case, it has a direct impact on how the tree search is explored. Although it is commonly expressed as a function of the objective function O, it is not clear how it can be shaped in order to drive the search to provide feasible solutions, the best one and to prove optimality, which often require different branching strategies. For this reason, the solver allows the user to define its own reward. That being said, a reward that gives a penalty of -1 at each step is integrated by default. This simple reward encourages the agent to conclude an episode as soon as possible. The end of an episode means that the solver reached optimality. Hence, giving such a penalty drove the agent to reduce the number of visited nodes for proving optimality. An alternative definition has been proposed in [11]. The reward signal consists in two terms, having a different importance. The first one is dedicated to find a feasible solution, whereas the second one drives the episode to find the best feasible solution. The reward is designed in order to prioritize the first term. The motivation is to drive the search to find a feasible solution by penalizing the number of non-assigned variables before a failure, and then, driving it to optimize the objective function.

Learning Agent. Once the environment has been defined, any RL agent can be used to train the model [36, 43]. The goal is to build a neural network NN that outputs an appropriate value to branch on at each node of the tree search. At the beginning of each new episode, an instance $\mathcal{Q}^p$ of the problem we want to solve is randomly taken from the training set and the learning is conducted on it. The training algorithm returns a vector of weights (θ) which is used for parametrizing the neural network. A recurrent issue in related works using learning approaches to solve combinatorial optimization problem is having access to enough data for training a model. It is not often the case, and this makes the design of new approaches tedious. To deal with this limitation, the solver integrates a support for generating synthetic instances, which are directly fed in the learning process. Then, the training is done using randomly generated instances sampled from a similar distribution to those we want to solve.

State Representation. In order to ensure the genericity of the framework, the neural architecture must be able to take any triplet $(\mathcal{Q}^p, \mathcal{C}_i^p, \mathcal{S}_i^p)$ as input, which requires to encode the RL state by a suitable representation. Doing so for the statistics $(\mathcal{S}_i^p)$ is trivial as they mainly consist of numerical values or categorical data. The information related to the instances $(\mathcal{Q}^p)$ is by definition problem-dependent, and several architectures are possible [11]. This information can also be omitted in our solver. However, a representation able to handle any CP model $(\mathcal{C}_i^p)$ is required. In another context, Gasse et al. [22] proposed a variable-constraint bipartite graph representation of mixed-integer linear programs in order to learn branching decisions inside a MIP solver using imitation learning. The representation they used is leveraged using a graph neural network. Following this idea, our solver adopted a similar architecture, referred to as a *tripartite graph* but tailored for CP models. A CP model $\langle X, D, C, O \rangle$ is represented by a simple and undirected graph $G(V_x, V_d, V_c, E)$ as follows. Each variable $x \in X$ is associated to a vertex from V_x, each possible value (union of all the domains) to a vertex from V_d, and each constraint to a vertex from V_c. Edges only connect either nodes from V_x to V_c if the variable x is involved in constraint c, or nodes from V_x to V_d if d is inside the current domain of x. Finally, each vertex and each edge can be labelled with a vector of features, corresponding to additional information of the model (arity of a constraint, domain size of a variable, type of a global constraint, etc.). The main asset of this representation is its genericity, as it can be used to represent any CP model. It is important to note that designing the best state representation is still an open research question and many options are possible. Two representations have been tested in this paper. The first one is the generic representation based on the tripartite graph, whereas the second one leverages problem-dependent features, as in [11].

Solving Algorithm. The solving process of `SeaPearl` is illustrated in Algorithm 1. It mostly works in the same manner as any modern CP solver. The main difference is the consistent use of a learned-heuristic for the value selection. While the search is not completed (lines 8–14), the fix-point algorithm is executed on the current node (line 9), and the features used as input of the neural network are extracted, both concerning the CP model (line 9) and the solving statistics (line 11). Using such information, the trained model is called in order to obtain the value on which the current variable must be branched on (line 12). Finally, the best solution found is returned (line 15). Note also that additional mechanisms, such as prediction caching [11], can be added to speed-up the search. The architecture of the network considered (line 12) is proposed in Fig. 2. It works as follows: (1) the GNN computes a latent d-dimensional vector representation of the features related to the current variable the tripartite graph, (2) the vector is used as input of a fully-connected neural network in order to obtain a score for each possible value (resp. action), and (3) this score is passed through a mask in order to keep only the values that are inside the domain.

Algorithm 1: Solving process of `SeaPearl`

---

1 ▷ **Pre:** $\mathcal{Q}^p$ is a specific instance of combinatorial problem p.
2 ▷ **Pre:** $\mathcal{C}_0^p$ is the state of the CP model $\langle X, D, C, O \rangle$ at the root node.
3 ▷ **Pre:** NN is a neural architecture giving a value at a node of the tree.
4 ▷ **Pre:** $\mathbf{w}$ is a trained weight vector parametrizing the neural network.
5 $\mathcal{C}_0^p := \mathtt{CPEncoding}(\mathcal{Q}_p)$
6 $\Psi := \mathtt{CP\text{-}search}(\mathcal{C}_0^p)$
7 $i := 0$
8 **while** Ψ **is not completed do**
9 | $\quad$ $\mathtt{fixPoint}(\mathcal{C}_i^p)$
10 | $\quad$ $\mathcal{S}_i^p := \mathtt{getSearchStatistics}(\Psi)$
11 | $\quad$ $x := \mathtt{selectVariable}(\mathcal{C}_i^p)$
12 | $\quad$ $v := \mathtt{NN}(\mathbf{w}, x, \mathcal{Q}^p, \mathcal{C}_i^p, \mathcal{S}_i^p)$
13 | $\quad$ $\mathcal{C}_{i+1}^p := \mathtt{branch}(\Psi, x, v)$
14 | $\quad$ $i := i + 1$
15 **return** $\mathtt{bestSolution}(\Psi)$

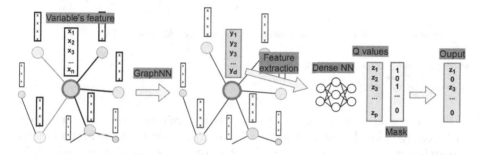

Fig. 2. Simplified representation of the neural network architecture

4 Modeling, Learning and Solving with SeaPearl

The goal of `SeaPearl` is to make most of the previous building blocks transparent for the end-user. This section illustrates how it is done with the graph coloring problem. Let $G(V, E)$ be an undirected graph. A *coloring* of G is an assignment of labels to each node such that adjacent nodes have a different label. The *graph coloring problem* consists in finding a coloring that uses the minimal number of labels. The programming language that we selected to develop `SeaPearl` is *Julia* [8], which is (1) efficient during runtime, and (2) rich in both mathematical programming [9,19,31,33] and machine learning libraries [25,54]. This resolves the issue encountered in [11] where an inefficient *Python* binding has been developed in order to call deep learning routines in C++ implementation of the solver `Gecode` [44]. More examples are also available on the Github repository of the solver. A regular CP model of the graph coloring problem is shown in Listing 1.1. The first step is to build a *trailer*, instantiating the trailing mechanisms and a *model*, instantiating the tuple $\langle X, D, C, O \rangle$. Variables, constraints, and the objective are then added to this object. Once the model is built, the solving process can then be run. At that time, no learning is done.

Listing 1.1. CP model for the graph coloring problem

```
## Preamble ##
n_vertex, n_edge, edges = getInput(instance)
trailer = SeaPearl.Trailer()
model = SeaPearl.CPModel(trailer)

## Variable declaration ##
k = SeaPearl.IntVar(0, n_vertex, trailer)
x = SeaPearl.IntVar[]

for i in 1:n_vertex
    push!(x, SeaPearl.IntVar(1, n_vertex, trailer))
    SeaPearl.addVariable!(model, last(x))
end

## Constraints ##
for v1, v2 in input.edges
    push!(model.constraints, SeaPearl.NotEqual(x[v1], x[v2], trailer))
end

for v in x
    push!(model.constraints, SeaPearl.LessOrEqual(v, k, trailer))
end

## Objective (minimizing by default) ##
model.objective = k
SeaPearl.solve!(model)
```

The next snapshot (Listing 1.2) shows how a reinforcement learning agent can be easily defined. The first instruction corresponds to the definition of the deep architecture, consisting of 4 graph attention layers (GATConv), and 5 fully-connected layers (Dense). To do so, a binding with the library Flux [25] has been developed. The second instruction defines the deep reinforcement learning agent. It requires as input (1) the neural architecture, (2) the optimizer desired (Adam), (3) the type of RL algorithm (DQN) and (4) hyper-parameters that depends on the specific algorithm selected (e.g., the discounting factor).

The last snapshot (Listing 1.3) shows how the training routines can be defined. The value-selection heuristic is first defined as a heuristic that will be trained using the previously defined RL agent. Then, a random generator, dedicated to construct new training instances, is instanciated. For the graph coloring case, this generator is based on the construction proposed by [13] and builds graphs of n vertices with density p that are k-colorable. Finally, the training can be run. To do so, the user has to provide the value-selection to be trained, the instance generator, the number of episodes, the search strategy, and the variable heuristic. Once trained, the heuristic can be used to solve new instances.

Listing 1.2. Reinforcement Learning agent for the graph coloring problem

```
## Neural network architecture ##
neuralNetwork = SeaPearl.FlexGNN(
    graphChain = Flux.Chain(
        GATConv(nInput => 10, heads=2),
        GATConv(20 => 10, heads=3),
        GATConv(30 => 10, heads=3),
        GATConv(30 => 20, heads=2),
    ),
    nodeChain = Flux.Chain(
        Dense(20, 20),
        Dense(20, 20),
        Dense(20, 20),
        Dense(20, 20),
    ),
    outputLayer = Dense(20, nOutput))

## Reinforcement learning agent ##
agent = RL.Agent(
    policy = RL.QBasedPolicy(
        learner = SeaPearl.CPDQNLearner(
            approximator = RL.Approximator(neuralNetwork, ADAM()),
            loss_function = huber_loss,
            discounting_factor = 0.9999,
            batch_size = 32,
            ...
        ),
        explorer = SeaPearl.CPEpsilonGreedyExplorer()))
```

Listing 1.3. Training a value-selection heuristic

```
## Defining the value selection heuristic as the RL agent ##
val_heuristic = SeaPearl.LearnedHeuristic(agent)

## Generating random instances ##
gc_generator = SeaPearl.GraphColoringGenerator()

## Training the model ##
SeaPearl.train!(
    valueSelectionArray = [val_heuristic],
    generator = gc_generator,
    nb_episodes = 1000,
    strategy = SeaPearl.DFSearch,
    variableHeuristic = SeaPearl.MinDomain())
```

We would like to highlight that these pieces of code illustrate only a small subset of the functionalities of the solver. Many other components, such as the reward, or the state representation can be redefined by the end-user for prototyping new research ideas. This has been made possible thanks to the multiple dispatching functionality of *Julia*, allowing the user to redefine types without requiring changes to the source code of `SeaPearl`.

5   Experimental Results

The goal of the experiments is to evaluate the ability of `SeaPearl` to learn good heuristics for value-selection. Comparisons against greedy heuristics on two NP-hard problems are proposed: *graph coloring* and *travelling salesman with time windows.* In order to ease the future research in this field and to ensure reproducibility, the implementation, the models and the results are released in opensource with the solver. Instances for training the models have been generated randomly with a custom generator. Training is done until convergence, limited to 13 h on AWS' EC2 with 1 vCPU of Intel Xeon capped to 3.0GHz, and the memory consumption is capped to 32 GB. The evaluation is done on other instances (still randomly generated in the same manner) on the same machine.

5.1   Graph Coloring Problem

The experiments are based on a standard CP formulation of the graph coloring problem (Listing 1.1), using the smallest domain as variable ordering. Instances are generated in a similar fashion as in [13]. They have a density of 0.5, and the optimal solutions have less than 5 colors. Comparisons are done with a heuristic that takes the smallest available label in the domain (min-value), and a random value selection. For each instance, 200 random trials are performed and the average, best and worst results are reported. The training phase ran for 600 episodes (execution time of 13 h) using DQN learning algorithm [36], a graph attention network has been used as deep architecture [50] upon the tripartite graph detailed in Sect. 3. A new instance is generated for each episode. The first experiment records the average number of nodes that has been explored before proving the optimality of the instances at different steps of the training, and using the default settings of `SeaPearl`. Results are presented in Fig. 3 for graphs with 20 and 30 nodes. Every 30 episodes, an evaluation is performed on a validation set of 10 instances. We can observe that the learned heuristic is able to reproduce the behaviour of the min-value heuristic, showing that the model is able to learn inside a CP solver. Results of the final trained model on 50 new instances are illustrated in Fig. 4 using performance profiles [18]. The metric considered is still the number of nodes explored before proving optimality. Results show that the heuristic performances can be roughly equaled.

5.2   Travelling Salesman Problem with Time Windows

Given a graph of n cities, The *travelling salesman problem with time windows* (TSPTW) consists of finding a minimum-cost circuit that connects a set of cities. Each city i is defined by a position and a time window, defining the period when it can be visited. Each city must be visited once and the travel time between two cities i and j is defined by $d_{i,j}$. It is possible to visit a city before the start of its time windows but the visitor must wait there until the start time. However, it is not possible to visit a city after its time window. The goal is to minimize the sum of the travel distances. This case study has been proposed previously as a proof of concept for the combination of constraint programming and reinforcement learning [11]. We reused the same design choices they did. The CP model is based on a dynamic programming formulation, the neural architecture is based on a graph attention network and the reward is shaped to drive the agent to find first a feasible solution, and then to find the optimal one. It is also noteworthy

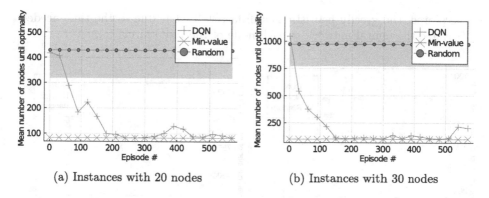

(a) Instances with 20 nodes   (b) Instances with 30 nodes

Fig. 3. Training curve of the DQN agent for the graph coloring problem

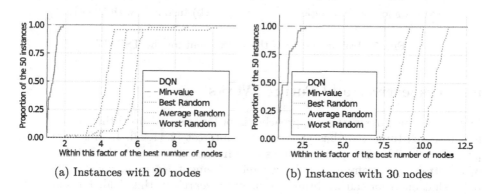

(a) Instances with 20 nodes   (b) Instances with 30 nodes

Fig. 4. Performance profiles (number of nodes) for the graph coloring problem

to mention that the default tripartite graph of our solver is not used in this experiment. A graph representing directly the current TSPTW instance is used instead, with the position and the time windows bounds as node features, and the distances between each pair of nodes as edge features.

Instances are generated using the same generator as in [11]. The training phase ran for 3000 episodes (execution time of 6 h) and a new instance is generated for each episode. The variable ordering used is the one inferred by the dynamic programming model, and the value selection heuristic consists of taking the closest city to the current one. The random value selection is also considered. As with the previous experiment, we record the average number of nodes that have been explored before proving optimality. Results are presented in Fig. 5 for instances with 20 and 50 cities. Once the model has been trained, we observe that the learned heuristic is able to outperform the heuristic baseline with a factor of 3 in terms of the number of nodes visited. This result is corroborated by the performance profiles in Figs. 6a-6b, which show the number of nodes explored before optimality for the final trained model. Execution time required to solve the instances is illustrated in Figs. 6c-6d. We can observe that even if less nodes are explored, the greedy heuristic is still faster. This is due to the time needed to traverse the neural network at each node of the tree search.

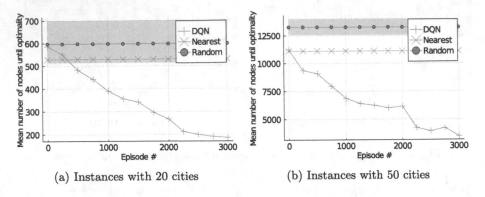

(a) Instances with 20 cities   (b) Instances with 50 cities

Fig. 5. Training curve of the DQN agent for the TSPTW

6 Perspectives and Future Works

Leveraging machine learning approaches in order to speed-up optimization solver is a research topic that has an increasing interest [2,11,22]. In this paper, we propose a flexible and open-source research framework towards the hybridization of constraint programming and deep reinforcement learning. By doing so, we hope that our tool can facilitate the future research in this field. For instance, four aspects can be directly addressed and experimented: (1) how to design the best representation of a CP state as input of a neural network, (2) how to select an an appropriate neural network architecture for efficiently learn value-selection heuristics, (3) what kinds of reinforcement learning algorithm are the most suited for this task, and (4) how the reward should be designed to maximize the performances. Besides, other research questions have emerged during the development of this framework. This section describes five of them.

Extending the Learning to Variable Selection. As a first proof of concept, this work focused on how a value-selection heuristic can be learned inside a constraint programming solver. Although crucial for the performances of the solver, especially for proving optimality [51], this has not been studied in this paper, and it has to be defined by the user. Integrating a learning component on it as well would be a promising direction.

Finding Solutions and Proving Optimality Separately. As highlighted by Vilim et al. [51], finding a good solution and exploring/pruning the search tree efficiently in order to prove optimality are two different tasks, that may require distinct heuristics. On the contrary, the reinforcement learning agent presented in this paper can hardly understand how and when both tasks should be prioritized. This leads to another avenue of future work: having two specialized agents, one dedicated to find good solutions, and the other one to prove optimality.

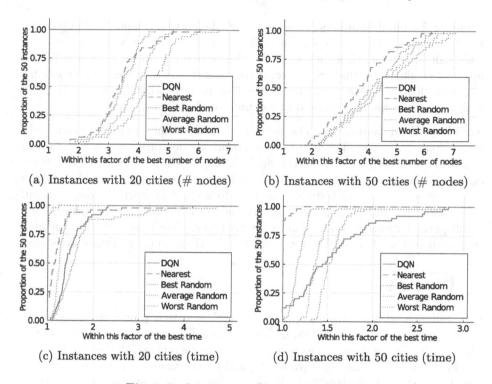

(a) Instances with 20 cities (# nodes)   (b) Instances with 50 cities (# nodes)

(c) Instances with 20 cities (time)   (d) Instances with 50 cities (time)

Fig. 6. Performance profiles for the TSPTW

Accelerating the Computation of the Learned Heuristics. As for any solving tool, efficiency is a primary concern. It is thus compulsory for the learned heuristic to be not only better than a man-engineered heuristic in terms of number of nodes visited but also in terms of execution time. As highlighted in the experiments, calling a neural neural network is time consuming compared to calling a simple greedy heuristic. Interestingly, this aspect has not been so considered in most deep learning works, as the trained model are only called few times, rendering the inference time negligible in practice. In our case, as the model has to be called at each node of the tree search (possibly more than 1 million times), the inference time becomes a critical concern. This opens another research direction: finding an appropriate trade-off between a large model having accurate prediction and a small model proposing worse prediction, but much faster. This has been addressed by Gupta et al. [24] for standard MIP solvers. Another direction is to consider the *network pruning* literature, dedicated to reduce heavy inference costs of deep models in low-resource settings [34, 35].

Reducing the Action Space. A recurrent difficulty is to deal with problems having large domains. On a reinforcement learning perspective, it consists of having a large action space, which makes the learning more difficult, and reduces the generalization to large instances. A possible direction could be to reduce the size of the action space using a dichotomy selection. Assuming a domain of n

values and a number of actions capped at k, the current domain is divided into k intervals, and the selection of an action consists in taking a specific interval. This can be done until a final value has been found. Another option is to use another architectures that are less sensitive to large action spaces, such as *pointer networks* [52], which are commonly used in natural language processing but which have also been considered in combinatorial optimization [16].

Tackling Real Instances. The data used to train the models are randomly generated from a specified distribution. Although this procedure is common in much of published research in the field [22,29], it cannot be used for solving real-world instances. One additional difficulty to consider in real-world instances is having access to enough data to be able to accurately learn the distribution. One way to do that is to modify slightly the available instances by introducing small perturbations on them. This is referred to as *data augmentation*, but may be insufficient as it can fail to represent the distribution of the future instances.

7   Conclusion

Combining machine learning approaches with a search procedure in order to solve combinatorial optimization problems is a hot topic in the research community, but it is still a challenge as many issues must be tackled. We believe that the combination of constraint programming and reinforcement learning is a promising direction for that. However, developing such hybrid approaches requires a tedious and long development process [11]. Based on this context, this paper proposes a flexible, easy-to-use and open-source research framework towards the hybridization of constraint programming and deep reinforcement learning. The integration is done on the search procedure, where the learning component is dedicated to obtain a good value-selection heuristic. Experimental results show that a learning is observed, and highlight challenges related to execution time. Many open challenges should be addressed for an efficient use of machine learning methods inside a solving process. We position this contribution not only as a new CP solver, but also as an open-source tool dedicated to help the community in the development of new hybrid approaches for tackling such challenges.

References

1. Anagnostopoulos, K.P., Mamanis, G.: A portfolio optimization model with three objectives and discrete variables. Comput. Oper. Res. **37**(7), 1285–1297 (2010)
2. Antuori, V., Hebrard, E., Huguet, M.-J., Essodaigui, S., Nguyen, A.: Leveraging reinforcement learning, constraint programming and local search: a case study in car manufacturing. In: Simonis, H. (ed.) CP 2020. LNCS, vol. 12333, pp. 657–672. Springer, Cham (2020). https://doi.org/10.1007/978-3-030-58475-7_38
3. Arulkumaran, K., Deisenroth, M.P., Brundage, M., Bharath, A.A.: Deep reinforcement learning: a brief survey. IEEE Signal Process. Mag. **34**(6), 26–38 (2017)
4. Bahdanau, D., Cho, K., Bengio, Y.: Neural machine translation by jointly learning to align and translate. arXiv preprint arXiv:1409.0473 (2014)

5. Baptiste, P., Le Pape, C., Nuijten, W.: Constraint-Based Scheduling: Applying Constraint Programming to Scheduling Problems, vol. 39. Springer Science & Business Media, Heidelberg (2012)
6. Bello, I., Pham, H., Le, Q.V., Norouzi, M., Bengio, S.: Neural combinatorial optimization with reinforcement learning (2017)
7. Bengio, Y., Lodi, A., Prouvost, A.: Machine learning for combinatorial optimization: a methodological tour d'horizon. Eur. J. Oper. Res. **290**, 405–421(2020)
8. Bezanson, J., Edelman, A., Karpinski, S., Shah, V.B.: Julia: a fresh approach to numerical computing. SIAM Rev. **59**(1), 65–98 (2017). https://doi.org/10.1137/141000671
9. Bromberger, S., Fairbanks, J., et al.: Juliagraphs/Lightgraphs. jl: an optimized graphs package for the julia programming language (2017)
10. Cappart, Q., Chételat, D., Khalil, E., Lodi, A., Morris, C., Veličković, P.: Combinatorial optimization and reasoning with graph neural networks. arXiv preprint arXiv:2102.09544 (2021)
11. Cappart, Q., Moisan, T., Rousseau, L.M., Prémont-Schwarz, I., Cire, A.: Combining reinforcement learning and constraint programming for combinatorial optimization. arXiv preprint arXiv:2006.01610 (2020)
12. Chami, I., Abu-El-Haija, S., Perozzi, B., Ré, C., Murphy, K.: Machine learning on graphs: a model and comprehensive taxonomy (2021)
13. Culberson, J.C., Luo, F.: Exploring the k-colorable landscape with iterated greedy. In: Dimacs Series in Discrete Mathematics and Theoretical Computer Science, pp. 245–284. American Mathematical Society (1995)
14. Dai, H., Dai, B., Song, L.: Discriminative embeddings of latent variable models for structured data. In: International conference on machine learning, pp. 2702–2711 (2016)
15. Dai, H., Khalil, E.B., Zhang, Y., Dilkina, B., Song, L.: Learning combinatorial optimization algorithms over graphs. In: Proceedings of the 31st International Conference on Neural Information Processing Systems, pp. 6351–6361 (2017)
16. Deudon, M., Cournut, P., Lacoste, A., Adulyasak, Y., Rousseau, L.-M.: Learning heuristics for the TSP by policy gradient. In: van Hoeve, W.-J. (ed.) CPAIOR 2018. LNCS, vol. 10848, pp. 170–181. Springer, Cham (2018). https://doi.org/10.1007/978-3-319-93031-2_12
17. Dewey, D.: Reinforcement learning and the reward engineering principle. In: AAAI Spring Symposia (2014)
18. Dolan, E.D., Moré, J.J.: Benchmarking optimization software with performance profiles. Math. Program. **91**(2), 201–213 (2002)
19. Dunning, I., Huchette, J., Lubin, M.: Jump: a modeling language for mathematical optimization. SIAM Rev. **59**(2), 295–320 (2017). https://doi.org/10.1137/15m1020575, http://dx.doi.org/10.1137/15M1020575
20. Dwivedi, V.P., Joshi, C.K., Laurent, T., Bengio, Y., Bresson, X.: Benchmarking graph neural networks. arXiv preprint arXiv:2003.00982 (2020)
21. Gambardella, L.M., Dorigo, M.: Ant-q: a reinforcement learning approach to the traveling salesman problem. In: Machine learning proceedings 1995, pp. 252–260. Elsevier (1995)
22. Gasse, M., Chételat, D., Ferroni, N., Charlin, L., Lodi, A.: Exact combinatorial optimization with graph convolutional neural networks. arXiv preprint arXiv:1906.01629 (2019)
23. Goodfellow, I., Bengio, Y., Courville, A., Bengio, Y.: Deep learning, vol. 1. MIT press, Cambridge (2016)

24. Gupta, P., Gasse, M., Khalil, E.B., Kumar, M.P., Lodi, A., Bengio, Y.: Hybrid models for learning to branch. arXiv preprint arXiv:2006.15212 (2020)
25. Innes, M., et al.: Fashionable modelling with flux. CoRR abs/1811.01457 (2018). https://arxiv.org/abs/1811.01457
26. Jussien, N., Rochart, G., Lorca, X.: Choco: an open source java constraint programming library. In: CPAIOR 2008 Workshop on Open-Source Software for Integer and Contraint Programming (OSSICP 2008), pp. 1–10 (2008)
27. Kearnes, S., McCloskey, K., Berndl, M., Pande, V., Riley, P.: Molecular graph convolutions: moving beyond fingerprints. J. Comput. Aided Mol. Des. **30**(8), 595–608 (2016)
28. Khalil, E., Dai, H., Zhang, Y., Dilkina, B., Song, L.: Learning combinatorial optimization algorithms over graphs. In: Advances in neural information processing systems, pp. 6348–6358 (2017)
29. Kool, W., Van Hoof, H., Welling, M.: Attention, learn to solve routing problems! arXiv preprint arXiv:1803.08475 (2018)
30. Michel, L., Schaus, P., Van Hentenryck, P.: MiniCP: a lightweight solver for constraint programming (2018). https://minicp.bitbucket.io
31. Legat, B., Dowson, O., Garcia, J.D., Lubin, M.: Mathoptinterface: a data structure for mathematical optimization problems (2020)
32. Li, H., Womer, K.: Modeling the supply chain configuration problem with resource constraints. Int. J. Proj. Manage. **26**(6), 646–654 (2008)
33. Lin, D., et al., other contributors: JuliaStats/Distributions.jl: a Julia package for probability distributions and associated functions (2019). https://doi.org/10.5281/zenodo.2647458
34. Lin, J., Rao, Y., Lu, J., Zhou, J.: Runtime neural pruning. In: Proceedings of the 31st International Conference on Neural Information Processing Systems, pp. 2178–2188 (2017)
35. Liu, Z., Sun, M., Zhou, T., Huang, G., Darrell, T.: Rethinking the value of network pruning. arXiv preprint arXiv:1810.05270 (2018)
36. Mnih, V., et al.: Playing atari with deep reinforcement learning (2013)
37. Monti, F., Frasca, F., Eynard, D., Mannion, D., Bronstein, M.M.: Fake news detection on social media using geometric deep learning. arXiv preprint arXiv:1902.06673 (2019)
38. Nejati, S., Le Frioux, L., Ganesh, V.: A machine learning based splitting heuristic for divide-and-conquer solvers. In: Simonis, H. (ed.) CP 2020. LNCS, vol. 12333, pp. 899–916. Springer, Cham (2020). https://doi.org/10.1007/978-3-030-58475-7_52
39. OscaR Team: OscaR: Scala in OR (2012). https://bitbucket.org/oscarlib/oscar
40. Refalo, P.: Impact-based search strategies for constraint programming. In: Wallace, M. (ed.) CP 2004. LNCS, vol. 3258, pp. 557–571. Springer, Heidelberg (2004). https://doi.org/10.1007/978-3-540-30201-8_41
41. Sanchez-Gonzalez, A., Godwin, J., Pfaff, T., Ying, R., Leskovec, J., Battaglia, P.W.: Learning to simulate complex physics with graph networks. arXiv preprint arXiv:2002.09405 (2020)
42. Schaus, P., Van Hentenryck, P., Régin, J.-C.: Scalable load balancing in nurse to patient assignment problems. In: van Hoeve, W.-J., Hooker, J.N. (eds.) CPAIOR 2009. LNCS, vol. 5547, pp. 248–262. Springer, Heidelberg (2009). https://doi.org/10.1007/978-3-642-01929-6_19
43. Schulman, J., Wolski, F., Dhariwal, P., Radford, A., Klimov, O.: Proximal policy optimization algorithms. arXiv preprint arXiv:1707.06347 (2017)
44. Schulte, C., Lagerkvist, M., Tack, G.: Gecode, pp. 11–13 (2006). http://www.gecode.org

45. Selsam, D., Bjørner, N.: Guiding high-performance sat solvers with unsat-core predictions. In: Janota, M., Lynce, I. (eds.) SAT 2019. LNCS, vol. 11628, pp. 336–353. Springer, Cham (2019). https://doi.org/10.1007/978-3-030-24258-9_24
46. Solnon, C.: Ant colony optimization and constraint programming. Wiley Online Library (2010)
47. Sutton, R.S., Barto, A.G.: Reinforcement Learning: An introduction. MIT press, Cambridge (2018)
48. Toth, P., Vigo, D.: The vehicle routing problem. SIAM (2002)
49. Tsolkas, D., Liotou, E., Passas, N., Merakos, L.: A graph-coloring secondary resource allocation for d2d communications in LTE networks. In: 2012 IEEE 17th international workshop on computer aided modeling and design of communication links and networks (CAMAD), pp. 56–60. IEEE (2012)
50. Veličković, P., Cucurull, G., Casanova, A., Romero, A., Lio, P., Bengio, Y.: Graph attention networks. arXiv preprint arXiv:1710.10903 (2017)
51. Vilím, P., Laborie, P., Shaw, P.: Failure-directed search for constraint-based scheduling. In: Michel, L. (ed.) CPAIOR 2015. LNCS, vol. 9075, pp. 437–453. Springer, Cham (2015). https://doi.org/10.1007/978-3-319-18008-3_30
52. Vinyals, O., Fortunato, M., Jaitly, N.: Pointer networks (2015)
53. Wu, Z., Pan, S., Chen, F., Long, G., Zhang, C., Yu, P.S.: A comprehensive survey on graph neural networks. IEEE Trans. Neural Netw. Learn. Syst. 32(1), 4–24 (2021). https://doi.org/10.1109/tnnls.2020.2978386
54. Yuret, D.: Knet: beginning deep learning with 100 lines of julia. In: Machine Learning Systems Workshop at NIPS, vol. 2016, p. 5 (2016)
55. Zhou, J., et al.: Graph neural networks: A review of methods and applications (2019)
56. Zhou, Y., Hao, J.K., Duval, B.: Reinforcement learning based local search for grouping problems: a case study on graph coloring. Expert Syst. Appl. 64, 412–422 (2016)

Learning to Sparsify Travelling Salesman Problem Instances

James Fitzpatrick[1]([envelope]) [ORCID], Deepak Ajwani[2] [ORCID], and Paula Carroll[1] [ORCID]

[1] School of Business, University College Dublin, Dublin, Ireland
james.fitzpatrick1@ucdconnect.ie, paula.carroll@ucd.ie
[2] School of Computer Science, University College Dublin, Dublin, Ireland
deepak.ajwani@ucd.ie

Abstract. In order to deal with the high development time of exact and approximation algorithms for NP-hard combinatorial optimisation problems and the high running time of exact solvers, deep learning techniques have been used in recent years as an end-to-end approach to find solutions. However, there are issues of representation, generalisation, complex architectures, interpretability of models for mathematical analysis etc. using deep learning techniques. As a compromise, machine learning can be used to improve the run time performance of exact algorithms in a matheuristics framework. In this paper, we use a pruning heuristic leveraging machine learning as a pre-processing step followed by an exact Integer Programming approach. We apply this approach to sparsify instances of the classical travelling salesman problem. Our approach learns which edges in the underlying graph are unlikely to belong to an optimal solution and removes them, thus sparsifying the graph and significantly reducing the number of decision variables. We use carefully selected features derived from linear programming relaxation, cutting planes exploration, minimum-weight spanning tree heuristics and various other local and statistical analysis of the graph. Our learning approach requires very little training data and is amenable to mathematical analysis. We demonstrate that our approach can reliably prune a large fraction of the variables in TSP instances from TSPLIB/MATILDA (>85%) while preserving most of the optimal tour edges. Our approach can successfully prune problem instances even if they lie outside the training distribution, resulting in small optimality gaps between the pruned and original problems in most cases. Using our learning technique, we discover novel heuristics for sparsifying TSP instances, that may be of independent interest for variants of the vehicle routing problem.

Keywords: Travelling Salesman Problem · Graph sparsification · Machine learning · Linear programming · Integer programming

1 Introduction

Owing to the high running time of exact solvers on many instances of NP-hard combinatorial optimisation problems (COPs), there has been a lot of research

© Springer Nature Switzerland AG 2021
P. J. Stuckey (Ed.): CPAIOR 2021, LNCS 12735, pp. 410–426, 2021.
https://doi.org/10.1007/978-3-030-78230-6_26

interest in leveraging machine learning techniques to speed up the computation of optimisation solutions. In recent years, deep learning techniques have been used as an end-to-end approach (see e.g., [30]) for efficiently solving COPs. However, these approaches generally suffer from (i) limited generalisation to larger size problem instances and limited generalisation from instances of one domain to another domain, (ii) need for increasingly complex architectures to improve generalisability and (iii) inherent black-box nature of deep learning that comes in the way of mathematical analysis [6]. In particular, the lack of interpretability of these models means that (1) we do not know which properties of the input instances are being leveraged by the deep-learning solver and (2) we cannot be sure that the model will still work as new constraints are required, which is typical in industry use-cases.

In contrast to the end-to-end deep learning techniques, there has been recent work (see e.g. [13]) to use machine learning as a component to speed-up or scale-up the exact solvers. In particular, Lauri and Dutta [19] recently proposed a framework to use machine learning as a pre-processing step to sparsify the maximum clique enumeration instances and scale-up the exact algorithms in this way. In this work, we build upon this framework and show that integrating features derived from operations research and approximation algorithms into the learning component for sparsification can result in reliably pruning a large fraction of the variables in the classical Travelling Salesman Problem (TSP). Specifically, we use carefully selected features derived from linear programming relaxation, cutting planes exploration, minimum-weight spanning tree (MST) heuristics and various other local and statistical analysis of the graph to sparsify the TSP instances. With these features, we are able to prune more than 85% of the edges on TSP instances from TSPLIB/MATILDA, while preserving most of the optimal tour edges. Our approach can successfully prune problem instances even if they lie outside the training distribution, resulting in small optimality gaps between the pruned and original problems in most cases. Using our learning technique, we discover novel heuristics for sparsifying TSP instances, that may be of independent interest for variants of the vehicle routing problem.

Overall, our approach consists of using a pruning heuristic leveraging machine learning (ML) as a pre-processing step to sparsify instances of the TSP, followed by an exact Integer Programming (IP) approach. We learn which edges in the underlying graph are unlikely to belong to an optimal TSP tour and remove them, thus sparsifying the graph and significantly reducing the number of decision variables. Our learning approach requires very little training data, which is a crucial requirement for learning techniques dealing with NP-hard problems. The usage of well analysed intuitive features and more interpretable learning models means that our approach is amenable to mathematical analysis. For instance, by inserting the edges from the Christofides and double-tree approximations in our sparsified instances, we can guarantee the same bounds on the optimality gap. We hypothesise that our approach, integrating features derived from operations research and approximation algorithms into a learning component for sparsification, is likely to be useful in a range of COPs, including but not restricted to, vehicle routing problems.

Outline. The paper is structured as follows: Sect. 2 describes the related works. In Sect. 3 we outline the proposed sparsification scheme: the feature generation, the sparsification model and post-processing techniques. Section 4 contains the experimental setup, exposition on the computational experiments and results. Discussion and conclusions follow in Sect. 5.

2   Notation and Related Work

Given a graph $G = (V, E; w)$ with a vertex set $V = \{1, ..., n\}$, an edge set $E = \{(u, v) | u, v \in V, u \neq v\}$ and a weighting function $w_G(e) \to \mathbb{Z}^+$ that assigns a weight to each edge $e \in E$, the goal of the TSP is to find a tour in G that visits each vertex exactly once, starting and finishing at the same vertex, with least cumulative weight. We denote by m the number of edges $|E|$ and by n the number of vertices $|V|$ of the problem.

2.1   Exact, Heuristic and Approximate Approaches

The TSP is one of the most widely-studied COPs and has been for many decades; for this reason, many very effective techniques and solvers have been developed for solving them. Concorde is a well-known, effective exact solver which implements a branch and cut approach to solve a TSP IP [1], and has been used to solve very large problems. An extremely efficient implementation of the Lin-Kernighan heuristic is available at [15], which can find very close-to-optimal solutions in most cases. Approximation algorithms also exist for the metric TSP that permit the identification of solutions, with worst-case performance guarantees, in polynomial time [7,26]. Many metaheuristic solution frameworks have also been proposed, using the principles of ant colony optimisation, genetic algorithms and simulated annealing among others [5,9,16]. In each of these traditional approaches to the TSP, there are lengthy development times and extensive problem-specific knowledge is required. If the constraints of the given problem are altered, the proposed solution method may no longer be satisfactory, possibly requiring further development.

2.2   Learning to Solve Combinatorial Optimisation Problems

Inspired by the success of deep learning to solve natural language processing and computer vision tasks, the question of how effective similar techniques might be in COP solution frameworks arises. Interest in this research direction emerged following the work of Vinyals et al. [30], in which sequence-to-sequence neural networks were used as heuristics for small instances of three different COPs, including the TSP. This was quickly followed by other ML-based approaches [3,18,21] that solve larger problem instances and avoid the need for supervised learning where access to data is a bottleneck. Graph neural networks [25] and transformer architectures [29] lead to significant speedups for learned heuristics, and have been demonstrated to obtain near-optimal solutions to yet larger TSP and vehicle routing problem (VRP) instances in seconds. Although they can produce competitive solutions to relatively small problems, these learning approaches appear to fail to generalise well to larger instance sizes, and most of these approaches require that the instance is Euclidean, encoding the coordinates of the vertices for feature computation. In cases of failure and poor solution quality, however, there is little possibility of interpreting why mistakes were made, making it difficult to rely on these models.

2.3 Graph Sparsification

Graph sparsification is the process of pruning edges from a graph $G = (V, E)$ to form a subgraph $H = (V, E' \subset E)$ such that H preserves or approximates some property of G [4,11,23]. Effective sparsification is achieved if $|E'| \ll |E|$. The running time of a TSP solver can be reduced if the underlying complete graph K_n can be sparsified such that the edges of at least one optimal Hamiltonian cycle are preserved. The work of Hougardy and Schroeder [17] sparsifies the graph defining symmetric TSP instances exactly, removing a large fraction of the edges, known as "useless" edges, that provably cannot exist in an optimal tour. Another heuristic approach due to Wang and Remmel [31] sparsifies symmetric instances by making probabilistic arguments about the likelihood that an edge will belong to an optimal tour. Both of these approaches have proven successful, reducing computation time significantly for large instances, but are unlikely to be easy to modify for different problem variants.

Recently, the sparsification problem has been posed as a learning problem, for which a binary classification model is trained to identify edges unlikely to belong to an optimal solution. Grassia et al. [12] use supervised learning to prune edges that are unlikely to belong to maximum cliques in a multi-step sparsification process. This significantly reduces the computational effort required for the task. Sun et al. [28] train a sparsifier for pruning edges from TSP instances that are unlikely to belong to the optimal tour. These approaches have the advantage that they can easily be modified for similar COP variants. The use of simpler, classical ML models lends them the benefits of partial interpretability and quick inference times. However, in the latter case, it is assumed that a very large number of feasible TSP solutions can be sampled efficiently, which does not hold for all routing-type problems, and in neither case are guarantees provided about the quality of the optimal solutions for sparsified problem instances.

3 Sparsification Scheme

The sparsification problem is posed as a binary classification task. Given some edge $e \in E$, we wish to assign it a label 0 or 1, where the label 0 indicates that the associated edge variable should be pruned and the label 1 indicates that it should be retained. We acquire labelled data for a set of graphs $\mathcal{G} = \{G_1, ..., G_n\}$ corresponding to TSP problem instances. For each graph $G_i = (V_i, E_i)$ we compute a set of p_i optimal tours $\mathcal{T}_i = \{t_i^1, ..., t_i^{p_i}\}$, as many as can be found within the tolerance level of the IP solver, and for each edge $e \in E_i$ we compute a feature representation $\vec{q}_e$. Each tour t_i has an implied set of edges $t_i \implies \epsilon_i \subset E_i$. Labelling each $e \in \bigcup_{j=1}^{p_i} \epsilon_i^j = \mathcal{E}_i$ with 1 and each edge $e \in E_i \setminus \mathcal{E}_i$ as 0, we train an edge classifier to prune edges that do not belong to an optimal tour. An optimal sparsifier would prune all but those edges belonging to optimal tours (potentially also solving the TSP). This classifier represents a binary-fixing heuristic in the context of an IP. In the following sections, we describe the feature representation that is computed for each edge and post-processing steps that are taken in order to make feasibility and approximation guarantees.

3.1   Linear Programming Features

We pose the TSP as an IP problem, using the DFJ formulation [8]. Taking A_{ij} as the matrix of edge-weights, we formulate it as follows:

$$\text{minimize} \qquad z = \sum_{i,j=1; i \neq j}^{n} A_{ij} x_{ij} \qquad (1)$$

$$\text{subject to} \qquad \sum_{i; i \neq j}^{n} x_{ij} = 1, \qquad j = 1, ..., n \qquad (2)$$

$$\sum_{j; j \neq i}^{n} x_{ij} = 1, \qquad i = 1, ..., n \qquad (3)$$

$$\sum_{(i,j) \in W} x_{ij} \leq |W| - 1, \qquad W \subseteq V; \quad |W| \geq 3 \qquad (4)$$

$$x_{ij} \in \{0, 1\}, \qquad i, j = 1, ..., m; \quad i \neq j \qquad (5)$$

Useful information about the structure of a TSP problem can be extracted by inspecting solution vectors to linear relaxations of this IP; in this way we can obtain insights into the candidacy of edges for the optimal tour. In fact, in several cases, for the MATILDA problem set, the solution to the linear relaxation at the root of the Branch and Bound (B&B) tree is itself an optimal solution to the TSP.

We denote the solution to the linear relaxation z_{LP} of the integer programme at the root node of the B&B tree by $\hat{\vec{x}}^0$. At this point, no variables have been branched on and no subtour elimination cuts have been introduced. That is, the constraints (4) are dropped and the constraints (5) are relaxed as:

$$x_{ij} \in [0, 1], \qquad i, j \in \{1, ..., m\}, i \neq j. \qquad (6)$$

One can strengthen this relaxation by introducing some subtour elimination constraints (4) at the root node. In this case, the problem to be solved remains a linear programming problem but several rounds of subtour elimination cuts are added. One can limit the computational effort expended in this regard by restricting the number of constraint-adding rounds with some upper bound $k = \lceil \log_2(m) \rceil$. The solution vector for this problem after k rounds of cuts is denoted by $\tilde{\vec{x}}^k$. We can also use as features the associated reduced costs $\vec{r}^k$ of the decision variables, which are computed in the process of a Simplex solver, standardising their values as $\hat{\vec{r}}^k = \vec{r}^k / \max \vec{r}^k$.

In order to capture broader information about the structure of the problem, stochasticity is introduced to the cutting planes approach. Inspired by the work of Fischetti and Monaci [10], the objective of the problem is perturbed. Solving the perturbed problem results in different solution vectors, which can help us to explore the solution space. We solve the initial relaxation z_{LP} in order to obtain a feasible solution, which can sometimes take a significant amount of computing effort. Subtour-elimination constraints are added to the problem for each subtour in the initial relaxation. Following this, $k = \lceil \log_2(m) \rceil$ copies of this problem are initialised. For each new problem, the edge weights A_{ij} are perturbed, and the problem is re-solved. The normalised reduced costs are obtained from each perturbed problem and for each edge (i, j) the mean reduced cost $\tilde{r}_{ij}$ is computed. We use the vector $\tilde{\vec{r}}$ of such values as a feature vector[1].

3.2 Minimum Weight Spanning Tree Features

The MST provides the basis for the Christofides–Serdyukov [7,26] and double-tree approximation algorithms, which give feasible solutions with optimality guarantees for a symmetric, metric TSP. Taking inspiration from these approximation algorithms, we use multiple MSTs to extract edges from the underlying graph thereby allowing us to compute features using them.

First, a new graph $H = (V, \emptyset)$ is initialised with the vertex set but not the edge set of the complete graph $G = (V, E) = K_n$. For $j = \lceil \log_2(n) \rceil \ll n$ iterations the MST $T = (V, E')$ of G is computed and the edges E' are removed from E and added to the edge set of H. Then, at

---
Algorithm 1. MST Features

Input: $G = (V, E), j$

1: $H \leftarrow (V, E'' - \emptyset)$
2: **for** $j \in \{1, 2, ..., \lceil \log n \rceil\}$ **do**
3: $\quad T = (V, E' \subset E) \leftarrow \text{MST}(G)$
4: $\quad G = (V, E) \leftarrow (V, E \setminus E')$
5: $\quad H = (V, E'') \leftarrow (V, E'' \cup E')$
6: $\quad w_H(e) = 1/j \quad \forall e \in E'$

Output: $H = (V, E'')$

---

each step, $G(V, E) \leftarrow G(V, E \setminus E')$, giving a new MST at each iteration with unique edges. The iteration at which edges are added to the graph H is stored, so that a feature $\hat{q}_{il}^j = 1_{il}^p / p$ may be computed, where 1_{il}^p is the indicator, taking unit value if the edge $e = (i, l) \in E$ was extracted at iteration p and zero otherwise.

---

[1] For additional detail regarding computation of these features, please see: https://arxiv.org/abs/2104.09345.

Since the value of j should be small, the vast majority of the edges will have zero-valued feature-values. This edge transferal mechanism can be used as a sparsification method itself: the original graph can be pruned such that the only remaining weighted edges are those that were identified by the successive MSTs, with the resulting graph containing $j(n-1)$ edges.

3.3   Local Features

The work of Sun et al. [28] constructs four local features on the graph $G = (E, V)$, comparing weights of an edge $(i, j) \in E$ to the edge weights $(k, j), k \in V$ and $(i, k), k \in V$. While relatively inexpensive to compute, yet less expensive features can be computed, comparing a given edge weight (i, j) to the maximum and minimum weights in E. That is, for each $(i, j) \in E$ we compute a set of features q_{ij} as:

$$q_{ij}^a = (1 + A_{ij})/(1 + \max_{(l,k)\in E} A_{lk}) \tag{7}$$

$$q_{ij}^b = (1 + A_{ij})/(1 + \max_{l\in V} A_{il}) \tag{8}$$

$$q_{ij}^c = (1 + A_{ij})/(1 + \max_{l\in V} A_{lj}) \tag{9}$$

$$q_{ij}^d = (1 + \min_{(l,k)\in E} A_{lk})/(1 + A_{ij}) \tag{10}$$

$$q_{ij}^e = (1 + \min_{l\in V} A_{il})/(1 + A_{ij}) \tag{11}$$

$$q_{ij}^f = (1 + \min_{l\in V} A_{lj})/(1 + A_{ij}) \tag{12}$$

The motivation for the features (7) and (10) is to cheaply compute features that relate a given edge weight to the weights of the entire graph in a global manner. On the other hand, motivated by the work of Sun et al. [28], the features (8), (9), (11), (12), compare a given edge weight to those in its direct neighbourhood; the edge weight associated with edge (i, j) is related only to the weights of the associated vertices i and j.

3.4   Postprocessing Pruned TSP Graphs

In this setting, sparsification is posed as a set of m independent classification problems. The result of this is that there is no guarantee that any feasible solution exists within a pruned problem instance. Indeed, even checking that any tour exists within a sparsified graph is itself an NP-hard problem. One can guarantee feasibility of the pruned graph by ensuring that the edges belonging to some known solution exist in the pruned graph; this forces both connectivity and Hamiltonicity (see Fig. 1). The pruned graph has at least one feasible solution and admits an optimal objective no worse than that of the solution that is known. For the TSP we can construct feasible tours trivially by providing any permutation of the vertices, but it is likely that such tours will be far from optimal. Assuming the problem is metric, we can use an approximation algorithm to construct a feasible solution to the problem that also gives a bound on the quality of the solution.

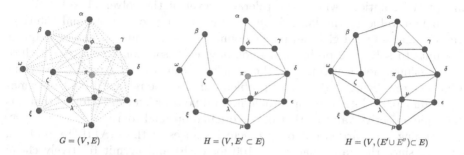

$$G = (V, E) \qquad\qquad H = (V, E' \subset E) \qquad\qquad H = (V, (E' \cup E'') \subset E)$$

Fig. 1. The pruned graph H does not admit a feasible solution. *Given a known solution, we add the edges E'' of the solution to the pruned graph in order to guarantee the feasibility of the pruned instance. If we obtain E'' using an approximation algorithm, then we can make guarantees about the quality of the solutions obtained from H.*

4   Experiments and Results

All experiments were carried out in Python[2]. Graph operations were performed using the NetworkX package [14] and the training was carried out with the Scikit-Learn package [22]. The linear programming features were computed using the Python interfaces for the Xpress and SCIP optimisation suites [2,20]. Training and feature computation was performed on a Dell laptop running Ubuntu 18.04 with 15. 6 GB of RAM an, Intel® Core™ i7-9750H 2.60GHz CPU and an Nvidia GeForce RTX 2060/PCIe/SSE2 GPU.

4.1   Learning to Sparsify

First we train a classification model to prune edges that are unlikely to belong to an optimal tour. That is, given the feature representation

$$\vec{q}_{il} = [q_{il}^a, q_{il}^b, q_{il}^c, q_{il}^d, q_{il}^e, q_{il}^f, \hat{\tilde{r}}_{il}^k, \tilde{r}_{il}, \hat{q}_{il}^j]^T$$

for the edge (i, l), we aim to train a machine learning model that can classify all the edges of a given problem instance in this manner. In each case we let the parameter $k = \lceil \log_2(m) \rceil$ and $j = \lceil \log_2(n) \rceil$, since these numbers grow slowly with the size of the graph and prevent excessive computation. In order to address the effects of class imbalance, we randomly under-sample the negative class such that the classes are equally balanced, and adopt class weights $\{0.01, 0.99\}$ for the negative and positive classes respectively to favour low false negative rates

---

2 All code available at: https://github.com/JamesFitzpatrickMLLabs/optlearn.

for the positive class, favouring correctness over the sparsification rate. Sample weights $w_{il} = A_{il} / \max_{il} A_{il}$ are applied to each sample associated with an edge (i, l) to encourage the classifier to reduce errors associated with longer edges.

In many instances there exists more than one optimal solution. We compute all optimal solutions within the tolerance level of the solver (1×10^{-8}) and assign unit value to the label for edges that belong to any optimal solution. Any optimal solution gives an upper bound on the solution for any subsequent attempts to solve the problem. To compute each solution, one can introduce a tour elimination constraint to the problem formulation to prevent previous solutions from being feasible. The problem can then be re-solved multiple times to obtain more solutions to the problem, until no solution can be found that has the same objective value as that of the original optimal tour.

Training was carried out on problem instances of the MATILDA problem set [27], since they are small ($n = 100$ for each) and permit relatively cheap labelling operations. One third (63) of the CLKhard and LKCChard problem instances were chosen for the training set, while the remaining problem instances are retained for the testing and validation sets. These two problem categories were selected for training once it was identified through experimentation that sparsifiers trained using these problems generally performed better (with fewer infeasibilities). This is in line with the findings of Sun et al.[28]. Since each of the linear programming features must be computed for any given problem instance, it is worthwhile checking if any solution $\hat{\tilde{x}}^0$ or $\tilde{x}^k$ is an optimal TSP solution, which would allow all computation to terminate at this point. All of the edges of a given graph belong to exactly one of training, test or validation sets, and they must all belong to the same set. Each symmetric problem instance in the TSPLIB problem set [24] for which the order $n \leq 3038$ is retained for evaluation of the learned sparsifier only.

Following pruning, edges of a known tour are inserted, if they are not already elements of the edge set of the sparsified graph. In this work, we compute both the double-tree and Christofides approximations, since they can be constructed in polynomial time and guarantee that the pruned graph will give an optimality ratio $\hat{\ell}_{opt} / \ell_{opt} \leq 3/2$, where $\hat{\ell}_{opt}$ is the optimal tour length for a pruned problem and ℓ_{opt} is the optimal tour length for the original problem. The performance of the classifier is evaluated using the optimality ratio $\tilde{\ell} = \hat{\ell}_{opt} / \ell_{opt}$ and the retention rate $\tilde{m} = \hat{m}/m$, where $\hat{m}$ is the number of edges belonging to the pruned problem instance. Since the pruning rate $1 - \tilde{m}$ implies the same result as the retention rate, we discuss them interchangeably. Introducing any known tour guarantees feasibility at a cost in the sparsification rate no worse than $(\hat{m} + n)/\hat{m}$.

4.2   Performance on MATILDA Instances

The sparsification scheme was first evaluated on the MATILDA problem instances, which are separated into seven classes, each with 190 problem instances. Logistic regression, random forest and linear support vector classifiers were evaluated as classification models. Although no major advantage was displayed after a grid search over these models, here, for the sake of comparison with Sun et al., an L1-regularised SVM with an RBF kernel is trained as the sparsifier. We can see from Table 1 that the optimality ratios observed in the pruned problems are comparable to those of Sun et al., with a small deterioration in the optimality gap for the more difficult problems (around 0.2%) when the same problem types were used for training. However, this comes with much greater sparsification rates, which leads to more than 85% of the edges being sparsified in all cases and almost 90% on average, as opposed to the around 78% observed by Sun et al. We similarly observe that the more difficult problems are pruned to a slightly smaller extent than the easier problem instances.

Table 1. Evaluation of the trained sparsifier against the problem subsets of the MATILDA instances. Here each cell indicates the mean value of the optimality ratio or pruning rate over each subset of problems (not including training instances)

Statistic (Mean)	Problem Class						
	CLKeasy	CLKhard	LKCCeasy	LKCChard	easyCLK-hardLKCC	hardCLK-easyLKCC	random
$\tilde{l}$	1.00000	1.00176	1.00000	1.00385	1.00049	1.00035	1.00032
$1 - \hat{m}$	0.90389	0.89021	0.91086	0.88298	0.88874	0.90357	0.89444

4.3 Pruning with and Without Guarantees

This section describes experiments carried out to test how much the sparsifier would need to rely on inserted tours to produce feasibly reduced problem instances. The optimality ratio statistics are shown in Table 2. Before post-processing, pruning was achieved with a maximum rate of edge retention $\max\{\hat{m}/m\} = 0.139$, a minimum rate $\min\{\hat{m}/m\} = 0.075$ and a mean rate $\langle \hat{m}/m \rangle = 0.100$ among all problem instances. In the vast majority of cases, at least one optimal tour is contained within the pruned instance (column 2), or one that is within 2% of optimal (column 4). For the more difficult problem instances, there were several infeasible instances produced each time. In some cases the pruned instances admitting sub-optimal solutions with respect to the original problem had improved optimality ratios (columns 3 and 4). The worst-case optimality improves for each problem class (except CLKeasy, where all problems were sparsified without losing an optimal solution) after post-processing. The distribution of these values with post-processing is depicted in Fig. 2.

This sparsification scheme was also tested on the TSPLIB problem instances (see Table 3). Before post-processing, for the majority of the problem instances, an optimal tour is contained within the sparsified graphs (column 2), and the vast majority of the problem instances have optimality ratios no worse than $\tilde{\ell} = 1.05$ (column 4). Unlike with the MATILDA instances, the pruning rate varies significantly. This is in accordance with the findings of Sun et al. [28] and indicates that the sparsifier is less certain about predictions made, and fewer edges are therefore removed in many cases. For some smaller problem instances (smaller than the training set problems) the pruning rate approaches 0.5167.

Table 2. Optimality ratio statistics before and after approximate tour insertion following learned sparsification (MATILDA problem instances). *Each cell indicates the number of problem instances of each class for which the optimality ratio resided within the bounds stated before and after post-processing. For example, the cell in column 2 containing 174 → 176 indicates that 174 purely sparsified instances admitted unit optimality ratios, but 176 did so after the post-processing. The last column contains the maximum optimality gap for each class. If there are infeasibly sparsified graphs, this value is denoted by ∞.*

Problem Class	# of Problems With Below Condition True			Worst Case
	$\ell_{opt}/\ell_{opt} = 1.0$	$\ell_{opt}/\ell_{opt} < 1.01$	$\ell_{opt}/\ell_{opt} < 1.02$	$\max\{\ell_{opt}/\ell_{opt}\}$
CLKeasy	190 → 190	190 → 190	190 → 190	1 → 1
CLKhard	110 → 123	153 → 181	165 → 190	∞ → 1.018
LKCCeasy	189 → 190	189 → 190	190 → 190	1.016 → 1
LKCChard	35 → 65	82 → 168	104 → 185	∞ → 1.024
easyCLK-hardLKCC	172 → 181	180 → 186	180 → 188	∞ → 1.024
hardCLK-easyLKCC	167 → 181	175 → 188	176 → 189	∞ → 1.030
random	174 → 176	186 → 188	188 → 190	∞ → 1.016
Total	1037 → 1106	1155 → 1291	1193 → 1322	∞ → 1.030

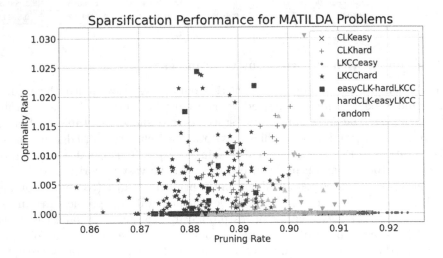

Fig. 2. Each point represents a problem instance, showing the pruning rate and the optimality ratio achieved for each problem in the MATILDA benchmark set.

The median pruning rate was 0.9109, with the mean pruning rate at 0.8904 and a standard deviation of 0.086. The highest pruning rate was for the problem *u2152.tsp*, for which 0.9758 m edges were removed, achieving an optimality ratio of 1.0.

Introducing the approximate tours brought the worst-case optimality ratio to 1.01186 (column 5) for *d657.tsp*, with median and mean pruning rates of, respectively, 0.8991 and 0.8718. The mean optimality ratio (excluding the infeasible instances) before approximate tour insertion was 1.01546 and (also without these same problems) 1.00073 after insertion, indicating that in most cases there is little reduction in solution quality as a result of sparsification. In Fig. 3 we can see depicted the relationship between the problem size (the order, n) and the pruning rate. Almost all of the instances retain optimal solutions after pruning, those that don't are indicated by the colour scale of the points. We can see from this plot that the drilling problem $d657.tsp$ and problems $pr[144, 226, 266].tsp$ $ts225.tsp$ have some of the largest optimality deviations (up to 1.2%), likely because in these problems the vertices are aligned in regular, often clustered patterns, which leads to problems with edge-weight distributions much different from the problems of the training set.

Table 3. Optimality ratio statistics before and after approximate tour insertion following learned sparsification (TSPLIB problem instances). *Results for each symmetric problem instance with $n \leq 1000$ using a logistic regression sparsifier. This set contained 76 problem instances, some of which are not metric TSPs. Although we cannot make provable guarantees for the non-metric problem instances, we can still use the approximate tours to ensure feasibility.*

Problem Class	# of Problems With Below Condition True			Worst Case
	$\hat{\ell}_{opt}/\ell_{opt} = 1.0$	$\hat{\ell}_{opt}/\ell_{opt} < 1.005$	$\hat{\ell}_{opt}/\ell_{opt} < 1.010$	$\max\{\hat{\ell}_{opt}/\ell_{opt}\}$
TSPLIB	$55 \rightarrow 56$	$69 \rightarrow 70$	$72 \rightarrow 73$	$\infty \rightarrow 1.01186$

4.4   Minimum Weight Spanning Tree Pruning

Empirical experiments demonstrated that using only successive MSTs, as outlined in Sect. 3.2, one can effectively sparsify symmetric TSP instances. This scheme proceeds without the need for a classifier, by building a new graph H with only the edges of the MSTs and their associated edge weights in G. This sparsified instance achieves a retention rate of jn/m, where j is the number of trees computed. In many cases, simply using this as a scheme for selecting edges for retention was sufficient to sparsify the graph while achieving a unit optimality ratio.

On the MATILDA problems, again with $j = \lceil \log_2(n) \rceil$, this achieves a worst-case optimality ratio of 1.0713 and a uniform pruning rate of 0.86, with the majority of the pruned instances containing an optimal solution. Under these conditions the easier problems (CLKeasy and LKCCeasy) had optimal tours preserved in every problem instance except for one. For the more difficult problem instances, in the vast majority of cases, the ratio does not exceed 1.02 (see Table 4). Insertion of approximate tours in this scenario does not lead to much improvement, since every instance of the MATILDA problem set is sparsified feasibly in this scheme.

For the TSPLIB problems, all but two can be sparsified feasibly without the insertion of approximate edges: $p654.tsp$ and $fl417.tsp$. Since k is a function of the problem order, $n = 100$, 86% of the edges of a graph will be removed, whereas at order $n = 1000$, the pruning rate reaches 98%. The worst optimality

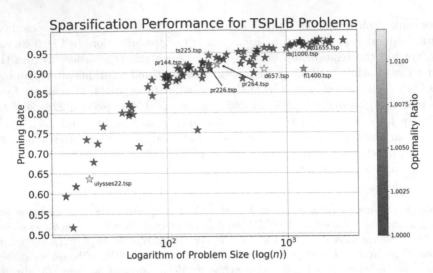

Fig. 3. Here each point represents a problem instance. The horizontal axis depicts the logarithm of the problem size in terms of the number of nodes. The vertical axis shows the sparsification rate and the colouring of the points indicates the optimality ratio observed.

Table 4. Optimality ratio statistics before and after approximate tour insertion following MST sparsification (MATILDA problem instances). *Comparison between the pure multiple MST pruning scheme and the same scheme with double-tree post-processing. In every case for the MATILDA problem set, the pruned instances are feasible.*

Problem Class	# of Problems With Below Condition True			Worst Case
	$\hat{\ell}_{opt}/\ell_{opt} = 1.0$	$\hat{\ell}_{opt}/\ell_{opt} < 1.02$	$\hat{\ell}_{opt}/\ell_{opt} < 1.05$	$\max\{\hat{\ell}_{opt}/\ell_{opt}\}$
CLKeasy	$190 \rightarrow 190$	$190 \rightarrow 190$	$190 \rightarrow 190$	$1 \rightarrow 1$
CLKhard	$85 \rightarrow 88$	$185 \rightarrow 185$	$190 \rightarrow 190$	$1.029 \rightarrow 1.029$
LKCCeasy	$189 \rightarrow 190$	$190 \rightarrow 190$	$190 \rightarrow 190$	$1.002 \rightarrow 1.002$
LKCChard	$77 \rightarrow 78$	$183 \rightarrow 183$	$189 \rightarrow 189$	$1.071 \rightarrow 1.054$
easyCLK-hardLKCC	$173 \rightarrow 190$	$190 \rightarrow 190$	$190 \rightarrow 190$	$1.012 \rightarrow 1.012$
hardCLK-easyLKCC	$171 \rightarrow 173$	$190 \rightarrow 190$	$190 \rightarrow 190$	$1.015 \rightarrow 1.015$
random	$176 \rightarrow 176$	$190 \rightarrow 190$	$190 \rightarrow 190$	$1.006 \rightarrow 1.006$
All	$1061 \rightarrow 1330$	$1318 \rightarrow 1330$	$1329 \rightarrow 1330$	$1.071 \rightarrow 1.054$

ratio for any feasible problem was for *pr226.tsp*, at $\tilde{l} = 1.106$, with a retention rate of 0.071. The majority of the sparsified instances retained an optimal tour (see Table 5), with the mean optimality ratio $\langle \tilde{\ell} \rangle$ for all feasible problems taking the value 1.0061. Including approximate tour edges in the sparsified graphs results in none having an optimality gap greater than 1.084 (*pr654.tsp*, which was previously infeasible, with a retention rate of 3.1%). The other previously infeasibly-sparsified problem, *fl417.tsp* admits an optimality ratio $\tilde{\ell} = 1.077$, retaining just 4.3% of its edges. The mean optimality ratio for all problems emerged as 1.0053.

Table 5. Optimality ratio statistics before and after approximate tour insertion following MST sparsification (TSPLIB problem instances). *For each problem instance the number of trees j differed, according to the order n.*

Problem Class	# of Problems With Below Condition True			Worst Case
	$\hat{\ell}_{opt}/\ell_{opt} = 1.0$	$\hat{\ell}_{opt}/\ell_{opt} < 1.02$	$\hat{\ell}_{opt}/\ell_{opt} < 1.05$	$\max\{\hat{\ell}_{opt}/\ell_{opt}\}$
TSPLIB	$51 \rightarrow 51$	$66 \rightarrow 67$	$71 \rightarrow 71$	$\infty \rightarrow 1.084$

4.5 Specifying the Pruning Rate

One of the advantages of including the post-processing step that guarantees the feasibility of the pruned instance is that we can exert greater control over the pruning rate. Training a sparsifier in the manner outlined above necessitates a trade-off between the optimality ratio and the pruning rate. Higher pruning rates result in sparsified problems that are easier to solve but that have typically poorer optimality ratios. Indeed, if the sparsification rate is too high, many of the sparsified instances will also become infeasible. Including approximate tours, however, allows us to choose effectively any decision threshold for the classifier, ranging from total sparsification and removing all edges, to choosing a threshold that results in no sparsification at all and the retention of all edges from the original graph. So long as there is at least one known feasible tour, regardless of its quality, then the pruned instance will be feasible after post-processing.

5 Discussion and Conclusions

In this work we have demonstrated that it is possible to learn to effectively sparsify TSP instances, pruning the majority of the edges using a small classification model while relying on LP and graph-theoretic features that can be efficiently computed. Providing optimality guarantees is possible by inserting edges of approximate tours, ensuring that even out-of-distribution problem instances can be tackled with this scheme. These features are supplemented with local statistical features, comparing edge weights to the global and neighbouring distributions. This scheme successfully generalises to larger problem instances outside of the training distribution. Where there is a lack of training data or where expediency is favoured, it has been shown that the MST extraction mechanism with inserted approximate tours performs well on most problem instances.

The motivation of this work has been to use the methods of ML to aid in the design of heuristics for solving combinatorial optimisation problems. This is in the hope that such an approach can reduce the development time required. Development time for ML solutions depends typically on the engineering of either features or problem-specific models and architectures. This can effectively transfer the engineering effort from one task to another, without producing tangible benefits. In this approach, classical ML models are used, which means that feature design is paramount to the success of the model. Here, LP features are designed that are not dependent upon the formulation of the problem; any TSP problem that requires formulation as binary IP with subtour-elimination (or capacity) constraints can have features produced in the same manner. This is attractive because it simply requires knowledge of IP formulations that give tight relaxations. Such relaxations may also be sufficient to solve the problem, potentially obviating the need for any further computation. Optimality ratio performance does not appear to depend on the problem size, which indicates that this pre-processing scheme might help extend the applicability of learned solving heuristics. Some preliminary experimental results indicated that this approach can be leveraged to speed up solving times for a custom-built Branch and Cut solver for the TSP using the DFJ formulation, but further experimentation and implementation in C will be necessary to determine the extent to which speedups can be achieved.

The use of the approximation algorithms guarantees the feasibility of the pruned problem instance, but it does not provide a tight bound on the optimality ratio. Tighter bounds can be achieved by using these in combination with effective heuristics. Alternatively, one could compute successive approximations by removing from consideration edges belonging to a tour once they have been computed, analogously to Algorithm 1. The observations from experimentation indicate that while these approximate tours rarely improve the optimality ratio at lower pruning rates, they are not only necessary for feasibility at higher pruning rates, but can help immensely to improve the final optimality ratio.

This scheme has the potential to be developed in a similar manner for other routing problems, in particular vehicle routing problems, for which solvers are not as effective in practice as they are for the TSP. To realise the benefits of scheme, an implementation would also have to be rewritten in a lower level language. Subsequent work could be done to evaluate the smallest problem sizes for which training can be carried out effectively and applicability to the VRP, where it might be possible to extend the use of the vehicle-flow IP formulation to larger problems, which typically need to be solved using a Branch, Cut and Price approach.

Acknowledgements. This publication has emanated from research supported in part by a grant from Science Foundation Ireland under Grant number 18/CRT/6183. For the purpose of Open Access, the author has applied a CC BY public copyright licence to any Author Accepted Manuscript version arising from this submission.

References

1. Applegate, D.L., et al.: Certification of an optimal TSP tour through 85,900 cities. Oper. Res. Lett. **37**(1), 11–15 (2009)
2. Ashford, R.: Mixed integer programming: a historical perspective with Xpress-MP. Ann. Oper. Res. **149**(1), 5 (2007)
3. Bello, I., Pham, H., Le, Q.V., Norouzi, M., Bengio, S.: Neural combinatorial optimization with reinforcement learning. arXiv preprint arXiv:1611.09940 (2016)
4. Benczúr, A.A.: Approximate s-t min-cuts in o (n^2) time. In: Proceedings of the 28th ACM Symposium on Theory of Computing (1996)

5. Braun, H.: On solving travelling salesman problems by genetic algorithms. In: Schwefel, H.-P., Männer, R. (eds.) PPSN 1990. LNCS, vol. 496, pp. 129–133. Springer, Heidelberg (1991). https://doi.org/10.1007/BFb0029743
6. Di Caro, G.A.: A survey of machine learning for combinatorial optimization. In: 30th European Conference on Operations Research (EURO) (2019)
7. Christofides, N.: Worst-case analysis of a new heuristic for the travelling salesman problem. Technical report, Carnegie-Mellon Univ. Pittsburgh Pa Management Sciences Research Group (1976)
8. Dantzig, G., Fulkerson, R., Johnson, S.: Solution of a large-scale traveling-salesman problem. J. Oper. Res. Soc. Am. **2**(4), 393–410 (1954)
9. Dorigo, M., Gambardella, L.M.: Ant colonies for the travelling salesman problem. Biosystems **43**(2), 73–81 (1997)
10. Fischetti, M., Monaci, M.: Exploiting erraticism in search. Oper. Res. **62**(1), 114–122 (2014)
11. Fung, W.-S., Hariharan, R., Harvey, N.J.A., Panigrahi, D.: A general framework for graph sparsification. SIAM J. Comput. **48**(4), 1196–1223 (2019)
12. Grassia, M., Lauri, J., Dutta, S., Ajwani, D.: Learning multi-stage sparsification for maximum clique enumeration. arXiv preprint arXiv:1910.00517 (2019)
13. Gupta, P., Gasse, M., Khalil, E.B., Kumar, M.P., Lodi, A., Bengio, Y.: Hybrid models for learning to branch. In: Larochelle, H., Ranzato, M.A., Hadsell, R., Balcan, M.-F., Lin, H.-T. (eds.) Advances in Neural Information Processing Systems 33: Annual Conference on Neural Information Processing Systems 2020, NeurIPS 2020, 6–12 December 2020, virtual (2020)
14. Hagberg, A., Swart, P., Chult, D.S.: Exploring network structure, dynamics, and function using NetworkX. Technical report, Los Alamos National Lab. (LANL), Los Alamos, NM (United States) (2008)
15. Helsgaun, K.: An effective implementation of the Lin-Kernighan traveling salesman heuristic. Eur. J. Oper. Res. **126**(1), 106–130 (2000)
16. Hopfield, J.J., Tank, D.W.: "Neural" computation of decisions in optimization problems. Biol. Cybern. **52**(3), 141–152 (1985). https://doi.org/10.1007/BF00339943
17. Hougardy, S., Schroeder, R.T.: Edge elimination in TSP instances. In: Kratsch, D., Todinca, I. (eds.) WG 2014. LNCS, vol. 8747, pp. 275–286. Springer, Cham (2014). https://doi.org/10.1007/978-3-319-12340-0_23
18. Kool, W., Van Hoof, H., Welling, M.: Attention, learn to solve routing problems! arXiv preprint arXiv:1803.08475 (2018)
19. Lauri, J., Dutta, S.: Fine-grained search space classification for hard enumeration variants of subset problems. In: The Thirty-Third AAAI Conference on Artificial Intelligence, AAAI 2019, The Thirty-First Innovative Applications of Artificial Intelligence Conference, IAAI 2019, The Ninth AAAI Symposium on Educational Advances in Artificial Intelligence, EAAI 2019, Honolulu, Hawaii, USA, 27 January–1 February 2019, pp. 2314–2321. AAAI Press (2019)
20. Maher, S., Miltenberger, M., Pedroso, J.P., Rehfeldt, D., Schwarz, R., Serrano, F.: PySCIPOpt: mathematical programming in Python with the SCIP optimization suite. In: Greuel, G.-M., Koch, T., Paule, P., Sommese, A. (eds.) ICMS 2016. LNCS, vol. 9725, pp. 301–307. Springer, Cham (2016). https://doi.org/10.1007/978-3-319-42432-3_37
21. Nazari, M., Oroojlooy, A., Snyder, L.V., Takác, M.: Reinforcement learning for solving the vehicle routing problem. In: Advances in Neural Information Processing Systems, pp. 9839–9849 (2018)

22. Pedregosa, F., et al.: Scikit-learn: machine learning in Python. J. Mach. Learn. Res. **12**, 2825–2830 (2011)

23. Peleg, D., Schäffer, A.A.: Graph spanners. J. Graph Theory **13**(1), 99–116 (1989)

24. Reinelt, G.: TSPLIB-a traveling salesman problem library. INFORMS J. Comput. **3**(4), 376–384 (1991)

25. Scarselli, F., Gori, M., Tsoi, A.C., Hagenbuchner, M., Monfardini, G.: The graph neural network model. IEEE Trans. Neural Netw. **20**(1), 61–80 (2008)

26. Serdyukov, A.I.: On some extremal walks in graphs. Upravlyaemye Sistemy **17**, 76–79 (1978)

27. Smith-Miles, K., van Hemert, J., Lim, X.Y.: Understanding TSP difficulty by learning from evolved instances. In: Blum, C., Battiti, R. (eds.) LION 2010. LNCS, vol. 6073, pp. 266–280. Springer, Heidelberg (2010). https://doi.org/10.1007/978-3-642-13800-3_29

28. Sun, Y., Ernst, A., Li, X., Weiner, J.: Generalization of machine learning for problem reduction: a case study on travelling salesman problems. OR Spectr. **2020**, 1–27 (2020). https://doi.org/10.1007/s00291-020-00604-x

29. Vaswani, A., et al.: Attention is all you need. In: Advances in Neural Information Processing Systems, pp. 5998–6008 (2017)

30. Vinyals, O., Fortunato, M., Jaitly, N.: Pointer networks. In: Advances in Neural Information Processing Systems, pp. 2692–2700 (2015)

31. Wang, Y., Remmel, J.: A method to compute the sparse graphs for traveling salesman problem based on frequency quadrilaterals. In: Chen, J., Lu, P. (eds.) FAW 2018. LNCS, vol. 10823, pp. 286–299. Springer, Cham (2018). https://doi.org/10.1007/978-3-319-78455-7_22

Optimized Item Selection to Boost Exploration for Recommender Systems

Serdar Kadıoğlu, Bernard Kleynhans[(✉)], and Xin Wang

AI Center of Excellence Fidelity Investments, Boston, USA
{serdar.kadioglu,bernard.kleynhans,xin.wang}@fmr.com

Abstract. Recommender Systems have become the backbone of personalized services that provide tailored experiences to individual users. Still, data sparsity remains a common challenging problem, especially for new applications where training data is limited or not available. In this paper, we formalize a combinatorial problem that is concerned with selecting the universe of items for experimentation with recommender systems. On one hand, a large set of items is desirable to increase the diversity of items. On the other hand, a smaller set of items enable rapid experimentation and minimize the time and the amount of data required to train machine learning models. We show how to optimize for such conflicting criteria using a multi-level optimization framework. Our approach integrates techniques from discrete optimization, unsupervised clustering, and latent text embeddings. Experimental results on well-known movie and book recommendation benchmarks demonstrate the benefits of optimized item selection.

Keywords: Recommender systems · Exploration-exploitation · Item selection · Set covering

1 Introduction

Recommender Systems have become central in our daily lives and are widely employed in the industry. Prominent examples include online shopping sites (e.g., Amazon.com [20]), music and movie services (e.g., YouTube [7], Netflix [30,44] and Spotify [13,26]), mobile application stores (e.g., iOS App Store and Google Play), and online advertising [32]. The primary goal of recommender systems is to help users discover relevant content such as movies to watch, articles to read, or products to buy. From the user's perspective, this creates a tailored digital experience, and from the business' perspective, it drives incremental revenue.

These systems learn users' preferences from historical observations to select the right content, at the right time, for the right channel. However, data sparsity is a common challenging problem, especially for newly launched recommender systems. The classical setting is composed of a set of users, U, and a set of items, I, from which top-k items are chosen (e.g., items with the highest probability to be clicked) and shown to the user at time t. For each recommendation, the reward

© Springer Nature Switzerland AG 2021
P. J. Stuckey (Ed.): CPAIOR 2021, LNCS 12735, pp. 427–445, 2021.
https://doi.org/10.1007/978-3-030-78230-6_27

is observed (e.g., whether the user clicked). The reward feedback is incorporated into the next decision at time $t + 1$, and the system proceeds.

While recommendation systems are concerned with selecting the top-k items at each step, there is an apriori decision that governs the entire process: *what should the universe of items I that can be recommended be?* In this paper, we focus on this problem, especially for new applications where training data is limited or not available.

To collect the necessary response data, recommender systems take advantage of randomized experimentation (i.e., *exploration*) to build personalization models (i.e., *exploitation*). However, randomized exploration incurs unwanted costs. From users' perspective, randomized experimentation is not the desired digital experience, in fact, the exact opposite of it. From business's perspective, randomization leads to missed opportunities. It has a cost on business KPIs from engagement metrics, e.g., the click-through rate, to take-action rates for positive outcomes, e.g., opening an account or buying a product.

It is therefore critical to speed-up exploration and switch to personalization as quickly as possible. However, collecting the amount of randomized response data to train efficient machine learning models takes considerable time and effort. This is a significant bottleneck, especially in low volume, low click-through rate applications, making the selection of the universe of items for randomized experimentation crucial.

In this paper, we present a multi-level optimization approach for selecting items to be included in randomized experimentation for recommender systems. Our selection procedure is designed to maximize knowledge transfer of users' responses and minimize the time-to-market for personalization. To that end, we jointly optimize the cardinality of the item universe and the diversity of items included in it. Our main contributions can be summarized as follows:

- By minimizing the cardinality of the item universe, we reduce the experimentation time window and mitigate the aforementioned negative impact on customer experience and business KPIs.
- We show how to use a latent embedding space to calculate diversity measures between items and maximize the diversity of the selected items.
- We propose a simple warm-start procedure based on item-to-item similarity to enable transfer learning from the randomized exploration phase to the personalized exploitation phase.
- More broadly, our hybrid approach serves as an integration block between modern recommender systems and classical discrete optimization techniques.

2 Problem Definition

Let us start with a formal description of our problem statement.

Definition 1 (Item Selection Problem (ISP)). *Given a set of items I, the goal of the Item Selection Problem (ISP) is to find the minimum subset $S \subseteq I$ that covers a set of labels L_c within each category $c \in C$ while maximizing the diversity of the selection S in the latent embedding space of items $E(I)$.*

Illustrative Example: To make our abstract ISP definition more concrete, we consider a running example based on movie recommendations.

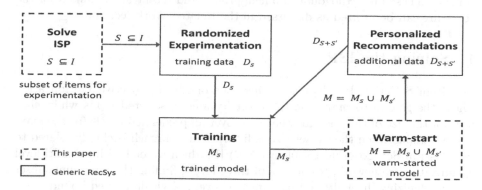

Fig. 1. Recommender System components from ISP to Personalization.

In a Movie Recommender System, the *items* I correspond to all available *movie titles* that could be recommended. Collecting data on user ratings for all available movies would be time-consuming. We are therefore interested in finding a subset S of movies to initialize the system for training data collection. We need to ensure that selected movies cover a wide range of variety. The *categories* of interest, C can include the genre, language and producer. Within each category $c \in C$, we can have a set of *labels*, such as action, comedy, thriller in the genre category, and English and French in the language category. Such metadata is commonly available in recommender systems. The ISP seeks to include at least one movie from each label L_c within the different categories $c \in C$. Notice that the selection can include multiple movies with the same label (e.g., comedy). Additionally, we want to maximize the diversity of selected movies in the *latent embedding space* $E(I)$. The latent representation can be based on textual data (e.g., synopses, movie reviews) or image data (e.g., cover art).

For unstructured text, the embeddings can be found using TFIDF [17], word embeddings such as Word2Vec [27], Doc2Vec [21], or language models like BERT [8]. For image data, the embedding can be based on convolutional neural networks. Finally, decomposition and dimensionality reduction techniques, such as NMF [22] and SVD [10] can be used over the latent representation.

3   Recommender System Components

The ISP is most relevant for recommender systems in new customer experiences for which there exists no historical data. As illustrated in Fig. 1, randomized experimentation is employed to collect training data D_S. This training data is later used to build personalization models M_S. The longer the exploration phase takes, the worse the customer experience and business outcomes are. To mitigate this, our strategy focuses on solving the ISP to guide the randomized exploration which is later augmented with warm-started models M_S'.

In the remainder, we focus on solving the ISP and the warm-start procedure. Training recommender models and creating item embeddings are generic components. These are orthogonal to our approach and recent algorithms from the literature can be utilized as discussed in the Related Work section.

4   Solving the ISP

The Item Selection Problem is an instance of Multi-Objective Optimization where the goal is to *maximize* the diversity among selected items while *minimizing* the size of the selection that can cover all predefined labels (or the maximum possible when the subset size is fixed). Our approach is closely related to the classical Set Covering Problem (SCP) [1] which we embed in a multi-level optimization framework. It consists of three levels; finding the minimum subset size, maximizing diversity and maximizing coverage within a fixed bound.

4.1   Minimizing the Subset Size

Selecting a subset of items that cover all predefined labels is a standard covering formulation. Let $M_{l,i}$ be the incident matrix where rows correspond to all predefined labels, $l \in L_c$, for each category $c \in C$, and columns correspond to item $i \in I$. We define $L_{c,i}$ as the label in category c for item i and set $M_{l,i}$ to 1 only if $M_{l,i} = L_{c,i}$. Let X be the set of decision variables where x_i is a binary variable denoting whether item $i \in I$ is included in the selection. Assume each selected item incurs a cost of 1 and let c represent the unit cost vector. Then formulating the unicost item selection problem, $P_{unicost}$, is straightforward:

$$
min \sum_{i}^{I} c_i x_i
$$

$$
\sum_{i \in I} M_{l,i} x_i \geq 1 \qquad \forall l \in L_c, \forall c \in C \qquad (P_{unicost})
$$

$$
x_i \in \{0,1\}, c_i = 1 \qquad \forall i \in I
$$

Assume *unicost_selection* $\subseteq I$ is the solution to $P_{unicost}$ where $k = |unicost\_selection|$ is the number of selected items.

4.2   Maximizing Diversity

Our simple mapping from ISP to SCP so far does not take diversity into account. To that end, we turn to the latent representation of items. The variety of the selected subset can be captured as the separation in the item embedding space $E(I)$. Given the minimum subset size k from the solution of $P_{unicost}$, we cluster the embedding space of items into k clusters. Let K denote the cluster centers. Then minimizing the total distance to centroids maximize the diversity among selected items. In other words, we are interested in a subset of items that are far away from each other in the latent space (i.e., maximum inter-distance) and closer to the cluster centers (i.e., minimum intra-distance).

Accordingly, we reformulate $P_{unicost}$ by changing its cost structure: the inclusion of item i incurs cost, c_i, based on its distance to the closest cluster.

$$c_i = \min \quad distance(i, k) \quad k \in K \quad \forall i \in I \qquad (P_{diverse})$$

The resulting problem formulation, $P_{diverse}$, is a reformulation of $P_{unicost}$ with the diversity cost structure. The diversity cost vector is normalized such that the total cost is the same as for $P_{unicost}$. To speed up the optimization process, we also initialize $P_{diverse}$ with the solution from $P_{unicost}$[1]. The solution of $P_{diverse}$, denoted $diverse\_selection$, is the minimum subset of items that are most spread out from each other in the embeddings space $E(I)$ while still maintaining the guarantee that all predefined labels are covered. For the distance metric, common choices are cosine distance or Euclidean distance.

4.3   Bounded Subset Size

While solving $P_{unicost}$ and $P_{diverse}$ successively leads to the smallest, most diverse set with coverage guarantees, it provides no control on the cardinality of the selection. However, remember that the time it takes to run randomized experiments is directly proportional to the number of items. Therefore, it is desirable to control the subset size and hence the time window of randomization using a predefined bound, t.

Given a constant t such that $t \leq |P_{diverse}|$ we are interested in selecting at most t items from the most diverse set of items, $diverse\_selection$, (not from the $entire\ set\ I$) such that the coverage is maximized.

Let X be the set of item selection variables as defined before. We introduce a set of binary decision variables, $is\_label\_covered_l$ to denote whether label l is covered with the selected items. Then $P_{max\_cover@t}$ can be formulated as:

$$max \sum_{l \in L_c, c \in C} is\_label\_covered_l$$

$$\sum_{i \in I} x_i \leq t$$

$$M_{l,i}x_i \leq is\_label\_covered_l \quad \forall l \in L_c, \ \forall c \in C \ \forall i \in I \qquad (P_{max\_cover@t})$$

$$\sum_{i \in I} M_{l,i}x_i \geq is\_label\_covered_l \quad \forall l \in L_c, \ \forall c \in C$$

$$x_i \in \{0, 1\} \quad \forall i \in I$$

$$is\_label\_covered_l \in \{0, 1\} \quad \forall l \in L_c, \ \forall c \in C$$

---

[1] Thanks to our anonymous reviewer for this suggestion.

Multi-Level Optimization for ISP(I, M, E, t)
In: Items: I
In: Incident Matrix: $M[label][item]$
In: Embedding Space: $E(I)$
In: Maximum Subset Size: t
Out: Selected Items: $S \subseteq I$

// First Level: Minimize the subset size
// Find the minimum set of items with full coverage
Formulate $P_{unicost}(I, M)$
$unicost\_selection \leftarrow$ **solve**$(P_{unicost})$

// Second Level: Maximize diversity
// Find the minimum set of items with full coverage that maximizes diversity
$k \leftarrow |unicost\_selection|$ ▷ Use the first level to decide the number of clusters
$K \leftarrow$ **cluster**$(E(I), num\_clusters = k)$ ▷ Find clusters in the embedding space
Initialize $cost \leftarrow$ zeros$(|I|)$ ▷ Set closest centroid distance as the diversity cost
for all item $\in I$ **do**
    $cost_{item} \leftarrow$ **min**(**distance**$(item, centroids \in K)$)
end for
Formulate $P_{diverse}(I, M, cost, unicost\_selection)$
$diverse\_selection \leftarrow$ **solve**$(P_{diverse})$ ▷ Solve for coverage and diversity

// Third Level: Maximize bounded coverage
// Find the maximum coverage within the diverse set subject to the bound
$t \leftarrow |diversity\_selection|$
Formulate $P_{max\_cover@t}(diverse\_selection, M, t)$
$S = max\_coverage \leftarrow$ **solve**$(P_{max\_cover@t})$

return S

Algorithm 1: Multi-Level Optimization for the Item Selection Problem (ISP).

In this formulation, the objective is to maximize the number of unique labels covered. The first constraint limits the total number of selected items with the given upper bound t. The second constraint links selection variables, X, with label coverage variables, $is\_label\_covered$. The constraint states that if a selected item exhibits the label, then that label is covered. The variable $is\_label\_covered$ can still be set to one even when no content is selected (since $0 \leq 1$), which is taken care of by the third constraint. It ensures that if a label is covered, at least one item offering it should be in the selection.

4.4   Multi-level Optimization

Bringing these components together, Algorithm 1 depicts our multi-level optimization framework that consist of solving $P_{unicost}$, $P_{diversity}$ and $P_{max\_cover@t}$. The framework is flexible to accommodate different latent embeddings, clustering techniques, and distance metrics that fit a given dataset the best. A concrete

---

Warm-Start Procedure(I, S, E, M$_S$, q)
In: Items: I
In: Selected Items: $S \subseteq I$
In: Embedding Space: $E(I)$
In: Trained Models for S: M_S
In: Distance Quantile: $q \in (0, 1)$
Out: Warm-started Model: M

$//$ Calculate item-to-item similarities using item embedding $E(I)$
for all $i \in I$ **do**
  $D_{i,j} \leftarrow$ **distance**(E_i, E_j) for $j \in I$
end for

$//$ Find distance threshold based on similarities of items in S
for all $s \in S$ **do**
  $D_s^{min} \leftarrow$ **min**$(D_{s,j})$ for $j \in S, j \neq s$
end for
$w \leftarrow$ **quantile**(D_s^{min}, q)

$//$ Warm-start untrained models
$M_{S'} \leftarrow \emptyset$
for all $s' \in S' = I \setminus S$ **do**
  $D_{s'}^{min} \leftarrow$ **min**$(D_{s,j})$ for $j \in S$
  if $D_{s'}^{min} \leq w$ **then**
    $p \leftarrow$ **argmin**$(D_{s,j})$ for $j \in S$
    $M_{S'} \leftarrow M_{S'} \cup$ **TransferLearning**(M_{S_p})
  end if
end for

return $M \leftarrow M_{S'} \cup M_S$

Algorithm 2: Warm-Start Procedure to transfer knowledge from the randomization (*exploration*) phase to the personalization (*exploitation*) phase.

instantiation we use in this paper is the TFIDF [17] featurization of item descriptions with the standard k-means clustering algorithm and cosine distance as the distance metric.

5   Warm-Starts

Given the solution of ISP, the experimentation phase can start. This yields the training data D_S which is used to build personalization model M_S. The idea behind transfer learning [3,4] is to leverage M_S such that, when the personalization phase starts, it is not restricted to the initial subset of items but can expand beyond the trained model M_S. To that end, we warm-start items $s' \in S' : I \setminus S$ to build $M_{S'}$ sharing knowledge from M_S.

As depicted in Algorithm 2, we take advantage of the item embedding $E(I)$ to calculate item-to-item similarities. Given pairwise distances, we find the closest item $s \in S$ for each *untrained* item s'. To use s for the warm-start of s', we enforce $distance(s, s') \leq w$ for $w > 0$ to ensure that the items are sufficiently similar. We obtain the distance threshold w from the distribution of pairwise distances within a certain quantile, e.g., the top decile $q = 10\%$.

Finally, for transfer learning between s and s', we can leverage the training data D_s or trained parameters of model M_s. After the initial warm-start, the training data for s and s' grows separately. This allows models to continue learning independently from each other after the warm-start.

6   Experiments

We experiment with well-known datasets from book and movie recommendations. The main goal of our experiments is to demonstrate the speed-up in the random experimentation phase enabled by our multi-level optimization framework while ensuring diversity and transfer learning capacity.

6.1   Evaluation Metrics and Questions

The exact time window, in the number of days/weeks required for experimentation, depends on several factors such as the expected interaction volume, the engagement level of users (e.g., average click-through rates) and the complexity of the learning algorithm to train (e.g., linear regression vs. wide&deep networks [5]). While these remain application-specific, to assess the effectiveness of our approach, we focus on the following evaluation metrics measured before and after warm-start:

- **Before warm-start:** The *number of items*, which serves as a proxy of exploration time (the lower, the better) and the *number of labels* covered, which measures the scope of exploration (the higher, the better).
- **After warm-start:** The *number of items*, which measures the capacity of transfer learning (the higher, the better) and *number of labels* covered, which is a proxy for the diversity of items that can be recommended (the higher, the better).

To demonstrate the potential speed-up in random experimentation and effectiveness of the warm-start procedure we consider the following specific questions:

Q1: What is the minimum number of items required to cover all labels?

Q2: How much speed-up is enabled in exploration phase when using optimized item selection to collect response data for training?

Q3: How effective is the warm-start procedure in increasing the number of items and the resulting coverage?

Q4: How sensitive is the ISP to the choice of latent embedding space of items?

Table 1. Summary statistics for Book and Movie Recommendation datasets.

Dataset	# Items	Categories	# Labels
GoodReads	1,000	{Genre, Publisher, Genre × Publisher}	574
	10,000		1,322
MovieLens	1,000	{Genre, Producer, Language, Genre × Language}	473
	10,000		1,011

6.2   Datasets: Book and Movie Recommendations

We use two well-known datasets from the recommender systems literature: the GoodReads Book Reviews [41,42] with 11,123 books (items) and the MovieLens (ml-25m) Movie Recommendations [14] with 62,423 movies (items). We consider two randomly selected subsets, small and large versions with 1,000 and 10,000 items, respectively. These datasets provide category and label metadata used in our ISP formulations. Table 1 summarizes the statistics of our datasets.

For book recommendations, there are 11 different genres (e.g., fiction, non-fiction, children), 231 different publishers (e.g., Vintage, Penguin Books, Mariner Books), and genre-publisher pairs. This leads to 574 and 1,322 unique book labels for the small and large datasets, respectively.

For movie recommendations, there are 19 different genres (e.g., action, comedy, drama, romance), 587 different producers, 34 different languages (e.g., English, French, Mandarin), and genre-language pairs. This leads to 473 and 1,011 unique movie labels for the small and large datasets, respectively.

In the ISP, we are interested in selecting movies (books) for exploration that cover all (or maximum) genres, producers (publishers), languages, and genre-language (genre-publisher) combinations.

6.3   Setup and Parameters

All our experiments were run on a machine with Linux RHEL7 operating system, a 16-core 2.2 GHz CPU, and 64 GB of RAM. To solve the optimization problems, we use the PYTHON-MIP [38] package, which comes with the precompiled COIN-OR CBC SOLVER [16]. For clustering, we employ the default k-means algorithm from the SKLEARN [2] library. To generate embeddings from unstructured text, we utilize the TEXTWISER [18] library. For the warm-start procedure we use a distance quantile of $q = 0.1$. Experiments for non-deterministic methods are repeated $n = 50$ times using different seeds and the results are averaged.

6.4   Embedding Space

The embedding space is based on textual descriptions of movies and books. To convert unstructured text data into meaningful vector representations, we use Term Frequency Inverse Document Frequency (TFIDF) [17], ignoring terms with a document frequency lower than the cut-off threshold of $min\_df = 20$.

To reduce dimensionality, we transform these vectors using non-negative matrix factorization [22] and generate 30-dimensional feature vectors for each item. In Sect. 6.9, we also experiment with other strategies to understand the sensitivity of ISP as a function of the embedding space.

6.5 Comparisons

We compare $P_{unicost}$, $P_{diverse}$, $P_{max\_cover@t}$ on each dataset against the following challenger algorithms:

1. *Random*: Uniform random selection as a simple baseline. The *Random* method uses the subset size k from the solution for $P_{unicost}$.
2. *Greedy*: The classical greedy heuristic for set covering that adds items itera-tively, whereby at each step, the item with the best $\frac{cost}{coverage}$ ratio is selected. This is a competitive baseline with a polynomial-time approximation scheme with worst-case guarantees [40].
3. *KMeans*: Unsupervised clustering approach that operates on the same embedding space. As in *Random*, it uses the subset size k from the solution of $P_{unicost}$ as the number of clusters. This method first clusters the latent space into k centers and then selects items closest to the centroids.

While *Greedy* maximizes coverage, it does not take diversity into account. This helps us assess the effectiveness of $P_{diverse}$. Analogously, while *KMeans* maximizes diversity, it omits label coverage. This in turn helps us determine the effectiveness of coverage constraints. Both *Random* and *KMeans* select k items for which we have an optimality certificate from $P_{unicost}$ that covers all labels.

Table 2. [Q1] Comparison of our solution to ISP and challenger approaches on the GoodReads dataset before and after warm-start in terms of the number of items selected and label coverage.

Dataset	Method	Before Warm-Start			After Warm-Start		
		Items	Labels	Coverage	Items	Labels	Coverage
GoodReads 1K Items 574 Labels	Random	374	325	57%	463	367	64%
	Greedy	374	574	100%	460	574	100%
	$P_{unicost}$	374	574	100%	470	574	100%
	KMeans	374	333	58%	741	431	75%
	$P_{diverse}$	446	574	100%	523	574	100%
GoodReads 10K Items 1,322 Labels	Random	1,080	606	46%	2,226	771	58%
	Greedy	1,080	1,322	100%	2,227	1,322	100%
	$P_{unicost}$	1,080	1,322	100%	2,433	1,322	100%
	KMeans	1,080	589	45%	2,834	838	64%
	$P_{diverse}$	1,165	1,322	100%	1,602	1,322	100%

Table 3. [Q1] Comparison of our solution to ISP and challenger approaches on the MovieLens dataset before and after warm-start in terms of the number of items selected and label coverage.

Dataset	Method	Before Warm-Start			After Warm-Start		
		Items	Labels	Coverage	Items	Labels	Coverage
	Random	243	220	46%	624	274	58%
MovieLens	Greedy	249	473	100%	648	473	100%
1K Items	$P_{unicost}$	243	473	100%	647	473	100%
473 Labels	KMeans	243	206	43%	659	276	58%
	$P_{diverse}$	248	473	100%	652	473	100%
	Random	523	298	29%	2,479	561	55%
MovieLens	Greedy	703	1,011	100%	3,031	1,011	100%
10K Items	$P_{unicost}$	523	1,011	100%	2,659	1,011	100%
1,011 Labels	KMeans	523	317	31%	1,801	542	54%
	$P_{diverse}$	558	1,011	100%	1,971	1,011	100%

6.6   Analysis of Coverage [Q1]

To answer Q1 and find the minimum set of items covering all labels, we solve $P_{unicost}$ and compare the number of selected items, the resulting label coverage before and after the warm-start procedure.

Book Recommendations: Table 2 summarizes our results for the GoodReads dataset. Solving $P_{unicost}$ before warm-start returns 374 items covering all 574 labels in the small dataset and 1,080 items that cover all 1,322 labels in the large dataset. This represents reductions of 63% and 89% compared to selecting all items. We then use $|P_{unicost}|$ for *Random* and *Greedy*. The *Greedy* algorithm is also competitive on both datasets in terms of label coverage. As expected, the number of labels covered by *Random* and *KMeans* is markedly lower. The solution for $P_{diverse}$ only requires 72 and 85 more items than $|P_{unicost}|$ demonstrating the slight pay-off to maximize the diversity of the selected content. After the warm-start procedure using Algorithm 2, *KMeans* yields the highest number of warm-started items. This is expected since clustering purely targets the diversity of the space, but unfortunately, its label coverage is no different than *Random*.

Movie Recommendations: Table 3 summarizes our results on the MovieLens datasets. $P_{unicost}$ achieves complete coverage with almost a 90% reduction compared to selecting all items. In this dataset, *Greedy* cannot achieve the quality of the optimum solution. Its optimality gap (249 vs. 243) for the small dataset is 2% and is significantly worse (703 vs. 523) at 34% for the large dataset. *Random* and *KMeans* continue performing poorly in terms of coverage before and after warm-start.

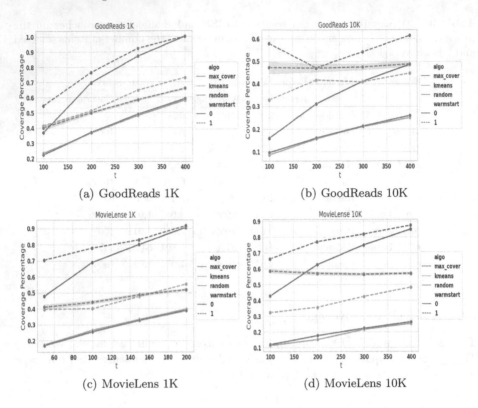

(a) GoodReads 1K   (b) GoodReads 10K

(c) MovieLens 1K   (d) MovieLens 10K

Fig. 2. [Q2] Bounded coverage of labels with varying number of selected items t.

Lastly, in terms of runtime, solving the multi-level optimization with $P_{unicost}$, $P_{diversity}$ and $P_{max\_cover@t}$ takes 20 min at most. This shows the efficiency of optimization technology when faced with recommendation benchmarks.

6.7   Analysis of Bounded Coverage [Q2]

To answer Q2 and demonstrate potential speed-up in random experimentation, we vary the subset bound t and analyze the label coverage before and after warm-starts for $P_{max\_cover@t}$, $KMeans$ and $Random$. We keep the range of t the same between datasets, with the exception of MovieLens 1K due to the smaller number of required items to cover all labels. In practice, the bound t is application-driven governed by time constraints, expected volumes, and user engagement. Figure 2 presents our results.

Before the warm-start, we see that, for each method, coverage increases consistently as t increases. Critically, for a given coverage level, the required number of items t is always lower for $P_{max\_cover@t}$ compared to other methods, indicating potential speed-up. For example, on the small GoodReads dataset, a coverage of 50% can be achieved at $t = 140$, while $t = 320$ is required to obtain the same

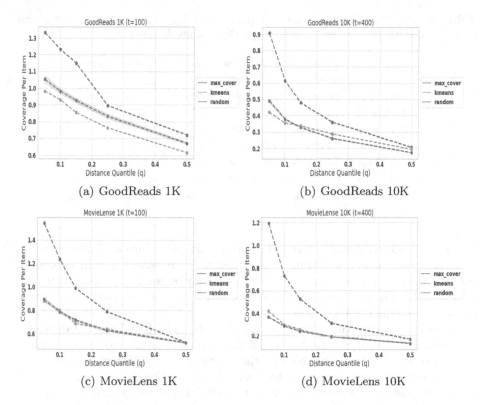

Fig. 3. [Q3] Warm-Start analysis of unit coverage with varying distance quantile q.

coverage using the $KMeans$ and $Random$ methods. There is a 2X reduction in the number of items, again demonstrating time savings in exploration.

After the warm-start, coverage increases for each method at each t, and notably, the coverage for $P_{max\_cover@t}$ continues to rank highest in both datasets. $KMeans$ and $Random$ results in similar coverage, but neither is capable of passing 50% (60%) with 200 (400) items on small and large sets whereas $P_{max\_cover@t}$ reaches 80% (85%) within the same bound.

It is worth noting that the number of items warm-started is not the same for different methods. In the next section, we analyze the efficiency of the warm-start procedure in terms of the number of labels covered per item.

6.8   Analysis of Warm-Start [Q3]

To answer Q3 and assess the effectiveness of the warm-start procedure in Algorithm 2 we perform sensitivity analysis on the distance quantile q and evaluate the average number of labels covered per item after warm-start. We keep the number of selected items fixed at $t = 100$ for small dataset and $t = 400$ for large dataset and find the selected items using $P_{max\_cover@t}$, $KMeans$ and $Random$.

Table 4. [Q4] Comparison of different item embeddings in terms of coverage and capacity to warm-start unseen items on the GoodReads datasets.

Dataset	Embedding	Items	Labels	Coverage	Unit Coverage
	TFIDF	254	313	54%	1.2
GoodReads	Word2Vec	233	318	55%	1.4
1K Items	GloVe	208	299	52%	1.4
574 Labels	Byte-Pair	235	300	52%	1.3
	TFIDF	1,743	723	55%	0.4
GoodReads	Word2Vec	900	608	46%	0.7
10K Items	GloVe	1,098	641	48%	0.6
1,322 Labels	Byte-Pair	940	553	42%	0.6

Using the selected items, we run the warm-start procedure at different values of q. As q increases, the distance constraint to warm-start an item is relaxed, thereby increasing the number of items that can be feasibly warm-started. In parallel, this possibly reduces the relevance of these items given the already collected training data.

Figure 3 presents the results for GoodReads and MovieLens datasets. For each method, the charts show the unit coverage (the number of covered labels divided by the number of items) after the warm-start. Notice that, as q increases, unit coverage decreases across the board for all methods and datasets. This clearly demonstrates the diminishing returns in label coverage as more items are included. Consistent with the coverage analysis, $P_{max\_cover@t}$ is the most effective approach in terms of the number of labels covered per item, significantly better than *Random* and *KMeans* especially for the top (semi-) decile, i.e., $q \leq 0.1$.

6.9   Analysis of Embedding Space [Q4]

To answer Q4 and evaluate the sensitivity of ISP with respect to the underlying item embeddings, we solve $P_{max\_cover@t}$ on the books dataset with a fixed $t = 100$ and $q = 0.1$. We experiment with several complementary embeddings using TEXTWISER [18]. Besides our baseline TFIDF, we employ FastText Word2Vec to learn word vectors [11,27], GloVe [29] embedding to learn global word representations, and Byte-Pair [35] embedding to learn character level information. In all cases, we apply Singular Value Decomposition (SVD) [10] to generate a fixed size 30-dimensional latent representation.

Table 4 reports the total number of items and percentage of label coverage after warm-start. The coverage is similar for the different embeddings hinting at the robustness of our multi-level framework. Nevertheless, more complex embeddings provide better unit coverage compared to TFIDF. In particular, the Word2Vec embedding achieves the best unit coverage in both datasets, closely followed by the GloVe embedding. This is thanks to the recent advances in NLP and the efficiency of pretrained models in capturing text semantics.

7   Related Work

Our work at the intersection of Operations Research (OR), Natural Language Processing (NLP), and Recommender Systems is related to several other approaches. From OR perspective, while cover formulations are standard in the literature, we show that optimization solvers can tackle problems derived from widely used recommendation datasets. Our multi-level optimization framework can be seen as an example of Hybrid Optimization [15] as it combines strengths of the cover formulation with unsupervised clustering on the embedding space. From NLP and Transfer Learning perspective, we take advantage of the recent advances in pre-trained word embeddings such as FastText [11], GloVe [29], and Byte-Pair [35]. From a Recommenders perspective, our framework leaves the choice of personalization algorithm open to a wide range of options, such as matrix factorization, collaborative filtering [19,34], nearest-neighbors [9], factorization machines [31], and deep learning models including Wide&Deep[5], DeepFM [12], and DCN [43] (see [23,45] for a survey).

While we considered an approach that starts with exploration followed by exploitation, these steps can be blended together [32]. For instance, multi-armed bandit learning policies [36] such as ϵ-Greedy, Thompson Sampling [37], and Upper Confidence Bounds [39] mixes exploration and exploitation. We can still incorporate ISP in these settings to guide the exploration component. Traditionally, statistical power analysis [6] and experimental design [33] methods offer a formal treatment to identify significant effects for a given statistical power. However, given the combinatorial nature of the problem and limited data context, these methods are rarely suited for item selection.

Finally, our warm-start procedure shares similarities with [3,4] which builds ensembles based on item similarity. A more involved approach would be to transfer models between different applications as in cross-system recommendations [28,46], e.g., cross-referencing between different systems such as book and movie recommendations.

8   Interactive Exploration of ISP

Finally, let us mention that the selection of the item universe is not purely an algorithmic problem. There are other criteria beyond coverage and diversity, such as preferences, time-sensitive and seasonal information, and regulatory or legal constraints. The selection of items can be viewed as an instance of human-in-the-loop optimization problem. For that purpose, we built an exploratory analysis tool as shown in Fig. 4. The tool allows interaction with our optimization algorithms. It uses UMAP [25] and t-SNE [24] to visualize the 2D embedding space (a) where centroids and selected items are color-coded. The tool provides hover-over display cues to view item metadata (a) and detailed tables (b). It shows relevant statistics of the response data (c & d) and presents the network structure (e) behind the warm-start procedure to surface item similarities.

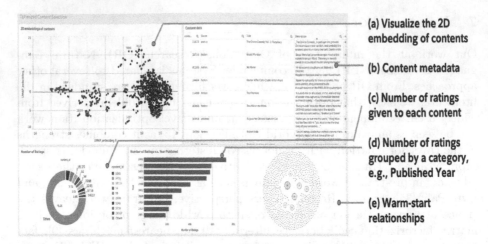

(a) Visualize the 2D embedding of contents

(b) Content metadata

(c) Number of ratings given to each content

(d) Number of ratings grouped by a category, e.g., Published Year

(e) Warm-start relationships

Fig. 4. Interactive visualization tool for item selection and exploratory analysis. Plot (a) visualizes the 2D embedding of items with trained items in blue, warm-started items in red and other untrained items in black; (b) shows the original content metadata such as Genre and Title; (c) and (d) show the feedback for each item or a group of items; (e) visualizes the warm-start relationships, such as how warm-started items are connected to trained items. (Color figure online)

9   Conclusion

We introduced a new combinatorial optimization problem, called the Item Selection Problem (ISP) for Recommender Systems in new and data-sparse applications. Our multi-level optimization framework combining OR, NLP, and Unsupervised Learning achieved significant speed-up in the exploration phase as demonstrated in our experiments. By speeding up the exploration, we alleviate issues stemming from randomization in customer experience and business outcomes. We hope that our ISP formalism not only facilitates further integration between these complementary fields but also helps practitioners design new recommendation system experiences.

References

1. Beasley, J.E.: An algorithm for set covering problem. Eur. J. Oper. Res. **31**(1), 85–93 (1987)
2. Buitinck, L., et al.: API design for machine learning software: experiences from the scikit-learn project. In: ECML PKDD Workshop: Languages for Data Mining and Machine Learning, pp. 108–122 (2013)
3. Caruana, R., Munson, A., Niculescu-Mizil, A.: Getting the most out of ensemble selection. In: Proceedings of the 6th IEEE International Conference on Data Mining (ICDM 2006), Hong Kong, China, 18–22 December 2006, pp. 828–833. IEEE Computer Society (2006)

4. Caruana, R., Niculescu-Mizil, A., Crew, G., Ksikes, A.: Ensemble selection from libraries of models. In: Brodley, C.E. (ed.) Machine Learning, Proceedings of the Twenty-first International Conference (ICML 2004), Banff, Alberta, Canada, 4–8 July 2004. ACM International Conference Proceeding Series, vol. 69. ACM (2004)
5. Cheng, H., et al.: Wide & deep learning for recommender systems. In: Karatzoglou, A., et al. (eds.) Proceedings of the 1st Workshop on Deep Learning for Recommender Systems, DLRS@RecSys 2016, Boston, MA, USA, 15 September 2016, pp. 7–10. ACM (2016)
6. Cohen, J.: Statistical Power Analysis for the Behavioral Sciences. Academic Press, Cambridge (2013)
7. Covington, P., Adams, J., Sargin, E.: Deep neural networks for Youtube recommendations. In: Sen, S., Geyer, W., Freyne, J., Castells, P. (eds.) Proceedings of the 10th ACM Conference on Recommender Systems, Boston, MA, USA, 15–19 September 2016, pp. 191–198. ACM (2016)
8. Devlin, J., Chang, M.W., Lee, K., Toutanova, K.: BERT: pre-training of deep bidirectional transformers for language understanding. arXiv preprint arXiv:1810.04805 (2018)
9. Goldberg, D., Nichols, D.A., Oki, B.M., Terry, D.B.: Using collaborative filtering to weave an information tapestry. Commun. ACM **35**(12), 61–70 (1992)
10. Golub, G.H., Reinsch, C.: Singular value decomposition and least squares solutions. In: Bauer, F.L. (ed.) Linear Algebra, Handbook for Automatic Computation. HDBKAUCO, vol. 2, pp. 134–151. Springer, Heidelberg (1971). https://doi.org/10.1007/978-3-662-39778-7_10
11. Grave, E., Bojanowski, P., Gupta, P., Joulin, A., Mikolov, T.: Learning word vectors for 157 languages. In: Proceedings of the International Conference on Language Resources and Evaluation (LREC 2018) (2018)
12. Guo, H., Tang, R., Ye, Y., Li, Z., He, X.: DeepFM: a factorization-machine based neural network for CTR prediction. In: Sierra, C. (ed.) Proceedings of the Twenty-Sixth International Joint Conference on Artificial Intelligence, IJCAI 2017, Melbourne, Australia, 19–25 August 2017, pp. 1725–1731. ijcai.org (2017)
13. Hansen, C., et al.: Contextual and sequential user embeddings for large-scale music recommendation. In: Santos, R.L.T., et al. (eds.) RecSys 2020: Fourteenth ACM Conference on Recommender Systems, Virtual Event, Brazil, 22–26 September 2020, pp. 53–62. ACM (2020)
14. Harper, F., Konstan, J.: The MovieLens datasets: history and context. ACM Trans. Interact. Intell. Syst. **5**(4), 1–19 (2015)
15. Hooker, J.N.: Integrated Methods for Optimization. International Series in Operations Research and Management Science, vol. 100. Springer, Boston (2007). https://doi.org/10.1007/978-0-387-38274-6
16. Forrester, J., et al.: coin-or/Cbc: Version 2.10.5, March 2020
17. Jones, K.S.: A statistical interpretation of term specificity and its application in retrieval. J. Document. **28**, 11–21 (1972)
18. Kilitcioglu, D., Kadioglu, S.: Representing the unification of text featurization using a context-free grammar. In: Proceedings of the AAAI Conference on Artificial Intelligence (2021)
19. Koren, Y., Bell, R.M., Volinsky, C.: Matrix factorization techniques for recommender systems. Computer **42**(8), 30–37 (2009)
20. Lake, T., Williamson, S.A., Hawk, A.T., Johnson, C.C., Wing, B.P.: Large-scale collaborative filtering with product embeddings. CoRR abs/1901.04321 (2019)
21. Le, Q., Mikolov, T.: Distributed representations of sentences and documents. In: International Conference on Machine Learning, pp. 1188–1196 (2014)

22. Lee, D.D., Seung, H.S.: Algorithms for non-negative matrix factorization. In: Advances in Neural Information Processing Systems, pp. 556–562 (2001)

23. Lops, P., de Gemmis, M., Semeraro, G.: Content-based recommender systems: state of the art and trends. In: Ricci, F., Rokach, L., Shapira, B., Kantor, P.B. (eds.) Recommender Systems Handbook, pp. 73–105. Springer, Boston (2011). https://doi.org/10.1007/978-0-387-85820-3_3

24. van der Maaten, L., Hinton, G.: Visualizing data using t-SNE. J. Mach. Learn. Res. **9**, 2579–2605 (2008)

25. McInnes, L., Healy, J., Melville, J.: UMAP: uniform manifold approximation and projection for dimension reduction. arXiv preprint arXiv:1802.03426 (2018)

26. Mehrotra, R., Shah, C., Carterette, B.A.: Investigating listeners' responses to divergent recommendations. In: Santos, R.L.T., et al. (eds.) RecSys 2020: Fourteenth ACM Conference on Recommender Systems, Virtual Event, Brazil, 22–26 September 2020, pp. 692–696. ACM (2020)

27. Mikolov, T., Sutskever, I., Chen, K., Corrado, G.S., Dean, J.: Distributed representations of words and phrases and their compositionality. In: Advances in Neural Information Processing Systems, pp. 3111–3119 (2013)

28. Pan, S.J., Yang, Q.: A survey on transfer learning. IEEE Trans. Knowl. Data Eng. **22**(10), 1345–1359 (2010)

29. Pennington, J., Socher, R., Manning, C.D.: Glove: global vectors for word representation. In: Empirical Methods in Natural Language Processing (EMNLP), pp. 1532–1543 (2014)

30. Quadrana, M., Karatzoglou, A., Hidasi, B., Cremonesi, P.: Personalizing session-based recommendations with hierarchical recurrent neural networks. In: Cremonesi, P., Ricci, F., Berkovsky, S., Tuzhilin, A. (eds.) Proceedings of the Eleventh ACM Conference on Recommender Systems, RecSys 2017, Como, Italy, 27–31 August 2017, pp. 130–137. ACM (2017)

31. Rendle, S.: Factorization machines. In: Webb, G.I., Liu, B., Zhang, C., Gunopulos, D., Wu, X. (eds.) ICDM 2010, The 10th IEEE International Conference on Data Mining, Sydney, Australia, 14–17 December 2010, pp. 995–1000. IEEE Computer Society (2010)

32. Ricci, F., Rokach, L., Shapira, B. (eds.): Recommender Systems Handbook. Springer, Boston (2015). https://doi.org/10.1007/978-1-4899-7637-6

33. Ryan, T.P., Morgan, J.: Modern experimental design. J. Stat. Theory Pract. **1**(3–4), 501–506 (2007)

34. Salakhutdinov, R., Mnih, A.: Probabilistic matrix factorization. In: Platt, J.C., Koller, D., Singer, Y., Roweis, S.T. (eds.) Advances in Neural Information Processing Systems 20, Proceedings of the Twenty-First Annual Conference on Neural Information Processing Systems, Vancouver, British Columbia, Canada, 3–6 December 2007, pp. 1257–1264. Curran Associates, Inc. (2007)

35. Sennrich, R., Haddow, B., Birch, A.: Neural machine translation of rare words with subword units. In: Proceedings of the 54th Annual Meeting of the Association for Computational Linguistics (Volume 1: Long Papers), pp. 1715–1725. Association for Computational Linguistics, Berlin, Germany, August 2016

36. Strong, E., Kleynhans, B., Kadioglu, S.: MABWiser: a parallelizable contextual multi-armed bandit library for Python. In: 2019 IEEE 31st International Conference on Tools with Artificial Intelligence (ICTAI 2019), pp. 885–890. IEEE (2019). https://github.com/fidelity/mabwiser

37. Thompson, W.R.: On the likelihood that one unknown probability exceeds another in view of the evidence of two samples. Biometrika **25**, 285–294 (1933)

38. Toffolo, T.A.M., Santos, H.G.: Python-MIP: Version 1.9.1. https://www.python-mip.com/

39. Valko, M., Korda, N., Munos, R., Flaounas, I., Cristianini, N.: Finite-time analysis of kernelised contextual bandits. In: Proceedings of the Twenty-Ninth Conference on Uncertainty in Artificial Intelligence, pp. 654–663 (2013)

40. Vazirani, V.V.: Approximation Algorithms. Springer, Heidelberg (2001). https://doi.org/10.1007/978-3-662-04565-7

41. Wan, M., McAuley, J.J.: Item recommendation on monotonic behavior chains. In: Pera, S., Ekstrand, M.D., Amatriain, X., O'Donovan, J. (eds.) Proceedings of the 12th ACM Conference on Recommender Systems, RecSys 2018, Vancouver, BC, Canada, 2–7 October 2018, pp. 86–94. ACM (2018)

42. Wan, M., Misra, R., Nakashole, N., McAuley, J.J.: Fine-grained spoiler detection from large-scale review corpora. In: Korhonen, A., Traum, D.R., Màrquez, L. (eds.) Proceedings of the 57th Conference of the Association for Computational Linguistics, ACL 2019, Florence, Italy, 28 July–2 August 2019, Volume 1: Long Papers, pp. 2605–2610. Association for Computational Linguistics (2019)

43. Wang, R., Fu, B., Fu, G., Wang, M.: Deep & cross network for ad click predictions. CoRR abs/1708.05123 (2017)

44. Wu, C., Alvino, C.V., Smola, A.J., Basilico, J.: Using navigation to improve recommendations in real-time. In: Sen, S., Geyer, W., Freyne, J., Castells, P. (eds.) Proceedings of the 10th ACM Conference on Recommender Systems, Boston, MA, USA, 15–19 September 2016, pp. 341–348. ACM (2016)

45. Zhang, S., Yao, L., Sun, A.: Deep learning based recommender system: a survey and new perspectives. CoRR abs/1707.07435 (2017)

46. Zhao, L., Pan, S.J., Xiang, E.W., Zhong, E., Lu, Z., Yang, Q.: Active transfer learning for cross-system recommendation. In: DesJardins, M., Littman, M.L. (eds.) Proceedings of the Twenty-Seventh AAAI Conference on Artificial Intelligence, Bellevue, Washington, USA, 14–18 July 2013. AAAI Press (2013)

Improving Branch-and-Bound Using Decision Diagrams and Reinforcement Learning

Augustin Parjadis[1]([✉]), Quentin Cappart[1], Louis-Martin Rousseau[1], and David Bergman[2]

[1] École Polytechnique de Montréal, Montreal, Canada
{augustin.parjadis-de-lariviere,quentin.cappart,
louis-martin.rousseau}@polymtl.ca
[2] University of Connecticut, Storrs, CT 06260, USA
david.bergman@uconn.edu

Abstract. Combinatorial optimization has found applications in numerous fields, from transportation to scheduling and planning. The goal is to find an optimal solution among a finite set of possibilities. Most exact approaches use relaxations to derive bounds on the objective function, which are embedded within a branch-and-bound algorithm. Decision diagrams provide a new approach for obtaining bounds that, in some cases, can be significantly better than those obtained with a standard linear programming relaxation. However, it is known that the quality of the bounds achieved through this bounding method depends on the ordering of variables considered for building the diagram. Recently, a deep reinforcement learning approach was proposed to compute a high-quality variable ordering. The bounds obtained exhibited improvements, but the mechanism proposed was not embedded in a branch-and-bound solver. This paper proposes to integrate learned optimization bounds inside a branch-and-bound solver, through the combination of reinforcement learning and decision diagrams. The results obtained show that the bounds can reduce the tree search size by a factor of at least three on the maximum independent set problem.

Keywords: Decision diagrams · Branch-and-bound · Reinforcement learning.

1 Introduction

Historically introduced for encoding Boolean functions and used for circuit design and verification [10,21], *Decision Diagrams* (DDs) have recently been reapplied in the field of combinatorial optimization [2,8,17], for example to sequencing problems [13] or the multidimensional bin packing problem [18]. Assuming a maximization objective, the optimal solution can be obtained in polynomial time in respect to the size of the decision diagram by following the

© Springer Nature Switzerland AG 2021
P. J. Stuckey (Ed.): CPAIOR 2021, LNCS 12735, pp. 446–455, 2021.
https://doi.org/10.1007/978-3-030-78230-6_28

longest path from the root to the terminal node of an exact DD. However, the size of exact DDs grows exponentially with the number of variables, which make them unsuitable for solving large problems. Recently, decision diagrams provided new means of obtaining bounds for combinatorial optimization problems that can be significantly better than those obtained via a traditional linear programming relaxation [6]. Bergman et al. [9] proposed to use DDs to encode a parametrizable and tractable approximation of the solution set. Such structures are referred to as *approximate* DDs.

The performance of this procedure highly depends on the ordering of the variables used to create the DDs. Better ordering can lead to tighter optimization bounds, which results in fewer nodes explored during the BnB search and an expected solution time reduction. Nevertheless, finding an ordering that yields the best bound is NP-hard and difficult to model. Typically heuristics have been considered for defining variable ordering [5] and only a few limited studies have proposed exact approaches for specific problem classes (see, e.g. [4]). Recently, Cappart et al. [12] proposed a deep reinforcement learning (DRL) [3,20] approach for computing the variable ordering. The idea is to train an agent to build a DD, with the incentive (i.e., the reward) to obtain bounds as tight as possible. However, this procedure has not been integrated in a BnB algorithm.

The contribution this paper makes is to illustrate the benefits and challenges of using both primal and dual bounds obtained with DRL-guided DDs within a BnB procedure. As a first experiment we apply this algorithm to the *maximum independent set problem* (MISP). Our preliminary results show that the proposed approach is able to prove optimality with significantly fewer nodes explored, due to the better bounds obtained. However, it is done at the expense of a significant increase in the execution time, as a deep neural network has to be called multiple times at each BnB node. An improved algorithm which uses caching and restricts the use of the DRL agent is also presented, and proves efficient on larger problems.

This paper introduces some preliminaries for DDs and reinforcement learning before presenting the BnB algorithm, followed by experimental results. Finally, the limitations of the current approach and directions for future research are discussed.

2 Learning Bounds Inside Branch-and-Bound

2.1 Decision Diagram-Based Branch-and-Bound

The DDs used in this work are Binary Decision Diagrams (BDDs), although the generalization to multi-valued decision diagrams is immediate. A BDD $B = (U, A, d)$ is a directed acyclic multi-graph in which the node set U is partitioned into layers $L_1, \ldots, L_{n+1}$. The set of solutions for a problem can be represented by a BDD in which each path from the root node r to the terminal node t encodes a feasible solution: each arc along the path gives the value of the variable associated with the layer the arc starts from, and a longest path in the diagram gives an optimal solution to the problem. Conversely, each feasible solution can be found in the BDD. Each layer $L_1, \ldots, L_n$ is associated with a unique variable x_k of the problem with $k \in \{1, \ldots, n\}$. Each arc $a \in A$ goes from one layer to the next and has label $d(a) \in \{0, 1\}$ that encodes the values of the layer's associated binary variable.

For larger problems in combinatorial optimization, BnB algorithms are widely used to generate a search tree of manageable size using optimization bounds. Such bounds can be obtained via feasible solutions and linear relaxation, and we focus here on how approximated DDs provide a simple alternative for computing bounds. An exact DD can be *relaxed* by merging nodes of a layer to narrow the diagram without removing any solutions, but adding unfeasible solutions [9]. A relaxed DD provides a dual bound when solved, the quality of which depends on the amount of merging operations done. Likewise, a *restricted* DD can be created by removing nodes of an exact DD; some solutions are deleted but none are created, yielding a subset of solutions that can provide a primal bound. A branching process can also be conducted on DD nodes, which provides a complete BnB scheme based on DDs as an alternative to the classic linear programming-based BnB. A DD-based BnB algorithm is described in Algorithm 2. Detailed information is proposed by Bergman et al. [5].

2.2 Variable Ordering and Reinforcement Learning

Recently, a machine learning approach has been proposed to address the NP-hard problem of variable ordering for DDs [12] using reinforcement learning (RL) [24]. The goal of a RL agent is to maximize the expected sum of rewards obtained by learning a behavior policy. In the case of DD construction, the agent incrementally builds an approximated DD and observes the bound given by the partially built diagram. The rewards given as feedback are defined by the evolution of the bound obtained (a reduction of an upper bound is encouraged, whereas an increase is penalized and vice versa for a lower bound). The state observed by the agent is a function of the problem instances and of the variables already added in the current DD. An action consists in selecting a new variable to add to the next layer of the DD. Finding a policy maximizing the action-value function for all states and actions is hard and a practical solution is to compute an approximation of Q using a Q-learning algorithm [25] applied to a graph neural network. The exact procedure can be found in can be found in [12].

2.3 The Branch-and-Bound Algorithm with a RL Agent

This section presents the main contribution of the paper: how to implement a BnB algorithm using DDs together with RL. The process is as follows. At each node explored during a DD-based BnB algorithm, a relaxed and a restricted DD are built and an ordering for the variables left at this stage of the search has to be determined for each DD. Currently, heuristics that try to greedily limit the width of the DD are used, and perform much better than simple lexicographic or random orderings.

We propose to work on the ordering of the approximate DDs by using a RL agent on a graph embedding rather than handcrafted heuristics. Two steps are to be considered for building this agent: (1) a training phase, consisting of learning a good policy for the problem at hand, and (2) a solving phase, corresponding to the execution of the BnB algorithm.

Vectorized Representation. A graph neural network (GNN) [11,19] is used to obtain a vectorized representation of a graph by computing node embeddings. The vectorization is built as follows: (1) the problem instance is represented as a graph (e.g., a MISP instance), (2) each vertex/edge of the nodes are decorated with relevant features (e.g. the weight of each vertex), (3) A GNN is used to obtain an d-dimensional embedding for each vertex of the graph, and (4) the embedding is given as input to a fully-connected neural network in order to obtain the final Q-values that are used for the prediction, indicating the quality of each vertex to be inserted in the current embedding.

Training Phase. The training is based on neural fitted Q-learning with a set of randomly generated graphs, and returns the weights $\mathbf{w}$ for the approximated action-value function $\hat{Q}$. The Q-value approximation is obtained with the use of a GNN with the library `structureToVec` [14]. The standard Q-learning algorithm can be enriched in several classic ways, among which are mini-batches to guarantee a better gradient descent based on several examples instead of one, reward scaling for a better handling of large reward quantities and an adaptive ϵ-greedy policy to move away from a local optimum, balancing exploration and exploitation. Details on the learning phase can be found in [12]. As the training uses the bound obtained via the partially built DD, the RL agent is dependent on the type of bound used. Therefore we consider two agents, one trained for relaxed DDs for which a low value upper bound is desired, and the other for restricted DDs for which the highest possible lower bound is desired. Once the parameters $\mathbf{w}$ for the Q-value have been learned, a simple policy can be derived as described below.

Solving Phase. The solver uses a trained RL agent each time a DD is developed. Algorithm 1 describes how an ordering is obtained by calling the agent several times, and Algorithm 2 describes the BnB algorithm using the DDs constructed with the RL agents. New variable orderings are computed at each node of the BnB to build the restricted and relaxed DDs. The RL agent is designed to obtain high-quality bounds, which then allows to update the lower bound or prune the node if applicable. Let B_{ut} be the DD induced by all nodes and paths going from u to t (which gives in particular $B_{rt} = B$). If B_{ut} is an exact DD, B'_{ut} designates a restriction and $\overline{B}_{ut}$ a relaxation of B_{ut}. If a restricted DD B'_{ut} is exact, no restriction was needed and thus no branching is performed. The variable orderings are obtained as described below.

In Algorithm 1, the state of the problem is given to the agent that responds with the Q-values for each variable. The variable u that maximizes the Q-value

---

Algorithm 1. DD construction with RL agent

1: $B = (\{L_1, \ldots, L_n\}, A, d)$ an empty DD, to be built with the variables V
2: **for** $j \in \{1, n\}$ **do**
3:    $u = \text{argmax}_{a \in V} \hat{Q}(s, a, \mathbf{w})$
4:    *build layer L_j with the variable u*
5:    $V \leftarrow V \backslash \{u\}, b_u = 1$
6: **end for**
7: **return** B

---

learned by the GNN is chosen and added to the partially built DD. A feature b_u for each $u \in V$ indicates if a variable has already been selected for the DD construction. After the selection of the variable u, the feature is updated. This modifies the state of the problem, so the Q-values are computed again to select the next variable. The process continues until all the variables have been considered and the DD is complete.

Algorithm 2. DD based branch-and-bound, 2 RL agents

1: $z_{\text{opt}} = -\infty$, $\mathcal{Q} = \{r\}$
2: **while** $\mathcal{Q} \neq \emptyset$ **do**
3:    select node $u \in \mathcal{Q}$
4:    create restricted DD B'_{ut} with Algorithm 1
5:    **if** $v^*(B'_{ut}) > z_{\text{opt}}$ **then** $z_{\text{opt}} = v^*(B'_{ut})$ **end if**
6:    **if** B'_{ut} non exact **then**
7:       create relaxed DD $\overline{B_{ut}}$ with Algorithm 1
8:       if not pruned, select exact cutset to add to $\mathcal{Q}$
9:    **end if**
10: **end while**

Complexity. Obtaining a variable ordering by calling a GNN is expensive because of the nature of the steps involved. The node embeddings are created by a message passing algorithm and fed to a fully connected neural network. This has to be done for each variable in the ordering, resulting in a time complexity of $\mathcal{O}(|E| \times n^2)$. In the next section, Algorithm 1 for variable ordering is compared against a classic heuristic called *MinState*, which has a time complexity of $\mathcal{O}(w \times n)$; with w the width of the DD. The additional complexity brought by the GNN should thus have to be justified with an efficient node reduction during the BnB.

3   Experimental Results

The goal of the first set of experiments is to show that a significant reduction of nodes explored in the tree search is possible with a well trained GNN. However, it is done at the expense of the computation time. The second experiment shows how this issue can be mitigated using a caching mechanism, and a hybrid heuristic. Similar to previous works [7,12], the case study considered is the maximum independent set problem (MISP). The solver is written in C++ and is implemented upon the code of Cappart et al. [12] for the RL agent part and Bergman et al. [5] for the DD-based BnB. The learning and the solving were ran on a Dell XPS 15 9570 with a Intel i7-8750H 2.20 GHz CPU. The training time was limited to eight hours. All instances considered have been randomly generated using a Barabasi-Albert scheme with the density parameter $\nu = 4$ [1].

Definition 1 (Maximum Independent Set Problem). *An independent set* I *in a weighted graph* $G = (V, E)$ *is a subset* $I \subseteq V$ *in which 2 vertices cannot be adjacent. The problem consists of finding an independent set of maximum weight, i.e. maximizing the function* $f(x) = \sum_{k=1}^{|V|} w_k x_k$ *such that* $\forall (i, j) \in E, x_i + x_j \leq 1$ *with* x_k *indicating whether the vertex* k *is selected or not, and* w_k *being the weight associated to this vertex.*

3.1   Performances of the Learned Variable Ordering

We plot in Fig. 1 the number of nodes explored during the DD-based BnB over 100 random instances. For the variable orderings, the *GNN* algorithm uses a GNN trained for relaxation over random graphs with the same distribution. Restriction is still done with a MinState ordering. The GNN consists of 4 iterations of belief propagation and a neural network with 2 layers of size 64. The *heuristic* algorithm uses MinState for relaxation and restriction. The results can be found in Table 1, with the average time in seconds. We can see that even if the number of nodes is drastically reduced, the execution time remains an important concern compared to MinState. Figure 1 shows the percentage of the graphs solved under a given number of nodes in performance profile [22]: the higher the curve, the better. Overall, the number of nodes needed to solve all the MISP instances is typically reduced by a factor of three to four with the RL agent for graphs over 100 nodes, noting that 90% of the instances were solved in fewer nodes.

Table 1. Average time and nodes explored for 100 MISP instances of size n

Algorithm	$n = 80$		$n = 100$		$n = 120$	
	Nodes	Time	Nodes	Time	Nodes	Time
GNN	1416	16.70	4628	79.90	34434	878.10
GNN+	2131	2.40	13152	10.00	81015	41.10
MinState	2802	0.32	22680	5.30	100822	40.40

3.2   Caching to Save Computation Time

To avoid calling the agent too often and speed-up the resolution, the GNN-computed orderings are stored and re-used for subsequent DDs that might have

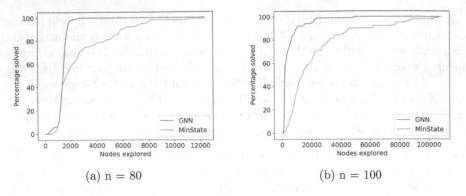

(a) n = 80   (b) n = 100

Fig. 1. Performance profile of a GNN agent and MinState for variable ordering over 100 instances of the MISP

a similar structure. To re-use an ordering, the number of variables for the current DD has to be close to the number of variables of the stored ordering and be included in it. When the number of variable differs too much, the stored ordering can be quite poor for the current DD (similar to a random ordering). This sacrifices some precision in favor of computational efficiency. Furthermore, the computation time bottleneck is expressed the most when it is needed the least, which is at the end of the BnB tree for small DDs. Bound improvements there are small and do not have major consequences, and small DDs are very fast to build with little need for a GNN. We experiment on the use of the RL agent before a given threshold for the number of nodes (here, 100 BnB nodes), and the use of the MinState heuristic after this threshold to close the search faster. Those improvements are natural given the cost of a GNN feed-forward operation, and we observe in Table 1 and in Fig. 2 with the algorithm GNN+ that the average number of nodes explored during a search can in fact be effectively reduced while keeping a competitive solving time for large problems.

3.3   Discussion

As highlighted in the experiment, the learned bounds can be successfully leveraged in the solver in order to reduce the number of nodes explored. However, the execution time is an important bottleneck, which makes the hybridization challenging to design, and additional mechanisms should be considered in order to improve the efficiency. *Caching* is one of them but is still not enough. We made other attempts to get improvements, such as (1) using the same agent for the restriction and the relaxation, (2) using a fixed ordering during the complete BnB process instead of dynamically recomputing it, (3) calling the GNN only every n steps or at random intervals, but without much success. We believe that finding other ways to improve is a promising research direction. A similar question has been addressed by Gupta et al. [16] for standard MIP solvers. They show that expressive and costly GNNs can be combined with inexpressive but cheap fully connected neural networks in order to obtain a better trade-off between prediction quality and computational efficiency.

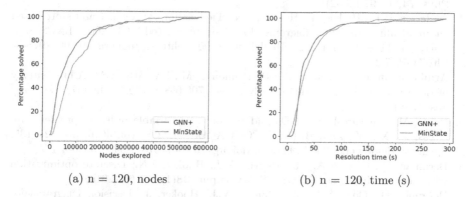

(a) n = 120, nodes   (b) n = 120, time (s)

Fig. 2. Performance profile of an improved GNN agent and MinState for variable ordering over 100 instances of the MISP

4 Conclusion

DDs provide a new flexible and general methodology for generating tight bounds for optimization problems, and DD-based BnB algorithms are often competitive with recent integer programming solvers. This paper presents a method that takes advantages of the flexibility of DDs for bound generation in BnB through machine learning. The reduction of the number of nodes indicates that this is a promising research direction, but subject to important challenges (e.g. the execution time). The generic nature of DD construction and BnB solving for discrete optimization problems makes for an interesting combination, with possible application to other problem classes like the maximum cut problem [6], the set covering problem [9] and the traveling salesperson problem [15,23].

The next step to better bridge the gap between machine learning and optimization would involve a general framework that takes advantage of the graph structure and recursive formulation of a problem by learning to construct DDs with high quality bounds for the problem, and solving it with an efficient BnB algorithm. Data generation for the training step would remain a challenge.

The use of machine learning to generate both lower and upper bounds for combinatorial optimization proved efficient and opens the door to multiple possibilities for future research on the integration of these disciplines. Applying deep learning tools to optimization faces practical bottlenecks that have to be overcome for practical applications and broad adoption in exact optimization solvers.

References

1. Albert, R., Barabási, A.L.: Statistical mechanics of complex networks. Rev. Mod. Phys. **74**, 47–97 (2002)
2. Andersen, H.R., Hadzic, T., Hooker, J.N., Tiedemann, P.: A constraint store based on multivalued decision diagrams. In: Bessière, C. (ed.) CP 2007. LNCS, vol. 4741, pp. 118–132. Springer, Heidelberg (2007). https://doi.org/10.1007/978-3-540-74970-7_11
3. Arulkumaran, K., Deisenroth, M.P., Brundage, M., Bharath, A.A.: A brief survey of deep reinforcement learning. CoRR abs/1708.05866 (2017). http://arxiv.org/abs/1708.05866
4. Behle, M.: On threshold BDDs and the optimal variable ordering problem. In: Dress, A., Xu, Y., Zhu, B. (eds.) COCOA 2007. LNCS, vol. 4616, pp. 124–135. Springer, Heidelberg (2007). https://doi.org/10.1007/978-3-540-73556-4_15
5. Bergman, D., Cire, A.A., van Hoeve, W.J., Hooker, J.N.: Discrete optimization with decision diagrams. INFORMS J. Comput. **28**(1), 47–66 (2016)
6. Bergman, D., Cire, A.A., van Hoeve, W.J., Hooker, J.: Decision Diagrams for Optimization. Springer, Cham (2016). https://doi.org/10.1007/978-3-319-42849-9
7. Bergman, D., Cire, A.A., van Hoeve, W.-J., Hooker, J.N.: Variable ordering for the application of BDDs to the maximum independent set problem. In: Beldiceanu, N., Jussien, N., Pinson, É. (eds.) CPAIOR 2012. LNCS, vol. 7298, pp. 34–49. Springer, Heidelberg (2012). https://doi.org/10.1007/978-3-642-29828-8_3
8. Bergman, D., Cire, A.A., van Hoeve, W.J., Yunes, T.: BDD-based heuristics for binary optimization. J. Heuristics **20**(2), 211–234 (2014). https://doi.org/10.1007/s10732-014-9238-1
9. Bergman, D., van Hoeve, W.-J., Hooker, J.N.: Manipulating MDD relaxations for combinatorial optimization. In: Achterberg, T., Beck, J.C. (eds.) CPAIOR 2011. LNCS, vol. 6697, pp. 20–35. Springer, Heidelberg (2011). https://doi.org/10.1007/978-3-642-21311-3_5
10. Bryant, R.E.: Graph-based algorithms for Boolean function manipulation. IEEE Trans. Comput. **100**(8), 677–691 (1986)
11. Cappart, Q., Chételat, D., Khalil, E., Lodi, A., Morris, C., Veličković, P.: Combinatorial optimization and reasoning with graph neural networks. arXiv preprint arXiv:2102.09544 (2021)
12. Cappart, Q., Goutierre, E., Bergman, D., Rousseau, L.M.: Improving optimization bounds using machine learning: decision diagrams meet deep reinforcement learning. Proc. AAAI Conf. Artif. Intell. **33**, 1443–1451 (2019)
13. Cire, A.A., van Hoeve, W.J.: Multivalued decision diagrams for sequencing problems. Oper. Res. **61**(6), 1411–1428 (2013). https://doi.org/10.1287/opre.2013.1221
14. Dai, H., Dai, B., Song, L.: Discriminative embeddings of latent variable models for structured data. In: International Conference on Machine Learning, pp. 2702–2711 (2016)
15. Deudon, M., Cournut, P., Lacoste, A., Adulyasak, Y., Rousseau, L.-M.: Learning heuristics for the TSP by policy gradient. In: van Hoeve, W.-J. (ed.) CPAIOR 2018. LNCS, vol. 10848, pp. 170–181. Springer, Cham (2018). https://doi.org/10.1007/978-3-319-93031-2_12
16. Gupta, P., Gasse, M., Khalil, E., Mudigonda, P., Lodi, A., Bengio, Y.: Hybrid models for learning to branch. In: Advances in Neural Information Processing Systems, vol. 33 (2020)

17. Hadzic, T., Hooker, J.: Postoptimality analysis for integer programming using binary decision diagrams. In: GICOLAG Workshop (Global Optimization), Vienna. Technical report, Carnegie Mellon University (2006)
18. Kell, B., van Hoeve, W.-J.: An MDD approach to multidimensional bin packing. In: Gomes, C., Sellmann, M. (eds.) CPAIOR 2013. LNCS, vol. 7874, pp. 128–143. Springer, Heidelberg (2013). https://doi.org/10.1007/978-3-642-38171-3_9
19. Kipf, T.N., Welling, M.: Semi-supervised classification with graph convolutional networks. arXiv preprint arXiv:1609.02907 (2016)
20. LeCun, Y., Bengio, Y., Hinton, G.: Deep learning. Nature **521**, 436–44 (2015). https://doi.org/10.1038/nature14539
21. Lee, C.Y.: Representation of switching circuits by binary-decision programs. Bell Syst. Tech. J. **38**(4), 985–999 (1959)
22. Moré, J.J., Dolan, E.D.: Benchmarking optimization software with performance profiles. Math. Program. **91**, 201–213 (2002). https://doi.org/10.1007/s101070100263
23. O'Neil, R.J., Hoffman, K.: Decision diagrams for solving traveling salesman problems with pickup and delivery in real time. Oper. Res. Lett. **47**(3), 197–201 (2019)
24. Sutton, R.S., Barto, A.G.: Reinforcement Learning: An Introduction. MIT Press, Cambridge (2018)
25. Watkins, C.J., Dayan, P.: Q-learning. Mach. Learn. **8**(3–4), 279–292 (1992). https://doi.org/10.1007/BF00992698

Physician Scheduling During a Pandemic

Tobias Geibinger[1], Lucas Kletzander[1(✉)], Matthias Krainz[2], Florian Mischek[1],
Nysret Musliu[1], and Felix Winter[1]

[1] Christian Doppler Laboratory for Artificial Intelligence and Optimization
for Planning and Scheduling, DBAI, TU Wien, Vienna, Austria
{tgeibing,lkletzan,fmischek,musliu,winter}@dbai.tuwien.ac.at
[2] St. Anna Children's Hospital, Medical University of Vienna, Vienna, Austria
matthias.krainz@stanna.at

Abstract. At the beginning of the pandemic last year some hospitals
had to change their physician schedules to take into account infection
risks and potential quarantines for personnel. This was especially important for hospitals that care for high-risk patients, like the St. Anna Children's Hospital in Vienna, which is a tertiary care center for pediatric
oncology. It was very important to develop solving methods for this complex problem in short time. We relied on constraint solving technology
which proved to be very useful in such critical situations. In this paper we
present a constraint model that includes the variety of requirements that
are needed to ensure day-to-day operations as well as the additional constraints imposed by the pandemic situation. We introduce an innovative
set of grouping constraints to partition the staff, with the intention to
easily isolate a small group in case of an infection. The produced schedules also keep part of the staff as backup to replace personnel in quarantine. In our case study, we evaluate and compare our proposed model on
several state-of-the-art solvers. Our approach could successfully produce
a high-quality schedule for the considered real-world planning scenario,
also compared to solutions found by human planners with considerable
effort.

Keywords: Physician scheduling · COVID-19 · Constraint
programming

1 Introduction

Scheduling staff in a hospital can usually be seen as a rostering problem, where
personnel are assigned to shifts to meet daily demands. In this paper we consider
a real-world problem occurring in the St. Anna Children's Hospital in Vienna,
where the normal scheduling procedures had to be rapidly adapted due to an
ongoing pandemic. In addition to ordinary operational requirements, we also
need to take additional constraints about infection risk into account. For example, it would be unwise for doctors to change their assigned stations frequently,
as they would risk coming into contact with more of their colleagues and more

© Springer Nature Switzerland AG 2021
P. J. Stuckey (Ed.): CPAIOR 2021, LNCS 12735, pp. 456–465, 2021.
https://doi.org/10.1007/978-3-030-78230-6_29

patients. Also, an important goal is to maintain a low total number of working doctors to reduce unnecessary infection risk and keep some physicians in reserve to cover for colleagues in quarantine.

Recently, the emergency scenarios during the pandemic have been considered in the pharmaceutical industry and for nurse rostering in [1,2]. However, in this paper we focus on a new physician scheduling problem [3], which to the best of our knowledge includes unique features and pandemic related constraints.

To produce schedules that can deal with these new and complex requirements, we developed a Constraint Programming (CP) model that allowed us to quickly prototype a model and adapt it to changing requirements. CP approaches have been applied for other physician scheduling and related problems [4–7], although without consideration of pandemic-related constraints. We also included several redundant constraints to the model and evaluated it on several state-of-the-art CP and mixed integer programming (MIP) solvers. We were able to quickly find high-quality solutions under several different configurations.

Our model was used to produce the schedule for the physicians in two hospital wards of the St. Anna Children's hospital. We show that the additional constraints can help in reducing the contacts both between physicians and with their patients, compared to work schedules generated under normal conditions.

The rest of this work is structured as follows: In Sect. 2 we describe the problem in detail. In Sect. 3 we present experimental results and briefly go over different usage scenarios in practice.

2   Problem Description and Constraint Model

2.1   Input Parameters and Decision Variables

Table 1 provides an overview of all problem input parameters. Given is a set of *departments* D and a set of *stations* S. Each station s is associated with a department $v_s \in D$. If work on a station s requires contact and collaboration with the whole department it is referred to as a *common station* $s \in C \subseteq S$. A regular station can be considered as a sub-area of a department.

The aim is to schedule a set of *physicians* P to a set of *skills* K on the different stations. Physicians have preferences for every combination of a station and a skill. These preferences are given by $p_{i,s,k} \in \{0, 1, 2, 3, 4\}$ with $i \in P$, $s \in S$, $k \in K$, where lower values mean a higher preference and a value of 4 indicates that the particular station and skill cannot be assigned to the respective physicians.

A subset of the personnel $R \subseteq P$ is at high risk for COVID-19 and should work in departments with lower risk of exposure to COVID-19 patients as stated by r_v for department v. Another subset $A \subseteq P$ must work in the current planning period (usually because they were part of the reserve previously). Each physician i has a corresponding maximum number of working hours per week w_i and a (possibly empty) set of days F_i where they are not allowed to work (e.g. vacation).

Table 1. Input parameters of the physician scheduling problem

Description	Parameter
Length of scheduling horizon in days	n $(n \equiv 0 \mod 7)$
Scheduling horizon	$L = \{1, \ldots, n\}$
Set of departments	D $(D^+ = D \cup \{0\})$
Set of stations	S $(S^+ = S \cup \{0\})$
Department of station	v_s $\forall s \in S^+$ $(v_0 = 0)$
Set of common stations	$C \subseteq S$
Set of physicians	P
Set of skills	K $(K^+ = K \cup \{0\})$
Station and skill preferences	$p_{i,s,k} \in \mathbb{N}$ $\forall i \in P, s \in S, k \in K$
Set of high-risk physicians	$R \subseteq P$
Department risk score	r_v $\forall v \in D$ $(r_0 = 0)$
Set of physicians that have to work	$A \subseteq P$
Maximum working hours per week	$w_i \in \mathbb{N}$ $\forall i \in P$
Set of forbidden working days	$F_i \subseteq L$ $\forall i \in P$
Set of shifts (0 is a dummy denoting a day off)	T $(T^+ = T \cup \{0\})$
Length of a shift in hours	$l_t \in \mathbb{N}$ $\forall t \in T^+$ $(l_0 = 0)$
Forbidden shift successors	$Q_t \subseteq T$ $\forall t \in T^+$ $(Q_0 = 0)$
Cover requirements (shift demands)	$d_{s,t,j,k} \in \mathbb{N}$ $\forall s \in S, t \in T, j \in L, k \in K$
Set of subsuming shifts	U
Department of subsuming shift	$dp_u \in D$ $\forall u \in U$
Cover requirements of subsuming shifts	$ds_{u,j} \in \mathbb{N}$ $\forall u \in U, j \in L$
Skillset of subsuming shift	$V_u \subseteq K$ $\forall u \in U$

Note that a history of previous assignments is also used for a seamless transition. For brevity it can be found in the appendix.[1]

The *planning horizon* (or length of the schedule) n is the number of days in the current scheduling period L. The set of possible *shifts* is T, where each $t \in T$ is associated with a length l_t in hours and a set of forbidden successors Q_t that must not be assigned on the following day.

Demand is given as a 4-dimensional matrix where $d_{s,t,j,k}$ denotes the number of employees required to work shift t in station s using skill k on day j. This demand matrix is supplemented by *subsumed demands*, which indicate that some regular shifts (1) in any single (non-common) station of a department should be replaced by a 24-h shift (2). The reasoning for this is that regular shifts are distributed over different stations of a department, while for the night a qualified physician can cover multiple stations simultaneously. Therefore, this demand can neither be fixed to one particular station nor one particular skill. Instead, we define a requirement for 24-h shifts that replace (subsume) a given number of regular shifts. The total set of these replacements is U. The parameter dp_u references the department affected by replacement u. For each day j in the

---

[1] See https://cdlab-artis.dbai.tuwien.ac.at/papers/pandemic-scheduling/.

planning horizon, $ds_{u,j}$ denotes the number of shifts that should be replaced across the whole department. These shifts must be chosen from those using any of the skills listed in V_u.

We demonstrate the use of subsumed demands via a simple example (Fig. 1) for one department on a single day. Here, one of the regular shifts of a senior physician or physician in any of the three non-common stations must be replaced by a 24-h shift. In the example schedule on the right, this is the shift of physician P4 in station S2. Alternatively, the shift of physician P2 in station S1 could also have been chosen.

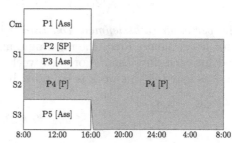

Station	SP	P	Ass
Cm	0	0	1
S1	1	0	1
S2	0	1	0
S3	0	0	1

Subsumed 1 of {SP, P}

(a) Demand matrix. One shift of skill SP or P must be subsumed by a 24-hour shift.

(b) A possible schedule fulfilling the demands.

Fig. 1. Example demand for a single department and day. The department has 3 stations, plus a common station (Cm) and demand during the day shift for three different skills: Senior physician (SP), physician (P), and Assistant (Ass).

A solution either assigns a day off (value 0) or a shift, skill, and station to all physicians on each day of the horizon using the decision variables in Table 2.

Table 2. Decision variables of the physician scheduling problem

Description	Variable
Assigned shift per physician and day	$x_{i,j} \in T^+$ $\forall i \in P, j \in L$
Assigned skill per physician and day	$y_{i,j} \in K^+$ $\forall i \in P, j \in L$
Assigned station per physician and day	$z_{i,j} \in S^+$ $\forall i \in P, j \in L$

2.2 Hard Constraints

The first set of constraints ensures that a physician either has a day off (all corresponding variables are 0) or has a proper shift, skill, and station assignment:[2]

$$([x_{i,j} = 0] + [y_{i,j} = 0] + [z_{i,j} = 0]) \in \{0, 3\} \quad \forall i \in P, j \in L \tag{1}$$

---

[2] We make use of the Iverson brackets: $[P] = 1$, if $P = true$ and $[P] = 0$ if $P = false$.

The *sequence constraints* handle limitations for work assignments on consecutive days. The maximum number of consecutive shifts is limited to 6, thus at least one out of every 7 consecutive days needs to be off.

$$\bigvee_{k=0}^{6} (x_{i,j+k} = 0) \quad \forall i \in P, j \in \{1, \ldots, n-6\} \tag{2}$$

Forbidden shift sequences are excluded by checking after assignment of shift t on day j that the shift on day $j+1$ is not in the forbidden set Q_t.

$$x_{i,j+1} \notin Q_{(x_{i,j})} \quad \forall i \in P, j \in \{1, \ldots, n-1\} \tag{3}$$

The *personalized assignment constraints* handle individual personnel limitations for assignments. First, the maximum number of working hours per week is checked for each physician.[3] For each week the sum of shift lengths l_t for all t that are assigned in this week is bounded by w_i.

$$\sum_{k=1}^{7} l_{(x_{i,(7 \cdot j+k)})} \leq w_i \quad \forall i \in P, j \in \{0, \ldots, \frac{n}{7} - 1\} \tag{4}$$

Secondly, forbidden days are handled by fixing $x_{i,j} = 0$ for all $j \in F_i$.

$$x_{i,j} = 0 \quad \forall i \in P, j \in F_i \tag{5}$$

The given demand needs to be covered exactly. The relevant *demand constraints* check that for each combination of station s, shift t and skill k the number of assigned physicians on each day j matches the demand $d_{s,t,j,k}$.

$$\sum_{i \in P} [x_{i,j} = t \wedge z_{i,j} = s \wedge y_{i,j} = k] = d_{s,t,j,k}$$
$$\forall s \in S, t \in T, j \in L, k \in K \text{ where } t > 2 \vee s \in C \tag{6}$$

However, there is an exception for shift types 1 (regular) and 2 (24-h) on non-common stations. Here, assignments of 24-h shifts also count towards the demand for regular shifts, since a 24-h shift functions as an extension of a regular shift. The additional constraint for subsumption u checks that the number of physicians assigned a 24-h shift on any non-common station of the correct department with any skill in V_u needs to match $ds_{u,j}$ on each day j.

$$\sum_{i \in P} [x_{i,j} \in \{1, 2\} \wedge z_{i,j} = s \wedge y_{i,j} = k] = d_{s,1,j,k} \quad \forall s \in S \setminus C, j \in L, k \in K \tag{7}$$

$$\sum_{i \in P} [x_{i,j} = 2 \wedge v_{(z_{i,j})} = dp_u \wedge z_{i,j} \notin C \wedge y_{i,j} \in V_u] = ds_{u,j} \quad \forall u \in U, j \in L \tag{8}$$

---

[3] Weeks are assumed to start on the first day of the schedule.

Two additional limits are defined for these assignments: Some assignments are impossible, when $p_{i,s,k} = 4$ for physician i, station s, and skill k on each day j.

$$p_{i,(x_{i,j}),(y_{i,j})} < 4 \quad \forall i \in P, j \in L \tag{9}$$

Changing the department is a major risk and therefore prohibited. For physician i the assignment of each station s in the current schedule needs to map to the same department v via v_s (days off are mapped to the dummy department 0). We model this requirement using the *nvalue* global constraint.

$$\text{nvalue}(\{v_{(z_{i,j})} | j \in L\}) \leq 2 \quad \forall i \in P \tag{10}$$

Physicians in $A \subseteq P$ are required to work, e.g., because they stayed at home for the whole previous period. At least one shift needs to be assigned to these physicians. Note that due to the minimization of the number of working personnel, this typically is enough to enforce a regular schedule for these physicians.

$$\bigvee_{j \in L} (x_{i,j} > 0) \quad \forall i \in A \tag{11}$$

2.3 Soft Constraints

The objective function is defined as the following weighted sum, minimizing the number of working physicians wp ($w_1 = 50$), the sum of preference scores pr ($w_2 = 5$), the sum of risk penalties r ($w_3 = 10$), and critically the number of station changes sc ($w_4 = 500$):[4]

$$\text{minimize } w_1 \cdot wp + w_2 \cdot pr + w_3 \cdot r + w_4 \cdot sc \tag{12}$$

The number of working physicians wp is defined as physicians with at least one shift assignment in the current schedule.

$$wp = \sum_{i \in P} \left[\bigvee_{j \in L} (x_{i,j} > 0) \right] \tag{13}$$

The total preference score pr sums the preference scores of all assignments.

$$pr = \sum_{i \in P} \sum_{j \in L} p_{i,(x_{i,j}),(y_{i,j})} \tag{14}$$

For the next objectives the last station assignment to a non-common station is tracked by $ls_{i,j}$. The initial value $ls_{i,0}$ is 0, the value is updated each day $z_{i,j}$ is set to a non-common station.

$$ls_{i,j} = \begin{cases} ls_{i,j-1} & \text{if } z_{i,j} \in C \\ z_{i,j} & \text{otherwise} \end{cases} \quad \forall i \in P, j \in L \tag{15}$$

---

[4] These weights were determined by the hospital staff.

The sum of risk penalties is calculated by using the department risk score of the assigned department (there can be at most one) for each physician $i \in R$.

$$r = \sum_{i \in R} r_{(v_{(ls_{i,n})})} \tag{16}$$

The number of station changes is obtained by counting occurrences of $ls_{i,j-1} \neq ls_{i,j}$ for all physicians i (unless $ls_{i,j-1} = 0$, in case of a previously unused physician).

$$sc = \sum_{i \in P} \sum_{j \in L} [ls_{i,j-1} > 0 \wedge ls_{i,j} \neq ls_{i,j-1}] \tag{17}$$

In addition to the hard and soft constraints we experimented with two sets of redundant constraints that impose redundant restrictions on the shift requirements. Furthermore, we evaluated two custom search strategies plus the solvers' default search strategy with our model. For details about the redundant constraints and search strategies see the appendix of this paper.

3   Experimental Evaluation

We implemented our model using the solver-independent modeling language MiniZinc 2.5.3 and applied it to create the schedule of a ward of the St. Anna Children's hospital for two weeks in April 2020.[5] The ward consists of two departments with three stations (plus the common station) each. There are 26 physicians to be scheduled in the ward, with four different skills. Four configurations of redundant constraints (none, only group 1, only group 2, and both), combined with three search strategies, result in a total of twelve different configurations of our approach.

We evaluated each variant using the state-of-the-art CP and MIP solvers: CPLEX 20.1 [8], CP Optimizer 20.1 [9], Gecode 6.3.0 [10], Gurobi 9.10 [11], Chuffed 0.10.4 [12] and OR-Tools 7.8 [13]. Each solver was executed using a single thread on an Intel Xeon CPUs E5-2650 v4 (2.90 GHz) with a time limit of one hour and a memory limit of 20 GB.

Neither Chuffed nor Gecode were able to find feasible solutions to the problem under any configuration within the runtime limit. CP Optimizer found feasible solutions using 6 out of the 12 variants, exactly those that included the second redundant constraint set. The best solution had an objective value of 5880. OR-Tools managed to find solutions under 10 configurations, achieving the optimal objective value of 1880 in 8 runs. Both Gurobi and CPLEX found feasible solutions with an objective value of 1880 using all 12 configurations. However, Gurobi was consistently faster in finding the best solution, and proved optimality of this solution under 9 configurations, whereas CPLEX and OR-Tools could not prove optimality. Since Gurobi was already able to find optimal solutions

---

[5] The anonymized instances are available at https://cdlab-artis.dbai.tuwien.ac.at/papers/pandemic-scheduling/.

within the one hour time limit, we did not perform experiments using a longer runtime.

Regarding the impact of the search strategies and redundant constraints, while the 3 configurations for which Gurobi could not prove optimality within an hour all included redundant constraint group 2, at the same time the fastest optimality proof did include these constraints (175 s, using only group 2, with default search strategy). Other than that, no distinctive trend could be seen with regards to the redundant constraints or used search strategies.

We note that the performance of the CP solvers may be improved by more extensive use of global constraints. Based on this, we also formulated constraint (2) with global *at least* constraints, but saw no performance improvement. Another possible improvement would be to formulate the demand requirements (6) using *global cardinality* constraints. However, this would require a major reformulation of the used decision variables. A solution of optimal quality was in any case found by Gurobi within at most 75 s regardless of the configuration, which indicates that using our model with the solver Gurobi is useful to quickly generate high-quality schedules in practice.

To evaluate the performance on a larger dataset, we constructed a second instance by doubling all cover requirements and duplicating each employee (leading to a total of 52 physicians). Both Gurobi and CPLEX were not able to prove optimality for the second instance within the runtime limit. However, Gurobi managed to find solutions with an objective value of 3660 in all configurations. CPLEX found feasible solutions for 11 out of 12 configurations, with 9 of them reaching the objective value 3660.

3.1   Impact of Pandemic-Related Constraints

We also compared the optimal solution found by our model with the one produced under "non-pandemic" conditions. Table 3 shows the differences between the two variants. While the assignments for the physicians in the high-risk group could not be reduced, two other physicians could be taken off work and placed in reserve to cover for absences. The main difference however is visible in the number of stations changes. While in the "non-pandemic" variant physicians changed their assigned stations 69 times, in the full model only a single change is necessary due to the additional constraints, which greatly decreases the potential for infection spreads. Notably, this was possible without compromising the qualifications and preferences of the physicians at all.

Table 3. Values of the different objectives in an optimal solution.

Objective	Full model	"Non-pandemic"
Station changes	1	69
Total working	24	26
Risk group working	5	5
Preferences	26	26

4   Conclusion

In this short paper we introduced a new physician scheduling problem. Novel constraints were proposed based on the needs of the St. Anna Children's Hospital in Vienna during the pandemic. Although our solution approach was developed for this particular hospital, the ideas considered in this paper are also useful for generating schedules in other healthcare institutions that must decrease infection risks. We provided a constraint model that includes pandemic specific constraints and investigated additional redundant constraints and search strategies.

State-of-the-art solvers were able to obtain very good solutions in a reasonable amount of time, whereas providing such solutions by a human planner requires scheduling experience and takes much more time. From the practical point of view, the investigation of solver independent formulations was very useful, as various solvers could be evaluated quickly. The constraint technology showed to be very convenient to develop a rapid solution.

For future work it would be interesting to evaluate our approach for instances of other healthcare institutions and to extend our approach for related problems such as the nurse rostering problem.

Acknowledgments. The financial support by the Austrian Federal Ministry for Digital and Economic Affairs, the National Foundation for Research, Technology and Development and the Christian Doppler Research Association is gratefully acknowledged.

References

1. Zucchi, G., Iori, M., Subramanian, A.: Personnel scheduling during COVID-19 pandemic. Optim. Lett. **15**(4), 1385–1396 (2021)
2. Seccia, R.: The nurse rostering problem in COVID-19 emergency scenario. Technical report (2020)
3. Erhard, M., Schoenfelder, J., Fügener, A., Brunner, J.O.: State of the art in physician scheduling. Eur. J. Oper. Res. **265**(1), 1–18 (2018)
4. Weil, G., Heus, K., Francois, P., Poujade, M.: Constraint programming for nurse scheduling. IEEE Eng. Med. Biol. Mag. **14**(4), 417–422 (1995)
5. Bourdais, S., Galinier, P., Pesant, G.: hibiscus: a constraint programming application to staff scheduling in health care. In: Rossi, F. (ed.) CP 2003. LNCS, vol. 2833, pp. 153–167. Springer, Heidelberg (2003). https://doi.org/10.1007/978-3-540-45193-8_11
6. Rousseau, L.-M., Pesant, G., Gendreau, M.: A general approach to the physician rostering problem. Ann. Oper. Res. **115**(1), 193–205 (2002)
7. White, C.A., White, G.M.: Scheduling doctors for clinical training unit rounds using tabu optimization. In: Burke, E., De Causmaecker, P. (eds.) PATAT 2002. LNCS, vol. 2740, pp. 120–128. Springer, Heidelberg (2003). https://doi.org/10.1007/978-3-540-45157-0_8
8. IBM and CPLEX. 20.1 IBM ILOG CPLEX Optimization Studio CPLEX User's Manual (2020). https://www.ibm.com/analytics/cplex-optimizer
9. IBM and CPLEX. 20.1 IBM ILOG CPLEX Optimization Studio CP Optimizer User's Manual (2020). https://www.ibm.com/analytics/cplex-cp-optimizer

10. Schulte, C., Lagerkvist, M., Tack, G.: Gecode 6.30 reference documentation (2020). https://www.gecode.org
11. Gurobi Optimization LLC. Gurobi Optimizer Reference Manual (2020). http://www.gurobi.com
12. Chu, G.: Improving combinatorial optimization. Ph.D. thesis, University of Melbourne, Australia (2011)
13. Laurent Perron and Vincent Furnon. Google OR-Tools 7.8 (2020). https://developers.google.com/optimization/

Author Index

Ajwani, Deepak 410
Akgün, Özgür 348
Åstrand, Max 365
Avgerinos, Ioannis 315

Baudoui, Vincent 179
Beck, J. Christopher 115
Bergman, David 106, 446
Bjørner, Nikolaj 1
Bleidorn, Dominik R. 133
Bracher, Adrian 283

Cappart, Quentin 392, 446
Cardonha, Carlos 106
Carroll, Paula 410
Chalumeau, Félix 392
Cire, André Augusto 231
Cohen, Eldan 115
Coppé, Vianney 231
Coulon, Ilan 392

Demirović, Emir 62

Engell, Sebastian 133
Enright, Jessica 348

Fabris, Irene 89
Feyzmahdavian, Hamid Reza 365
Fitzpatrick, James 410
Formenti, Enrico 196
Frohner, Nikolaus 283

Geibinger, Tobias 456
Gillard, Xavier 231

Hanen, Claire 214
Hill, Alessandro 26
Horn, Matthias 72
Hoshino, Richard 89

Jabbour, Said 163
Jefferson, Christopher 348

Johansson, Mikael 365
Jung, Victor 332

Kadıoğlu, Serdar 427
Kamel, Nadjet 163
Karlsson, Emil 45
Klanke, Christian 133
Kletzander, Lucas 456
Kleynhans, Bernard 427
Kordon, Alix Munier 214
Krainz, Matthias 456
Kronqvist, Jan 299

Lambers, Roel 149
Levatich, Maxwell 1
Lombardi, Michele 266
Lopes, Nuno P. 1
Lozano, Leonardo 106

McCreesh, Ciaran 348
Meel, Kuldeep S. 248
Milano, Michela 266
Mischek, Florian 456
Misener, Ruth 299
Mohammadalitajrishi, Mahshid 248
Mourtos, Ioannis 315
Musliu, Nysret 456

Nekkache, Ikram 163

Parjadis, Augustin 446
Pedersen, Theo 214
Pesant, Gilles 248
Porrmann, Till 383
Pralet, Cédric 179
Prosser, Patrick 348

Raidl, Günther R. 72, 283
Régin, Jean-Charles 196, 332
Riva, Sara 196

Römer, Michael   383
Rönnberg, Elina   45
Rothuizen, Laurent   149
Rousseau, Louis-Martin   392, 446
Rybalchenko, Andrey   1

Sais, Lakhdar   163
Schaus, Pierre   231
Silvestri, Mattia   266
Spieksma, Frits C. R.   149
Stanczak, Marvin   179

Ticktin, Jordan   26
Tsay, Calvin   299

van Driel, Ronald   62
Vidal, Vincent   179
Vossen, Thomas W. M.   26
Vuppalapati, Chandrasekar   1

Wang, Keliang   106
Wang, Xin   427
Winter, Felix   456

Yfantis, Vassilios   133
Yorke-Smith, Neil   62

Zois, Georgios   315
Zschaler, Steffen   348

Printed in the United States
by Baker & Taylor Publisher Services